Springer Water

For further volumes:
http://www.springer.com/series/13419

Anik Bhaduri · Janos Bogardi
Jan Leentvaar · Sina Marx
Editors

The Global Water System in the Anthropocene

Challenges for Science and Governance

GWSP

Springer

Editors
Anik Bhaduri
Janos Bogardi
Sina Marx
Global Water System Project
Bonn
Germany

Jan Leentvaar
UNESCO-IHE
Delft
The Netherlands

ISBN 978-3-319-07547-1 ISBN 978-3-319-07548-8 (eBook)
DOI 10.1007/978-3-319-07548-8
Springer Cham Heidelberg New York Dordrecht London

Library of Congress Control Number: 2014944339

Printed on acid-free paper

Springer is part of Springer Science+Business Media (www.springer.com)

Foreword

This book is one of the three main outputs of the "Conference on Water in the Anthropocene: Challenges for Science and Governance" organized in Bonn, in May 2013, by the Global Water Systems Project, a project developed along a decade of relevant research work for the study of the complex interactions and feedbacks occurring in the global water system. The other two main outputs are the Bonn Declaration on Global Water Security, adopted at end of the 2013 Conference, and the Special Issue of the journal "Current Opinion in Environmental Sustainability," published in December 2013, under the title *Water and the Anthropocene: New Perspectives for Global Sustainability*.

The chapters of the book are directly related to aspects of the very rapid evolution of the world since the onset of the second half of the twentieth century. This evolution was due to an enormous progress of science and technology in the developed countries and to a remarkable increase in the production and supply of a vast range of goods and services. In this period, the political and social conditions of the world have, in general, changed in a positive manner, with a significant improvement in the standard of living of hundreds of millions of people. There was, also, an explosion of human activity, with major impacts at a global scale on the Earth's system. This unprecedented event has been named "the Great Acceleration."

During the second half of the twentieth century, the world population has doubled reaching 6 billion in 2000, while the global economy grew 15-fold. On the other hand, the percentage of the world's urban population increased, during the same period, from 30 to 50 %, and it is estimated that it might grow to close to 70 % by 2050. The driving forces pushing the Great Acceleration constitute an interlinked system, characterized by population growth, increased consumption, the abundance of cheap energy and the implementation of liberalizing economic policies.

At the end of the 1980s, the Brundlandt Report launched the concept of Sustainable Development, which became a popular flag amongst academics, politicians and common citizens. However, the implementation of this concept proved to be difficult. It became clear that the era of prosperity recently experienced by the more developed societies began to undergo a slowdown, largely as a result of the foreseeable depletion of oil reserves, the occurrence of climate change, the devastating pollution, the noticeable impoverishment of biodiversity and the growing lack of natural resources. Water situation is viewed with particular concern, because it is the life support of humans and ecosystems, as well as an essential ingredient of the economic and social development process.

From the dawn of the industrial revolution, two centuries ago, industrialisation has carried with it a new model of civilisation. Its principal agents—technological development, intensive agriculture, expansion of transport and growth of urbanisation, responding to the increased population and the growth of individual demand for goods and services—have led to a level of water consumption that is not currently sustainable in many regions of our world of 7 billion people. Climate change further exacerbates this situation.

We are, in reality, depleting our natural capital, placing at risk our future prosperity and even our very survival. Humanity is no longer able to control some of the feedback effects derived from its own action. The impacts of climate change are indeed a good example of this. Over merely two centuries, we have transferred to the atmosphere, in the form of gases or heat, a substantial part of the hydrocarbons which took millions of years to accumulate on Earth.

Concerned with the significant modifications created on planet Earth during the Great Acceleration, the American biologist Eugene Stroemer and the Dutch Nobel laureate geochemist Paul Crutzen, proposed, in 2000, the consideration of a new geological epoch that they called "the Anthropocene." This new epoch is characterized by global environmental changes considered sufficiently significant to distinguish the Anthropocene from the Holocene, that corresponds to, approximately, the last 12,000 years in which humanity has lived on the Earth, benefiting from a relatively stable climate.

In 2009, a proposal was submitted to the Stratigraphic Commission of the Geological Society of London to consider the Anthropocene as a formal distinct unit of the geological epoch divisions. This proposal has been seriously considered, but in spite of the intense activity of independent working groups, it may take years or even decades until the International Union of Geological Sciences officially formalizes the acceptance this new epoch.

The climate's reaction to the variation of the concentration of greenhouse gases and the magnitude of the biosphere changes associated to it, lend credibility to the idea that, in fact, we have already entered in a new era, not comparable with any interglacial episode of the Quaternary. Events, such as the extinction or migration of species and the replacement of natural vegetation by crops developed under monoculture, are warnings of persistent biostratigraphic signs.

Stroemer, Crutzen and their followers believe that it has become evident that we are no longer living in the Holocene. It cannot but be considered worrying that, over merely two centuries, human beings, who in the early Holocene changed progressively from hunter-gatherers to agricultural and sedentary populations, have recently transformed into major predators and squanderers of resources, assuming the role of a global geophysical force on a scale similar to some of the great forces of nature, such as earthquakes or volcanoes. Population growth is not the most important issue. The real problem is that we are becoming too wealthy and consuming exponentially more resources than can be renewed at adequate rates. In this way, human beings have transformed themselves into the principal force capable of interfering with the natural balances of the Earth. Our planet is currently faced with an entirely new situation, with an unexpected danger: the proliferation of an endemic and invasive species, the human species, whose influence has transformed the atmosphere, impoverished the biosphere, altered the lithosphere and greatly modified the hydrosphere.

In the case of the hydrosphere, the transformations involve not only freshwater and estuarine water, but also the oceans, whose changes, in turn, influence inland waters. The impact of climate change affects mainly the quantity of water (available and necessary) and the quality of the water, but it also causes a sea level rise, with important consequences for inland, surface and ground waters, as well as estuarine and coastal waters. Moreover, climate change leads to changes in rainfall patterns and tends to increase extreme water-related phenomena, in particular floods and droughts. Clearly, the Anthropocene raises novel challenges regarding water management and governance.

This book contributes to the ongoing evolution of the debate on water issues, from conventional wisdom to new forms of thinking and reflection, as required in the Anthropocene where we are already living. The book is, indeed, a step forward in the accomplishment of this objective. Moreover, the book has also the merit of helping to establish links between science and practice in the area of water resources management and governance and of identifying in which ways research and innovation can favour water resources sustainability. As a consequence, the new type of knowledge required in the Anthropocene will be, progressively, built up.

The challenges of the twenty-first century—imposed by the limits of natural resources, financial instability, social inequity (within countries and between countries and regions of the world) and environmental degradation—are a clear sign that 'business as usual' cannot continue. We are acceding to new phase of human experience and entering into a new world, qualitatively and quantitatively different from the one we know.

Competitive demand for water, food and energy may cause geopolitical conflicts capable of triggering major social and political instability and irreversible environmental damage. Any strategy focused merely on parts of the water-food-energy-climate system, which does not consider the strong interactions at play, may have unexpected and serious consequences. The consideration of the water-food-energy nexus demands that sectorial decisions respect an inter-sectorial perspective. Such consideration of this perspective clearly fosters synergies and requires management of trade-offs between the sectors.

In the Anthropocene, it is indispensable to find the best ways to safeguard the future of water and, with it, the future of humanity. However, these paths are full of difficulties, which will occur not within a millennium, nor even a century. In fact, we face the risk that intolerable situations may arise within a few decades.

It should also be noted that we are the first generation in history to possess a vast knowledge of the way our activities influence the Earth's system. We have built our past, we are building our present and we can build our future. We are part of the history of the Earth. For the first time in history, we are really becoming aware of what is going on, and, therefore, we are the first fully responsible for changing our relationship with the planet.

Clearly, it is only possible to pursue the world's development in a sustainable manner, if we adopt new paradigms of development, implying radical changes in the human behaviour of the more developed societies as well as in the emerging economies. Human societies must be aware of the urgent need to change past trends, in order to prevent the risk of being confronted with drastic, swift and irreversible negative developments.

In the first place, we must endeavour to acquire a stronger awareness of water management and governance problems and of their importance. We must, also, create conditions to identify those problems in due time and to provide the necessary solutions. Finally, we must be aware that innovation and technological developments, while crucial, cannot alone solve the complex and multifaceted problems of water management and governance, in a world that is undergoing a change that is increasingly global.

The current situation may require, amongst other things, a proper commitment and responsibility for the implementation of an appropriate governance of the Earth's system, based on the creation of new national and international institutions concerned with sustainable development. This may also require an active stewardship towards a reform of the intergovernmental system of environment, as well as a global vision of water governance.

The future we want, as proclaimed in the outcome document of the Rio+20 Conference, will certainly require that man adopts a much more prudent behaviour in his relationship with water. Only in this way, it will be possible to prevent that water becomes the source of conflicts, so many fear and, instead, acts as a forceful driver of social cohesion.

Luis Veiga da Cunha

Preface

Our constantly evolving planet has witnessed many biological and geological events in the past, marked by epochs that have altered its functioning in fundamental ways, through major changes in climate regime, tectonics and volcanism or by the mass extinction of species. We know that humans have influenced the environment in many ways in the past, but since the industrial revolution and even more during the "great acceleration in the human enterprise" following World War II, humans have assumed the role of a dominating force in changing the biosphere, geosphere, atmosphere, hydro- and cryosphere and hence affecting crucial functioning of the Earth system. We are exhausting resources, causing multiple changes without understanding their interrelated and complex outcomes. We have accelerated major processes (e.g. erosion, nitrogen applied to the land mass) while decelerating others (e.g. loss in delivery of river water and sediments to the world's oceans) in a very short period, and rapidly altered our relationship with the environment in the beginning of a new geological epoch, termed "the Anthropocene". This implies the significant role of human activities in creating a lasting impact on and in codetermining the future evolution of the planet.

Human activities also impact the global water system as part of the Earth system in a significant way and change the way water moves around the globe like never before. Thus, understanding and managing the global cycle of water, an irreplaceable resource vital to all aspects of both environmental and social systems on this planet, is fundamental for achieving global environmental sustainability.

Since its inception, the Global Water System Project (GWSP) has coordinated and supported a broad research agenda to study the complex global water system with its interactions of environmental and social components as a continuum and coupled system, and helped to understand its complex feedback processes. The GWSP Conference "Water in the Anthropocene: Challenges for Science and Governance. Indicators, Thresholds and Uncertainties of the Global Water System" held in Bonn in 2013, synthesized the major achievements in global water research within the last decade. It presented global as well as regional perspectives of the water system's responses at different scales and explored its management vis-a-vis globally relevant change.

This book is an important outcome of the conference, identifying how research can assist policy and practice of sustainable freshwater management in the era of the Anthropocene. The book covers global, regional and local perspectives and addresses issues, such as water resource management and governance, variability in supply, increasing demands for water, environmental flows, and land use change.

The book comprises of 28 chapters that are classified into four broad themes:

Global Water System: Current State and Future Perspectives; Dimensions of Change in River Basins and Regions; Ecosystem Perspectives in Water Resources Management; and Governing Water in the Anthropocene.

The chapters under "Global Water System: Current State and Future Perspectives" present assessments of global water resource availability, deal with earth observations and the role of indicators, data and models of the global water system. They discuss aspects of how to account for water and uncertainties globally, covering both physical processes and socially mediated water fluxes, water withdrawals and uses as well as virtual water trade.

The theme "Dimensions of Change in River Basins and Regions" focuses on adapting to global changes at the river basin and regional scale. This part includes contributions about adaptive resource management towards water security in river basins, chapters addressing institutions and governance challenges in water scarce regions as well as chapters bringing in historical perspectives to understand river systems in the Anthropocene.

The third theme "Ecosystem Perspectives in Water Resources Management" presents different approaches to ecologically sustainable water management drawing on various case studies. The part focuses on how to mitigate the negative impacts of anthropogenic activities on the resilience of social-ecological systems.

The fourth part, "Governing Water in the Anthropocene" concentrates on the crosscutting issue of global water governance, acknowledging the fact that the global "water crisis" is in fact a governance crisis. Case studies in water governance and management under global change from different parts of the world are complemented by contributions dealing with issues like water law, ethics and institutions in water governance.

Acknowledgments

The book "The Global Water System in the Anthropocene" is one of the main outputs of the conference "Water in the Anthropocene" held in Bonn during May 2013, and financed by German Federal Ministry of Education and Research (BMBF). BMBF also supports GWSP-IPO which played a significant role in this book editing. The decade long support of GWSP through BMBF is gratefully acknowledged.

Contents

Part I Global Water System: Current State and Future Perspectives

1 **Balancing the Needs of All Services Provided by Global Water Resources** 3
Elizabeth Curmi, Keith Richards, Richard Fenner, Grant M. Kopec and Bojana Bajželj

2 **Performance Indicators in the Water Resources Management Sector** 15
Balázs M. Fekete and Eugene Z. Stakhiv

3 **Virtual Water and Trade: A Critical Economic Review** 27
Erik Gawel

4 **Data, Models and Uncertainties in the Global Water Cycle** 45
R. J. Harding, A. J. Dolman, D. Gerten, I. Haddeland, C. Prudhomme and P. van Oevelen

5 **Integrated Assessments of Water Scarcity: Knowns, Unknowns and Ways Forward** 59
Tabea K. Lissner, Caroline A. Sullivan, Dominik E. Reusser and Jürgen P. Kropp

6 **A Global Approach to Estimating the Benefit-Cost Ratio of Water Supply Measures in the Agricultural Sector** 73
A. K. Schürkmann, A. Biewald and S. Rolinski

Part II Dimensions of Change in River Basins and Regions

7 **The Role of Institutions and Water Variability in Food Security in Uzbekistan: The Case of Rice Markets in the Khorezm Region** 91
Yadira Mori-Clement, Anik Bhaduri and Nodir Djanibekov

8 **Dams on Shared Rivers: The Concept of Benefit Sharing** 105
Waltina Scheumann, Ines Dombrowsky and Oliver Hensengerth

9 **Challenges and Solutions for Urban-Tourist Water Supply on Mediterranean Tourist Islands: The Case of Majorca, Spain** 125
Angela Hof, Macià Blázquez-Salom, Margalida Comas Colom and Alfredo Barón Périz

10 **Unconventional Water Resources of Agricultural Origin and Their Re-utilization Potential for Development of Desert Land Aquaculture in the Aral Sea Basin** 143
B. K. Karimov, M. Matthies and B. G. Kamilov

11 **A Historian's Perspective on Rivers of the Anthropocene** 161
Philip V. Scarpino

12 **Transdisciplinarity, Human-Nature Entanglements, and Transboundary Water Systems in the Anthropocene** 173
Jason M. Kelly

13 **Eastern European Perspective on the Environmental Aspects in Current Flood Risk Management: The Example of the Czech Republic** 183
Blanka Loučková

14 **Adaptive Irrigation Management in Drought Contexts: Institutional Robustness and Cooperation in the Riegos del Alto Aragon Project (Spain)** 197
Sergio Villamayor-Tomas

Part III Ecosystem Perspectives in Water Resources Management

15 **The Missing Piece in the Conservation Puzzle: Cohesion Among Environmental, Economic and Social Dimensions** 215
Francisco A. R. Barbosa, Paulina M. Maia-Barbosa, Diego G. F. Pujoni and Lorena T. Oporto

16 **Integrating Input-Output Modeling with Multi-criteria Analysis to Assess Options for Sustainable Economic Transformation: The Case of Uzbekistan** 229
Maksud Bekchanov, Anik Bhaduri, Manfred Lenzen and John P. A. Lamers

17 **Sustaining Freshwater Biodiversity in the Anthropocene** 247
Jaime Garcia-Moreno, Ian J. Harrison, D. Dudgeon, V. Clausnitzer, W. Darwall, T. Farrell, C. Savy, K. Tockner and N. Tubbs

18 **Water Governance and Management Systems and the Role of Ecosystem Services: Case Study Insights—Groundwater Management in the Sandveld Region, South Africa** 271
Kathrin Knüppe and Claudia Pahl-Wostl

19 **Tackling the 'How' Question: Enabling and Enacting Practical Action for Managing the Wicked Problem of Nonpoint Source Pollution in Catchments** 289
James J. Patterson, Jennifer Bellamy and Carl Smith

20 **Experiences with a Transdisciplinary Research Approach for Integrating Ecosystem Services into Water Management in Northwest China** 303
Tuck Fatt Siew, Petra Döll and Hamid Yimit

Part IV Governing Water in the Anthropocene

21 **Multilevel Governance of Irrigation Systems and Adaptation to Climate Change in Kenya**........................... 323
Jampel Dell'Angelo, Paul F. McCord, Elizabeth Baldwin, Michael E. Cox, Drew Gower, Kelly Caylor and Tom P. Evans

22 **Transboundary Water Management in Federal Political Systems: A Story of Three Semi-arid Rivers** 343
Dustin Garrick, Lucia De Stefano, Jamie Pittock and Daniel Connell

23 **Legal Plurality in Mekong Hydropower: Its Emergence and Policy Implications**.............................. 355
Diana Suhardiman and Mark Giordano

24 **International River Basin Organizations Lost in Translation? Transboundary River Basin Governance Between Science and Policy** 369
Susanne Schmeier

25 **The Human Right to Water and Sanitation: Reflections on Making the System Effective** 385
Pedi Obani and Joyeeta Gupta

26 Patterns of Water Law 401
Joseph W. Dellapenna

27 The Notion of the Global Water Crisis and Urban Water Realities 415
Antje Bruns and Fanny Frick

28 The Need for a Value-Reflexive Governance of Water in the Anthropocene 427
Simon Meisch

Part I
Global Water System: Current State and Future Perspectives

Perspectives

There is no doubt that occurrence and distribution of water on Earth is governed by a unique and globally functional hydrological cycle. This water cycle is intertwined with other biochemical cycles. Scientist are not only probing the global interconnections between hydrometeorological phenomena, stream flow, and ground water movement, but also that of water use and stewardship discovering amazing links, and hydrological dependencies and vulnerabilities. Regions where land use change, deforestation, or other relevant developments may occur influence hydrometeorological phenomena thousands of kilometers away.

During the last decade, the Global Water System Project (GWSP) pioneered research to assess these global interdependencies, to estimate global water availability, management, and also potential water stress due to population growth, economic development, and climate change/variability. Several dimensions of the problem like water and global trade, water quality deterioration, and disappearance of aquatic biodiversity have been analyzed. Irrespective of spectacular results and timely warning issued, global scale hydrology and water resources management are still in their infancy. It would also be premature to claim that policy makers regularly followed scientific advice in this regard.

While water is a global concern, it is also considered as a profoundly local issue. Water governance and management, while widely advocated to be integrated and basin/aquifer-based, is still rather exercised within jurisdictional (national, provincial, and municipal) contexts. Water conflicts, while occurring at different scales, happen to be most pronounced in local disputes among competing local users and sectors.

Several contributions in the Conference "Water in the Anthropocene: challenges for science and governance" analyzed global phenomena and critically evaluated methods applied to estimate water availability and its natural, or socially mediated movements in the form of physically measurable fluxes or virtual water transfers through international trade.

Resource assessments may not look at the supply side only. Therefore, Curmi et al. suggest an alternative approach to evaluate the demands for goods and services provided by water for both humans and ecosystems and to map these demands back onto resource flows. This approach allows very informative visualization of resource cycling through different continents, water use sectors, and hydrological processes (precipitation, evaporation, stream flow, etc).

Large scale hydrology and water resources management are inherently important for policy making. However, the gap between scientific presentations and results and the information needed for political and policy making processes is quite wide. Much effort, especially the development of meaningful indicators is needed for policy relevant communication of scientific findings and recommendations. Two contributions in this book—that by Lissner et al. and by Fekete and Stakhiv—address the question of comprehensive and reliable indicator development.

There is a broad agreement that integrated approaches are needed to allow the consideration of multiple determinants together to provide the basis for informed decisions. Yet, it would be premature to claim success through wide scale practical applications of the proposed indicators. Recent model developments show considerable achievements in capturing the complexities of interlinked hydrological processes within the global water cycle. These models are needed to assess our potential water future(s). How much is the anticipated growth of water scarcity? How far will climate change aggravate water availability and distribution in space and time? How far are we able to constrain and quantify model and data uncertainty? The comprehensive paper by Harding et al. reviews these questions and juxtaposes the inherent challenges with the research agenda of the Global Energy and Water Exchanges (GEWEX) project that is part of the World Climate Research Programme (WCRP).

On a global scale and by a large margin, agriculture is the biggest (and most inefficient) user of water. Hence, the food and consequently water security of a growing population depends on how agricultural water use (mainly irrigation) will develop. In their contributions, Schürkmann et al. estimate the global benefit/cost ratio for new irrigation infrastructure by using model-generated shadow prices for water.

Another economic aspect of the present and future water management is addressed by Gawel in his critical review of the virtual water and trade concept. While virtual water and the water footprint concepts have attractive informative value, it was concluded that the virtual water concept is limited in its capacity to serve as policy advice or to guide economic decision-making. The problems that hamper water resources management in general like distorted pricing, bad governance, and lack of capacity cannot be rectified by virtual water-related schemes or trade rules.

These contributions highlight some crucial aspects of what may be called global water system. Many of its problems, policy relevance, and research needs were addressed. The fascinating duality of water as a local and simultaneously global issue is likely to remain in the focus of research for quite some time.

Chapter 1
Balancing the Needs of All Services Provided by Global Water Resources

Elizabeth Curmi, Keith Richards, Richard Fenner, Grant M. Kopec and Bojana Bajželj

Abstract Global assessments of water use tend to focus on the supply side, where data on physical hydrology provide an apparently (but often questionable) secure underpinning. However, one difficulty with this approach is that it struggles to deal with the issues of multiple uses of water and of treatment and recycling. Another is that global analysis offers little guidance to water policy and management, which invariably and necessarily act at more local scales. An alternative approach is therefore to evaluate demand for the goods and services offered by water, to both human beings and to ecosystems, and then to map these demands back onto resource flows. This paper describes the sources (precipitation, surface water and groundwater) and the uses of water in delivering *all* of its services (including its provisioning of environmental services), and uses two Sankey diagrams to visualise this system. The results stress the need for an integrated assessment of all water sources and services, simultaneously considering human and ecosystem needs, and highlighting the need to improve human water-use efficiency and productivity rather than lazily invading further the needs of ecosystems on whose additional services humans rely.

Background and Introduction

Water is essential in supporting a number of important human-related and environmental services, and includes water needed to produce food, energy and industrial products as well as that needed to maintain terrestrial and aquatic

E. Curmi (✉) · R. Fenner · G. M. Kopec · B. Bajželj
Department of Engineering, University of Cambridge, Trumpington Street, Cambridge CB2 1PZ, UK
e-mail: ec459@cam.ac.uk

K. Richards
Department of Geography, University of Cambridge, Downing Place, Cambridge CB2 1PZ, UK

A. Bhaduri et al. (eds.), *The Global Water System in the Anthropocene*,
Springer Water, DOI: 10.1007/978-3-319-07548-8_1,

environmental systems that deliver ecosystem service of indirect benefit to humans. Many global water studies have focussed on understanding how rivers, lakes and groundwater are used to supply various economic sectors, including agriculture, industry and domestic use, and have also assessed future trends in both availability and use of freshwater resources in these sectors (Gleick 2011; Shiklomanov 2000; Postel et al. 1996). Postel et al. (1996) interpolated climate, vegetation and soil information for different geographical zones to estimate the water balance for different regions, concluding that 54 % of geographically-accessible global runoff is currently being used to supply human-related services. High-resolution global water models have also been developed to estimate current and future water at the river basin level (Alcamo et al. 2003; Vörösmarty 2000; Hanasaki et al. 2008). However the availability of *all* water (including precipitation) and its use in supplying *all* services (including terrestrial and aquatic ecosystems) are usually omitted in these assessments (Hoff 2009).

Managing *All* Water for *All* Services: Including the Environment

Precipitation falls on many different land cover types, not only supporting rain-fed crop production, but also sustaining terrestrial ecosystems whose services include the production of timber, carbon sequestration, water quality maintenance, flood alleviation, and many more (Rockström and Gordon 2001). Large-scale changes in land cover have occurred as a result of increased conversion to land uses such as crop and pasture lands (for food production), at the expense of natural forest and grassland (terrestrial ecosystems) (Defries et al. 2004). This has not only resulted in a loss of important natural systems but also has had large scale impacts on the quantity and quality of water (Scanlon et al. 2007). Increases in rain-fed crop and pasture land, replacing forest and grassland have decreased terrestrial evapotranspiration and increased streamflow. However, they have also degraded the quality of water due to salinisation and fertiliser leaching (Scanlon et al. 2007). Approximately 60–70 % of the world's food production is grown in rain-fed farming systems (Rost et al. 2008). More than 95 % of farmland in sub-Saharan Africa is rain-fed, whilst this figure is 90 % and 60 % in Latin America and South Asia, respectively (Molden 2009). Rain-fed agriculture is expected to grow faster than irrigated cultivation in the coming decade as many countries with a shortage of renewable groundwater and/or surface water resources realise the potential to improve the productivity of rain-fed agriculture (Hoff et al. 2010).

Groundwater and surface water (Renewable FreshWater Resources; RFWR) are essential in supporting several economic sectors (i.e. agriculture, domestic water supply and industry), but are also critical for maintaining aquatic ecosystems. However, the variability of available RFWR during the year has led to the creation of reservoir storage to enable a stable flow over the year for agriculture, domestic and industrial uses (including hydropower). It is estimated that approximately

16,000 km^3 to 19,000 km^3 per year of global river runoff is controlled by reservoirs (Shiklomanov 2000; Jones 2010). There is strong evidence that the resultant flow alteration has had a detrimental effect to the ecology in river systems (Richter and Thomas 2007). In some areas the extraction of freshwater resources for direct economic benefit is so extreme that major river systems—the Colorado, Indus, Huang Ho and Ganges—have ceased to flow to the sea for periods (Sophocleous 2007). Fortunately, there is now recognition that freshwater is important not only for economic sectors, but also for aquatic ecosystems that provide a variety of essential goods and services such as fisheries, wildlife, flood protection etc. (Sophocleous 2007; Smakhtin et al. 2004). This has led to the concept of 'environmental flows', whose main aim is to mimic aspects of the natural flow variability of river systems, which requires understanding of the role of that variability in sustaining ecosystems (Arthington et al. 2006). Calculating environmental flows is a considerable challenge for water managers as it requires a participatory and interdisciplinary approach, where stakeholders, ecologists, hydrologists, and water managers work together to ensure a balance is maintained between water allocation for industry, domestic and agriculture sectors, *and* the aquatic and riparian ecosystem in order to maintain them in good condition (Pahl-Wostl et al. 2013).

From this brief review we should conclude that water is important not only in providing direct human-related services (i.e. agriculture, domestic water supply and industry), but is also in maintaining terrestrial and aquatic ecosystems that provide critical indirect services. An improved balance must be reached between these direct and indirect water services, especially because sustaining the indirect environmental services is an essential element of inter-generational stewardship. To achieve this goal, better understanding of the allocation and use of current global water resources is needed, to help identify inefficiencies and potential trade-offs. This paper therefore aims to highlight the source, use, service and sinks of all global water resources, and to emphasise that to meet future global water demand more efficient management of *all* water resources is needed, with an improved balance between water allocated for direct human-related uses bringing immediate economic benefit, and for the sustenance of environmental services.

Visualising *All* the Services Provided by Global Water Resources

Curmi et al. (2013) have produced two linked diagrams that show how both precipitation (Fig. 1.1) and the surface and groundwater renewable freshwater resources (Fig. 1.2) are distributed and used. These are in the form of Sankey diagrams, that trace the flow of global water, from its sources (i.e. precipitation, surface water, groundwater), to their uses (i.e. agriculture, industry, domestic, environment) and the services they provide (food, energy, drinking water, sanitation, ecosystems etc.), and finally to their sinks (i.e. evapotranspiration, outflow). The two diagrams distinguish water and its use in a manner similar to the 'green'

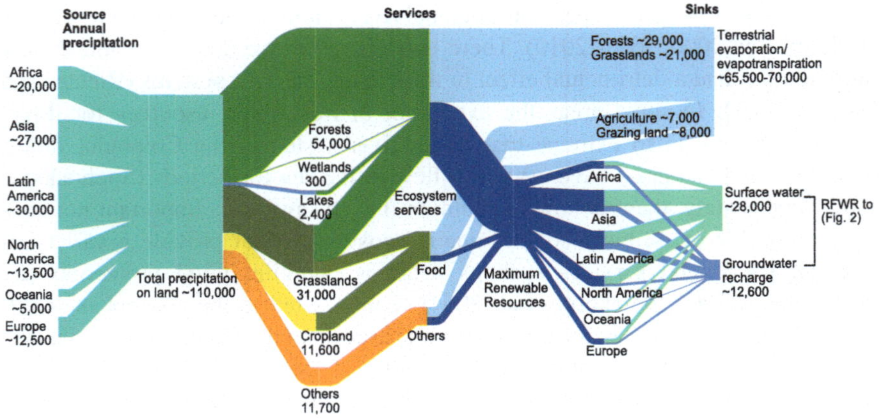

Fig. 1.1 The services provided by precipitation. Figures are in km^3/year (Reprinted from Journal of Environmental Management, 129, Curmi E, Richards K, Fenner R, Allwood J.M, Kopec G.M, Bajzelj B, 456–462, Copyright (2013), with permission from Elsevier)

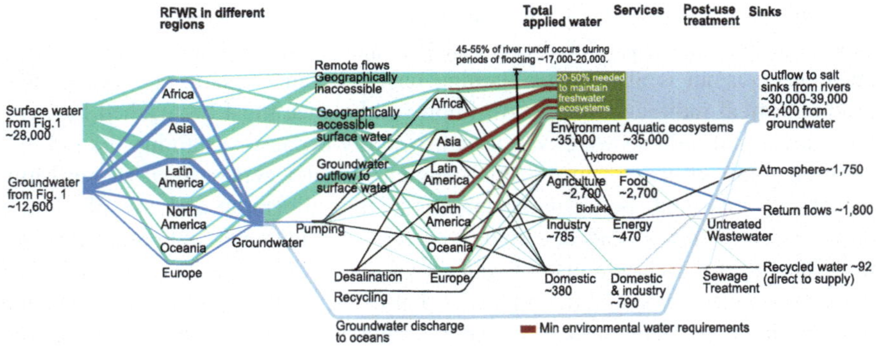

Fig. 1.2 The services provided by surface and groundwater (Reprinted from Journal of Environmental Management, 129, Curmi E, Richards K, Fenner R, Allwood J.M, Kopec G.M, Bajzelj B, 456–462, Copyright (2013), with permission from Elsevier)

(Fig. 1.1) and 'blue' (Fig. 1.2) distinction made by Falkenmark and Rockström (2006). Figure 1.1 summarises how precipitation is consumed by evapotranspiration ('green' water) during plant growth in rain-fed crop, managed grassland production systems and in terrestrial ecosystems (e.g. forests). The average annual supply of excess precipitation in sustaining river flow and recharging groundwater (RFWR) appears at the right-hand side of Fig. 1.1, and then enters the left of Fig. 1.2. This is the managed 'blue' water. However, in common with Jewitt (2006), we prefer to retain the distinction between rainfall and RFWR, because this provides a more practical and physical framework for understanding the linkages and trade-offs involved in assessing and managing water resources as a whole, and the services they provide. It also distinguishes the role of rainfall and streamflow in sustaining, respectively, terrestrial and aquatic ecosystems and *their* services; and facilitates recognition that the major service of food production involves joint appraisal of land and water management in order to optimise the varying balance of rainfall infiltration into the soil, and application of irrigation flows, in sustaining crop yields.

Figure 1.1 shows how precipitation is differentially distributed amongst the continents and falls on different land cover types (i.e. forests, grasslands, croplands) to support services that include terrestrial ecosystem services (~79,000 km^3/year) and food production (~18,000 km^3/year). Approximately 60 and 80 % of the water these two services use, respectively, is 'lost' to the atmosphere through evapotranspiration. The balance is the renewable freshwater resource; this is also shown distributed amongst the continents, and is separated into renewable surface flows and groundwater at the right of Fig. 1.1.

The services provided by surface and groundwater (RFWR) are shown in Fig. 1.2. Some surface water is remote flows that are largely unusable (~7,700 km^3/year), while geographically accessible surface water amounts to ~20,000 km^3/year. Groundwater includes water that is pumped (~1,000 km^3/year) for use in different economic sectors, water that outflows into surface water (~10,000 km^3/year) to provide ecologically-important baseflow in rivers, and groundwater discharged directly into the oceans (~2,400 km^3/year) (Curmi et al. 2013). Services provided by RFWR include a contribution to terrestrial food production, energy production, industrial and domestic supply, and the services of aquatic ecosystems. The diagram shows that aquatic ecosystems receive most water, but much of this is available for other uses; and a significant proportion runs off during flood periods (Curmi et al. 2013). The environmental flows shown in red are the minimum flows needed to maintain river systems in fair condition, according to Smatkhin et al. (2004). Most RFWR eventually reaches a sink, either when it flows into the ocean, or when it is lost to the atmosphere through evapotranspiration. Water used in energy production and for domestic and industrial use is usually discharged back into river systems as return flows, often at a higher temperature or polluted and untreated (return flows); some is treated and reused (recycled). However, the diagram shows that recycled (and desalinated) water represent a very small fraction of the total RFWR used.

Managing Global Water for *All* Services

These diagrams clearly show how global water resources are currently allocated amongst different services. However, population growth and climate change are together expected to increase future global water demand, and as the total supply is unlikely to change markedly, this increase can only be met by supply-side policies that change the balance of allocation of water between different services (most obviously, transferring water from its role in sustaining ecosystems to supporting direct, economic services such as agriculture and domestic supply), or by more efficient use of water through demand management and increased multiple use (demand-side policies). The former strategy is not sustainable, as it will require diverting water away from the maintenance of terrestrial and aquatic ecosystems, and will degrade their indirect services, perhaps irreversibly and in ways that will cost more in the long term. To avoid these implications, all of the services provided by all water will need to be managed conjunctively and effectively, which will require effective management of both water and land resources.

Land Management and the Effective Management of All Water Resources

Land management has an effect not only on the hydrological cycle (increasing or decreasing evapotranspiration as land cover changes), but also on the availability and quality of water resources. The productivity of the land resource also depends on the amount of water it receives. The importance of integrated management of land and water resources has been highlighted in many studies (Calder et al. 2005; Jewitt 2006; Curmi et al. 2013). Calder et al. (2005) suggested two criteria for evaluating land and water management practices; whether precipitation exceeds evapotranspiration (or not), and whether river flow exceeds an agreed minimum (or not). This results in four 'quadrants', each representing different water management consequences of land use change, and each suggesting options for avoiding negative consequences. A similar 'quadrant' approach suggested by Curmi et al. (2013) highlights more directly different measures for the effective joint management of land and water resources, distinguishing approaches that improve the efficiency of precipitation use and RFWR use.

Effective land management practices can improve the efficiency with which both precipitation and RFWR are used. Better management of soil moisture during crop growth e.g. through mulching or cover cropping (Curmi et al. 2013; Molden 2007), and more appropriate crop selection could reduce the non-productive evaporation associated with crop growth. Agroforestry, incorporating trees into agricultural systems, and silvipasture, that combines forestry and animal grazing, are examples that are mutually beneficial for both food production and terrestrial ecosystems. Similar land measures could be adopted in irrigated agriculture, together with integrated pest management and breeding and biotechnology.

Water Management for the Effective Management of All Water Resources

We also need to effectively manage the supply and use of available water resources (both precipitation and RFWR) and find a balance between water that is needed for economic activities and for the environment. Investments in rain-water harvesting, on field storage and small scale irrigation can help alleviate some of the problems of water availability in dry seasons and dry spells in rain-fed agriculture production systems. This is well-known, but involves supporting small-scale farming with good agricultural advisory services, and has often been undervalued in development projects relative to larger-scale capital-intensive infrastructure investment.

Effective Management of Groundwater for All Services

Groundwater supplies water to billions of people, and plays an important part in supplying water to the agricultural sector and to the maintenance of freshwater ecosystem services (Gleeson et al. 2012). On average, the global annual use of groundwater for human-related services is estimated at 1,000 km^3 of which approximately 67 % is used for irrigation, 22 % for domestic and 11 % for industrial purposes (UNESCO 2012) (Fig. 1.2). This seems to be quite small when compared to the available renewable groundwater supply (left of Fig. 1.2), however the excessive use of groundwater is occurring in many regions including North China, India and North America (Konikow and Kendy 2005; Wada et al. 2012) as shown in Fig. 1.3 (an expansion of the 'pumping' line in Fig. 1.2). Global groundwater depletion was estimated at 283 ($\pm$40) km^3 for the year 2000 (Wada et al. 2012).

Excessive pumping in aquifers close to surface water bodies can capture some of the groundwater flow that would have been discharged, without pumping, as baseflow into surface water (Sophocleous 2002). Under natural conditions before the drilling of boreholes and wells, aquifers attain an approximate equilibrium, where over time recharge exceeded discharge during wet years and vice versa in dry years. Extracting water from wells disrupts this equilibrium, and a decrease in groundwater outflow to surface water due to excessive pumping can lead to the drying up of springs, marshes and riverine-riparian systems, which is detrimental

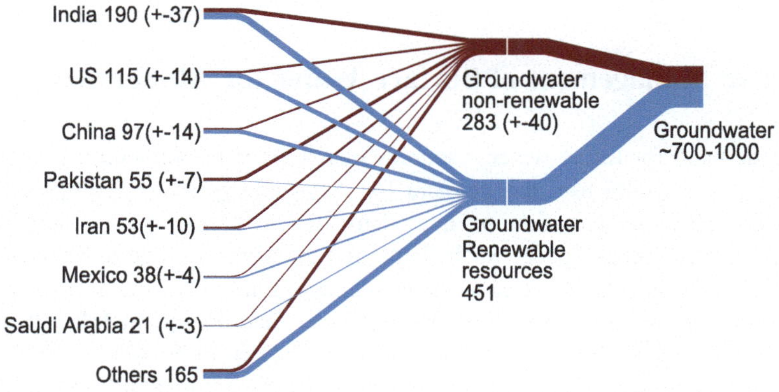

Fig. 1.3 Use of groundwater in selected countries, highlighting renewable and non-renewable (depletion) quantities

to freshwater ecosystems (Sophocleous 2002); this has already happened in many regions.

It is not possible to stop using groundwater, as it is an important resource for many people. However, a balance needs to be struck between allocating water directly for immediate human use (domestic supply, industry and agriculture), and ensuring flows to the environment that sustain ecosystems so that their indirect services can continue to be realised. Long term goals should be established to improve the sustainability of groundwater resource management, because of the long time scales relating to groundwater processes and their impacts (Gleeson et al. 2012). The ‘safe yield’ which defines the long-term balance between the amount of groundwater withdrawn and recharged should be reviewed to recover a balance between the immediate groundwater needs of humans, and the sustainable requirement to maintain recharge and river baseflow (Sophocleous 2000). Groundwater and surface water should also be viewed as a coupled resource, rather than two separate resources. Legislation should therefore not focus only on the sustainable use of groundwater, or the protection of rivers or lakes, but include all water resources in integrated assessments. Examples of such legislation include the conjunctive stream-aquifer management policy in Kansas which includes baseflow requirements as part of the safe-yield regulations (Sophocleus 2000). The National Water Act (NWA) in South Africa also adopts an integrated water resources management approach, and recognises that precipitation, surface water and groundwater are part of a common resource. The legislation uses the term ‘ecological reserve’ to establish a connection between the use of surface and groundwater and the maintenance of freshwater ecosystems (Levy and Xu 2011). Of course, calculating the groundwater quantities needed to maintain acceptable baseflow levels is challenging, and extractions from groundwater will always have an effect on the state of equilibrium of the aquifers. Perhaps an ‘optimal yield’ (Seward et al. 2006) or ‘managed yield’ (Meyland 2011) concept needs to be used more widely, since both recognise the trade-offs between short-term human use and longer-term needs for sustainable environmental health.

Effective Management of Surface Water for All Services

As the demand for future water resources increases, with population growth and increasing per capita use, the demand for clean energy from hydropower also increases. Thus, construction of new dams for irrigation, public supply and hydropower continues apace. Such dams and reservoir storage systems have a negative effect on freshwater ecosystems (changing the downstream flow regime, and withholding sediment and nutrients from rivers and floodplains). It is therefore important that these structures are built or re-engineered to provide a combination of services, including effective management of the environmental flows needed to maintain freshwater ecosystems. Several studies have assessed the possibility of adjusting current dam operational procedures to provide not only a stable supply of water resources, but also to maintain a complex pattern of environmental flows (Richter and Thomas 2007; Olden and Naiman 2010; Suen 2011). Hydropower dams can be operated to release water on a daily basis at a rate that is similar to natural river flow so that the impacts on ecosystems can be reduced (Richter and Thomas 2007). Therefore forecasts for hydropower operations would include not only the hydropower operator's ability to meet the energy demands, but also fluctuating environmental flow objectives. Optimising this use of water for multiple purposes is increasingly understood to be feasible. Guo et al. (2011), for example, have built an ecological operational model for the Three Gorges Project in China to maximise hydropower production subject to ecological constraints related to flow releases that satisfy the minimum and optimal environmental flow demands for fish spawning and reproduction. According to Guo et al. (2011), the hydropower benefit of the reservoir will not decrease dramatically, whilst ecological demands are met. This example shows that reservoir systems can manage multiple uses of water, and a balanced water demand for energy, agriculture, public supply and environmental flows can be achieved.

Multiple Use of Water and Reducing Food Waste

Only 92 km^3 of water is recycled and reused, mostly in the agricultural sector. Investing in more treatment facilities and reusing water several times for different services before it flows to a final sink could also increase the available supply of water resources without affecting other water services. If all industrial and domestic return flows were treated and reused then the amount of water that could be recycled and reused would increase by $\sim$300 km^3 which is equal to approximately 20 % of the RFWR consumed by the agricultural sector. This would require significant investment, but wastewater (defined as all water that is returned to the system from agriculture, industry and domestic) should be seen as an essential resource not only because it is a source of water but also for its nutrient load, which can be reused by the agricultural sector. Typically, nutrients in treated wastewater effluent from sewage treatment plants include nitrogen, phosphorous and potassium, nutrients that can be available for crop production if the water is re-used (Corcoran et al. 2010). There are emerging technologies that can recover nutrients from wastewater that can be used as fertiliser, and although this implies an energy cost in the production process, such methods could provide a hidden benefit in reducing water demand (de-Bashan and Bashan 2004).

Another measure to use RFWR efficiently is to reduce food waste; when food is wasted the water used to grow that food is also wasted. Kummu et al. (2012) estimate that freshwater used in food supply-chain losses are approximately 215 km^3/year. These losses are particularly high in countries that are considered dry such as North Africa and West-Central Asia. Reducing this waste would improve the efficiency of water use in food production, or could be used to maintain other services such as those provided by aquatic ecosystems.

Discussion and Conclusion

The importance of managing *all* global water resources (precipitation and RFWR) for *all* services (direct and indirect services) is emphasised in this paper. There have been many assessments that have focused on calculating the RFWR use and have largely ignored the importance of precipitation for crop production, or for the maintenance of terrestrial ecosystems. Zhao et al. (2010) undertake an input and output analysis of the water footprint and virtual water trade of products in the Haihe river system in China. Their analysis focused only on the amount of RFWR used for the production and trade of these products; the use of precipitation is completely ignored. The justification was that if there is no substitution for the water use of a product with some other water use, then this is irrelevant in the local decision process. But even where irrigation water is applied, precipitation contributes to the plant productivity, and ignoring this misses part of the picture. There is in any event a substitute for this water use, which is the maintenance of terrestrial ecosystem services which provide important indirect services to humans, and that can if necessary be valued.

Global water resource demand is expected to increase in the future. Historically, water resource analysis has focussed on increasing supply of water to different economic sectors, often without proper consideration for the role of water in sustaining the services of terrestrial and aquatic ecosystems. Now that there is a language of ecosystem service assessment, land and water management must incorporate this into its thinking, so that water supply is considered in relation to both short term direct use and longer-term sustainability of these services. It is also imperative that demand-related studies are at the forefront of water management plans, through an emphasis on efficient management of global water resources. This paper highlights global water use for multiple services, and also stresses the importance of integrated management of water and land resources; after all, a land-use decision is also a water-use decision (Duda 2003).

Acknowledgements The research on which this paper is based is financially supported by BP as part of its Energy Sustainability Challenge Initiative, and we gratefully acknowledge this. The authors would also like to thank Dr. Julian Allwood, Professor Chris Gilligan, Professor John Dennis, Professor Paul Linden, Professor Danny Ralph, Professor John Pyle and Dr. Richard McMahon for their excellent contributions. The views expressed in the paper are solely those of the authors.

References

Alcamo J, Doll P, Henrichs T, Kaspar F, Lehner B, Rosch T, Siebert S (2003) Development and testing of the WaterGAP 2 global model of water use and availability. Hydrol Sci J 48(3): 317–337. doi:10.1623/hysj.48.3.317.45290

Arthington AH, Bunn SE, Poff NL, Naiman RJ (2006) The challenge of providing environmental flow rules to sustain river ecosystems. Ecol Appl 16(4):1311–1318

Calder I, Gosain A, Rama Mohan Rao MS, Batchelor C, Snehalatha S, Bishop E(2005) FAWPIO-India Progress Report- August 2005. Forest, Land and Water Policy, Improving outcomes (FAWPIO) programme. DFID

Corcoran E, Nellemann C, Baker E, Bos R, Osborn D, Savelli H (2010) Sick water? The central role of wastewater management in sustainable development. A rapid response assessment. United Nations Environmental Programme, UN Habitat, GRID-Arenal, Norway

Curmi E, Richards K, Fenner R, Allwood JM, Kopec G, Bajzelj B (2013) An integrated representation of the services provided by global water resources. J Environ Manage 129: 456–652. doi:10.1016/j.jenvman.2013.07.031

de-Bashan LE, Bashan Y (2004) Recent advances in removing phosphorus from wastewater and its future as fertilizer (1997–2003). Water Res 38:4222–4246. doi:10.1016/j.watres.2004.07.014

Defries RS, Foley JA, Asner GP (2004) Land-use choices: balancing human needs and ecosystem function. Front Ecol Environ 2(5):249–257

Duda AM (2003) Integrated management of land and water resources based on a collective approach to fragmented international conventions. Philos Trans R Soc Lond B Biol Sci 358(1440): 2051–2062. doi:10.1098/rstb.2003.1410

Falkenmark M, Rockström J (2006) The new blue and green water paradigm: breaking new ground for water resources planning and management. J Water Resour Manage 32:129–132

Gleeson T, Alley WM, Allen DM, Sophocleous M, Zhou Y, Taniguchi M, VanderSteen J (2012) Towards sustainable groundwater use: setting long-term goals, backcasting, and managing adaptively. Ground Water 50(1):19–26. doi:10.1111/j.1745-6584.2011.00825.x

Gleick PH, Allen L, Cohen MJ, Cooley H, Christian-Smith J, Heberger M, Morrison J, Palaniappan M, Schulte P (2011) The World's water, vol 7, The biennial report on freshwater resources. Island Press, Washington

Guo W, Wang H, Xu J, Xia Z (2011) Ecological operation for three Gorges reservoir. Water Sci Eng 4(2):143–156. doi:10.3882/j.issn.1674-2370.2011.02.003

Hanasaki N, Kanae S, Oki T, Masuda K, Motoya K, Shirakawa N, Tanaka K (2008) An integrated model for the assessment of global water resources—Part 2: applications and assessments. Hydrol Earth Syst Sci 12(4):1027–1037. doi:10.5194/hess-12-1027-2008

Hoff H (2009) Global water resources and their management. Curr Opin Environ Sustain 1(2):141–147. doi:10.1016/j.cosust.2009.10.001

Hoff H, Falkenmark M, Gerten D, Gordon L, Karlberg L, Rockström J (2010) Greening the global water system. J Hydrol 384(3–4):177–186. doi:10.1016/j.jhydrol.2009.06.026

Jewitt G (2006) Integrating blue and green water flows for water resources management and planning. Phys Chem Earth 31:753–762

Jones JAA (2010) Water sustainability, a global perspective. Hodder Education, London

Konikow LF, Kendy E (2005) Groundwater depletion: a global problem. Hydrogeol J 13(1):317–320. doi:10.1007/s10040-004-0411-8

Kummu M, de Moel H, Porkka M, Siebert, Varis O, Ward PJ (2012) Lost food, wasted resources: global food supply chain losses and their impacts on freshwater, cropland and fertiliser use. Sci Total Environ 438:477–489

Levy J, Xu Y (2011) Review: groundwater management and groundwater/surface-water interaction in the context of South African water policy. Hydrogeol J 20(2):205–226. doi:10.1007/s10040-011-0776-4

Meyland SJ (2011) Examining safe yield and sustainable yield for groundwater supplies and moving to managed yield as water resources limits become a reality. Paper presented at the WIT Transaction on Ecology and the Environment, 6th Conference on sustainable water resources management, Riverside, California, 23rd–25th May, 2011

Molden D, (2007) Water for food, water for life, a comprehensive assessment of water management in agriculture. Earthscan, London; International Water Management, Columbo

Molden D (2009) Comprehensive assessment of water management in agriculture series;7. In: Wani SP, Rockström J, Oweis T (eds) Rainfed agriculture: unlocking the potential. Cab International, MA, USA

Olden JD, Naiman RJ (2010) Incorporating thermal regimes into environmental flows assessments: modifying dam operations to restore freshwater ecosystem integrity. Freshw Biol 55(1):86–107. doi:10.1111/j.1365-2427.2009.02179.x

Pahl-Wostl C, Arthington A, Bogardi J, Bunn SE, Hoff H, Lebel L, Tsegai D (2013) Environmental flows and water governance: managing sustainable water uses. Curr Opin Environ Sustain 5:1–11 doi:10.1016/j.cosust.2013.06.009

Postel SL, Daily GC, Ehrlich PR, Daily C (1996) Human appropriation of renewable fresh water. Science 271(5250):785–788

Richter BD, Thomas GA (2007) Restoring environmental flows by modifying dam operations. Ecol Soc 12(1):12

Rockström J, Gordon L (2001) Assessment of green water flows to sustain major biomes of the World: implications for future ecohydrological landscape management. Phys Chem Earth Part B 26(1):843–851

Rost S, Gerten D, Bondeau A, Lucht W, Rohwer J, Schaphoff S (2008) Agricultural green and blue water consumption and its influence on the global water system. Water Resour Res 44:W09405

Scanlon BR, Jolly I, Sophocleous M, Zhang L (2007) Global impacts of conversions from natural to agricultural ecosystems on water resources: quantity versus quality. Water Resour Res 43(3). doi:10.1029/2006WR005486

Seward P, Xu Y, Brendonck L (2006) Sustainable groundwater use, the capture principle and adaptive management. Water SA 32(4):473–482

Shiklomanov I (2000) Appraisal and assessment of World water resources. Water Int 25(1): 11–32. doi:10.1080/02508060008686794

Smakhtin V, Revenga C, Döll P (2004) A pilot global assessment of environmental water requirements and scarcity. Water Int 29(3):307–317

Sophocleous M (2000) From safe yield to sustainable development of water resources—the Kansas experience. J Hydrol 235(1–2):27–43. doi:10.1016/S0022-1694(00)00263-8

Sophocleous M (2002) Interactions between groundwater and surface water: the state of the science. Hydrogeol J 10(1):52–67. doi:10.1007/s10040-001-0170-8

Sophocleous M (2007) The science and practice of environmental flows and the role of hydrogeologists. Ground Water 45(4):393–401. doi:10.1111/j.1745-6584.2007.00322.x

Suen JP (2011) Determining the ecological flow regime for existing reservoir operation. Water Resour Manage 25(3):817–835. doi:10.1007/s11269-010-9728-3

Vörösmarty CJ (2000) Global water resources: vulnerability from climate change and population growth. Science 289(5477):284–288. doi:10.1126/science.289.5477.284

UNESCO (2012) Managing water under uncertainty and risk. The United Nations World Water Development Report 4, UNESCO, Paris

Wada Y, van Beek LPH, Bierkens MFP (2012) Nonsustainable groundwater sustaining irrigation: a global assessment. Water Resour Res 48(November 2011), W00L06. doi:10.1029/2011 WR010562

Zhao X, Yang H, Yang Z, Chen B, Qin Y (2010) Applying the input and output method account for water footprint and virtual water trade in the Haihe River Basin in China. Environ Sci Technol 44(23):9150–9156

Chapter 2
Performance Indicators in the Water Resources Management Sector

Balázs M. Fekete and Eugene Z. Stakhiv

Abstract Governance, i.e., the institutional administrative component of water resources management, in particular and natural resources management in general, is an increasingly complex endeavor that forms the basis of integrated water resources management. Hence, it is very difficult to assess the performance of numerous adopted policy decisions and regulations that guide future water management adjustments. A myriad of interconnected aspirational goals, embedded within various well-intentioned UN Declarations, address desired water resources management improvements (efficiency, productivity) as a mechanism to support a wide variety of economic, social, ecological and cultural objectives. A wide array of performance indicators has been developed to track the relative effectiveness of these policies on water use productivity and efficiency. Any useful water resources management performance index must start with an accurate specification of available resources at various water resources management accounting level (country, region, river basins, etc.). Traditionally, water resources accounting was carried out via statistical surveys (e.g., FAO-AQUASTATS). A major step forward is deploying high resolution hydrological data assimilation along with geographical information systems to develop water resources assessment and link those data to spatially distributed socio-economic information.It is difficult enough to accurately describe the core state variables of a water management system, such as annual renewable water resources at national and river basin scales. The complexity and uncertainty magnifies when these state variables are used in composite indices to assess the performance of a diverse assortment of water related

B. M. Fekete (✉)
Department of Civil Engineering, The City College of New York, CUNY Environmental CrossRoads Initiative, CUNY CREST Institute, City University of New York, Steinman Hall Room #188, 160 Convent Avenue, New York, NY 10031, USA
e-mail: bfekete@ccny.cuny.edu

E. Z. Stakhiv
UNESCO International Center for Integrated Water Resources Management (ICIWaRM), Department of Geography and Environmental Engineering, Johns Hopkins University, Baltimore, MD, USA

A. Bhaduri et al. (eds.), *The Global Water System in the Anthropocene*,
Springer Water, DOI: 10.1007/978-3-319-07548-8_2,

investments, or the relative effectiveness of institutional reforms on various components of the water sector (irrigation, water supply, hydropower, etc.). Single indicators are clearly inadequate to guide different aspects of the water management. Composite indices, such as the Water Poverty Index (WPI) or the Environmental Sustainability Index (ESI), which combine aspirational goals with state variables, skew the evaluation outcomes. Furthermore, indices applied uniformly to regions or countries with vastly different hydro-climatological, geographical and socio-economical conditions are prone to lead to misleading comparison.This chapter offers an overview of some of the indicator systems used in the past and discusses some of the challenges in producing resources management indicators. The paper outlines a conceptual framework for indicators that are suitable to guide both water management planning and evaluation of the implementations.

Introduction

The search for meaningful indicators to track progress of various UN initiatives such as the Millennium Development Goals has a long history going back to the United Nations Conference on Environment and Development (UNCED) or Rio Conference (1992). Numerous composite indices have been developed with the intention of assessing the current state of a nation's status with respect to a set of desired objectives—e.g., the Human Development Index (HDI) (UNDP 1990); the Water Poverty Index (WPI) (Sullivan 2002; Sullivan and Meigh 2003) or the Environmental Sustainability Index (ESI). Several variants of these indices have been developed and tested in various countries and river basins (Kemp-Benedict et al. 2011; Cai et al. 2011).

The genesis for this monitoring evolved from the 1992 United Nations (UN) Dublin Conference on Water and the Environment, which established the main principles of modern water management, and which served as the basis of Chap. 18 of the UNCED Agenda 21 (Rio de Janeiro, 1992). These principles included the imperative necessity of reliable information for water resources planning and management. This issue has been repeatedly reemphasized in a series of subsequent UN conferences, such as the World Summit on Sustainable Development (WSSD) (Johannesburg, August–September 2002), the Commission on Sustainable Development (CSD) (2004 and 2005) as well as other international meetings such as the Istanbul Fifth World Water Forum (2009) and the Sixth World Water Forum in Marseilles (2012).

One of the main outcomes of the United Nations Conference on Sustainable Development or Rio + 20 Conference (2012) was the agreement by member States to launch a process to develop a set of Sustainable Development Goals (SDGs), which will build upon the Millennium Development Goals and converge with the post 2015 development agenda. It was decided to establish an "*inclusive and transparent intergovernmental process open to all stakeholders, with a view to developing global sustainable development goals to be agreed by the General Assembly*". It was further agreed that SDGs must be:

- Action-oriented
- Concise
- Easy to communicate
- Limited in number
- Aspirational
- Global in nature
- Universally applicable to all countries while taking into account different national realities.

The development of SDGs will be followed by calls for developing a corresponding set of performance indicators to track the implementation of various SDG initiatives. It is one thing, however, to ask for basic information for status reports on a set of key indicators, as a way of conveying a 'snapshot' overview of a nations' status as a comparative exercise. It is quite another matter to use these indicators and composite indices to determine the relative effectiveness of various UN targets and initiatives, particularly complex ones. The basic reasons are that:

- water resources management comprises many different sectors (irrigated agriculture; municipal water supply, hydroelectric power, flood control, ecosystems, navigation, etc.), each with a different array of management objectives and guiding legislation and regulations;
- many of the water sector reforms deal with a complex array of governance and institutional reform issues, whose direct effects are difficult, if not impossible to evaluate collectively, much less individually;
- national level, aggregate, time-averaged indicators are simply too coarse to be of use for evaluating the relative effectiveness of individual policies on a particular water sector;
- there are too many higher-order national or exogenous global factors that dominate and often catalyze water sector changes, such as trade agreements, globalization, political change or instabilities and lack of financial resources, which are not taken into account in the composite indicators.

Utility of Indicators

Indicators by definition are simplified abstractions of the real world. Some indicators, such as the GDP (Gross Domestic Product) or CPI (Consumer Product Index) are used routinely to summarize the changes in a nation's production and consumption patterns. A great deal of technical effort and resources are devoted to these composite indices so they are often deemed to conceptually and practically reflect key economic trends at the national level.

Traditional one dimensional indicators expressed as a simple ratio of some resource with respect to socioeconomic conditions often lack correlation in a statistical sense (Chenoweth 2008) with policy relevant states of the managed

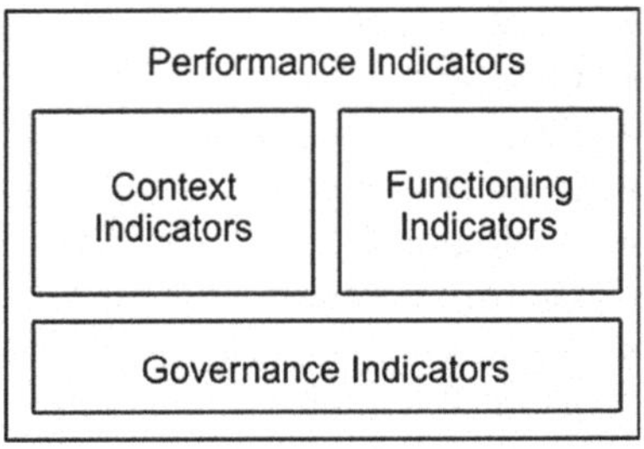

Fig. 2.1 Performance indicators structured around context and function indicators underlined by governance indicators

resources. Indicators structured to address efficiency or productivity in the sense of getting more for the investment ('crop per drop', or an increase in yield) have a bit more success, but there is a great degree of variability between different countries (Kemp-Benedict et al. 2011; Cai et al. 2011).

In a broader sense, resources management needs to consider the '**context**' with respect to the '**function**' of the management activity and its alignment with the prevailing '**governance**' approach (e.g., federal systems vs centralized control). These three dimensions, when structured appropriately may enable a more practical assessment of the relative **performance** of the managed sector (Fig. 2.1).

Context Indicators relate to the natural context (e.g., water availability, rainfall), to infrastructure (such as water treatment capacity, or storage), or to human and economic capitals. 'Context indicators' are required to act as benchmarks when assessing the achievements of another territory with a comparable context. Chenoweth (2008) demonstrated that simple context indicators (e.g., Falkenmark (1986) Water Scarcity Index), and other comparable indices that use per capita water availability as key metric) as a means to evaluate performance are inadequate. "The common sense definition of water scarcity being a state of insufficient water to satisfy normal requirements is of little use to policy makers as it fails to acknowledge degrees of water scarcity and how different societies adapt to this scarcity." In other words, those indices that describe predetermined thresholds are essentially inadequate for policy formulation purposes, or even as descriptors of the current state of scarcity. Chenoweth (2008) also noted, the problem arises because composite indices such as the WPI or the ESI (with the exception of the HDI) often try to link vaguely defined concepts such as IWRM or sustainability with particular desired water resources management outcomes.

Function Indicators relate to inputs, outputs and outcomes (e.g., water use intensity). A number of indicators relate to describing the dynamic functioning of the water sector at the national level (e.g., water withdrawals, water depletion or wastewater actually treated). The WPI and ESI are examples of such formulations. The WPI (Sullivan 2002) intended to provide links between poverty, social deprivation, environmental integrity, access, water availability and health) was applied to 147 countries in 2002. It showed that some of the world's richest nations such as the United States (32nd) and Japan (34th) fare poorly in water ranking, while two developing countries (Guyana, Suriname) score in the top ten. The WPI grades 147 countries to show where the best and worst water situations exist. The

Environmental Sustainability Index (ESI) has similar puzzling outcomes. In the 2012 survey, the US ranked 49th, below Spain (32), Greece (33), Nicaragua (35), and just above Cuba (50), Zambia (57) and Egypt (60).

These composite indices focus on abstract goals and lack the ability to take into account improvements in management, technological adaptation, or globalization (trade and economic changes throughout the globe). The absence of a significant correlation between water availability—either *internal* renewable water resources (the amount of water generated entirely within national boundaries), or with *total* renewable water resources, contradicts the expectation embedded in these indicators that per capita water resources availability play a significant role in determining the ability of a country to satisfy basic water and other human needs (Chenoweth 2008).

Governance Indicators offer possible explanations behind the different levels of performance achieved through the intervention of various policies, programs and regulations intended to improve water use effectiveness, and between a given territory or river basin in comparison with different benchmark territories. Management and governance are sometimes mistakenly considered synonyms, their distinctions is important (Pahl-Wostl 2009). While management refers to activities to keep the state of the resources within desired bounds, governance takes into account the different actors in charge of regulating those who carry out resource management decisions. Depending on the complexity of the inter-relationship between formal and informal institutional hierarchies ranging from state to non-state actors steering resource management objectives faces different challenges. Multi-level, polycentric governance that is characteristic for Western democracies leads to complex interaction between the various governing actors with different level of jurisdiction and degree to influence on management decisions (Pahl-Wostl 2009). The breadth of governance indicators must embrace territorial water resources and a wide variety of water use management improvements to provide an insightful diagnosis of possible weak spots in need of further investigation and possible improvement or reforms (Saleth and Dinar 2004).

Performance Indicators synthesize the three core indicators (context, function, governance) as a targeted consideration of the functioning of a particular sector in relation to its objectives, within a given context. Issues of efficiency/productivity, effectiveness and impact can be considered (e.g., access to water supply and sanitation or value added in agriculture or industry). Only performance indicators can serve an evaluation function—i.e., to provide insights as to general effectiveness or cost-effectiveness of a particular course of action or a set of policies or combination of investments. Performance indicators are not 'state variables' that describe some existing or desired condition. Rather they are meant to illuminate whether a strategy of prescribed actions achieve their intended effect—i.e., they meet the desired outcomes. Water productivity (WP) defined as the ratio of net (economic) benefits from water use (crop, forestry, fishery, livestock or other mixed agricultural systems) and the amount of water utilized in its production (Cai et al. 2011), appears to be a more relevant metric that measures how the system converts water and other resources to goods and services. It is a performance metric relating two state variables (economic output and resources utilization) and assesses existing conditions that could be tracked over time to determine relative changes and improvements.

Cai et al. (2011) assessed WP across 10 major river basins, in several continents, at different stages of development and in various hydro-climatic zones with varying proportion of irrigated land from a high (78 %) to low (1 %). Crop yields were highest in the river basins with the highest proportion of irrigation and lowest in rain-fed systems. The authors ultimately concluded that there was too much variability in the river basins and factors other than water availability were at least as important in WP—especially the availability of markets and highly variable prices for commodities. Kemp-Benedict et al. (2011) studying the same 10 river basins from a different perspective arrived to similar conclusion. They tried to make the link between water availability and poverty. This approach differs from the obverse of the water availability—water-specific forms of deprivation, which forms the core of the Water Poverty Index (Sullivan and Meigh 2003). Kemp-Benedict et al. (2011) found that defining poverty itself is difficult, as it covers such tangible concepts as income, assets, as well as sociological conditions of relative deprivation and well-being (education, mortality, life expectancy) and a sense of empowerment and control over one's life.

The Water Performance index appears to be a step in the right direction, but it still misses the essence of acknowledging the differences in the value of water. Water rich countries like Canada or Japan, will rank poorly in terms of return on each "drop of water", since their agriculture tends to produce crops that are water intensive. In contrast, countries with limited water resources such as Israel are likely to invest in high value crops (e.g., fruits, vegetables) providing high return on investment in water infrastructure.

Indicators developed, so far, rarely take into account the inherent temporal variability of the underlying natural, human induced and socioeconomic processes, although a considerable number of studies were carried out to assess how certain indicators would change under future climate conditions (Arnell 1999, 2010; Schewe et al. 2013)

Since performance indicators are expected to be able to track the changes over time and evaluate if the overall performance is heading in the right directions, all the underlying indicators have to be expressed in a time-varying manner. The core (context, function and governance) indicators need to provide some metrics reflecting those changes that occur outside of the control of the policy makers or managers, in order to distinguish changes in the performance indicators over time that are due to exclusively to policy and management decisions.

Governance and IWRM

One of the most difficult evaluations is the performance assessment of institutional changes (laws, policies, regulations) that are considered key to effective water resources management. *Saleth and Dinar* (2004), in their pioneering study of institutional performance in the water sector, note that "*…water scarcity whether quantitative, qualitative, or both—originates more from inefficient use and poor management than from any real physical limits on supply augmentation.*" According to their analysis, the crisis in the water sector is mainly a function of limitations of contemporary institutions, which allocate and manage water and they advocate a series of institutional reforms that are key for successful implementation of IWRM.

Governance reforms are very difficult to implement, and more difficult still to evaluate their relative effectiveness on improvements in water use efficiency, economic productivity, poverty reduction or improvements in environmental sustainability. *Saleth and Dinar* (2004) tried to develop a mechanism for evaluating these important institutional changes. The quantitative results were quite mixed and in most cases inconclusive. The reason was that effective water governance is predicated on a series of prerequisites, beginning with a well-defined system of water rights or water law and enforcement. Other related water policies, administrative changes and water extraction regulations, including those for privatization, cost recovery, water transfers—were all intertwined with the fundamental requirement to define and enforce water rights.

Governance can be defined as the web of policies, institutional arrangement and management instruments mobilized by the actors making decisions impacting the functioning of the production system on a territory. These instruments, or measures include:

- *Technical* measures used in resource assessment and design of structures used to control, store and supply water for different purposes.
- *Economic* measures used to encourage efficient and responsible allocation and use of water resources including pricing, charges, subsidies and penalties.
- *Administrative* information systems, maps/models, plans, guidelines and other decision support and management tools.
- *Legal* measures, which prescribe restrict or prohibit different water uses including abstraction/discharge permits, codes of conduct and minimum standards.
- *Institutional* regulatory bodies, management arrangements, planning procedures, coordination and partnership mechanisms
- *Social/Participatory* measures to increase awareness of water issues and mobilize users to participate in planning, management and financing of water resource development

IWRM, of course, is the fundamental comprehensive management platform for attaining what is termed 'water security' and sustainable development. The most widely acceptable definition of water security would read as "the availability of an acceptable quantity and quality of water for health, livelihoods, ecosystems and production, coupled with an acceptable level of water-related risks to people, environments and economies" (Grey and Sadoff 2007). The definition is firmly embedded in the concept of sustainable development, with its aim to ensure a triple bottom line of social, environmental, and economic development outcomes. A somewhat more useful categorization for 'water security' was proposed by Turton

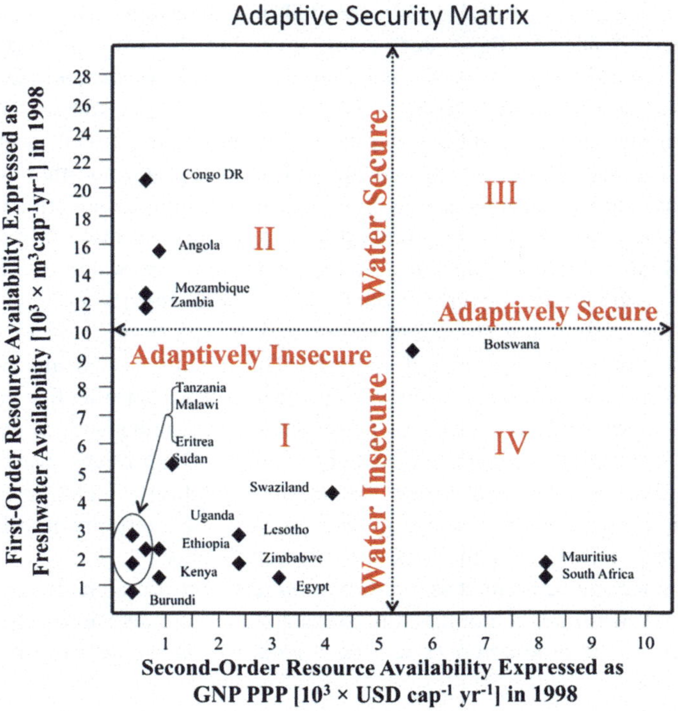

Fig. 2.2 Adaptive security matrix adopted from Turton and Warner (2002) and populated with a wider range of countries

and Warner (2002). They formulated a four quadrant graph of water security/insecurity that is represented by the per capita availability of total renewable water resources versus the relative adaptability of a nation to deal with water scarcity, as reflected by GDP per capita (Fig. 2.2). Both are gross measures, but at least there is a recognition that nations can effectively deal with relative water resources scarcity through improved performance and institutional adaptation. Generally speaking, nations cannot create new water (except for desalination), and cannot readily move from Quadrant I to quadrant II. However, they can become adaptively secure by increasing their GDP. Economic growth will allow investments in technology and infrastructure, and is the key to adaptive water security (towards the quadrants III and IV).

Water Resources Accounting

Reliable water resource accounting is a fundamental basis for establishing any water related indicators. The lack of adequate hydrological monitoring (Hannah et al.

Table 2.1 Global freshwater fluxes to ocean estimates in km^3/year

Source	Discharge
Baumgartner and Reichel (1975)	37,713
Korzoun et al. (1978)	46,900
L'vovich et al. (1990)	39,700
Oki et al. (1995)	22,311
Postel et al. (1996)	40,700
Grabs et al. (1996)	42,700
Nijssen et al. (2001)	36,103
Fekete et al. (2002)	38,320
Dai et al. (2009)	37,288
Schlosser and Houser (2007)	36,000
Syed et al. (2010)	32,851
Wisser et al. (2010)	37,405
Haddeland et al. (2011)	42,000–66,000

2010; Fekete et al. 2012) severely hinders optimal design and operation of the water infrastructure and allocation of water resources. Global water resource assessments that could guide international efforts to promote sustainable human developments are in particularly difficult due to inconsistencies in data reporting, obstacles in data sharing and declining monitoring capabilities (Vörösmarty et al. 2001).

Water resource planners often use various statistics (long term mean, extremes, exceedance probability, etc.) in lieu of the time varying records partly as a convenience to reduce the amount of data needed, but as a necessity, when observation records are only available as statistical summaries. Traditionally, these statistics were assumed to be stationary, but this assumption was challenged (Milly et al. 2008) arguing that with changing climate these statistics will no longer remain steady and water managers will need to consider Global Circulation Model predictions to prepare for anticipated changes.

The first global assessments of the global freshwater resources (Baumgartner and Reichel 1975; Korzoun et al. 1978; L'vovich et al. 1990) were solely based on water balance considerations. Later studies included country statistics (Shiklomanov 2000; Shiklomanov and Rodda 2002) or discharge gauge records (Grabs et al. 1996; Fekete et al. 2002). Up until the first decade of the 21st century the different assessments appeared to narrow around the 36,000–39,000 km^3/year (Table 2.1) with some notable exception. Both Oki et al. (1995) and Syed et al. (2010) based their water budget assessments on atmospheric water budget derived from the weather forecast model reanalysis from NCAR–NCEP (Kalnay et al. 1996; Kistler et al. 2001) which has known deficiencies in representing the amount of water participating in the water cycle (Fekete et al. 2004).

Before Haddeland et al. (2011), the global estimates appeared within the 36,000–40,000 km^3/year that Fekete et al. (2002) suggested as the most plausible range based on the recognition that only 50 % is monitored for discharge recording 20,700 km^3/year on average based on the Global Runoff Data Centre's data (Fekete et al. 2002).

The discharge to ocean estimates from Haddeland et al. (2011) were generated as a model inter-comparison exercise under the EU WATCH[1] program where global scale hydrological models were tested with the same bias corrected Global Circulation Model climate forcings, therefore the deviation from previous reported values is not surprising, but the wide range among the tested models is disturbing. Recent effort under the Inter-sectoral Impact Model Intercomparison Project (ISI-MIP) (Schiermeier 2012) arrived to similar spread while testing different large scale impact assessment models with hydrological component (Davie et al. 2013; Schewe et al. 2013).

The large uncertainties in water resource estimates based on hydrological modeling combined with the similar uncertainties in GCM future climate predictions makes (Milly et al. 2008) suggestion to rely more on climate and hydrological simulations for numerical water resources planning questionable. Fekete and Stakhiv (2013) argued that water managers best guidance will remain to use hydrological observations and continuously update of the long-term statistics. Since water infrastructures are typically designed for 30–40 years, when major repairs and upgrades are inevitable, past records with sufficient extra safety buffer will remain viable basis for specifying design criteria, while the continuous update of the long-term statistics allows for adaptation of changing climate (Fekete and Stakhiv 2013).

Conclusions

Improving availability of spatially specific hydro-meteorological, bio-geophysical and socioeconomic data opened new opportunities in supporting water managers and decision makers with up-to-date and comprehensive information. Given the plethora of diverse data available, policy makers will need aggregated information in the form of performance indicators that combine context, function and governance indicators as a single metric providing insights into the effectiveness of a particular course of action or a set of policies or investments. Comprehensive performance indicators need to objectively assess whether given policy decisions lead to desired outcomes without preconditioning or biasing the policy options to a narrowed subset that are preferred "a priori".

Simplistic context indicators such as the Falkenmark index or function indicators such as the Water Poverty Index or the Environmental Sustainability Index, promoted in the past, are insufficient to substitute for true performance indicators, because they are limited to identifying potential shortages of water without recognizing the capabilities of individual societies to adapt to water resources limitations via the deployment of technology, better water resource management, etc.

In this chapter, a conceptual framework for performance suitable to evaluate water resources management was proposed that distinguishes three underlying core indicators (**context**, **function** and **governance**) to characterize the boundary

[1] http://www.eu-watch.org.

conditions (context) within which water resources management needs to operate, the objectives (functioning) of the water resources utilization and the regulatory, institutional and management environment (governance) dictating how policy decisions and implementations are carried out.

Instructive **performance indicators** need to synthesize **context**, **function** and **governance** indicators in order to provide an integrated metric that can guide water managers and policy makers. Context, function and governance indicators need to be able the characterize spatial and temporal variability, while performance indicators are expected to factor in the spatial differences and temporal variations in a manner that would allow performance evaluations irrespective of space and time.

Regardless of the simplicity or complexity of various indicators intended to help evaluation for policy making and management decisions, the validity of any indicators ultimately hinges on the underlying data. The huge uncertainties in water resources estimates severely hinder the application of any indicators. Without adequate monitoring and reliable data, 'tweaking' the existing array of indicators remains a futile exercise that won't improve decision making.

References

Arnell NW (1999) Climate change and global water resources. Glob Environ Change 9:S31–S49

Arnell NW (2010) Hydrology and global environmental change. In: Malone TF, Roederer JG (eds) Global environment change. Pearson, Cambridge, UK, pp 4–6

Baumgartner A, Reichel E (1975) The world water balance. Elsevier, New York

Cai XL, Molden D, Mainuddin M et al (2011) Producing more food with less water in a changing world: assessment of water productivity in 10 major river basins. Water Int 36:42–62. doi:10.1080/02508060.2011.542403

Chenoweth J (2008) A re-assessment of indicators of national water scarcity. Water Int 33:5–18. doi:10.1080/02508060801927994

Davie JCS, Falloon PD, Kahana R et al (2013) Comparing projections of future changes in runoff from hydrological and biome models in ISI-MIP. Earth Syst Dyn 4:359–374. doi:10.5194/esd-4-359-2013

Dai A, Qian T, Trenberth KE (2009) Changes in continental freshwater discharge from 1948 to 2004. J Climat 22:2773–2792

Falkenmark BYM (1986) Fresh water-time for a modified approach. AMBIO 15:192–200

Fekete BM, Looser U, Robarts RD (2012) Rationale for monitoring discharge on the ground. J Hydrometeorol 13:1977–1986

Fekete BM, Stakhiv EZ (2013) Water management preparation strategies for adaptation to changing climate. In: Charles R, Kumagai M, Robarts RD (eds) Climate change and global warming of inland waters: Impacts and mitigation for ecosystems and societies, 3rd edn. John Wiley & Sons

Fekete BM, Vörösmarty CJ, Grabs W (2002) High-resolution fields of global runoff combining observed river discharge and simulated water balances. Glob Biogeochem Cycles 16(15):1–6. doi:10.1029/1999GB001254

Fekete BM, Vörösmarty CJ, Roads JO, Willmott CJ (2004) Uncertainties in precipitation and their impacts on runoff estimates. J Climatol 17:294–303

Grabs W, de Couet T, Pauler J (1996) Freshwater fluxes from the continents into the world oceans based on data of the global runoff data base. Global Runoff Data Centre, Koblenz

Grey D, Sadoff CW (2007) Sink or swim? Water security for growth and development. Water Policy 9:545. doi:10.2166/wp.2007.021

Haddeland I, Clark DB, Franssen W et al (2011) Multimodel estimate of the global terrestrial water balance: setup and first results. J Hydrometeorol 12(5):869–884. doi: 10.1175/2011JHM1324.1
Hannah DM, Demuth S, van Lanen HAJ et al (2010) Large-scale river flow archives: importance, current status and future needs. Hydrol Process. doi:10.1002/hyp.7794
Kalnay E, Kanamitsu M, Kistler R et al (1996) The NCEP/NCAR 40-year reanalysis project. Bull Am Meteorol Soc 77:437–471. doi: 10.1175/1520-0477(1996)077
Kemp-Benedict E, Cook S, Allen SL et al (2011) Connections between poverty, water and agriculture: evidence from 10 river basins. Water Int 36:125–140
Kistler R, Kalnay E, Collins WD et al (2001) The NCEP/NCAR 50-year reanalysis: Monthly means CD-ROM and Documentation. Bull Am Meteorol Soc 82:247–267. doi: 10.1175/1520-0477(2001)082
Korzoun VI, Sokolov AA, Budyko MI et al (1978) Atlas of the world water balance. Gidrometeoizdat, Leningrad
L'vovich MI, White GF, Belyaev AV et al (1990) Use and transformation of terrestrial water systems. In: Turner BL, Clack WC, Kates RW et al (eds) Earth as transformed by human action. Cambridge University Press, Cambridge, USA, pp 235–252
Milly PCD, Betancourt J, Falkenmark M et al (2008) Stationarity is dead: whither water management? Science 319:573–574
Oki T, Musiake K, Matsuyama H, Masuda K (1995) Global atmospheric water balance and runoff from large river basins. Hydrol Process 9:655–678
Pahl-Wostl C (2009) A conceptual framework for analysing adaptive capacity and multi-level learning processes in resource governance regimes. Glob Environ Change 19:354–365. doi:10.1016/j.gloenvcha.2009.06.001
Postel SL, Daily GC, Ehrlich PR (1996) Human appropriation of renewable fresh water. Science 271(2):785–788
Saleth RM, Dinar A (2004) The institutional economics of water: a cross-country analysis of institutions and performance. Edward Elgar, Cheltenham, UK
Schewe J, Heinke J, Gerten D et al (2013) Multimodel assessment of water scarcity under climate change. Proc Nat Acad Sci U.S.A. doi:10.1073/pnas.1222460110
Schiermeier Q (2012) Models hone picture of climate impacts. Nature 482:286. doi:10.1038/482286a
Schlosser CA, Houser PR (2007) Assessing a satellite-era perspective of the global water cycle, J Climate 20:1316–1338
Shiklomanov IA (2000) Appraisal and assessment of world water resources. Water Int 25:11–32
Shiklomanov IA, Rodda JC (2002) World water resources at the beginning of the 21st century. UNESCO, Paris
Sullivan CA (2002) Calculating a water poverty index. World Dev 30:1195–1210
Sullivan CA, Meigh J (2003) Considering the water poverty index in the context of poverty alleviation. Water Policy 5:513–528
Syed TH, Famiglietti JS, Chambers DP et al (2010) Satellite-based global-ocean mass balance estimates of interannual variability and emerging trends in continental freshwater discharge. Proc Nat Acad Sci U.S.A 107:17916–17921
Turton AR, Warner JF (2002) Exploring the population/water resources nexus in the developing world. The Woodrow Wilson Institute, Washington, DC, p 52–82
UNDP (1990) Human Development Report 1990. United Nations Development Programme (UNDP)
Vörösmarty CJ, Askew A, Barry RG et al (2001) Global water data: a newly endangered species. AGU EOS Trans 82:54, 56, 58
Wisser D, Fekete BM, Vörösmarty CJ, Schumann AH (2010) Reconstructing 20th century global hydrography: A contribution to the Global Terrestrial Network- Hydrology (GTN-H). Hydrol Earth Syst Sci 14(1):1–24. doi: 10.5194/hess-14-1-2010

Chapter 3
Virtual Water and Trade: A Critical Economic Review

Erik Gawel

Abstract In recent years many scholars have dealt with aspects of a "globalisation of water resources" implicating the need for a global approach to governing scarce water resources. Especially the concepts of virtual water and water footprints have garnered increasing attention due to their pledge to disclose the linkages of local water consumption and global agricultural trade. In response, trade-restricting policy instruments have been promoted by some authors in response to seemingly inefficient, unfair or unsustainable "virtual water"-trade patterns. To shed some light on the link between food trade, water and sustainability this paper discusses the informative value of the virtual water and water footprint concepts from an economic point of view, including various refinements of these indicators which have been suggested in the literature. Additionally, the performance of trade-related global water governance arrangements based on virtual water will be considered, bringing up again the debate about the environmental benefits of free trade. It must be concluded that the virtual water concept is limited in terms of its usefulness in providing policy advice or guiding economic decision-making. Specific sustainability problems (distorted pricing, bad governance, trade performance) should be solved in the respective arenas and not by virtual water-related global governance schemes or even trade barriers.

E. Gawel (✉)
Department of Economics, Helmholtz Centre for Environmental Research – UFZ,
Permoserstr. 15, 04318 Leipzig, Germany
e-mail: erik.gawel@ufz.de

A. Bhaduri et al. (eds.), *The Global Water System in the Anthropocene*,
Springer Water, DOI: 10.1007/978-3-319-07548-8_3,

Virtual Water and Trade: Aspects of a Discussion

In recent years the concepts of virtual water (VW) and water footprints (WF) have garnered increasing attention as a quantitative assessment of global water use. The catchword *virtual water* was coined by the geographer *Anthony Allan* in the 1990s and refers to the water used in the production process of a good. Subsequently, the concept has been widely linked to trade issues, particularly for agricultural products. The corresponding term *virtual water trade* refers to the circumstance that virtual water crosses international borders as a result of trade in water-intensive goods (Allan 1998). These trade flows and their implications have been discussed controversially, in particular since the beginning of the millennium, when the concept of WF (e.g. of a country) was additionally introduced (Hoekstra and Hung 2002).[1] Thus, virtual water can be accounted for by various measures and indicators such as virtual water contents, footprints and trade flows (Chapagain and Hoekstra 2008). Apart from mere accounting, VW concepts also have been interpreted *normatively*, that is to say that a certain footprint level or trade direction might be to a certain extent "right" or "wrong" or could be "improved" in a way. If this is the case, the concept might also be used as a scientific basis for policy advice on water-related trade or consumption patterns in order to tackle water scarcity. While many scholars from natural sciences consider the concept appropriate for normative purposes there is also serious critique against it (Perry 2014), particularly from economists (e.g. Wichelns 2004, 2011; Meran 2011; Gawel and Bernsen 2013).

As normative criteria both *resource efficiency* and *justice* have been introduced: It is remarkable that the discussion on VW and WF in this regard has changed its perspective. In the 1990s virtual water trade was originally meant to increase the global water use efficiency (Allan 1998; Hoekstra 2006), which is attained by producing water-intensive goods in the most water-abundant or water-productive regions (Hoekstra and Hung 2002). Later, WF calculations have been established to provide an accounting framework to implement resource fairness. This way, WF quotas might contribute to an allocation not according to natural water endowment, but according to the philosophy of fair shares ("virtual water for all"—Zehnder 2010).

The role of trade is somewhat ambivalent in this debate: On the one hand trade is most welcome to achieve a global water use efficiency and to contribute to global water savings by making water available for global food demand where it is the least scarce and most productive. On the other hand, VW volumes and trade flows at times are seen to have problematic moral implications for fairness and justice, which seemingly can only be addressed by restraining free trade on a global scale (Hoekstra 2011). VW trade seems to instigate fairness concerns, especially in the context of trade relations between industrial and developing countries (World Water Council 2004). In this perspective trade is no longer a supporting tool to meet global demand for food by using less water, rather it is considered now a powerful means to

[1] While usually VW and WF are just seen as different ways of looking at the same thing (Gawel and Bernsen 2013), Velazquez et al. 2009 try to elaborate some relevant differences.

access and to take possession of scare resources all over the world for particular needs in a possibly unfair way (Wackernagel and Rees 1997).

What is important here is that in *every* case free trade might not always work towards achieving these aims—be it water use efficiency or a certain idea of water justice (Hoekstra 2006). Thus, a normative interpretation of VW always implies to reorganise trade patterns according to a given normative criterion related to VW accounting. Accordingly, global water governance arrangements have been deemed necessary by some authors (Verkerk et al. 2008) to counter the detrimental impacts of trade on the sustainable and "fair" use of water resources (Hoekstra 2006, 2011).

Two crucial research questions arise from this: What can virtual water (VW) analysis tell us about the "right" direction or volume of foreign trade (Sect. "What Can Virtual Water Analysis Tell Us About the "Right" Trade Patterns?")? And is there a need for a global water governance reorganising trade patterns with respect to water availability in order to meet sustainability requirements (Sect. "Virtual-Water Oriented Trade as a Means to Obtain Water Resource Efficiency?")?

What Can Virtual Water Analysis Tell Us About the "Right" Trade Patterns?

Virtual-Water-Oriented Trade as a Means to Obtain Water Resource Efficiency?

According to the first normative criterion VW analysis might reveal "efficiency gains" through virtual water trading (global water savings—Fader et al. 2011) and uncover "efficiency" deficits in current trade patterns. In this perception, virtual water has to flow from water-abundant to water-scarce regions. Thus, *strengthening* of VW trade and a strict alignment to a "global water use efficiency" should be pursued (Allan 1998; Hoekstra and Hung 2003).

However, from an economic perspective, the idea of reorganising trade patterns according to the availability of one single production factor appears to be weird in a sense. First, having a look at the empirical evidence water availability is without any clear impact on trade volumes and trade flows (Kumar and Singh 2005).[2] Rather there is a more significant impact by the availability of (arable) land: Often trade flows can be observed from "water-scare but land-abundant" countries to "water-abundant, but land-scarce" countries. For instance, the Netherlands as a water-abundant but land-scarce country turns out to be a net importer of VW whereas Kazakhstan (water-scarce but land-abundant) is a net exporter (Kumar and Singh 2005). Theoretically, these findings are hardly surprising: Water is just one single trade determining production factor besides land, labour, capital and

[2] An early statistical and modelling analysis on the water-food trade relations has been conducted by Yang et al. (2003).

knowledge. It can be shown in a simple Heckscher-Ohlin-model (HO-model) representing the standard economic textbook theory for foreign trade between countries with different resource endowments that 1. relative endowment of water is not decisive for optimal trade direction with respect to mutual welfare and that 2. relative endowment *should not* be relevant either, again for reasons of welfare (Meran 2011). According to the HO-theorem in a 2-goods-2-factors (say water W and land L)-2-countries-model a country turns out to be an exporter of the very good that is using the "relative abundant" factor more intensively. But this does not necessarily imply that the water-abundant countries are net exporters of virtual water: The HO-theorem is based on *relative abundance* with respect to the *factor proportion* (i.e. $W_1/L_1 > W_2/L_2$) between countries, not on absolute factor scarcity. Therefore, a relatively water-rich country ($W_1 > W_2$) may be at the same time a net importer of VW (Ansink 2010). There is no distinct relation between water availability and preferable trade position (Meran 2011). This is why any attempt to bring trade patterns into line with mere VW numbers run the risk of decreasing overall welfare. Furthermore, foreign trade may result in global water savings but does not necessarily have to: Theoretically, a mere (but welfare-increasing) redistribution of the same total amount of water used might be a possible solution as well.

Virtual-Water Oriented Trade as a Means to Achieve Resource Fairness?

Recently, normative concepts of *justice* have also been associated with virtual water trading. This way, virtual water trade now may appear as a problem to be tackled by VW analysis revealing fairness deficits in both consumption and trade patterns and suggesting a *restriction* of VW trade with respect to fairness norms leading to a certain "anti-trade bias" (van den Bergh and Verbrüggen 1999). The manifold arguments brought forward in this regard (see Table 3.1) can be mainly divided into three dimensions (Gawel and Bernsen 2013): 1. imperfections in the current regime of world trade; 2. the wide regional and interpersonal disparities in income and (water) consumption; and 3. potential adverse local impacts as a consequence of using water resources to produce tradable goods.

The most apparent flaw of this approach is the arbitrariness of the distribution norms to come into question and the consequential ambiguity of their implications: Again, no clear conclusion can be drawn from the calculations of virtual water indicators. Rather, nearly everything seems to be problematic: a trade flow from the North to the South (dependencies of the poor) as well as a trade flows from the South to the North (exploitation of scarce resources, pollution export). Moreover, a dilemma becomes evident considering a water-rich country such as the USA with respect to what it should do with its abundant water resources. An above-average consumption of water would appear to be just as "unfair" as a hoarding strategy concerning the own (sufficient) endowment (Hoekstra 2011; Verkerk et al. 2008).

Table 3.1 Normative criticisms to virtual water trade and their related policy dimensions

Indicators of virtual water		Corresponding set of concerns	Dimension of the problem		
			Trade regime	Consumption patterns	Adverse local impacts
Virtual water specific consumption	*Virtual water content of a commodity*	Implies a high water consumption of the respective product, which is why the water intensive meat consumption in industrialised countries is questioned (Lotze-Campen and Welp 2007)		X	X
	Water footprint of a product	Is meant to show consumers their negative impact on (external) water resources (Hoekstra et al. 2009)		X	X
	Water footprint of a person	An indicator of the "wasteful" behaviour of people in the industrialised world (Hoekstra et al. 2009). May lie below or above the average "fair share" a person is entitled to (Hoekstra 2006)		X	
	Water footprint of a company	Indicator for the awareness and risk of a company of its water use throughout all of its operations (The CEO Water Mandate 2009)		X	X
	Water footprint of a country	The extent to which a nation is depleting its own water resources, as well as water resources in other parts of the world (Hoekstra and Chapagain 2007). May differ from a nation's "reasonable share" (Verkerk et al. 2008)		X	
	Global virtual water budget	May not coincide with the "maximum human water footprint" (Verkerk et al. 2008)		X	X

(continued)

Table 3.1 (continued)

Indicators of virtual water		Corresponding set of concerns	Dimension of the problem		
			Trade regime	Consumption patterns	Adverse local impacts
Virtual water trade flows	*Magnitude of virtual water trade*	Shows the rising interdependencies between countries, which are oftentimes thought to be at the detriment of the poorest countries (Neubert, 2008).	X		X
	Water savings through virtual water trade	Shows the global water savings of Virtual Water trade that may or may not occur (de Fraiture et al. 2004; Chapagain et al. 2005)			X
	Virtual water trade balance	Water scarce countries may have a trade surplus while water rich countries may have a trade deficit (Chapagain and Hoekstra 2008). Large trade surpluses in industrial countries result in dependencies of importing countries (Warner 2003), while trade deficits point to wasteful consumption patterns (Sonnenberg et al. 2009). Adverse local impacts may result from virtual water net imports, as well as net exports	X	X	X
	Virtual water import dependency	Import dependency is often seen as problematic from the viewpoint of poor arid countries (World Water Council 2004). Water-abundant industrial countries are enabled to pressure importers politically (Roth and Warner 2008). Water import dependency and water scarcity are usually not correlated (Hoekstra and Hung 2002)	X		
	Top exporters and top importers	Virtual Water exports are dominated by industrialised countries (Zehnder 2003)	X		
	Interregional trade flows	Dry regions may be net exporters or on the other hand be too dependent on Virtual Water imports from another region (World Water Council 2004)	X	X	
	External water footprint of a country	A country can be blamed for certain environmental problems in the exporting nation (Hoff 2009). Industrial countries are generally seen to have a too high external water footprint		X	X

Source Gawel and Bernsen (2013)

The only other option would be to export VW, which (as illustrated above) would also entail "unacceptable" states of dependence for developing countries, according to trade critics. Whatever this country might do it could be condemned for the sake of justice. This reveals the contradictory nature of these normative concepts.

Do Water Footprints Reveal Relevant Information on Sustainability?

WFs are intended to be a transfer of previous footprint concepts such as the carbon footprint to the realm of water resources, assessing amongst other things the virtual water content of people, nations, firms or products (Hoekstra et al. 2009). WF are considered to be in line with previous developed carbon footprint concepts (Gerbens-Leenes et al. 2007). However, greenhouse gases are a real *global* problem, since their emission contributes to global warming in a *homogeneous* way, regardless of where and when exactly the emission takes place. Thus, the carbon footprint always reveals strictly comparable quantitative information about an activity's impact on climate change. However, final decisions about abatement strategies have to take into account not only the magnitude of the carbon footprint but additional (economic) indicators like abatement costs and the values that people place on different climate-relevant activities. In contrast, water is a *heterogeneous* resource with diverse local impacts. While a significant carbon footprint always indicates a high impact on climate change (even though still no direct policy conclusion can be derived), a considerable WF does not even provide information about whether environmental harm has actually occurred (CEO Water Mandate 2009). In a similar vein, the concept of "water neutrality" (Hoekstra 2008) is more difficult to substantiate than the corresponding "carbon neutrality" since water depletion and water pollution are site-specific problems in particular.

Therefore, no valid information concerning the sustainability of water use can be provided by WF. And no economically relevant information is given either on where best to reduce water input.

Suggested Remedies: Is There a Case for a VW-Based Global Governance?

Many scholars have stated that the majority of water crises around the world do not actually relate to absolute water scarcity, but to poor water management and the lack of appropriate water prices resulting in water management crises (e.g. Rogers and Hall 2003; OECD 2003). On which scale should these management problems be addressed? The traditional view in hydrology and water resource management is that the mobility of water is confined within river basins, and thus "the management of water on one continent has no direct bearing on the management of water on another continent" (Young et al. 1994, p. 18). The Agenda 21 explicitly stipulates that water resources should be managed at the river basin level, as does the European Union's Water Framework Directive.

However, this traditional view has been contested by some (e.g. Hoekstra 2011), who point to global linkages ("teleconnections") induced by natural and anthropogenic forces, which might lead to a "globalisation of water resources" and imply the need for a global governance approach. Apart from physical "teleconnections" throughout the biosphere such as large-scale irrigation impacts on intercontinental climate patterns, moisture feedback effects and water-related climate change due to global GHG emissions, a second kind of interconnection "originating from economic globalisation and agricultural trade" is seen to be at work (Hoff 2009, p. 141). Thus, "globalisation" and international trade could change the location of production and water use and "[transform] water into a global issue" (World Water Assessment Programme 2009, p. 35). The need for global water governance is therefore put forward because many driving forces behind water-related problems and conflicts are beyond the scope of national, local, or water catchment–oriented governance (Pahl-Wostl et al. 2008; Schnurr 2008; Moss and Newig 2010; Hoekstra 2011). Following this perception, water might appear as a "major global public good" (Pahl-Wostl et al. 2008).

A coordinated global water policy framework, however, does not exist today (Dellapenna and Gupta 2009), and global water governance still remains an academic concept (Ünver 2008). However, ideas for regulating foreign trade with respect to water resources are rampant in this debate: Petrella (2001) even suggests the introduction of a world water contract that would declare water a global good and the common patrimony of humanity, which should not be subject to trade transactions or purchased by foreign investors. Ethical aspects are also emphasised by McKay (2003) who proposes a VW Trading Council within the WTO, which would be concerned with the redistribution of VW on ethical grounds. Hoekstra (2006, 2011) and Verkerk et al. (2008) suggest a whole range of global institutional arrangements to promote "fairness" and "resource efficiency" in water use around the world. These include an international water-pricing protocol, international business agreements, a pollution tax on internationally traded goods that cause water pollution in their waste stage, the labeling of water-intensive products, and a scheme of WF quotas. WF quotas have the objective of assigning a "reasonable" or "fair" share of the world's water resources to every country and person, in the face of widely diverging per-capita consumption rates between industrial and developing countries (Hoekstra 2011). For a country to remain within its "reasonable bounds", a tax on water-intensive (import) goods is recommended (Hoff 2009).

However, from an economic point of view, it appears to be rather questionable whether trade restrictions really make sense in order to promote sustainability (Gawel and Bernsen 2011a; LeVernoy and Messerlin 2011): The contentious issue whether trade is always welfare-increasing or might particularly aggravate environmental problems (and should therefore be regulated) has been discussed already since the 1990s. From the ecological economics point of view, trade might have a tendency toward overexploitation of resources and also imply the risk of discriminating developing countries (Daly 1993; Daly and Cobb 1994). In contrast, neoclassical economists emphasise the economic benefits of trade in general and the ineptitude of trade regulation to specifically address regional environmental problems (Bhagwati 1993; Schulz 1996; Siebert 1996). Even in the face of market failure it is local environmental policy that is needed rather than trade restricting rules. Although local governance and water pricing are far from being perfect, to say nothing of the trading rules that determine international trade, one should be cautious in deciding whether a particular global governance approach can really address the existing problems. Firstly, the conception of water as a global public good (theoretically characterized as non-rivalrous and non-excludable) is questionable, since the scope of its benefits and externalities is still mostly local or regional (Mehta 2002). Market-based global drivers or impacts should not be confused with global commons! Therefore, even though many water use-related impacts are widespread around the globe, they are, in a strict economic sense, not truly global in nature as is the case for climate protection (Mehta 2003; Vörösmarty et al. 2004; Gawel and Bernsen 2011c).[3] This does not affect global water governance approaches in general but reminds us that teleconnections "originating from economic globalization and agricultural trade" (Hoff 2009, p. 141) should not be mixed up with global externalities (that may indeed occur in the global hydrological cycle). Secondly, global trade-flow regulations or the imposition of average water consumption levels will lead to nothing but distortions and losses in wealth. Trade restricting policies would be highly arbitrary and even paternalistic since they represent ideas of global equity and fairness without taking into consideration the needs and preferences of individuals or even "poor countries". Furthermore, they deny developing countries the capability to decide on production and trade patterns in their own best interests. Hence, policies and instruments aiming at reorganising global trade patterns according to merely quantitative VW calculations run the risk of being inefficient (neglecting costs and preferences), ineffective (not solving local environmental problems) and at the same time even patronising (restricting local production and trade decisions).

May Recent Refinements Overcome the Flaws?

May VW analysis deliver more relevant information if it is refined (for an overview see Lillywhite et al. 2010) and takes more aspects of scarcity into account?

[3] Population growth and changing consumption patterns will differ among regions, and this will affect the associated impact of water scarcity, which is why "water is far from having the properties of a global public good" (Mehta 2003, p. 556). By comparison, the global climate system *can* be characterized as a global public good, because no person on earth can be excluded from its benefits or from the negative consequences of climate change.

The most prominent recent development in this field might be the distinction between *green* and *blue water*. According to Falkenmark (2003), blue water refers to the water found in rivers, lakes and groundwater aquifers, while green water denotes the water stored the unsaturated zone of soils stemming directly from rainfall and thus being used for biomass production of rainfed agro-ecological systems. Only one third of global precipitation becomes runoff in rivers and recharges aquifers, whereas two thirds infiltrate into the soil and form green water resources (Hoff et al. 2010). The supposed relevance of the water's colour stems from the assumption, that in terms of supply green water is a "free good", which bears low or zero opportunity costs (Schubert 2011), while blue water causes high opportunity costs due to its many alternative uses, and deserves specific attention in VW and WF accounting (Yang et al. 2006).

However, the colour approach first and foremost reveals that unspecific calculations of VW disregarding heterogeneity of resource use, commonly used so far, have obviously been deficient. But what do we gain by taking additionally into account the water's colour? The over-simplifying assumption that green water always has lower opportunity costs than blue water neglects that soil moisture can indeed have substantial opportunity costs, while ground- or surface water can have low opportunity costs under certain conditions (Wichelns 2010a). Assuming that green water bears no opportunity costs reveals a great deal of anthropocentrism, because cultivated land might alternatively serve as a habitat for other species and contribute to biodiversity—even though biodiversity in itself serves humankind (Biewald 2011), which is why the existence of negative or missing opportunity costs for green water is doubtful. To assess appropriately water abstraction for human needs we need to take into account the respective full opportunity costs which is a continuous variable. A simple dichotomous distinction between "blue" and "green" cannot serve as a sound base for an economic assessment of resource use—even more if we consider that opportunity costs of blue water are unclear and context-sensitive and that green water is needed for ecosystem services competing with agricultural use and thus is not at all an economically "free good" (that is free of opportunity costs). Using green water for a non-commercial habitat (instead of cultivation of crops) might be of the same total economic value as using blue water for shipping (instead of irrigation). Thus, the introduction of colours does not solve the main flaw of the concept—to address heterogeneity of water resource use in a sound way that allows for discriminating products or trade flows to be "good" or "bad".

Moreover, already the distinction of water into a blue and green category runs into serious difficulties, because these two are not necessarily distinct (Wichelns 2011). In the hydrological cycle, water which transpires from plants and evaporates from surface water or soils comes back as rainfall and interacts with rivers, lakes and groundwater reservoirs, while certain plants and land uses can have significant impacts on blue water resources (Ridoutt and Pfister 2010).

On the aggregated level of trade flows, the colour of water is irrelevant from the perspective of both trading partners and consumers in the importing countries, because decisions about efficient water use are still made on the local level and have to take into account opportunity costs depending on the socio-economic context without any distinct relation to "colours".

Beyond "colouring" water there have recently been introduced several other concepts in order to improve the explanatory power of VW analysis. To make VW and WF assessments more informative as to the actual local impact of water use, some authors have attempted to weight explicitly water footprints with indicators of scarcity, sustainability or even shadow prices.

The concept of an "unsustainable WF" (Schubert 2011) only considers blue water, which has been "unsustainably" extracted in the place of production, or which has been polluted to "some inacceptable degree". WF calculations according to this concept (just like scarcity-weighted WF—see below) usually lead to different figures than traditional WF analysis, which reveals a great deal of arbitrariness and contributes to the fragmentation of the WF concept (Lenzen et al. 2013). Just as arbitrary is the qualification "to a certain degree", which gives no idea about the external costs of agricultural production. On the other hand, the definition of "unsustainable" might not be universally accepted, and therefore cannot be simply prescribed to any country. In conclusion, the concept would demand a huge amount of data collection while being only a controversial and unnecessary loop way in assessing (locally already well-known) problems of water scarcity and pollution.

Another upcoming strand of literature aims at weighting the VW flows using explicit information on economic scarcity. This is relevant since it allows for more accurate addressing of heterogeneity of resource uses in terms of values (instead of colours or dichotomous variables like "non-sustainability").

Ridoutt and Pfister (2010) introduced a scarcity-weighted water footprint, which again includes blue and grey water only. Here, the water use at every production step is weighted with a scarcity indicator from the producing region. Additionally, the characterisation of blue virtual water consumption in a specific river basin with the methods of lifecycle impact assessment (LCIA) has been proposed by Pfister et al. (2009). Others (Aldaya and Llamas 2009; Aldaya et al. 2010; Garrido et al. 2010) take an economic approach by including water productivity into virtual water analysis. Garrido et al. (2010) introduce a new indicator of apparent water productivity, the price of a good divided by its virtual water content, as well as the terms of trade of virtual water, the value of virtual water imports divided by the value of virtual water exports. Thus, virtual water flows here are valued by the prices of the goods for whose production the water was used. Biewald et al. (2011) attempt to evaluate blue water savings induced by trade by a weighting with shadow prices. The authors assess the green and blue water use with and without international trade, and come to the conclusion that globally less water is used as a result of trade, although the use of blue water slightly increases, while regionally especially arid countries save large amounts of blue water. The blue water savings are then weighted with shadow prices, which leads to an index depicting the value of water savings (difference in blue water consumption [trade and no trade] times water shadow price).

May these enhancements of VW and WF analysis really overcome the fundamental flaws described above? The newly introduced "economic" approach of assigning value-oriented informations (e.g. Garrido et al. 2010) has already been commented on by Wichelns (2010c) who observes that the inclusion of economic productivities, product prices and exchange terms is "of questionable value" (p. 692) because methods to assess agricultural water productivities already exist. The perspective that such an approach will add credibility and stature to the application of virtual water to policy questions is therefore misplaced (p. 694) and is in any case too "water-centric" ignoring all other factors of production.

The approach of Biewald et al. (2011), for instance, is to weight the (blue) water savings resulting from trade with shadow prices to assess the savings' monetary value. Hence, instead of quantitative flows we obtain value flows. Does this really remedy the mentioned shortcomings? First of all, it is not obvious how fictional shadow prices, even if these could correctly be calculated, can give information about the welfare gains from free trade, because they do not consider changes in other activities and factor uses. Weighted VW calculations still do not contain the information which actually is of interest, that is, the efficiency of local decisions on water use. Thus, the problems of traditional VW analysis remain. Just like it doesn't matter to a country whether it imports blue or green VW resources (Wichelns 2010b), it should not matter whether A imports greater euro-amounts in VW than B, as long as this water has been employed sustainably in the place of production, and water has been remunerated according to its scarcity. Finally, shadow prices still give no information as to what would be a "good" trade flow, or where water savings of which colour are especially desirable (Gawel and Bernsen 2011b).

The various efforts to enhance the water footprint's informative value all strive in completely different directions, which adds to the impression of a widespread confusion. Unfortunately, while none of the new concepts is really convincing, the great awareness which is supposed to be created among consumers will be degraded because VW and WF analysis will be ever more fragmented, since every methodology will lead to different VW contents, values and implications.

Conclusions

With respect to VW, foreign trade is subject to either expectations (water savings, water use efficiency) or concerns (fairness, participation)—in either case it is often suggested that trade should be (re-) organised according to (physical) water availability. However, VW accounting unfortunately does not at all provide reliable information neither for economic decisions on water-related trading nor a global governance regulating trade or consumption patterns.

First, VW accounting lacks relevant economic scarcity information concerning heterogeneous water resources and thus to provide suitable policy advice. The mere counting of water quantities does not offer specific information on values, particularly whether trading VW really illustrates or even causes an unsustainable exploitation of water resources. Since no information on local costs and benefits of water extraction and water use is given, and the influence of other production factors as well as preferences for import and export goods is neglected, it is not possible to the determine the "right" direction of VW trade flows this simple way. There is no stable relation between a country's welfare and its net position of trading VW. Taking up trade can lead to a negative net position of a water-scarce country while increasing welfare at the same time. This is due to the fact that trade patterns (volumes and directions) are economically based on both preferences and full comparative production costs not only on relative water endowment. For the same reason, the welfare enhancing effect of trading water-intensive goods does not necessarily depend on concurrent global water savings. Moreover, VW net positions are no sound indicators for fairness of water resources distribution.

The various efforts to enhance the water footprint's informative value all strive in completely different directions. Ultimately, however, these refinements mainly confirm the impression that the concept in its current form is not useful in giving reasonable policy advice. It may be noted that the concepts have succeeded in creating a kind of qualitative awareness to the great amounts of water which are at times "hidden" in our food, but until now has not offered specific information for any further policy-relevant conclusions.

Second, to evaluate trade flows numerous normative criteria are used, be it "water resource efficiency" or fairness of global water access. Hence, the normative framework as well as the conclusions drawn appear to be highly contradictory. It remains unsettled whether the aspired goal consists of a realignment of trade flows according to either principles of equity and justice or concepts of scarcity or "global water use efficiency". Applying these (contradictory) normative criteria almost every conceivable trade pattern could be animadverted on.

When addressing globally sustainability problems of regional water use, a "problem of fit" must be stated, that is, problems related to water depletion and pollution are addressed at the wrong scale. To achieve sustainability in resource use we have to take into account local and regional economic scarcity as well as relevant externalities induced by water use. This should not be mixed up in (restricting) trade policies. The prerequisites for a sustainable regional water management (cost-covering water prices, good governance etc.) and the challenge of fairness in global trade regimes have to be addressed in their respective arenas. For that reason, environmental and trade policies should not be based on mere VW calculations and their predominantly misleading policy implications.

Acknowledgments The author would like to thank two anonymous referees for helpful comments.

References

Aldaya MM, Llamas MR (2009) Water footprint analysis (hydrologic and economic) of the Guadiana River Basin. Accompanying report to the U.N. world water development report no. 3. United Nations Educational, Scientific and Cultural Organization, Paris

Aldaya MM, Garrido A, Llamas MR, Varelo-Ortega C, Novo P, Casado RR (2010) Water footprint and virtual water trade in Spain. In: Garrido A, Llamas MR (eds) Water policy in Spain. CRC Press, Leiden, pp 49–59

Allan JA (1998) Virtual water: a strategic resource, global solutions to regional deficits. Groundwater 36(4):545–547

Ansink E (2010) Refuting two claims about virtual water trade. Ecol Econ 69(10):2027–2032

Bhagwati J (1993) The case for free trade. Sci Am 269:18–23

Biewald A (2011) Give virtual water a chance! An attempt to rehabilitate the concept. GAIA 20(3):168–170

Biewald A, Lotze-Campen H, Rolinski S, Hoff H (2011) The impact of trade on local green and blue water availability. Discussion paper. Potsdam Institute for Climate Impact Research, Potsdam, 15 Apr 2011

CEO Water Mandate (2009) From footprint to public policy—the business future for addressing water issues. Pacific Institute DP, Pacific Institute of Public Policy, Port Vila, Vanuatu. http://ceowatermandate.org/files/research/From_Footprint_to_Public_Policy.pdf. Accessed 18 Jan 2014

Chapagain AK, Hoekstra AY (2008) The global component of freshwater demand and supply: an assessment of virtual water flows between nations as a result of trade in agricultural and industrial products. Water Int 33(1):19–32

Chapagain AK, Hoekstra AY, Savenije HHG (2005) Saving water through global trade. Value of water research report series 17. Institute of Water Education, Delft, Netherlands. http://www.waterfootprint.org/Reports/Report17.pdf. Accessed 18 Jan 2014

Daly HE (1993) The perils of free trade. Sci Am 269:24–29

Daly HE, Cobb J (1994) For the common good—redirecting the economy toward community, the environment and a sustainable future. Beacon Press, Boston

Dellapenna JW, Gupta J (eds) (2009) The evolution of the law and politics of water. Springer, Berlin

Fader M, Gerten D, Thammer M, Heinke J, Lotze-Campen H, Lucht W, Cramer W (2011) Internal and external green-blue agricultural water footprints of nations, and related water and land savings through trade. Hydrol Earth Syst Sci Discuss 8(1):483–527

Falkenmark M (2003) Freshwater as shared between society and ecosystems: from divided approaches to integrated challenges. Philos Trans R Soc Lond B 358(1440):2037–2049

de Fraiture C, Cai X, Amarasinghe U, Rosegrant M, Molden D (2004) Does international cereal trade save water? The impact of virtual water trade on global water use. In: Comprehensive assessment research report 4, Colombo, Sri Lanka. http://www.iwmi.cgiar.org/Assessment/files/pdf/publications/ResearchReports/CARR4.pdf. Accessed 30 Jun 2011

Garrido A, Llamas MR, Varela-Ortega C, Novo P, Rodríguez-Casado R, Aldaya MM (2010) Water footprint and virtual water trade in Spain—policy implications. Springer, New York

Gawel E, Bernsen K (2011a) Do we really need a water footprint? global trade, water scarcity and the limited role of virtual water. GAiA 21(3):162–167

Gawel E, Bernsen K (2011b) The colour of water: what does it tell us about scarcity? GAiA 21(4):224–228

Gawel E, Bernsen K (2011c) Globalisation of water: the case for global water governance? Nature + Culture 6(3):205–217

Gawel E, Bernsen K (2013) What is wrong with virtual water trading? On the limitations of the virtual water concept. Environ Plann C (Government and Policy) 30(6):168–181

Gerbens-Leenes W, Hoekstra A, Holland R, Koch G, Moss J, Ndebele P, Orr S, Ronteltap M, de Ruyter van Stevenink E (2007) Water neutrality: a concept paper. http://www.indiaresource.org/campaigns/coke/2008/Waterneutrality.pdf. Accessed 18 Jan 2014

Hoekstra AY (2006) The global dimension of water governance: nine reasons for global arrangements in order to cope with local water problems. Value of water research report series 20. Institute for Water Education, Delft, Netherlands. http://doc.utwente.nl/58371/1/Report_20.pdf. Accessed 18 Jan 2014

Hoekstra AY (2008) Water neutral: reducing and offsetting the impacts of water footprints. Value of water research report series 28. Institute for Water Education, Delft, Netherlands. http://www.waterfootprint.org/Reports/Report28-WaterNeutral.pdf. Accessed 18 Jan 2014

Hoekstra AY (2011) The global dimension of water governance: why the river-basin approach is no longer sufficient and why cooperative action at global level is needed. Water 2011(3): 21–46

Hoekstra AY, Chapagain AK (2007) Water footprints of nations: water use by people as a function of their consumption pattern. Water Resour Manag 21(1):35–48

Hoekstra AY, Chapagain AK, Aldaya MM, Mekonnen MM (2009) Water footprint manual, state of the art 2009. Water footprint report. Water Footprint Network, Enschede, Netherlands. http://www.waterfootprint.org/?page=files/Publications. Accessed 18 Jan 2014

Hoekstra AY, Hung PQ (2002) Virtual water trade: a quantification of virtual water flows between nations in relation to international crop trade. Value of water research report series 11. Institute for Water Education, Delft, Netherlands. http://www.waterfootprint.org/Reports/Report11.pdf. Accessed 18 Jan 2014

Hoekstra AY, Hung PQ (2003) Virtual water trade: a quantification of virtual water flows between nations in relation to international crop trade. Value of water research report series 11. Institute for Water Education, Delft. http://www.waterfootprint.org/Reports/Report11.pdf. Accessed 18 Jan 2014

Hoff H (2009) Global water resources and their management. Curr Opin Environ Sustain 1(2): 141–147

Hoff H, Falkenmark M, Gerten D, Gordon L, Karlberg L, Rockström J (2010) Greening the global water system. J Hydrol 384(3–4):177–186

Kumar MD, Singh OP (2005) Virtual water in global food and policy making: is there a need for rethinking? Water Resour Manag 19(6):759–789

Lenzen M, Moran D, Bhaduri A, Kanemoto K, Bekchanov M, Geschke A, Foran B (2013) International trade of scarce water. Ecol Econ 94:78–85

LeVernoy A, Messerlin P (2011) Water and the WTO: don't kill the messenger. Paper prepared for the workshop "accounting for water scarcity and pollution in the rules of international trade" Amsterdam, 25–26 Nov 2011

Lillywhite RD, Allan AJ, Chapagain A, Garnett T (2010) Water footprinting—a review in support of the defra project WU0120. Department of Environment Food and Rural Affairs, London

Lotze-Campen H, Welp M (2007) More food with less water: the role of efficiency gains, lifestyles and trade. In: Lozán JL, Graßl H, Hupfer P, Menzel L, Schönwiese D (eds) Global change: enough water for all? Hamburg, p 306–309

McKay J (2003) A proposal for an international virtual water trading council: building institutional frameworks at international level to reduce poverty. In: Ünver O, Gupta RK, Ayşegül K (eds) Water development and poverty reduction. Kluwer Academic Publishers, Norwell, pp 111–128

Mehta L (2003) Problems of publicness and access rights: perspectives from the water domain. In: Kaul I (ed) Providing global public goods, managing globalization. Oxford, New York, p 556–575

Meran G (2011) The unimportance of "virtual water" for environmental policy. http://www.infraday.tu-berlin.de/fileadmin/fg280/veranstaltungen/infraday/conference_2011/papers_presentations/presentation—meran.pdf. Accessed 18 Jan 2014

Moss T, Newig J (2010) Multi-level water governance and problems of scale: setting the stage for a broader debate. Environ Manag 46(1):1–6
Neubert S (2008) Strategic virtual water trade—a critical analysis of the debate. In: Scheumann W (ed) Water politics and development cooperation: local power plays and global governance. Springer, Berlin, pp 123–145
Organisation for Economic Co-operation and Development (OECD) (2003) Improving water management: recent OECD experience. OECD, Paris
Pahl-Wostl C, Gupta J, Petry D (2008) Governance and the global water system: a theoretical exploration. Glob Gov 14(4):419–435
Perry C (2014) Water footprints: path to enlightenment, or false trail? Agric Water Manag 134:119–125
Petrella R (2001) Wasser für alle. Ein globales Manifest, Rotpunktverlag, Zürich
Pfister S, Koehler A, Hellweg S (2009) Assessing the environmental impacts of freshwater consumption in LCA. Environ Sci Technol 43(11):4098–4104
Ridoutt BG, Pfister S (2010) A revised approach to water footprinting to make transparent the impacts of consumption and production on global freshwater scarcity. Glob Environ Change 20(1):113–120
Rogers P, Hall A (2003) Effective water governance. TEC Background Paper 7. Global Water Partnership, Stockholm. http://www.tnmckc.org/upload/document/bdp/2/2.7/GWP/TEC-7.pdf . Accessed 18 Jan 2014
Roth D, Warner J (2008) Virtual water: virtuous impacts? The unsteady state of virtual water. Agric Hum Values 25(2):257–270
Schnurr M (2008) Global water governance: managing complexity on a global scale. In: Scheumann W (ed) Water politics and development cooperation: local power plays and global governance. Springer, Berlin, pp 107–120
Schubert H (2011) The concepts of virtual water and water footprints. acatech Materialien no. 4. Acatech, Munich
Schulz C-E (1996) Trade policy and ecology. Environ Resource Econ 8(1):15–38
Siebert H (1996) Trade policy and environmental protection. Kiel Working Papers 730. https://www.econstor.eu/dspace/bitstream/10419/878/1/194370208.pdf. Accessed 18 Jan 2014
Sonnenberg A, Chapagain A, Geiger M, August D (2009) Der Wasser-Fußabdruck Deutschlands. Woher Stammt das Wasser, das in unseren Lebensmitteln steckt?, WWF Germany, Berlin
Ünver O (2008) Global governance of water: a practitioner's perspective. Glob Gov 14(4): 419–435
Van den Bergh J, Verbrüggen H (1999) Spatial sustainability, trade and indicators: an evaluation of the 'ecological footprint'. Ecol Econ 29(1):61–72
Velázquez E, Madrid C, Beltrán MJ (2011) Rethinking concepts of virtual water and water footprint in relation to the production-consumption binomial and the water-energy nexus. Water Resour Manag 25(2):743–761
Verkerk MP, Hoekstra AY, Gerbens-Leenes PW (2008) Global water governance: conceptual design of global institutional arrangements. Value of water research report series 26. Institute for Water Education, Delft, Netherlands. http://www.waterfootprint.org/Reports/Report26-Verkerk-etal-2008GlobalWaterGovernance.pdf. Accessed 18 Jan 2014
Vörösmarty C, Lettenmaier D, Leveque C, Meybeck M, Pahl-Wostl C, Alcamo J, Cosgrove W, Grassl H, Hoff H, Kabat P, Lansigan F, Lawford R, Naiman R (2004) Humans transforming the global water system. Eos, Trans Am Geophys Union 85(48):509–520
Wackernagel M, Rees W (1997) Unser Ökologischer Fußabdruck: Wie der Mensch Einfluss auf die Umwelt Nimmt. Birkhäuser Verlag, Basel
Warner J (2003) Virtual water – virtual benefits? Scarcity, distribution, security and conflict reconsidered. In: Hoekstra AY (ed) Virtual Water Trade–Proceedings of the International Expert Meeting on Virtual Water Trade. IHE Delft Value of Water Research Report Series No. 12
Wichelns D (2004) The policy relevance of virtual water can be enhanced by considering comparative advantages. Agric Water Manag 66(1):49–63

Wichelns D (2010a) Virtual water: a helpful perspective, but not a sufficient policy criterion. Water Resour Manag 24(10):2203–2219

Wichelns D (2010b) Virtual water and water footprint offer limited insight regarding important policy questions. Int J Water Resour Dev 26(4):639–651

Wichelns D (2010c) Virtual water and water footprints: policy relevant or simply descriptive? Int J Water Resour Dev 26(4):689–695

Wichelns D (2011) Virtual water and water footprints. Compelling notions, but notably flawed. GAIA 20(3):171–175

World Water Assessment Programme (2009) Water in a changing world. Third UN world water development report. UNESCO, Paris

World Water Council (2004) Virtual water trade—conscious choices. E-conference synthesis. World Water Council, Marseille. http://www.waterfootprint.org/Reports/virtual_water_final_synthesis.pdf. Accessed 18 Jan 2014

Yang H, Reichert P, Abbaspour K, Zehnder AJB (2003) A water resources threshold and its implications for food security. Environ Sci Technol 37(14):3048–3054

Yang H, Wang L, Zehnder AJB, Abbaspour KC (2006) Virtual water trade: an assessment of water use efficiency in the international food trade. Hydrol Earth Syst Sci 10:443–454

Young GJ, Dooge JCI, Rodda JC (1994) Global water resource issues. Cambridge University Press, Cambridge

Zehnder AJB (2003) Water issues: the need for action at different levels. Aquat Sci 65(1):1–20

Zehnder AJB (2010) Water food nexus. Presentation at the first water research horizon conference, Berlin, 13 July 2003

Chapter 4
Data, Models and Uncertainties in the Global Water Cycle

R. J. Harding, A. J. Dolman, D. Gerten, I. Haddeland, C. Prudhomme and P. van Oevelen

Abstract Growing water scarcity will be a major challenge for society in the 21st century. Tackling this challenge requires a multi-scale and interdisciplinary approach to water science in order to understand the complex and interlinked nature of the global water system and how it may change now and in future. There are still considerable uncertainties in our quantification and understanding of the global water cycle. One of the major themes for Global Energy and Water Exchanges (GEWEX, a core project of the World Climate Research Programme) in the coming years is to better understand and predict precipitation variability and changes, and to understand how changes in land surface and hydrology influence past and future changes in water resources and security. These questions focus on the exploitation of improved data sets of precipitation, soil moisture, evapotranspiration, and related variables to close the water budget over land, for providing improved information for products related to water quantity and quality for decision makers, and for initializing seasonal and long-term climate change projections. Through a number of case studies this paper explores newly available data sets and modelling initiatives describing the global water cycle and its associated uncertainties. These studies illustrate how the GEWEX science questions cover many of the challenges facing water science in the coming years, including the improvement of our modelling and prediction of precipitation and

R. J. Harding (✉) · C. Prudhomme
Centre for Ecology and Hydrology, Wallingford, Oxon, UK
e-mail: rjh@ceh.ac.uk

A. J. Dolman
VU University Amsterdam, Amsterdam, The Netherlands

D. Gerten
Potsdam Institute for Climate Impact Research, Potsdam, Germany

I. Haddeland
Norwegian Water Resources and Energy Directorate, Oslo, Norway

P. van Oevelen
International GEWEX Project Office, Silver Spring, MD, USA

A. Bhaduri et al. (eds.), *The Global Water System in the Anthropocene*,
Springer Water, DOI: 10.1007/978-3-319-07548-8_4,

evaporation, the development and use of new data sets, the better understanding of extremes and the representation of realistic land surface complexity, with all anthropogenic influences, into our analysis. The ultimate result should be better and more relevant tools to inform society of potential impacts and adaptation options to climate and environmental change.

Introduction

Humanity faces many challenges in the future—not the least of these is the possibly growing scarcity of freshwater and the interrelationship between water use and use of the land surface (Gerten et al. 2013). While most of the problems and solutions to future water scarcity lie in governance and ownership of water, all assessments must be underpinned by a thorough and quantified knowledge of the global water cycle and an understanding of the interrelationships between water, climate and land surface processes (see e.g. Dadson et al. 2013).

Water is crucial to most of the services our aquatic and terrestrial ecosystems provide: agricultural production, carbon budgets (and other biogeochemical cycles), biodiversity, energy generation, industrial production and human health. Particularly hydrologic extremes play an important role—floods and droughts are pressure points on water scarcity and environmental damage (IPCC 2012). For decades there has been increasing pressures on available water in many regions of the world due to increasing water demand because of a growing population and wealth (Kummu et al. 2010), which together with the potential impacts of climate change on water availability and water demand are likely to aggravate water scarcity in the future (Schewe et al. 2014).

Increasing greenhouse gases are likely to lead to increases in temperature, a trend already observed (IPCC 2007). Higher temperatures will increase evaporation, over the oceans in particular, and hence will increase water vapour in the atmosphere (the evidence suggests the absolute humidity has increased but the relative humidity has decreased slightly, continuing the increase in evaporative demand, IPCC 2013). The increase in humidity is likely to lead to overall higher rainfall globally and the likelihood of more intense rainfall regionally (e.g. Groisman et al. 2005). The changing patterns of precipitation are far more complex than those of temperature, depending on not just the thermodynamics of the atmosphere but also on the details of the atmospheric circulations. These circulations frequently depend on the small or regional scale processes (and are difficult to simulate with climate models). It can be concluded, however, that the mean picture is likely to be one of increasing rainfall, although there will be large regions where rainfall decreases. There is a consensus that wet areas (and particularly northern latitudes) are likely to get wetter and drier areas drier (see e.g. Allan et al. 2010). The variability in space and time of rainfall makes it difficult to establish trends, though. Overall we have yet to identify a statistically robust mean increase in global precipitation, but we do see in the last few decades a trend of increasing precipitation at high latitudes and an increase in intense rainfall in some regions (IPCC 2013).

The Global Energy and Water EXchanges (GEWEX) Project is one of four core projects under the World Climate Research Programme (WCRP). *GEWEX itself was set up in the early 1990s to co-ordinate international efforts to observe, understand and model the hydrological cycle and energy fluxes in the Earth's atmosphere and at the surface.* In the last three decades it has co-ordinated and encouraged the production of consistent global data sets describing the water and energy budget of the of the earth, it has sponsored a suite of extensive land surface campaigns to better characterise land-atmosphere interactions (of energy and water) and co-ordinated international modelling studies to better understand and describe these interactions and their impacts on water resources.

This paper summarizes some typical case studies undertaken by GEWEX scientists to contribute to our understanding of global water resources and presents the new GEWEX questions developed in the last 2 years to guide the future research direction of the global change research community.

Uncertainties in the Global Water Cycle: Past and Future

To provide an unambiguous analysis of the changes and uncertainties in the global water cycle we require global data sets produced using consistent methodologies and which make use of a wide range of in situ and satellite data and modelling products. GEWEX has co-ordinated and sponsored the development of a wide range of global products. A number of consistent data sets of precipitation covering the globe (or at least the land areas) have been produced (see Dadson et al. 2013 for an overview). These gridded estimates rely on the interpolation of the network rain gauges. While many regions, such as Europe and North America, have extensive rain gauge networks, many regions have sparse networks and the estimates are often supplemented by satellite estimates and analysis from the national weather forecasting centres. Biemans et al. (2009) analyzed seven datasets and found a spread of at least 10 % for the annual global mean (see Fig. 4.1), which propagates to runoff estimates if used in hydrological models. Regionally the range in estimates can be much larger, notably in the Arctic and mountainous regions, where the spread in estimates can be greater than 100 % (see Biemans et al. 2009, Fig. 3b). Even in the larger basins precipitation uncertainty is typically 30 % of the mean. Precipitation is poorly simulated by the climate models—the mean global precipitation from the CMIP5 runs has a 22 % spread and a bias, on average, of 10 % (Trenberth pers. comm.). Again, at the basin scale this spread will be even larger. Accurate precipitation estimates are obviously fundamental to assessing water resources but are also critical in understanding flows of energy through the climate system.

River discharge is monitored widely around the world. The Global Runoff Data Centre (GRDC http://www.bafg.de/) archives discharge data for almost 9,000

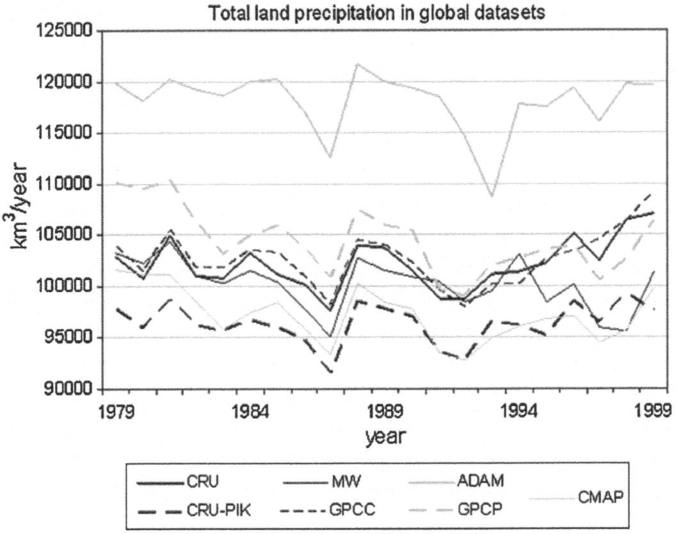

Fig. 4.1 Rainfall estimates from major global data sets (Biemans et al. 2009)

gauging stations worldwide, two-thirds of which have daily data. However, like the rainfall data, the spatial (and temporal) coverage is patchy with large gaps in Africa (excluding South Africa) and Southern Asia. River discharge is uniquely valuable as one of the few measures which integrate over large areas of the land surface (see e.g. Dai et al. 2009), unfortunately many discharge measurements have either been discontinued or are unavailable to the global change community. The number of stations in the GRDC data base peaks in the 1970s and many parts of Africa, south America and Asia have little new data since that time. The international community should continue strenuous effort to support countries to maintain their hydrometric networks and make their observations freely available. The interpretation of discharge measurements is complex because of the impact of human interventions: dams, extractions, transfers and land cover changes and also for some catchments 'naturalised' flows have been estimated.

In an attempt to provide globally consistent fields of discharge macro-scale hydrological models are often used. Global Water System Project (GWSP) and the EU funded WATCH project (e.g. Harding et al. 2011) have co-ordinated an intercomparison of hydrological models globally (WaterMIP, Haddeland et al. 2011), making use of a new global data set of meteorological data (the WATCH Forcing Data, WFD, Weedon et al. 2011). This ensured the models used consistent driving data and a consistent terrestrial grid including a common river routing network. Eleven models were included in the intercomparison, including Global Hydrological Models (GHMs) and stand-alone versions of the land surface models commonly used in climate models (Haddeland et al. 2011). The main distinction between these two classes of models is that GHMs solve the water balance alone whereas the land surface models solve both the energy and water balances. All but

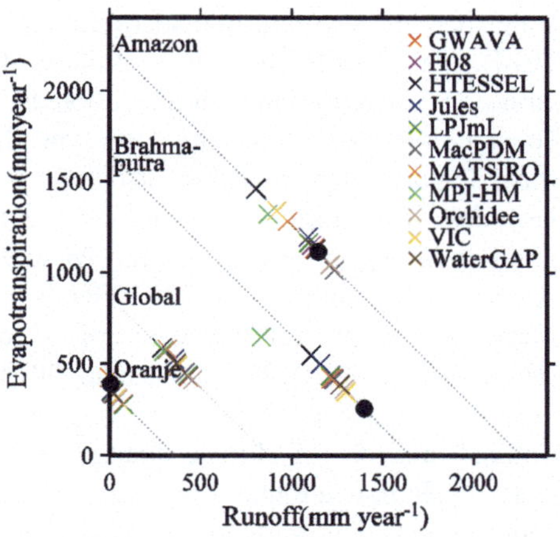

Fig. 4.2 Runoff and evapotranspiration for global terrestrial surface and three major basins calculated from a range of global hydrology and land surface models (see Haddeland et al. 2011 for details)

one of the models (WaterGAP) was run without calibration via observed discharge data. The initial analysis was for "naturalised" conditions (Haddeland et al. 2011) i.e. excluding human influences related to land cover changes, damming, water abstraction and irrigation.

The eleven models showed a significant spread of the partitioning of precipitation into evaporation and runoff. The average annual global land surface evaporation varied between models from 415 to 586 mm/year and runoff from 290 to 457 mm/year, Fig. 4.2. There was no single cause for the spread in model outputs, although the different model treatment of snow was a major factor explaining the different shapes of the simulated annual hydrographs. Because all models use the same precipitation the range in mean annual discharge must be a reflection of the wide range in evaporation calculations. There is clearly a need to improve evaporation simulations in large scale hydrological models and provide benchmarks to validate models.

Compared with observed discharge most models overestimate total annual runoff in semi arid regions (e.g. in the Oranje River, see Fig. 4.2)—probably a result of both water extractions not being included in this phase of WaterMIP, and wetland evaporation, typically not being included in these models. Interestingly the runoff for the Brahmaputra was under-estimated—this is probably a result of the underestimate of precipitation in the Himalayan region. In the Brahmaputra River basin temperature based evapotranspiration schemes resulted in less runoff than evapotranspiration schemes also taking radiation into account. In other basins, e.g. the Amazon River basin, the parameterization of the evaporation water intercepted on the canopy was found to be one of the factors causing the spread in evapotranspiration and runoff estimates.

The climate modelling community has a long history of systematic model intercomparison through the climate model intercomparison projects (CMIPs; Meehl et al. 2007). The impact modeling community has only recently started assessing future large-scale changes in land surface water fluxes and available water resources in multimodel frameworks (Hagemann et al. 2013; Schewe et al. 2014). The results show that climate change impacts do not only depend on emission scenarios and climate models, but that different impact models give considerably different results. However, the results also exhibit a number of large-scale features. In particular, discharge is projected to increase at high northern latitudes, in eastern Africa and on the Indian peninsula, and to decrease in a number of regions including the Mediterranean and parts of North and South America (Hagemann et al. 2013; Schewe et al. 2014). In other parts of the globe, however, the projections are subject to a large spread across the ensemble (Hagemann et al. 2013; Schewe et al. 2014). In addition to climate change, humans alter the water cycle by constructing dams and through water withdrawals. Multimodel analyses of climate change and direct human impacts on the terrestrial water cycle indicate that direct human impacts on the mean annual water fluxes in some regions, e.g. parts of Asia and in the western United States, are of the same order of magnitude, or even exceed impacts to be expected for moderate levels of global warming (+2° K) (Haddeland et al. 2013).

Evaporation: The Cinderella of the Global Water Cycle

Evaporation[1] (*ET*) is a key component of the global hydrological and energy cycle. Together with precipitation, *ET* determines to a large extent the availability of soil moisture. In turn land cover and the availability of soil moisture determine the evaporation. Evaporation is difficult to measure in a consistent way and so, despite its importance and unlike rainfall and discharge, there is not extensive network of evaporation measurements (the nearest is the FLUXNET network (http://fluxnet.ornl.gov/introduction), with just over 500 sites). It is only in the last decade that consistent global estimates of evaporation have become readily available, because of the lack of an extensive network of in situ measurement such estimates rely heavily on a combination of satellite and modelling products. GEWEX has sponsored the development of these products through its LandFlux project (http://wgdma.giss.nasa.gov/landflux.html) and Regional Hydrological Projects (http://www.gewex.org/projects.html).

Current Earth System models show a large variability in *ET* estimates (Mueller et al. 2011) and a good benchmark data set at the global scale is still unavailable. The key components of *ET* (i.e. transpiration through plant's stomata, bare soil evaporation and evaporation of vegetation-intercepted water) vary globally and in time. Interception loss (the evaporation directly from free water on vegetation following rainfall), is often not included explicitly in evaporation estimates, it will be particularly important in forest (responsible for the evaporation of

[1] In this paper we refer to Evaporation as the sum of all its components, transpiration, soil, open water, interception etc. This is sometimes also referred to as Evapo-transpiration.

approximately 13 % of the total incoming rainfall over broadleaf evergreen forests, 19 % in broadleaf deciduous forests, and 22 % in needleleaf forests—according to estimates using a novel satellite driven way of calculating the interception loss of vegetated land surfaces, Miralles et al. 2011). Thus in high rainfall regions the high interception losses from forest will lead to higher total evaporation compared with grassland or arable crops.

In the framework of the LandFlux-EVAL initiative (www.iac.ethz.ch/url/LandFlux-EVAL), several *ET* datasets based on observations, diagnostic datasets, land surface models and re-analyses are evaluated. We find that recent declining trends in terrestrial evaporation (Jung et al. 2010) are corroborated but suggest that these may be related to rainfall variability arising from El Nino/La Nina cycles (El Nino conditions are associated with negative anomalies of ET and soil moisture in most of the tropics and southern hemisphere), Miralles et al. (2013). Future climate scenarios suggest a possible increase in El Nino like activity, Collins (2005), further emphasising the possibilities of increased water stress in the future.

Overall the data sets evaluated suggest a mean global evaporation of 1.5 mm per day. This estimate is somehow lower than previously existing estimates. There is, however, still considerable uncertainty attached to all of these estimates. The input of reliable precipitation remains one of the key uncertainties in the current *ET* products. Further assimilation of soil moisture may help to reduce these uncertainties.

Global Droughts in the 21st Century

Droughts have profound impact not only on water supply but food production, biodiversity and human well being in general. Both GEWEX and WCRP have identified hydrological extremes as a major topic requiring research to understand better the underlying mechanisms and possibilities for future changes. Droughts take many forms: meteorological, hydrological, agricultural etc., but underlying all are a prolonged scarcity of rainfall, usually exacerbated by increased potential evaporation and water extractions (see e.g. Teuling et al. 2013). It is generally accepted that along with increasing rainfall intensity climate change will also bring increasing occurrence of drought (see e.g. IPCC 2013). Drought can be quantified in many ways, depending on a studies purpose and perspective. A simple index commonly used is the Palmer Drought Severity Index which makes a balance between precipitation and evaporation. Using this index there has been a suggestion that drought occurrence has already increased through the 20th century (Dai et al. 2004) but the magnitude of this change has been shown to depend on the methodology used to estimate evaporative losses (Sheffield et al. 2012), hence suggesting that prediction of future drought depends critically on its definition and method of calculation.

To assess the impact of climate change on hydrological droughts a multi-model experiment was undertaken, including seven Global Impact Models (GIMs) driven by climate data from five Global Climate Models (GCMs) from CMIP5 under four different Representative Concentration Pathways (RCPs), Moss et al. (2010). Drought severity was defined as the fraction of land under runoff deficit (runoff less than a drought threshold) and is a measure of the time-integrated effect of several interlinked processes and stores, including precipitation, evaporation and soil moisture storage. Results show a likely increase in the global severity of drought at the end of the 21st century, with systematically greater increases for the RCPs describing stronger radiative forcings. Under RCP8.5 (the most extreme), droughts exceeding 40 % of the non-arid parts of the land area are projected by nearly half of the simulations. This increase in drought severity has a strong signal-to-noise ratio at the global scale, but Southern Europe, Middle East, South East United States, Chile and South West Australia are identified as possible hotspots for future water scarcity issues. The uncertainty due to GIMs is greater than that from GCMs, particularly if including a GIM that accounts for the dynamic response of plants to CO_2 and climate, as this model simulates little or no increase in drought frequency. This analysis demonstrates that different representations of terrestrial water cycle processes in GIMs are responsible for a much larger uncertainty in the response of hydrological drought to climate change than previously thought. When assessing the impact of climate change on hydrology it is hence critical to consider a diverse range of GIMs to better capture the uncertainty associated with the models (Prudhomme et al. 2014).

Quantifying Multiple Pressures on Future Global Water Resources

The current era of the "Anthropocene" is characterized by multiple pressures on global freshwater resources. Especially the continuing population growth and associated growing demand for water-consumptive goods (such as food for humans and livestock) has already led to over-exploitation of surface and groundwater resources in many locations. As population growth and also lifestyle changes toward more water-demanding products are very likely to continue in the future, water resources will be exposed to even larger pressures, not least because anticipated climate change is about to reduce water availability in many regions, particularly some semi-arid regions (Mediterranean, Western USA, Southern Africa and North-eastern Brazil) where water is already scarce (Kundzewicz et al. 2007). A pressing question is whether there will be enough (both 'blue' and 'green') water resources to produce the food for a growing world population under conditions of ongoing global warming and associated precipitation changes (Rockström et al. 2007; Falkenmark and Lannerstad 2010; Gerten et al. 2011; Wada et al. 2013).

Besides data-based approaches, recent research makes use of global hydrological, vegetation and/or crop growth models to address questions of this kind (see e.g. Hoff et al. 2010; Elliott et al. 2014). Recent applications include assessments of future water resources and supply for different levels of mean global warming using either a single model (e.g. Gosling et al. 2010; Gerten et al. 2011, 2013), or comparing results from up to 12 global hydrological and land surface and ecosystem models (Davie et al. 2013; Schewe et al. 2014). Such studies show that already today there is high water scarcity in many regions (for example North Africa, the Middle—East and South Asia, Gerten et al. 2011), but that climate change—even if limited to a mean global warming of 2 °C above preindustrial levels—would increase the number of people living in water-scarce river basins or countries by several hundred millions. For example, using the dynamic global vegetation and water balance model LPJ mL (Rost et al. 2008), Gerten et al. (2013) found that solely due to climate change, an additional of 8 % of global population will live in water-scarce catchments for a +2 °C world, rising to 13 % for a +5 °C world. Projected increases in world population will increase this number strongly, which indicates the challenge to ensure water security, and food security alike. Gerten et al. (2011) found that the blue and green water resources (the latter defined as the evapotranspiration during the growing season on current cropland and partly on grazing land) of many countries is not sufficient to produce a 'balanced' diet of 3,000 kcal per capita per day. This was found for large areas of North Africa, the Middle East and the Indian subcontinent, where imports of food and underlying virtual water appear to be a necessity. Rising atmospheric CO_2 concentration, however, is potentially beneficial to plant water use efficiency and hence total yields, as has been shown e.g. for future projections of worldwide irrigation water demand (Konzmann et al. 2013; Elliott et al. 2014). An assessment of the potential of various options to close the emerging water-for-food gaps, among them more effective on-farm crop water management (such as harvesting runoff and suppression of bare soil evaporation) suggest that substantial increases in crop production on existing farmland is possible. However, it appears likely that further cropland expansion and virtual water trade is inevitable (Rost et al. 2009; Fader et al. 2013; Elliott et al. 2014).

GEWEX Questions

Throughout 2012 GEWEX undertook an extensive consultation with the scientific and policy community and identified four major scientific areas, or questions, which need attention in the coming 5 years:

1. Observations and Predictions of Precipitation: *how can we better understand and predict precipitation variability and changes?*
2. Global Water Resource Systems: *how do changes in land surface and hydrology influence past and future changes in water availability and security?*
3. Changes in Extremes: *how does a warming world affect climate extremes, especially droughts, floods, and heat waves, and how do land area processes, in particular, contribute?*

4. Water and Energy Cycles and Processes: *how can understanding of the effects and uncertainties of water and energy exchanges in the current and changing climate be improved and communicated?*

See also http://www.gewex.org/pdfs/GEWEX_Science_Questions_final.pdf.

GEWEX Science Question 1: Observations and Predictions of Precipitation

How Can We Better Understand and Predict Precipitation Variability and Changes?

This question focuses on the exploitation of improved data sets of precipitation as well as related variables, such as soil moisture, water storage, and sea surface salinity expected in the coming 5 years. These improvements will come from ongoing and planned satellite missions as well as greater use of in situ observations; their evaluation and analysis to document mean, variability, patterns, extremes and probability density functions; their use to confront models in new ways and to improve our understanding of atmospheric and land surface processes that in turn feed into improved simulations of precipitation; and new techniques of data assimilation and forecasts that can lead to improved predictions of the hydrological cycle. These results should all lead to improved understanding and prediction of precipitation variability and related climate services.

GEWEX Science Question 2: Global Water Resource Systems

How Do Changes in Land Surface and Hydrology Influence Past and Future Changes in Water Availability and Security?

There is a need to address terrestrial water storage changes and close the water budget over land through exploitation of new data sets, data assimilation, and improved physical understanding and modelling skill across scales, from catchments to regional to global with links to the entire hydrological cycle, including ground water. In particular need of attention is the use of realistic land surface complexity with all anthropogenic effects taken into account, instead of a fictitious natural environment. This encompasses all aspects of global change, including water management, land use change, and urbanization. Water quality and especially water temperature, both of which are greatly affected by industrial and power plant use, are of immediate concern, to be followed by nutrients. The ecosystem response to climate variability and responsive vegetation must be included, as must cryospheric changes such as permafrost thawing and changes in mountain glaciers. Feedbacks, tipping points, and extremes are of particular concern. The results should enhance the evaluation of the vulnerability of water systems, especially to extremes, which are vital for considerations of water security and can be used to increase resilience through good management and governance.

GEWEX Science Question 3: Changes in Extremes

How Does a Warming World Affect Climate Extremes, Especially Droughts, Floods, and Heat Waves, and How Do Land Area Processes, in Particular, Contribute?

A warming world is expected to alter the occurrence and magnitude of extremes such as droughts, heavy rainfalls, and floods, as well as the geographic distribution of rain and snow. Such changes are related to an acceleration of the hydrologic cycle and circulation changes, and include the direct impact of warmer conditions on atmospheric water vapor amounts, rainfall intensity, and snow-to-rain occurrence. How well are models able to handle extremes and how can we improve their capability? New improved and updated data sets at high frequency (e.g., hourly) are needed to properly characterize many of these facets of our climate and to allow for assessment against comparable model data sets. New activities are needed to promote analyses quantifying which changes are consistent with our expectations and how we can best contribute to improving their prediction in a future climate. Confronting models with new observationally-based products will lead to new metrics of performance and highlight shortcomings and developmental needs that will focus field programs, process studies, numerical experimentation, and model development. New applications should be developed for improved tracking and warning systems, and assessing changes in risk of drought, floods, river flow, storms, coastal sea level surges, and ocean waves.

GEWEX Science Question 4: Water and Energy Cycles and Processes

How Can Understanding of the Effects and Uncertainties of Water and Energy Exchanges in the Current and Changing Climate Be Improved and Conveyed?

This question includes goals of improved consistency between net solar and infrared radiation and sensible and latent heat fluxes at the surface to reveal processes that in turn must be replicated in climate models, at multiple scales. This question relates also to uncertainties introduced by incomplete understanding of cloud-aerosol-precipitation interactions and their feedbacks to the climate system. Only through a better understanding of the uncertainties in observations and models will it be possible to discriminate natural variability from longer-term trends of key variables such as temperature and precipitation. Possibilities of new satellite-based measurements, combined with observations at the surface and in the ocean, should enable improved reconciliation between observed changes in the radiative imbalance at the top-of-atmosphere (TOA) and the inventory of changes in energy throughout the Earth system. Upgraded GEWEX data sets, global reanalyses of atmosphere and ocean, and improved modelling together with advanced diagnostics being planned throughout the GEWEX Panels which play key roles in advancing this topic. The result is improved tools and products for climate services.

The case studies presented in this paper are a small sample of the studies of GEWEX scientists, they illustrate some of the advances in recent years but also some of the considerable uncertainties remaining. They also put into perspective the new GEWEX questions and priorities for future research: including the improvement of our modelling and prediction of precipitation (and evaporation), development and use of new data sets, the improvement of land surface models, the better understanding of extremes and the representation of realistic land surface complexity with all anthropogenic effects into our analysis. In the next decade GEWEX will work to bring together scientists from a wide range of disciplines to work towards these on these topics. The ultimate result should be better and more relevant tools and analyses to inform society of potential impacts and adaptation options to climate and environmental change.

Acknowledgments This paper was developed as part of the European Union (EU-FP6)-funded integrated program called WATCH (Contract 036946). We would also like to thanks the many scientists who have contributed to the development of the GEWEX aims and science questions and the organisers of the GWSP conference for providing a forum to discuss these issues.

References

Allan RP, Soden BJ, John VO, Ingram WI, Good P (2010) Current changes in tropical precipitation. Environ Res Lett 5:025205

Biemans H, Hutjes R, Kabat P et al (2009) Effects of precipitation uncertainty on discharge calculations for main river basins. J Hydrometeor 10:1011–1025

Collins M (2005) El Nino- or La Nina-like climate change? Clim Dyn 24:89–104. doi: 10.1007/s00382-004-0478-x

Dadson S, Acreman MC, Harding RJ (2013) Water security, global change and land-atmosphere feedbacks. Phil Proc Roy Soc A 371:20120412

Dai A, Trenberth KE, Qian T (2004) A global data set of Palmer Drought Severity Index for 1870-2002: Relationship with soil moisture and effects of surface warming. J Hydrometeorology 5:1117–1130

Dai A, Qian T, Trenberth KE, Milliman JD (2009) Changes in continental freshwater discharge from 1948-2004. J Clim 22:2773–2791

Davie JCS, Falloon PD, Kahana R et al (2013) Comparing projections of future changes in runoff and water resources from hydrological and ecosystem models in ISI-MIP. Earth Syst Dyn 4:279–315

Elliott J, Deryng D, Müller C et al (2014) Constraints and potentials of future irrigation water availability on global agricultural production under climate change. Proc Natl Acad Sci (in press)

Fader M, Gerten D, Krause M et al (2013) Spatial decoupling of agricultural production and consumption: quantifying dependence of countries on food imports due to domestic land and water constraints. Environ Res Lett 8:014046

Falkenmark M, Lannerstad M (2010) Food security in water-short countries—coping with carrying capacity overshoot. In: Martínez-Cortina L, Garrido A, López-Gunn E (eds) Rethinking water and food security. Taylor & Francis, London, pp 3–22

Gerten D, Heinke J, Hoff H et al (2011) Global water availability and requirements for future food production. J Hydrometeor 12:885–899

Gerten D, Lucht W, Ostberg S et al (2013) Asynchronous exposure to global warming: freshwater resources and terrestrial ecosystems. Environ Res Lett 8:034032

Gosling SN, Bretherton D, Haines K et al (2010) Global hydrology modelling and uncertainty: running multiple ensembles with a campus grid. Phil Trans Royal Soc A 368:4005–4021

Groisman PY, Knight RW, Easterling DR et al (2005) Trends in intense precipitation in the climate record. J Clim 18:1343–1367

Hoff H, Falkenmark M, Gerten D et al (2010) Greening the global water system. J Hydrol 384:177–184

Haddeland I, Clark DB, Franssen W et al (2011) Multimodel estimate of the terrestrial global water balance: setup and first results. J Hydrometeor 12:869–884

Haddeland I, Heinke J, Biemans H et al (2013) Global water resources affected by human interventions and climate change. Proc Nat Acad Sci 111(9):3251–3256. doi: 10.1073/pnas.1222475110

Harding R, Best M, Blyth E et al (2011) Preface to the "water and global change (WATCH) special collection: current knowledge of the terrestrial global water cycle". J Hydrometeorol 12:1149–1156

Hagemann S, Chen C, Clark DB et al (2013) Climate change impact on available water resources obtained using multiple global climate and hydrology models. Earth Syst Dyn 4:129–144. doi: 10.5194/esd-4-129-2013

IPCC (2013) Climate change 2013: the physical science basis. Contribution of working Group I to the fifth assessment report of the intergovernmental panel on climate change. In: Stocker TF, Qin D, Plattner G-K, Tignor M, Allen SK, Boschung J, Nauels A, Xia Y, Bex V, Midgley PM (eds). Cambridge University Press, Cambridge, p 1535

IPCC (2012) Managing the risks of extreme events and disasters to advance climate change adaptation. A special report of working Groups I and II of the intergovernmental panel on climate change. In: Field CB, Barros V, Stocker TF, et al (eds). Cambridge University Press, Cambridge, p 582

IPCC (2007) Climate change 2007: the physical science basis. Contribution of working Group I to the fourth assessment report of the intergovernmental panel on climate change. In: Solomon S, Qin D, Manning M et al (eds). Cambridge University Press, Cambridge, p 996

Jung M, Reichstein M, Ciais P et al (2010) Recent decline in the global land evapotranspiration trend due to limited moisture supply. Nature 467:951–954

Konzmann M, Gerten D, Heinke J (2013) Climate impacts on global irrigation requirements under 19 GCMs, simulated with a vegetation and hydrology model. Hydrol Sci J 58:1–18

Kummu M, Ward PJ, de Moel H et al (2010) Is physical water scarcity a new phenomenon? Global assessment of water shortage over the last two millennia. Environ Res Lett 5:034006

Kundzewicz ZW, Mata LJ, Arnell NW et al (2007) Freshwater resources and their management. Climate change 2007: impacts, adaptation and vulnerability. contribution of working Group II to the fourth assessment report of the intergovernmental panel on climate change. In: Parry ML, Canziani OF, Palutikof JP (eds). Cambridge University Press, Cambridge, pp 173–210

Meehl GA et al (2007) Global climate projections. Climate change 2007: the physical science basis. Contribution of working Group I to the fourth assessment report of the intergovernmental panel on climate change. In: Solomon S et al (eds). Cambridge University Press, Cambridge, pp 764–770

Miralles DG, Holmes TRH, De Jeu RAM et al (2011) Global land-surface evaporation estimated from satellite-based observations. Hydrol Earth Syst Sci 15:453–469. doi:10.5194/hess-15-453-2011

Miralles DG, van der Berg MJ, Gash JH et al (2013) El Niño–La Niña cycle and recent trends in continental evaporation. Nature. doi:10.1038/NCLIMATE2068

Moss RH et al (2010) The next generation of scenarios for climate change research and assessment. Nature 463:747–756

Mueller B, Seneviratne SI, Jiménez C et al (2011) Evaluation of global observations based evapotranspiration datasets and IPCC AR4 simulations. Geophys Res Lett 38:L06402. doi:10.1029/2010GL046230

Prudhomme C et al (2014) Hydrological droughts in the 21st century: hotspots and uncertainties from a global multimodel ensemble experiment. Proc Natl Acad Sci, USA

Rockström J, Lannerstad M, Falkenmark M (2007) Assessing the water challenge of a new green revolution in developing countries. Proc Natl Acad Sci U S A 104:6253–6260

Rost S, Gerten D, Bondeau A et al (2008) Agricultural green and blue water consumption and its influence on the global water system. Water Resour Res 44:W09405

Rost S, Gerten D, Hoff H et al (2009) Global potential to increase crop production through water management in rainfed agriculture. Environ Res Lett 4:044002

Schewe J et al (2014) Multi-model assessment of water scarcity under climate change. Proc Natl Acad Sci, USA (online early)

Sheffield J, Wood E F, Roderick M L (2012) Little change in global drought over the past 60 years. Nature 491:435–438. doi: 10.1038/nature11575

Teuling AJ et al (2013) Evapotranspiration amplifies European summer drought. Geophys Res Lett 40(10):2071–2075

Wada Y, Wisser D, Eisner S et al (2013) Multi-model projections of irrigation water demand under climate change. Geophys Res Lett 40:4626–4632

Weedon GP, Gomes S, Viterbo P et al (2011) Creation of the WATCH forcing data and its use to assess global and regional reference crop evaporation over land during the twentieth century. J Hydrometeorol 12:823–848

Chapter 5
Integrated Assessments of Water Scarcity: Knowns, Unknowns and Ways Forward

Tabea K. Lissner, Caroline A. Sullivan, Dominik E. Reusser and Jürgen P. Kropp

Abstract Inadequate access to water is already a problem in many regions of the world and processes of global change are expected to further exacerbate the situation. Many aspects determine the adequacy of water resources: beside actual physical water stress, where the resource itself is limited, economic and social water stress can be experienced if access to resource is limited by inadequate infrastructure, political or financial constraints. Further, water quality is an essential determinant of adequate water access. All human activities as well as ecosystems require water in sufficient quantity and quality for their functioning. To assess the adequacy of water availability for human use, integrated approaches are needed that allow to view the multiple determinants in conjunction and provide sound results as a basis for informed decisions. This contribution gives an overview of existing knowledge on different aspects to measure water scarcity and points out gaps in existing approaches. It then proposes two parts of an integrated approach to look at the multiple dimensions of water scarcity. It first outlines the AHEAD approach to

T. K. Lissner (✉) · D. E. Reusser · J. P. Kropp
Potsdam Institute for Climate Impact Research,
60 12 03, 14412 Potsdam, Germany
e-mail: lissner@pik-potsdam.de

D. E. Reusser
e-mail: reusser@pik-potsdam.de

J. P. Kropp
e-mail: kropp@pik-potsdam.de

T. K. Lissner
Geography Department, Humboldt-University, Berlin, Germany

C. A. Sullivan
School of Environment Science and Engineering, Southern Cross University, Lismore, NSW, Australia
e-mail: caroline.sullivan@scu.edu.au

J. P. Kropp
Department of Geological and Environmental Sciences, University of Potsdam, Potsdam, Germany

A. Bhaduri et al. (eds.), *The Global Water System in the Anthropocene*, Springer Water, DOI: 10.1007/978-3-319-07548-8_5,

measure Adequate Human livelihood conditions for wEll-being And Development. The approach allows viewing impacts of climate change, e.g. changes in water availability, within the wider context of AHEAD conditions. However, adequate water access is not determined by water availability alone. To assess the water requirements for different sectors in more detail, we present a second approach to assess the multiple determinants of water adequacy, including aspects of quantity, quality as well as access, in an integrated way.

Introduction

Water is an essential resource for human livelihoods in itself, but also a vital factor of production for food, energy and other industrial products. Much work has been devoted to measuring water scarcity and a range of data and approaches have been made available from a variety of sources (Eriyagama et al. 2009; Brown and Matlock 2011). Water scarcity is already present in many regions of the world, and processes of global change are expected to exacerbate such problems (WWAP 2012). Situations of water scarcity lead to competitions between water users, requiring informed management decisions to ensure all needs are being met in a sustainable manner. Beside actual physical water stress, where the resource itself is limited, economic and social water stress can be experienced if access to resource is limited by inadequate infrastructure, political or financial constraints, or if water quality is poor.

To adequately assess the state of development, consistent and meaningful measures of water availability and adequate access are fundamental. However, accounts of water stress often only focus on water resource availability, rather than on important additional dimensions (i.e. access, quality, costs) (Sullivan 2002). Especially in developing countries, however, often infrastructure and access pose a bigger problem to water supply than actual resource availability (Rijsberman 2006). Most indicators of water scarcity are based on some relationship between water use or withdrawal and actual (blue) water availability (Hoekstra et al. 2011). Though knowledge of the multiple drivers of inadequate water access exists, so far integrated approaches linking physical aspects of water scarcity to the socio-economic conditions are lacking. Quantified and integrated assessments are of high importance to inform the policy process. To assess the adequacy of water availability for human livelihoods, more differentiated approaches are needed, looking at user-specific determinants of water availability. Any methodology to address water availability relies on adequate data to depict water needs, water availability and water use. Uncertainties in data sources as well as in knowledge are an essential factor to be taken into account when aiming to produce policy relevant results.

The topic of water needs and scarcity is a cross-sectoral issue. It needs to be considered in the context of these multiple aspects and embedded into existing socio-economic realities. The article has two important goals. First, we give an overview of current knowledge about water scarcity and outline important aspects of the main determinants in Sect. "Determinants of Water Scarcity". We give an overview of current knowledge, but also highlight missing links. We then outline how processes of global change interact with water resources and introduce two approaches, which can help in overcoming some of the challenges faced in the analysis of water scarcity (Sect. "Global Change, Development and Water Resources: Ways Forward"). The AHEAD framework quantifies *Adequate Human livelihood conditions for wEll-being And Development*, with water playing a central role for many aspects of the index. We present a first quantification at global scale and outline, how the index can improve our understanding of water scarcity. We further present in detail how the multiple aspects affecting the adequacy of water resources can be formalized in holistic way. Approaches such as the Water Poverty Index (Sullivan 2002) and the Climate Vulnerability Index (Sullivan and Meigh 2005) give important insight into the multiple aspects of water poverty, identifying and integrating several dimensions. Building on these findings, we propose a measurable framework to integrate the determinants of water adequacy. We conclude the paper with a summary of the main findings and ways forward to improve our knowledge of water scarcity.

Determinants of Water Scarcity

Water scarcity describes a situation where available water resources are insufficient to meet needs, either due to quantity, quality or access reasons. Many indices and approaches exist to define and measure water scarcity, usually focussing on distinct aspects of the topic. This results in a extensive body of knowledge on the single determinants of water scarcity, however these are often assessed separately and several uncertainties remain. The following sections give a brief overview of the main aspects determining water scarcity, including water availability (Sect. "Water Availability"), water needs and use (Sect. "Water Needs and Use") as well as water quality and access (Sect. "Water Quality and Access").

Water Availability

Account of water availability usually describe resources at a given spatial and temporal resolution, e.g. total internal renewable water resources at country-level. Accounts often differentiate between e.g. internal and external, surface and subsurface or renewable and non-renewable (see e.g. FAO (2011)). It is thus extremely important to clarify the spatial and temporal scales of analysis, as well as the accessibility of the required resource. The differentiation into green and blue water resources presented by Falkenmark and Rockström (2004) is an important contribution to differentiate the accessibility of water resources for different users. Assessments of water scarcity usually take into account blue water availability or total renewable internal water resources, often at national scale and with annual averages (Hoekstra et al. 2011; Rijsberman 2006). These scales do not reflect variability at higher spatial and temporal scales, however such variability is often significant.

Information on water availability is available either from accounts of measured water availability, e.g. the FAO Aquastat Database or from hydrological models. As several drivers may impact water availability, including climate and land-use change as well as other human activities, hydrological models are essential in providing information on potential changes. Both, measured and modeled accounts of water availability are subject to uncertainties. As rainfall is highly heterogeneous across time and space, measuring spatial variability is difficult. Similarly, the measurement of actual evapotranspiration is difficult and strongly depends on soil and vegetation and river and groundwater measurements are spatially sparse (Heistermann and Kneis 2011).

Modeling uncertainties have several causes. Amongst other things, data to define boundary conditions is insufficient and no agreed upon structure for models exist (Beven 2009) and spatially heterogeneous processes cannot be resolved for global assessment. As a comparison of a number of water models showed, model results have a significant spread (Harding et al. 2011). A further comparison of these models using a range of climate models as input further showed, that uncertainty from hydrological models is higher than the uncertainty from climate models (Wada 2013). To meaningfully assess water availability for human use, it is thus essential to be clear about scale and scope of the analysis to be able to choose an adequate representation of water resources. Further, awareness of uncertainties associated with data sources is crucial to reach informed decisions.

Water Needs and Use

Looking at the demand side of water scarcity, one can differentiate between water withdrawals as well as water consumption and water needs (requirements). Withdrawals usually refer to the amount of water withdrawn for sector-specific utilization, of which some may be returned to the environment after use. Actual water use is usually lower, as it only describes water which actually consumed and not returned to the environment (Shiklomanov 2000). Both, withdrawals and consumption, describe the amounts of water currently used, but not necessarily whether the amount available fulfills needs. Influential approaches based on water use and consumption calculate the relationship between water use and water availability (e.g. Alcamo et al. 2003) or look at water consumption and the water footprint of lifestyle choices (e.g. Hoekstra 2006).

Opposed to this, descriptions of water requirements describe (sector-specific) water needs, though results of these assessments are quite diverging, as the overview in Table 5.1 shows, and are highly dependent on development status and lifestyle choices. As the large ranges show, it difficult to define generally

Table 5.1 Overview of sectoral water needs according to different sources, all converted to $m^3/cap^{-1}/yr^{-1}$

	Chenoweth (2008)	Falkenmark (1997)	Shuval (1992)	Range
Municipal	30.6	36	100	30.6–100
Industrial	12.6	36–432	–	12.6–432
Agricultural	–	504–1,584	25	25–1,584

applicable levels of water needs, nonetheless for direct country-comparisons such approaches can provide a useful contribution. One important example of looking at water scarcity from a user perspective is the widely used Falkenmark Indicator, which defines water stress at different levels of per capita availability. Based on an assessment of generally valid sectoral water needs, it defines levels of average water needs per capita to produce a sufficient diet in semi-arid countries (Falkenmark 1997). By accounting for water needs at country-scale, this indicator does not take into account important differences between countries or look at the potential of importing water through food imports, for example. Additionally, intra-annual variations, which play an important role for agriculture, or management capacities, are not taken into account.

When looking at water needs, environmental water requirements play an important role, as a significant fraction of the available water is needed for functioning ecosystems. Usually these flows are not taken into account for scarcity assessments (WWAP 2012). The fraction of water needed for environmental flows depends on the topography, vegetation and climate of the region and values between 20 and 50 % of total flows have been suggested (Smakhtin et al. 2004).

Water Quality and Access

While the availability of sufficient resources to meet water requirements is a prerequisite, the water also needs to be accessible and of sufficient quality for use (WWAP 2012). As the Millennium Development Goals (UN 2012) reflect, access is often the main impediment for adequate municipal water resources. Accessibility to water is determined through distance and time needed for its collection as well as through the reliability and costs (Howard and Bartram 2003; Sullivan 2002). Adequate water supply infrastructure also increases the likelihood of better water quality. Human water security is already highly stressed globally through high levels of water pollution (Vörösmarty et al. 2010a) and significant investments in water treatment are needed to make water usable. Water-borne diseases are a major threat to human health in developing countries (Bates et al. 2008). Current approaches to water scarcity usually focus on quantities, but much less at access and especially water quality.

Global Change, Development and Water Resources: Ways Forward

Processes of global change affect the availability of water resources in several ways. Climate change will alter temperatures and precipitation patterns, resulting in changes of seasonal and temporal variations in physical water availability, with a likely increase in climate extremes such as heavy rain and flooding as well as droughts (Bates et al. 2008). Population dynamics, especially in developing and water stressed countries, will likely lead to reductions in per capita availability, even if total water resources remain constant.

Water use is clearly determined by prevailing lifestyles. The highest fraction of water is used for agricultural purposes. Different dietary patterns have very different water-intensities in their production (Rijsberman 2006; Mekonnen and Hoekstra 2010b) and the production of animal products is especially water-intensive (Mekonnen and Hoekstra 2010a). Increasing prosperity and development across the world currently leads to shifts towards an overall increase in calorie intake as well as shift in dietary patterns towards more energy- and water-intensive food, with stronger increases in the consumption of animal products (Pradhan et al. 2013). Consequently, higher water requirements for the agricultural sector are likely with advancing development. Similarly, energy production already uses a significant portion of water resources. Electrification, especially in rural, developing regions, is urgently needed to raise living standards (AGECC 2010), but is also associated with increases in water use and potentially higher levels of pollution (Maheu 2009).

Human livelihoods are determined by several aspects, including resources such as food and water, essential infrastructure such as health care as well as aspects of social structure, such as political stability. Many of these aspects directly or indirectly require adequate access to sufficient water resources. As outlined in Sect. "Water Availability", water and climate models introduce a significant level of uncertainty in the quantification of available water resources in the coming decades. It is also clear, however, that the availability of such models is essential to assess current and future water availability at detailed spatial scales. Water resources will be affected by climate change and the predictions of potential changes are essential for anticipatory water resource planning.

While the overall availability of water resources needs to be sufficient, especially the accessibility as well as the quality of water resources pose major problems. Often, a lack of water is caused by poor quality an insufficient access, rather than actual resource scarcity. To address this water challenge, methods are needed that are able to view the multiple determinants of adequate water availability for essential sectors in an integrated way. At the same time, these methods need to be able to cope with the uncertainty, which are inevitable in projections of future developments.

The available approaches and concepts introduced in Sect. "Determinants of Water Scarcity" provide important aspects of water scarcity, however the available knowledge needs to be put into a perspective, which includes the different facets of water scarcity and relates them to human livelihood realities and the challenges of global change. The following section outlines two approaches, which can contribute to such an integrated view.

The AHEAD Index: Livelihood Conditions and Human Well-Being

The AHEAD index aims to measure *Adequate Human livelihood conditions for wEll-being And Development* in an integrated way (see Lissner et al. 2013a, full article in preparation). It is based on 16 elements, which were identified through a review of influential approaches on human needs and well-being. Three sub-indices are differentiated, namely the Subsistence, Infrastructure and Social Structure sub-indices, each containing relevant components, which cover a range of needs from the respective domain (see Fig. 5.1). The sub-indices as well as the full index are aggregated using fuzzy logic, which allows to take into account the properties of the variables and data and retain important aspects of their relationships in the aggregation process (Kropp et al. 2006; Lissner et al. 2012). The first step of a fuzzy logic approach is the fuzzification of the input variables. Here, the degree of membership to a linguistic category, in our case to the phrase ''conditions are adequate'', is calculated through the use of membership functions. Upper and lower threshold determine the membership range and fuzzified variables take continuous values between 0 and 1. Following the fuzzification, values are aggregated using context-specific decision rules (Lissner et al. 2012). To represent adequate water resources for the purpose of measuring adequate AHEAD conditions, we calculated water stress levels using the Falkenmark indicator (Falkenmark 1997). Though this indicator has several limitations and shortcomings (see Sect. "Water Needs and Use"), for the purpose of country-comparisons it provides a useful approach. Where average annual per capita water availability falls below 500 $m^3/cap^{-1}/yr^{-1}$, water stress is present, leading to a fuzzified value of 0 (lower threshold). The adequacy of water increases linearly to values of up to 1 for regions where water availability is above 1,700 $m^3/cap^{-1}/yr^{-1}$ (upper threshold). To describe the adequacy of water availability for AHEAD, additionally we take into account water access, represented by the variable 'access to an improved water source' WHO (2009), Howard and Bartram (2003), to account for the fact that water access is often not limited through resource availability, but through a lack of infrastructure and quality. For the aggregation process, decision rules are defined that reflect the specific properties of the elements. In the case of water availability, a MIN operator is used to reflect the fact, that water availability is a basic requirement and cannot be substituted by other elements. Consequently, inadequate water resources also implicate inadequate AHEAD conditions. The same is true for all elements of the Subsistence sub-index. Opposed to this,

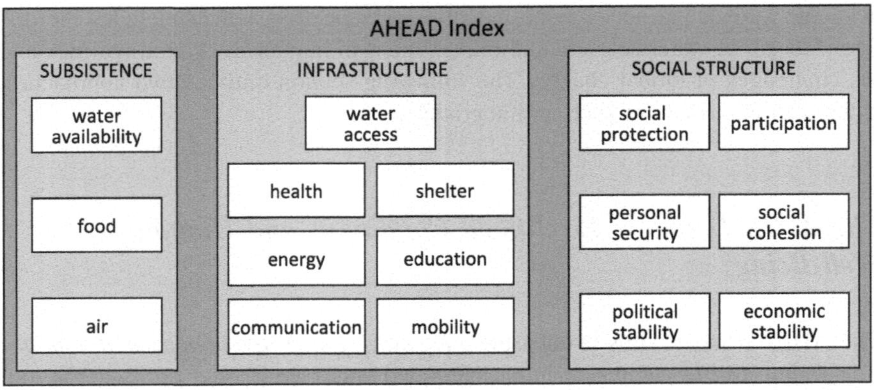

Fig. 5.1 Components and structure of the AHEAD index. Elements are grouped into three sub-indices according to their properties and main functions

elements from the Infrastructure or Social Structure sub-indices, may be substitutable to some extent and this is also reflected in the choice of aggregation rules. Figure 5.1 shows the components and structure of the AHEAD index.

The structured approach allows to refer back to those input factors, which are most important for the overall result. In a first calculation of the approach on a global scale, we used freely available data to calculate current AHEAD conditions at country scale. To exemplify how changes in water availability affect AHEAD, we used output from the WaterGAP model (Alcamo et al. 2003; Döll et al. 2003). Within the Integrated Project Water and Global Change (WATCH),[1] model results have been calculated using output from three climate models (IPSL, ECHAM and CNCM3 (Randall et al. 2007; Harding et al. 2011)) under assumptions of an SRES-A2 scenario (Nakićenović et al. 2000). Figure 5.2 shows selected results from the calculations: values show the aggregate AHEAD conditions and changes over time as a function of water availability in example countries (colored symbols) as well as the world mean (black).

While changes in water availability have little impact on global mean AHEAD conditions, changes at country scale are often significant, as for example in Iran, where conditions are projected to deteriorate over the course of the century. Other country examples underline model differences. Results for both, Slovakia and Senegal, for example, show that model results can differ substantially in terms of

[1] All data are available for download at ftp://ftp.iiasa.ac.at/. We summed surface and subsurface runoff (Qs and Qsb without human influence) (for details on the data convention see http://www.eu-watch.org/watermip/data-format) and converted the data from a grid based resolution in kg/s/m^2 to a per capita availability per country in m^3/year. The available data covers the years 1900–2100. We calculated 30-year average availability for a baseline period (1971–2000), further referred to as 1990, as well as for three scenario periods 2030 (2011–2040), 2060 (2041–2070) and 2090 (2071–2100). Population scenarios to calculate per capita availability were used consistent with the SRES climate scenario A2 (IIASA 1996).

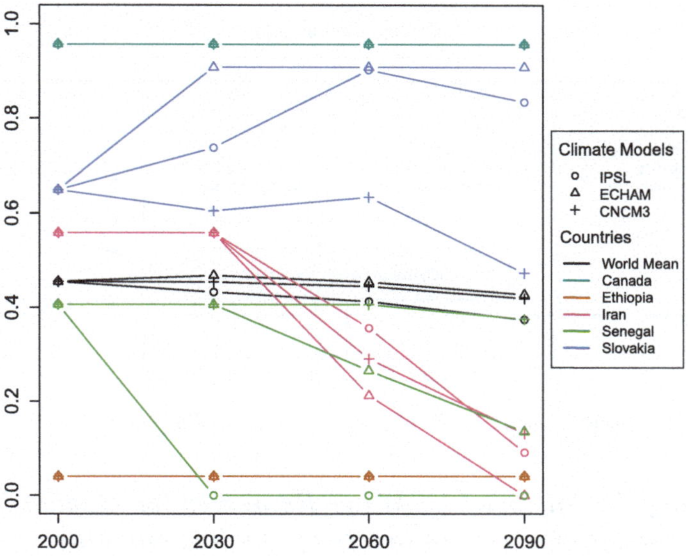

Fig. 5.2 Changes in the adequacy of AHEAD conditions over time, resulting from changes in water availability, for selected countries and the world average

magnitude as well as direction: for both countries the three models diverge strongly towards the end of the century. In some countries, such as Canada and Ethiopia, AHEAD conditions are not primarily determined by water resource availability. Canada has the overall highest adequacy of AHEAD conditions and water resource are adequate, also under future conditions. Ethiopia, opposed to this, shows the lowest adequacy of livelihood and well-being conditions. While water resources remain at a mean level of adequacy, also under climate change, nonetheless AHEAD conditions are inadequate due to other factors.

The results can give first indications to where water resource limitation affect AHEAD conditions and where detailed further analyses are important. By looking at water resources within the wider frame of the AHEAD framework, a prioritization of country specific limitations becomes possible. With the goal of increasing the overall quality of livelihoods and human well-being, the approach allows to prioritize between sectors and put limited resources towards sectors most in need. The approach can help in identifying regions, in which uncertainty in climate and hydrological models needs to be addressed further, but can also underline regions where results have a clear message and point towards efficient interventions to improve AHEAD conditions.

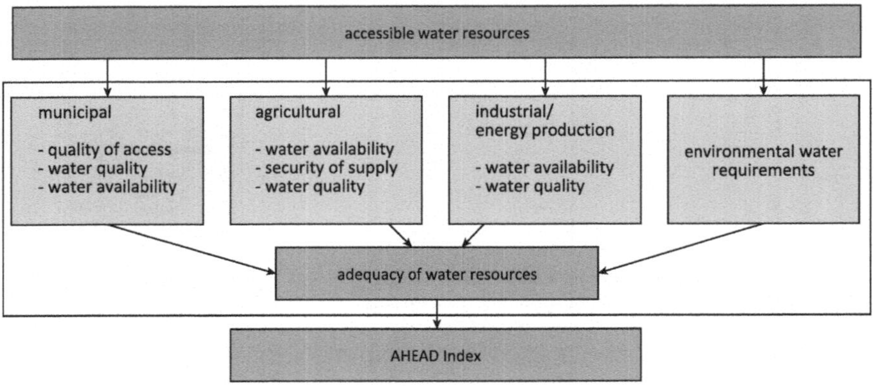

Fig. 5.3 Determinants of the adequacy of water resources for different users

Measuring the Adequacy of Water Resources for Human Use: Sector-Specific Determinants of Quality, Quantity and Infrastructure

The AHEAD index gives important insights into general water availability for human livelihoods and well-being. However, by only looking at water resource availability without a sectoral differentiation, important other aspects are not included. Different water users have different requirements regarding the quantity, quality and accessibility of water resources. Especially in developing countries, water quality and access often play a more important role in determining the adequacy of water than resource availability alone. For integrated assessments of the adequacy of water resources for human use, these various aspects need to be combined in a useful way. The Water Poverty Index provides an important approach towards a holistic view of water, including aspects such as access and affordability of water (Sullivan 2002). An important step forward is the operationalization of such an approach. To do so, we differentiate the major current water users municipal, agriculture and energy/industrial production. We further include environmental water requirements, as functioning ecosystems are a critical aspect for human livelihoods. Each sector has specific requirements regarding access infrastructure, water quality as well as quantity (see Lissner et al. 2013b, full article in preparation). As outlined in Fig. 5.3, water adequacy for municipal use is mainly determined by the type of access infrastructure and water quality. In developed countries, for example, high investments in water security ensure adequate municipal water access (Vörösmarty et al. 2010a), whereas water deprivation in developing countries is high due to a lack in such infrastructure (UN 2012). Agriculture as the largest water consumer mainly depends on sufficient water resource availability, which can be enhanced through supply security (e.g. through water storage and irrigation); water quality is especially an issue for irrigated agricultural production. Energy and industrial production depend on water of sufficient quality and quantity. Further, environmental water requirements are taken into account as a percentage of average water availability, combined with a measure of threat to biodiversity (Vörösmarty et al. 2010b). Again, fuzzy logic provides a useful method to aggregate the relevant factors to represent sectoral and overall water adequacy. Based on the review of sectoral water requirements (Table 5.1), we include sector-specific lower and upper thresholds of water needs. Further, we include sector-specific determinants of water quality as well as aspects of infrastructure. The final aggregate measure representing the "adequacy of water resources" thus includes sector-specific indicators of relevant aspects of water access, water quality as well as water availability.

The approach allows identifying the most decisive factor in determining water adequacy for each sector as well as the overall aggregate, showing which factor is least adequate in the region under analysis and where improvements are most urgently needed. The results of this analysis of decisive factors can give important information to policy makers, who want to most effectively improve the access to adequate water resources. It can identify, whether water availability is the limiting factor, or whether water quality or access factors play a more important role in limiting water access. Situations of water scarcity can lead to trade-offs, as sectors compete for access to sufficient resources (WWAP 2012). By using sector-specific thresholds of needs, the outlined approach gives the possibility to assess different sectoral allocation scenarios to reduce competition and find integrated solutions.

Conclusions

The assessment of water availability and determinants of water scarcity is of high scientific as well as political interest and relevance. A lot of knowledge is available, which also includes known unknowns and awareness about uncertainties. Water is an essential human need and the basis for socio-economic activities. Processes of global change affect both the side of water availability as well as the side of water consumption and water requirements. In many cases, climate change adds on to existing development deficiencies and current patterns of unsustainable water use. Available knowledge needs to be viewed in a context, which relates it to development realties and human livelihood conditions.

We have presented two parts of an integrated fuzzy logic analysis, which allows analyzing the multiple determinants for adequate water availability. The AHEAD framework allows to view the availability of water resources in a wider context of human well-being and livelihoods. The inclusion of a range of factors which affect AHEAD components provides a framework to view important aspects in conjunction and impacts of water availability on AHEAD conditions can be made visible. Zooming into the water sector in more detail, the proposed method to measure the adequacy of water resources outlined in Sect. "Measuring the Adequacy of Water Resources for Human Use: Sector-Specific Determinants of Quality, Quantity and Infrastructure" can improve understanding of the most decisive factors for adequate water access, including a sectoral differentiation. As anthropogenic aspects, such as access to and quality of water resources are often more important than resource availability alone, the approach is an important extension to the AHEAD approach. Together, the two approaches provide an important way forward to assess water scarcity in the anthropocene. Approaches such as the ones outlined here can help in making complex information more accessible and pave the way towards prioritizing between sectors.

There is high uncertainty in projections of water availability, stemming from both hydrological and climate models. Our approach can help in identifying where such uncertainties are high and relevant and further detailed analyses are needed. But it can also identify those regions, where it is quite certain that other aspects need to be improved to secure adequate water access. Actionable and policy-relevant information which is based on sound scientific findings, as is provided by the presented approaches, is much needed (WWAP 2012, Chap. 6). It can help in highlighting what we know and help determining the importance of the things we do not know.

Acknowledgements The work has been developed within a project funded by the German Federal Ministry of the Environment, Nature Conservation and Nuclear Safety, specifically through the International Climate Initiative (ICI).

References

AGECC (2010) Energy for a sustainable future. The Secretary-General's Advisory Group on Energy and Climate Change (AGECC), New York

Alcamo J, Döll P, Henrichs T, Kaspar F, Lehner B, Rösch T, Siebert S (2003) Global estimates of water withdrawals and availability under current and future business-as- usual conditions. Hydrolog Sci J 48(3):339–348

Bates B, Kundzewicz Z, Wu S, Palutikof J (2008) Climate change and water. Technical Paper of the Intergovernmental Panel on Climate Change. IPCC Secretaritat, Geneva

Beven K (2009) Environmental modelling: an uncertain future?. Routledge, New York

Brown A, Matlock MD (2011) A review of water scarcity indices and methodologies. Technical report, University of Arkansas, The Sustainability Consortium

Chenoweth J (2008) Minimum water requirement for social and economic development. Desalination 229(1–3):245–256

Döll P, Kaspar F, Lehner B (2003) A global hydrological model for deriving water availability indicators: model tuning and validation. J Hydrol 270(1–2):105–134

Eriyagama N, Smakhtin V, Gamage N (2009) Mapping drought patterns and impacts: a global perspective. IWMI Research Report 133

Falkenmark M (1997) Meeting water requirements of an expanding world population. Philos T Roy Soc B 352(1356):929–936

Falkenmark M, Rockström J (2004) Balancing water for humans and nature. The new approach in ecohydrology, Earthscan, London

FAO (2011) Climate change, water and food security. Technical report

Harding R, Warnaars T, Weedon G, Wiberg D, Hagemann S, Tallaksen L, van Lanen H, Blyth E, Ludwig F, Kabat P (2011) Executive summary of the completed WATCH Project. http://www.eu-watch.org/publications/technical-reports. Accessed 09 Jan 2014

Heistermann M, Kneis D (2011) Benchmarking quantitative precipitation estimation by conceptual rainfall-runoff modeling. Water Resour Res 47(6):W06,514

Hoekstra AY, Chapagain AK (2006) Water footprints of nations: Water use by people as a function of their consumption pattern. Water Resour Manag 21(1):35–48

Hoekstra AY, Chapagain AK, Aldaya MM, Mekonnen MM (2011) The water footprint assessment manual: setting the global standard. Earthscan, London

Howard G, Bartram J (2003) Domestic water quantity, service level and health. World Health Organisation, Geneva

IIASA (1996) Population projections for the IPCC special report on emission scenarios (1996). http://webarchive.iiasa.ac.at/Research/POP/IPCC/index.html. Accessed 09 Jan 2014

Kropp JP, Block A, Reusswig F, Zickfeld K (2006) Semiquantitative assessment of regional climate vulnerability: the North-Rhine Westphalia study. Clim Change 76:265–290

Lissner TK, Holsten A, Walther C, Kropp JP (2012) Towards sectoral and standardised vulnerability assessments: the example of heat wave impacts on human health. Clim Change 112(3–4):687–708

Lissner TK, Reusser DE, Kropp JP (2013a) Tackling climate change impacts in the context of sustainability: the livelihood index as an integrative framework. In: Geophysical research abstracts, vol 15, EGU2013-9480, 2013, Vienna, Austria

Lissner TK, Reusser DE, Sullivan CA, Kropp JP (2013b) Meeting water needs for sustainable development: an overview of approaches, measures and data sources. In: geophysical research abstracts, vol 15, EGU2013-13804, 2013, Vienna, Austria

Maheu A (2009) Energy choices and their impacts on demand for water resources: an assessment of current and projected water consumtion in global energy production. Unisfra and McGill University, Montreal

Mekonnen M, Hoekstra A (2010a) The green, blue and grey water footprint of farm animals and animal products. Value of water research report series No. 48, UNESCO-IHE, Delft

Mekonnen MM, Hoekstra AY (2010b) The green, blue and grey water footprint of crops and derived crop products. Value of water research report series No. 47, UNESCO-IHE, Delft

Nakićenović N, Alcamo J, Davis G, Vries BD (2000) IPCC special report on emissions scenarios (SRES). Technical report, Intergovernmental Panel on Climate Change

Pradhan P, Reusser DE, Kropp JP (2013) Embodied greenhouse gas emissions in diets. PLoS ONE 8(5):e62228

Randall D, Wood R, Bony S, Colman R, Fichefet T, Fyfe J, Kattsov V, Pitman A, Shukla J, Srinivasan J, Stouffer R, Sumi A, Taylor K (2007) Climate models and their evaluation. In: Solomon S, Qin D, Manning M, Chen Z, Marquis M, Averyt K, Tignor M, Miller H (eds) Climate change 2007: the physical science basis. Contribution of Working Group I to the Fourth Assessment Report of the Intergovernmental Panel on Climate Change. Cambridge University Press, Cambridge, New York

Rijsberman FR (2006) Water scarcity: fact or fiction? Agr Water Manage 80(1–3):5–22

Shiklomanov IA (2000) Appraisal and assessment of world water resources. Water Int 25(1):11–32

Shuval H (1992) Approaches to resolving the water conflicts between Israel and her neighbors—a regional water-for-peace plan. Water Int 17(3):37–41

Smakhtin V, Revenga C, Döll P (2004) A pilot global assessment of environmental water requirements and scarcity. Water Int 29(3):307–317

Sullivan CA (2002) Calculating a water poverty index. World Dev 30(7):1195–1210

Sullivan CA, Meigh J (2005) Targeting attention on local vulnerabilities using an integrated index approach: the example of the climate vulnerability index. Water Sci Technol 51(5):69–78

UN (2012) The Millennium development goals report 2012. United Nations, New York

Vörösmarty CJ, McIntyre PB, Gessner MO, Dudgeon D, Prusevich A, Green P, Glidden S, Bunn SE, Sullivan CA, Liermann CR, Davies PM (2010a) Global threats to human water security and river biodiversity. Nature 467(7315):555–561

Vörösmarty CJ, McIntyre PB, Gessner MO, Dudgeon D, Prusevich A, Green P, Glidden S, Bunn SE, Sullivan CA, Liermann CR, Davies PM (2010b) Supplementary information: global threats to human water security and river biodiversity. Nature 467(7315):555–561

Wada Y (2013) Future irrigation water demand under climate change: regional variability and uncertainties arising from GHMs and CMIP5 climate projections. In: Water in the Anthropocene. Challenges for Science and Governance. 21–24 May 2013, Bonn

WWAP (2012) The united nations world water development report 4: managing water under uncertainty and risk. WWAP (World Water Assessment Programme) UNESCO, Paris

WHO Global Health Observatory Data Repository. http://apps.who.int/gho/data/. Accessed Nov 2012

Chapter 6
A Global Approach to Estimating the Benefit-Cost Ratio of Water Supply Measures in the Agricultural Sector

A. K. Schürkmann, A. Biewald and S. Rolinski

Abstract This study assesses at the global scale the potential costs and benefits of new infrastructure needed for the additional supply of irrigation water, focusing on rainwater harvesting, desalination and groundwater extraction. The cost and applicability of each measure is assessed and estimated separately. The potential benefit of additional water supply infrastructure is given by the water shadow price, which is generated by the global land and water use model MAgPIE (Model of Agricultural Productivity and its Impact on the Environment). Based on these results the irrigation potential (in Mha) is calculated. We find that groundwater extraction is cost-efficient in the most places and therefore has the highest irrigation potential (152.5 Mha) followed by rainwater harvesting (61.5 Mha) and desalination (0.5 Mha). The results reflect the current practice of supplying irrigation water, and a sensitivity analysis shows that rainwater harvesting has the largest potential to alleviate irrigation water scarcity through decreasing prices. The sensitivity analysis also shows that if the price of desalinated water continues to decline as it has in the past, desalination could become cost efficient especially in arid, coastal regions of the world.

Introduction

World population is projected to reach a number of 9–10 billion by 2050 (Lutz and Samir 2010) while income levels are expected to increase (Rask and Rask 2010). Higher incomes lead to more food consumption in total and an increase in the

A. K. Schürkmann (✉)
Global Water System Project (GWSP), Walter-Flex-Str. 3, 53113 Bonn, Germany
e-mail: ASchuerkmann@gmail.com

A. Biewald · S. Rolinski
Potsdam Institute for Climate Impact Research (PIK), Telegrafenberg 31, 14473 Potsdam, Germany

A. Bhaduri et al. (eds.), *The Global Water System in the Anthropocene*, Springer Water, DOI: 10.1007/978-3-319-07548-8_6,

share of animal calories consumed (Kearney 2010; Valin et al. 2014, *Accepted*). To meet this increasing demand for agricultural products, agricultural production must also increase. Although production has grown continuously in the past, increases in crop production have slowed down recently (Foley et al. 2011).

Increasing agricultural production is mainly achieved in one of the following ways: expanding cultivated area, developing and implementing new and more productive crop varieties, and intensifying agriculture on currently cultivated land.

Agricultural intensification reduces the yield gap (Cassmann 1999; Ramankutty et al. 2002), which is defined as the difference between the potential maximum yield of a crop in a certain place and the actual yield (see: Global Yield Atlas 2013). For many crops, global yield gaps are substantial; if 95 % of the crops' harvested areas met their current climatic potential, up to 60 % higher yields could be achieved (Licker et al. 2010). In their study Licker et al. (2010) found that one of the most promising ways to close the yield gap is increasing irrigation. This is because water is frequently the limiting factor regarding plant growth, this is particularly true for the arid regions of the world where yield gaps are high.

Globally, irrigation agriculture accounts for 40 % of the world's food production while occupying only 20 % of cultivated land (Siebert et al. 2007; UN Water World Water Assessment Program 2009). Increasing irrigation and expanding the area under irrigation on currently cultivated but unirrigated land are major methods for increasing future food production. However, in many regions of the world water is scarce or access to water is limited (Rosegrant et al. 2009). Problems of water scarcity in agriculture can be addressed (1) by reducing the demand for irrigation water and (2) by increasing the amount of water available for irrigation.

Demand for irrigation water can be reduced by increasing irrigation efficiency, minimizing irrigation water losses or distributing water in a more productive way (Seckler et al. 2003; Rockström and Barron 2007; Molden et al. 2010). While many studies examine the potential to reduce irrigation water demand, the present study focusses on the expansion of the area under irrigation which requires additional water supply. Measures for increasing irrigation water supply include dam construction, water transfer, desalination of salt and brackish water, rainwater harvesting, groundwater extraction and waste water reuse. Even though implementing any of these measures is subject to difficult political and social constraints and decisions (Woolley et al. 2009), implementation depends largely on the cost and cost efficiency of the specific measure (Hussain et al. 2007).

Implementation of a water supply measure is cost efficient only when the benefits of increased agricultural production outweigh its construction, operation and maintenance costs. Costs and benefits of a measure vary substantially depending on local or regional conditions. To determine where the impact of a water supply measure is highest, it is necessary to compare the cost efficiency (1) for the same measure in different locations and (2) for different measures in the same location.

A common way to assess the cost efficiency of an investment is calculating the benefit-cost ratio (BCR). The benefit-cost ratio expresses the monetary benefit of a project or investment relative to its cost. Unfortunately, the cost of many water supply measures is lacking for many parts of the world. Despite their importance, two common methods for supplying freshwater, water transfer and dam construction, will not be assessed in the current study. Data for the cost of water transfer are lacking. While some data are available for the location and size of moderate and large dams globally, there are very few data available for the cost of supplied irrigation water. Furthermore, dams often have substantial and complex, negative social, political and environmental impacts that cannot be included in a global study (WCD 2000). Therefore, in the present study we focus on three methods for increasing agricultural water supply: rainwater harvesting, desalination, and groundwater extraction. To our knowledge there has been no global scale analysis of their cost efficiency so far.

These three measures represent different types of water sources, namely groundwater (groundwater extraction), surface water (rainwater harvesting) and unconventional water (desalination; after Siebert et al. 2010). They represent different parts of the water cycle and different methods for supplying additional water. In this context, different methods mean either spatial or temporal redistribution of water as done by groundwater extraction and rainwater harvesting or the generation of additional freshwater by desalination.

At the small to medium scales, rainwater harvesting is often cost efficient in developing countries and can benefit individual farmers and communities (for case studies on India, see: Panigrahi et al. 2005; Pandey 1991; Goel and Kumar 2004; Sharma et al. 2010; for China see Yuan et al. 2003; Liang and van Dijk 2011; and for Kenya see Ngigi et al. 2005). Furthermore, at the global scale, small scale rainwater harvesting structures can significantly increase yields and therefore improve food security (Wisser et al. 2010).

Although there has been broad interest in the development of new, cheaper technologies for extracting salt from seawater (Karagiannis and Soldatos 2008), using desalinated water for irrigation is not common practice, occurring only in Spain and some parts of the Arabian Peninsula (Al-Rashed and Sherif 2000; Mezher et al. 2011; Zarzo et al. 2013).

Groundwater is commonly extracted throughout most of the world and is often overexploited (e.g. Aeschberg-Hertig and Gleeson 2012; Werner et al. 2012 or ISARM Internationally Shared Aquifer Resources Management a collection of global groundwater related data available on http://www.isarm.org/publications/119). In some places groundwater contributes up to 90 % of irrigation water supply (Al-Rashed and Sherif 2000).

To assess the potential benefit of increased water supply, we use the water shadow price, which is calculated by the global land and water use model MAgPIE (Model of Agricultural Production and its Impact on the Environment; Lotze-Campen et al. 2008).

This study is the first attempt to assess globally the cost efficiency of different irrigation water supply measures. Using spatially explicit cost and benefit data, we determine where investments in different water supply measures may be most cost efficient.

To assess and compare the cost efficiency of the three water supply measures, we developed a conceptual framework that differentiates between the applicability,

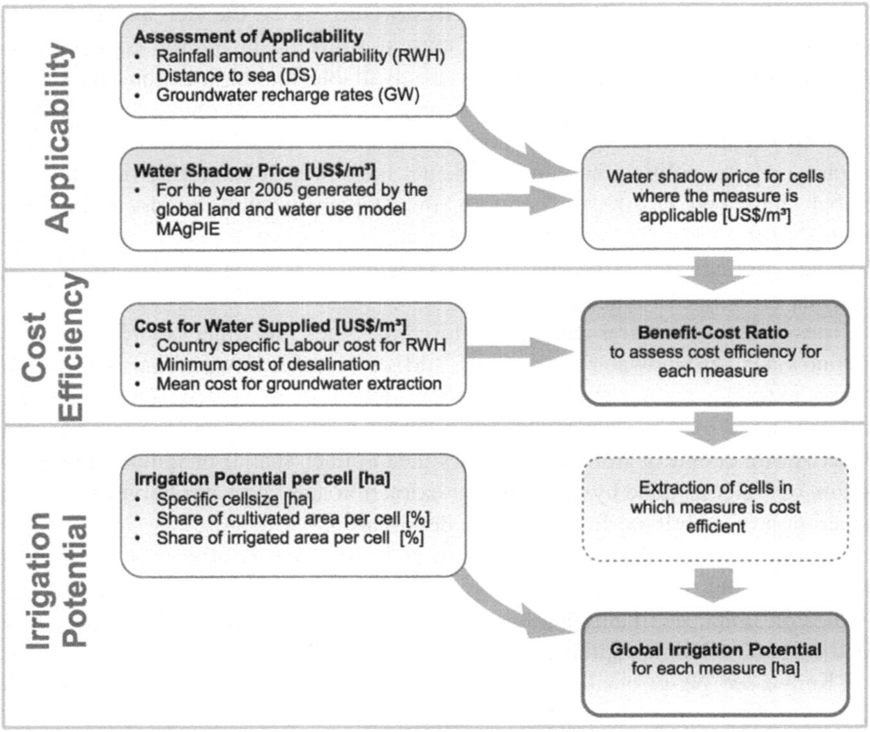

Fig. 6.1 Flowchart of the conceptual framework illustrating the different steps of the analysis. (*RWH* rainwater harvesting, *DS* desalination, *GW* groundwater extraction)

the cost efficiency, and the potential impact of each measure (Fig. 6.1). Applicability refers to the physical feasibility of implementing a measure depending environmental conditions and resource availability.

Where a measure was deemed applicable, the benefit-cost ratio was determined, using the cost and water shadow price data. A sensitivity analysis was conducted to address uncertainties in the collected data. Then the irrigation potential was calculated for each measure based on the amount of land under cultivation and the share of that land that was irrigated. In the final step, the applicability, cost efficiency, and irrigation potential for the three measures were compared.

Methods

The Lund-Potsdam-Jena dynamic global vegetation model with managed Lands (LPJmL) was used to generate vegetation growth, crop yields, and water consumption on a $0.5° \times 0.5°$ grid in daily time steps. This model uses twelve crop functional types and nine functional types for natural vegetation to simulate crop

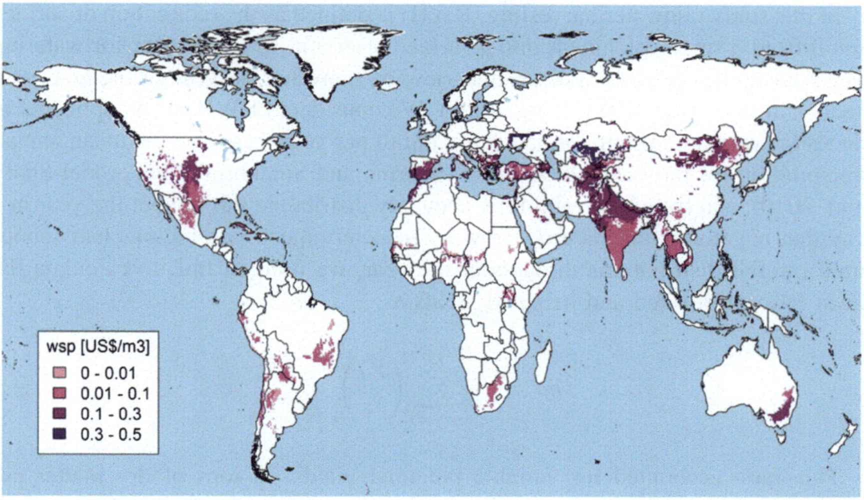

Fig. 6.2 Global map of the water shadow price (WSP) generated by MAgPIE for the year 2005 in US$/m^3. White cells indicate areas with no irrigation agriculture or where irrigation water is not scarce. Cell size is 0.5° × 0.5°

yields and land use based on observed land use patterns and climatic and biogeochemical conditions (Sitch et al. 2003; Bondeau et al. 2007). Results from LPJmL, which include the availability of irrigation water per grid cell, were combined with regional economic information, and used as input data for MAgPIE, a model of agricultural production and its impact on the environment, which calculates the water shadow price.

MAgPIE is a global, spatially explicit, economic land and water use model with a cost minimization function (Lotze-Campen et al. 2008). It minimizes global production costs for producing crops and livestock and requires that agricultural demands be met. MagPIE simulates time steps of 10 years starting in 1995 and uses the optimal land use pattern from the previous period as a starting point. In each time step, MAgPIE meets growing demand, driven by changes in population and income projections, by increasing agricultural production. Production can be increased by either increasing the area under cultivation or intensifying agriculture on cultivated lands. Intensification of agriculture can, besides others, be achieved through increased irrigation. The available amount of irrigation water is calculated by LPJmL and used as one constraint in MAgPIE.

When all available water in a cell is used for irrigation, the water shadow price (WSP) is calculated and an estimate of how much an additional unit of water would be worth in the model context (Fig. 6.2). The water shadow price is used as a water scarcity indicator and expresses the willingness to pay to increase irrigation water supply by one unit. Currently the water shadow price is computed only for cells where irrigation agriculture is already partially practiced (based on Döll and Siebert 2000).

In this study rainwater harvesting (RWH) is defined as the redirection of surface runoff from a small catchment into a surface reservoir, to store water for watering crops during dry periods (For an overview of rainwater harvesting methods see: Boers and Ben-Asher 1981; Ngigi 2003). We considered RWH to be applicable in places (a) with a minimum of 350 mm rainfall per year, based on the mean annual precipitation for the years 2000–2009 (Sharma and Smakhtin 2006; Abdel-Shafy et al. 2010), and (b) where rainfall is unevenly distributed throughout the year in a way that negatively affects crop growth. To determine which regions had suboptimal rainfall distributions throughout the year, we used LPJmL to calculate the mean ratio of rain fed and irrigated yields as:

$$mr = \frac{1}{15} \sum_{crop=1}^{15} \left(\frac{Y_{rf}}{Y_{ir}} \right)_{crop}$$

This ratio accounted for variable potential yields (in tons of dry matter per hectare [t/ha]) of the 15 main food crops implemented in MAgPIE. The reciprocal is the potential to increase crop yields through irrigation. RWH was deemed applicable for cells with at least 350 mm/year rainfall and where crop yields could be increased by more than 10 % through irrigation.

The cost of a RWH facility was considered to be the sum of construction, operation and maintenance costs over the expected lifespan of the structure. Over the lifespan of the structure, material and maintenance costs were considered to be negligible (Fox and Rockström 2000; Saha et al. 2007). We assumed that labour costs are proportional to the size of the reservoir, with ~3500 h being required to build a 150 m^3 reservoir (Fox and Rockström 2000) and that the reservoir lifespan is 25 years (Goel and Kumar 2004; Sturm et al. 2009). Over the projected lifespan of the reservoir, the total amount of water provided (w_{tot}) is 3750 m^3. The labour investment per unit water (L_w) can be calculated as:

$$L_w = \frac{L_t}{w_{tot}}$$

yielding an estimate of 0.93 h/m^3.

Hourly salaries (S_h) for workers in construction and agriculture were calculated based on data from the International Labour Organization (ILO), converted to US Dollars using historical exchange rates and then adjusted for inflation to the base year of 2005 using the conversion factors provided by Sahr (2012). For countries lacking data, we used the median hourly salary for countries with similar economies, based on the World Bank classifications. Finally a discount rate of 3 % over the lifespan of 20 years was incorporated to account for the increasing value of the initial investment over time.

Multiplication of labour investment per unit water (L_w) with hourly salaries (S_h) for a farm worker (in \$/h) gives the final cost (C_{RWH}) per m^3 water supplied by rainwater harvesting:

$$L_w * S_h = C_{RWH}$$

We considered desalination applicable in areas with direct access to the sea, where water can be extracted and processed without transportation over large distances. Therefore, we limited our analysis to MAgPIE cells with at least one neighboring cell belonging to one of the oceans.

We excluded all large inland water bodies, except the Caspian Sea because it holds saltwater with an average concentration of about one third that of sea water (Dumont 1998) and its water level has risen over the last four decades (Ozyavas and Khan 2012).

A variety of methods exist for desalinating seawater (El-Ghonemy 2012), and the cost of desalination (C_{DS}) can range from 0.45 to 11.0 US\$/m^3 based on the technology, facility size, energy source and salt content of the water (e.g. Karagiannis and Soldatos 2008; Froiui and Oumeddur 2008; Mezher et al. 2011).

Because desalination is a comparatively expensive method for supplying additional freshwater, we chose the minimum reported cost of 0.45 US\$/m^3 as a optimistic global estimate. This price of 0.45 US\$/m^3 comes from a desalination plant south of Tel Aviv, Israel, which uses reverse osmosis and is connected to the electricity grid. It produces about 330 000 m^3 of water per day and up to 110 million m^3/y mainly to secure the freshwater supply to surrounding towns (Dreizin 2006; Sauvet-Goichon 2007).

Groundwater is the largest unfrozen freshwater resource in the world and accounts for about 40 % of irrigation water worldwide (BGR/UNESCO 2008; Siebert et al. 2010).

Because groundwater overexploitation leads to declining groundwater levels (Aeschberg-Hertig and Gleeson 2012), groundwater extraction is applicable only in places with recharge rates high enough to support irrigation. We assumed that groundwater extraction is applicable only where recharge rates are higher than 20 mm/year for major groundwater basins and areas with complex hydrogeological structures and where recharge rates exceed 100 mm/year for local and shallow aquifers. Although saline water is sometimes used for irrigation (Flowers 2004), it can increase soil salinity and negatively impact plant growth (Shalhevet 1994), so we excluded areas with saline groundwater. We determined where groundwater extraction would meet these criteria for applicability, based on data from the Worldwide Hydrogeological Mapping and Assessment Program (WHYMAP; BGR/UNESCO 2008).

The cost of groundwater extraction (C_{GW}) is proportional to the distance between the groundwater level and the land surface. The cost of groundwater extraction is comparatively low, ranging from 0.01 US\$/m^3 to 0.08 US\$/m^3 (Water Resources Group 2009). We used the mean value (0.04 US\$/m^3) for the cost of groundwater extraction

We used the benefit-cost ratio (BCR) to determine whether an investment in water supply infrastructure was cost efficient. The benefit-cost ratio for each measure (BCR_{RWH}; BCR_{DS}; BCR_{GW}) was calculated as:

$$BCR_{measure} = \frac{WSP}{C_{measure}}$$

where *WSP* was the water shadow price for the year 2005 in US\$/m^3, C was the cost for water supplied by one of the three measures, rainwater harvesting (C_{RWH}), desalination (C_{DS}) and groundwater extraction (C_{GW}), also in US\$/m^3 for the year 2005.

The BCR is a dimensionless number with values <1 indicating that the costs are higher than benefits, and values >1 indicating that the economic benefits outweigh the costs. Being a dimensionless indicator, the benefit-cost ratio allows for comparison of the economic performance of different projects and investments regardless of their nature and possible incompatibility.

To determine how uncertainties in the costs impacted the BCR, we conducted a sensitivity analysis of the effects of the infrastructure costs for each measure on the BCR with costs ranging from 1–200 % of the calculated price. This sensitivity analysis shows how high subsidies for a measure would have to be to make the investment worthwhile.

The irrigation potential was calculated based on LPJmL data and the absolute size of MAgPIE cells in ha. LPJmL determines the share of cultivated and irrigated area per cell. Using results for 2005, we calculated the potential increase in irrigated area for each cell as the difference between the cultivated and irrigated areas per cell. The potential increase in irrigated area of all cells with BCR >1 were summed to calculate the global irrigation potential for each measure.

Results and Discussion

Overall, we found that GW was applicable in more places than either RWH or DS (green cells in Fig. 6.3), mainly because of the abundance of groundwater globally. However, in many places in which GW was applicable, water is not the predominant factor limiting crop growth (for comparison see Fig. 6.2). Thus, considering only water-limited localities, RWH is applicable in more areas than either GW or DS (351.6 Mha, 225.6 Mha, 31.6 Mha, respectively; blue cells in Fig. 6.3). Despite relatively widespread applicability of RWH and GW, the costs associated with implementing these measures make them economically inefficient in many areas. For example, we found RWH to be cost-efficient only in India and a small part of Ukraine (Fig. 6.3 top). Our results are supported by several case studies from India, in which RWH was cost-efficient (Pandey 1991; Sharma et al. 2010) with BCRs of 1.17 (Panigrahi et al. 2005) and 1.33 (Goel and Kumar 2005). However, in several places our results do not reflect the results of case studies from other parts of the world. For example, we underestimated the BCR of RWH in rural Beijing, which varies between 1.96 and 6.2 (Liang and van Dijk 2011).

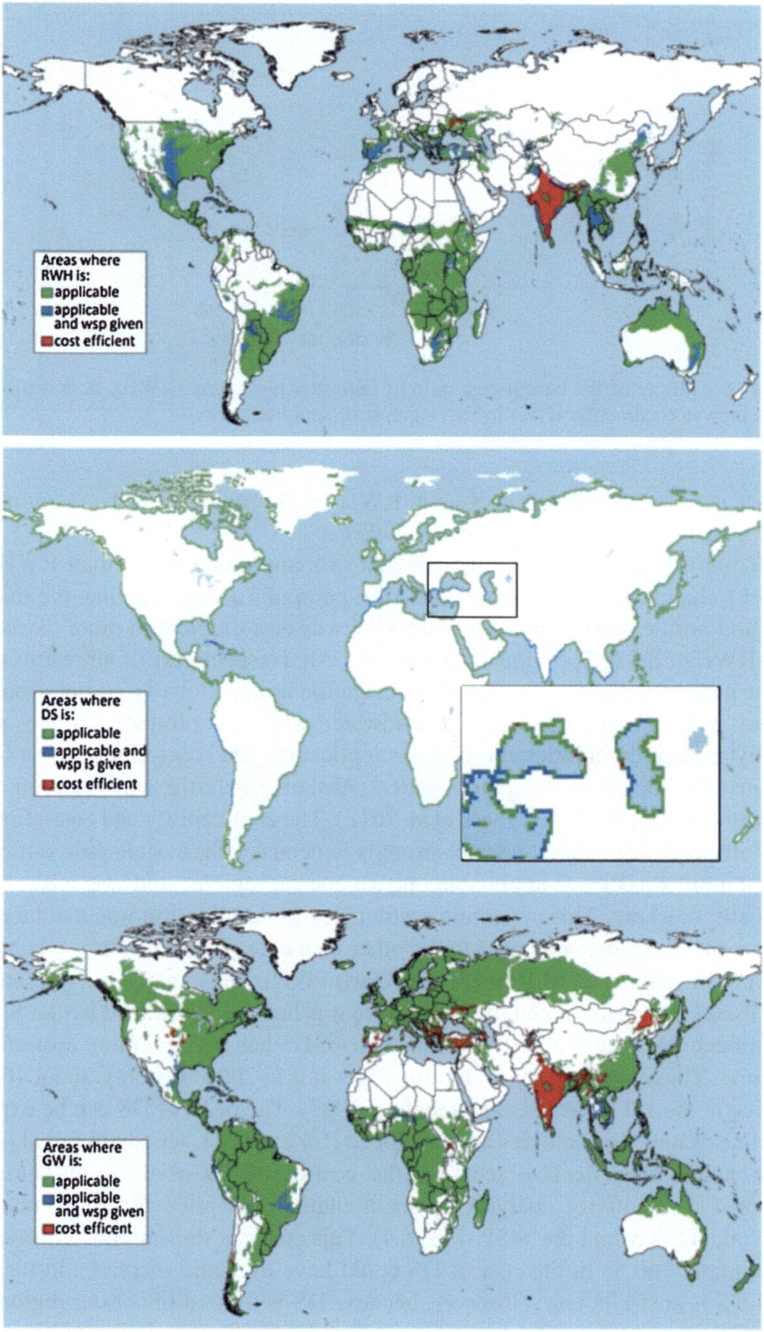

Fig. 6.3 Areas where measure is applicable (*green*), areas where a measure is applicable and a water shadow price is given (*blue*), and areas where a measure is cost-efficient (*red*)

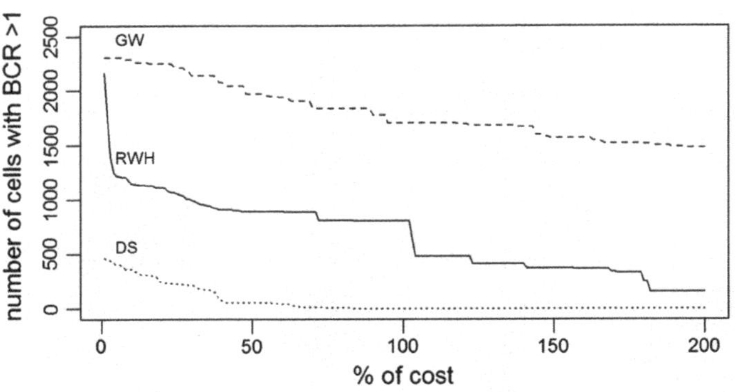

Fig. 6.4 Sensitivity of the benefit-cost ratio of rainwater harvesting (RWH), desalination (DS) and groundwater extraction (GW) to varying cost of supplied water

In other parts of China and in Kenya, RWH has been shown to be economically viable (Yuan et al. 2003; Ngigi et al. 2005).

GW, on the other hand, is cost-efficient in many more places than RWH (e.g. most of India and its neighboring countries, parts of Turkey, Ukraine, the midwest USA, and northeastern China). In total, GW was cost-efficient in more places than either RWH or DS (152.5 Mha, 61.5 Mha, 0.5 Mha respectively). Our results reflect current practices globally, as 40 % of irrigation water globally comes from GW (Siebert et al. 2010). However, groundwater is over-exploited in many places globally, especially in India, limiting the applicability and cost-efficiency of GW as a means for supplying irrigation water (Aeschberg-Hertig and Gleeson 2012; Glendenning et al. 2012; Varghese et al. 2012). The applicability and cost-efficiency of GW in supplying irrigation water strongly depends on the sustainability of its use.

By definition, DS was applicable only in coastal regions, and was cost-efficient practically nowhere. This economic inefficiency of desalination was due largely to its relatively high cost compared to the other measures. These results are consistent with current practices (Al-Rashed and Sherif 2000). Only in Spain is desalinated water used for irrigation, and that is because it is heavily subsidized by the Spanish government (Mezher et al. 2011). Nonetheless, DS has become more cost-efficient over time. The average price of desalinated water in 2000 was only about 10 % of the price in the 1960s (Reddy and Ghafour 2007). The price of DS can be expected to decline as new technologies are developed (for example, see Shaffer et al. 2012).

To explore the effects of cost on the cost-efficiency of these measures, we conducted a sensitivity analysis by manipulating the price for implementation while holding constant the WSP (Fig. 6.4). This analysis showed that a decrease of approximately 60 % in the cost of DS could have dramatic increases in the areas where DS is cost-efficient. However, because DS is limited to coastal regions, the role of DS in supplying irrigation water will undoubtedly remain limited compared to GW and RWH.

Varying the cost of GW from 1–200 % had relatively little effect on the cost-efficiency of GW in supplying irrigation water. Because the current cost of GW is so low (0.04 US\$/m^3), even with doubling the cost to 0.08 US\$/m^3, it is still well below the WSP in most places (Fig. 6.2). Similarly, the already low cost of GW cannot be reduced much more such that reducing the cost has little effect on increasing the area where it is cost-efficient.

Compared to GW, varying the cost of RWH had a stronger impact on cost-efficiency, due to the wide range in labor costs among countries, which vary from 0.04 US\$/m^3 to over 30 US\$/m^3. Reducing the high labor costs by over 95 % results in a large increase in the cost-efficiency of RWH. This seems reasonable because we have most likely over-estimated the labor cost associated with constructing a RWH facility for three main reasons. First, farm work is highly seasonal, which means that in many places, farmers have periods in the year in which their time is not fully occupied with any kind of work. During these times of the year, constructing a RWH facility would not come at a cost. Second, farm work, especially in remote locations, can be informal so that actual cost of farm labor is not accurately reported. By its nature, informal labor is cheaper than formal employment, such that reported salaries are probably higher than unreported salaries. Third, the assumption that a RWH facility is filled only once per year may not be true in places with multiple rainy periods throughout the year. In these places, more water is provided throughout the year (the facility can be filled multiple times) at the same labor cost as a facility that is only filled once.

The irrigation potential according to cost-efficiency is highest for GW (152.5 Mha; 10.1 % of global cultivated area) followed by RWH (61.5 Mha; 4.1 % of global cultivated area). RWH is applicable in the most places with an irrigation potential of 351.6 Mha while GW is only applicable in 225.6 Mha. For DS the numbers are up to orders of magnitudes smaller for applicability and cost-efficiency with 31.6 Mha and 0.5 Mha, respectively (Fig. 6.5). In places where RWH was found to be cost-efficient, the same was true for GW and due to the low cost for GW it was always the economically more viable option. However, taking into account the overestimation of applicability of GW and the underestimation of the cost-efficiency of rainwater harvesting we discussed before, this figure may change. Another aspect is the current implementation of these two measures, even though RWH has been practiced for centuries in some places, it was entirely unknown in other water scarce regions until recently (Boers and Ben-Asher 1981). This adds to the potential of RWH, as extracting groundwater has been for many years common practice in most parts of the world, which threatened groundwater resources in many places (Aeschbach-Hertig and Gleeson 2012).

A drawback of this study is that the water shadow price and therefore the benefit-cost ratio are only calculated for cells that are at least partially equipped for irrigation. Because of this areas with possibly high irrigation potential and the highest yield gaps, especially in Africa and Asia (Wisser et al. 2010; Rosegrant et al. 2002), are not included in this study.

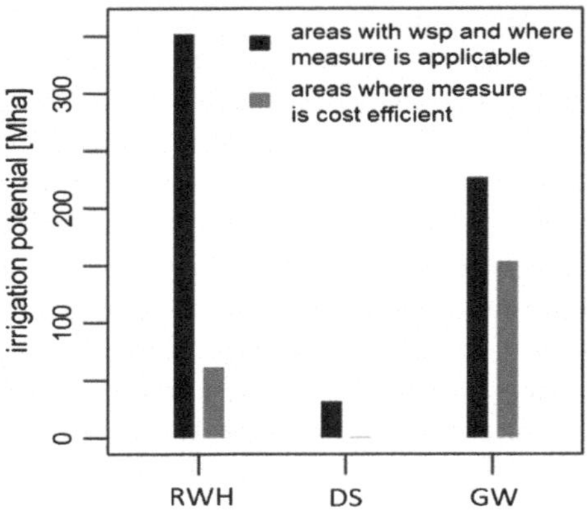

Fig. 6.5 Irrigation potential in Mha for each measure based on applicability and water shadow price (*black*) and cost-efficiency (*grey*) for rainwater harvesting (RWH), desalination (DS) and groundwater extraction (GW)

Conclusion

This study was a first attempt to assess and compare the cost efficiency of different irrigation water supply measures on a global level. Following the steps laid out in the conceptual framework, the current practice of irrigation water supply and its related cost efficiency was reflected in this global study. It could be shown that additional irrigation water can be supplied in a cost efficient way in many regions. Even when only focusing on places where irrigation is already partially practiced, the irrigation potential for the different measures is high for RWH and GW and may contribute substantially to closing the yield gap in many regions.

While RWH is widely applicable, it is cost-efficient for only 4.1 % of the global cultivated land area. With a more accurate assessment of the cost for water supplied by RWH this number is expected to increase. GW has a medium potential to increase irrigation water supply and is economically efficient for 10.1 % of the global cultivated land area. However, incorporating over-exploitation of groundwater, into future analyses will reduce applicability of GW and the area where we found it to be cost efficient. DS only plays a minor role as an irrigation water supply measure due to its high price and its spatial limitation. However, if prices continue to decline, DS may become more important for agriculture in coastal regions and where other freshwater sources are not available.

Future research should focus on better cost assessment (especially for RWH), include data on groundwater over-exploitation, and incorporate the option of expanding irrigation into regions where irrigation is not yet practiced. Calculation of the water shadow price for regions where irrigation is not yet practiced is particularly important for Africa, where yield gaps are high, and the impact of irrigation is expected to be highest (Wisser et al. 2010).

References

Abdel-Shafy HI, El-Saharty AA, Regelsberger M, Platzer C (2010) Rainwater in Egypt: quantity, distribution and harvesting. Mediterr Mar Sci 11:245–257

Aeschbach-Hertig W, Gleeson T (2012) Regional strategies for the accelerating global problem of groundwater depletion. Nat Geosci 5:853–861

Al-Rashed MF, Sherif MM (2000) Water resources in the GCC countries: an overview. Water Resour Manage 14:59–75

BGR/UNESCO (2008) Groundwater resources of the world, 1:25 000 000. http://www.whymap.org/whymap/EN/Products/products_node_en.html

Boers ThM, Ben-Asher J (1981) A review of rainwater harvesting. Agric Water Manage 5:145–158

Bondeau A, Smith P, Zeahle S, Schaphoff S, Lucht W, Cramer W, Gerten D, Lotze-Campen H, Müller C, Reichstein M, Smith B (2007) Modelling the role of agriculture for the 20th century global terrestrial carbon balance. Glob Change Biol 13:679–706

Cassman KG (1999) Ecological intensification of cereal production systems: yield potential, soil quality, and precision agriculture. Agronomy—Faculty Publications. Paper 105. http://digitalcommons.unl.edu/agronomyfacpub/105

Döll P, Siebert S (2000) A digital global map of irrigated areas. ICID J 49:55–66

Dumont HJ (1998) The Caspian lake: history, biota, structure, and function. Limnol Oceanogr 43:44–52

Dreizin Y (2006) Ashkelon seawater desalination project—off-taker's self costs, supplied water costs, total costs and benefits. Desalination 190:104–116

El-Ghonemy AMK (2012) Water desalination systems powered by renewable energy sources: review. Renew Sustain Energy Rev 16:1537–1556

Flowers JT (2004) Improving crop salt tolerance. J Exp Bot 55:307–319

Fox P, Rockström J (2000) Water-harvesting for supplementary irrigation of cereal crops to overcome intra-seasonal dry-spells in the Sahel. Phys Chem Earth 25:289–296

Foley JA, Ramankutty N, Brauman KA, Cassidy ES, Gerber JS et al (2011) Solutions for a cultivated planet. Nature 478:337–342

Frioui S, Oumeddour R (2008) Investment and production costs of desalination plants by semi-empirical method. Desalination 223:457–463

Glendenning CJ, van Ogtrop FF, Mishra AK, Vervoort RW (2012) Balancing watershed and local scale impacts of rain water harvesting in India—a review. Agric Water Manage 107:1–13

Global Yield Atlas (2013) http://www.yieldgap.org/web/guest/home

Goel AK, Kumar R (2004) Economic analysis of water harvesting in a mountainous watershed in India. Agric Water Manage 71:257–266

Hussain I, Turral H, Molden D, Ahmad MUD (2007) Measuring and enhancing the value of agricultural water in irrigated river basins. Irrig Sci 25:263–282

Karagiannis IC, Soldatos PG (2008) Water desalination cost literature: review and assessment. Desalination 223:448–456

Kearney J (2010) Food consumption trends and drivers. Philos Trans R Soc B Biol Sci 365:2793–2807

Liang X, van Dijk MP (2011) Economic and financial analysis on rainwater harvesting for agricultural irrigation in the rural areas of Beijing. Resour Conserv Recycl 55:1100–1108

Licker R, Johnston M, Foley JA, Barford C, Kucharik CJ, Monfreda C, Ramankutty N (2010) Mind the gap: how do climate and agricultural management explain the 'yield gap' of croplands around the world? Glob Ecol Biogeogr 19:769–782

Lotze-Campen H, Müller C, Bondeau A, Rost S, Popp A, Lucht W (2008) Global food demand, productivity growth, and the scarcity of land and water resources: a spatially explicit mathematical programming approach. Agric Econ 39:325–338

Lutz W, Samir KC (2010) Dimensions of global population projections: what do we know about future population trends and structures? Philos Trans R Soc B Biol Sci 365:2779–2791

Mezher T, Fath H, Abbas Z, Khaled A (2011) Techno-economic assessment and environmental impacts of desalination technologies. Desalination 266:263–273
Molden D, Oweis T, Steduto P, Bindraban P, Hanjra MA, Kijne J (2010) Improving agricultural water productivity: between optimism and caution. Agric Water Manag 97:528–535
Ngigi SN (2003) What is the limit of up-scaling rainwater harvesting in a river basin? Phys Chem Earth 28:943–956
Ngigi SN, Savenije HHG, Rockström J, Gachene CK (2005) Hydro-economic evaluation of rainwater harvesting and management technologies: farmers investment options and risks in semi-arid Laikipia district of Kenya. Phys Chem Earth 30:772–782
Ozyavas A, Khan SD (2012) The driving forces behind the Caspian Sea mean water level oscillations. Environ Earth Sci 65:1821–1830
Pandey S (1991) The economics of water harvesting and supplementary irrigation in the semi-arid tropics of India. Agric Syst 36:207–220
Panigrahi B, Panda SN, Agrawal A (2005) Water balance simulation and economic analysis for optimal size of on-farm reservoir. Water Resour Manage 19:233–250
Ramankutty N, Foley JA, Norman J, McSweeney K (2002) The global distribution of cultivable lands: current patterns and sensitivity to possible climate change. Glob Ecol Biogeogr 11:377–392
Rask KJ, Rask N (2010) Economic development and food production-consumption balance: a growing global challenge. Food Policy 36:186–196
Reddy KV, Ghaffour N (2007) Overview of the cost of desalinated water and costing methodologies. Desalination 205:340–353
Rockström J, Barron J (2007) Water productivity in rainfed systems: overview of challenges and analysis of opportunities in water scarcity prone savannahs. Irrig Sci 25:299–311
Rosegrant MW, Cai X, Cline SA, Nakagawa N (2002) The role of rainfed agriculture in the future of global food production. International Food Policy Research Institute (IFPRI), Washington, DC, USA
Rosegrant MW, Ringler C, Zhu TJ (2009) Water for agriculture: maintaining food secutity under growing scarcity. Annu Rev Environ Resour 34:205–222
Saha R, Ghosh PK, Mishra VK, Bujarbaruah KM (2007) Low-cost micro-rainwater harvesting technology (Jalkund) for new livelihood of rural hill farmers. Curr Sci 92:1258–1266
Sahr RC (2012) Inflation conversion factors for years 1774 to estimated 2022. Political science department, Oregon State University, Corvallis, OR 97331, USA. http://oregonstate.edu/cla/polisci/download-conversion-factors
Sauvet-Goichon B (2007) Ashkelon desalination plant—a successful challenge. Desalination 203:75–81
Seckler D, Molden D, Sakthivadivel R (2003) The concept of efficiency in water-resources management and policy. In: Kijne JW Barker R, Molden D (eds) Water productivity in agriculture: limits and opportunities for improvement, vol 1. CABI, Walingford, UK, p 37–51(Book series title: Comprehensive assessment of water management in agriculture series)
Shaffer DL, Yip NY, Gilron J, Elimelech M (2012) Seawater desalination for agriculture by integrated forward and reverse osmosis: improved product water quality for potentially less energy. J Membr Sci 415–416:1–8
Shalhevet J (1994) Using water of marginal quality for crop production—major issues. Agric Water Manage 25:233–269
Sharma BR, Smakhtin VU (2006) Potential of water harvesting as a strategic tool for drought mitigation. International Water Management Institute (IWMI)
Sharma BR, Rao KV, Vittal KPR, Ramakrishna YS, Amarasinghe U (2010) Estimating the potential of rainfed agriculture in India: prospects for water productivity improvements. Agric Water Manage 97:23–30
Siebert S, Döll P, Feick S, Hoogeveen J, Frenken K (2007) Global map of irrigation areas version 4.0.1. Johann Wolfgang Goethe University, Frankfurt am Main, Germany/Food and Agriculture Organization of the United Nations, R Source: GWSP Digital Water Atlas (2008). Map 31: Irrigated Areas (V1.0). http://atlas.gwsp.org

Siebert S, Burke J, Faures JM, Frenken K, Hoogeveen J, Döll P, Portmann FT (2010) Groundwater use for irrigation—a global inventory. Hydrol Earth Syst Sci 14:1863–1880

Sitch S, Smith B, Prentice IC, Arneth A, Bondeau A, Cramer W, Kaplan J, Levis S, Lucht W, Sykes M, Thonicke K, Venevsky S (2003) Evaluation of ecosystem dynamics, plant geography and terrestrial carbon cycling in the LPJ dynamic vegetation model. Glob Change Biol 9:161–185

Sturm M, Zimmermann M, Schütz K, Urban W, Hartung H (2009) Rainwater harvesting as an alternative water resource in rural sites in central northern Namibia. Phys Chem Earth 34:776–785

Valin H, Sands RD, van der Mensbrugghe D, Ahammad H, Bodirsky BL, Hasegawa T, Havlík P, Kyle P, Mason-D'Croz D, Paltsev S, Tabeau A, Blanc E, Fujimori S, Heyhoe E, van Meijl H, Rolinski S, Willenbockel D, von Lampe M, Nelson GD (2014) Global economic models and food demand towards 2050: an intercomparison exercise. Agric Econ 45(1):51–67

Varghese SK, Buysse J, Frija A, Speelman S, van Huylenbroeck G (2012) Are investments in groundwater irrigation profitable? A case of rice farms from south India. J Am Water Resour Assoc 49:52–66

Water Resources Group (2009) Charting our water future—economic frameworks to inform decision-making. http://www.mckinsey.com/App_Media/Reports/Water/Charting_Our_Water_Future_Full_Report_001.pdf

Werner AD, Zhang Q, Xue L, Smerdon BD, Li X, Zhu X, Yu L, Li L (2012) An initial inventory and indexation of groundwater mega-depletion cases. Water Resour Manage 27:507–533

Wisser D, Frolking S, Douglas EM, Fekete BM, Schumann AH (2010) The significance of local water resources captured in small reservoirs for crop production—a global-scale analysis. J Hydrol 384:264–275

Woolley J, Cook SE, Molden D, Harrington L (2009) Water, food and development: the CGIAR challenge program on water and food. Water Int 34:4–12

World Commission on Dams (2000) Dams and developement—a new framework for decision-making, report of the world commision on dams. Earthscan, London

World Water Assessment Programme (2009) The United Nations world water development report 3: Water in a changing world. Report No. 978-92-3-104235-5, 407 (UNESCO, 2009)

Yuan T, Fengmin L, Puhai L (2003) Economic analysis of rainwater harvesting and irrigation methods, with an example from China. Agric Water Manage 60:217–226

Zarzo D, Campos E, Terrero P (2013) Spanish experience in desalination for agriculture. Desalin Water Treat 51:53–66

Part II
Dimensions of Change in River Basins and Regions

Today it is quite clear that feeding of more than 10 billion people in the near future will have serious implications on the carrying capacity of our planet. Future water demands need to be satisfied, while resource availability in space and time is subject to increasing uncertainties due to climate variability and change. Throughout history, humans have intensively used natural resources, unknowingly induced changes in water and land use patterns, which are reflected in stresses of the water and nutrient cycles. All of the above have wide scale ramification effects, and could cause widespread instabilities in ecosystems upon which humanity relies.

There is a justified concern that following a business-as-usual trajectory, we would transgress the boundaries of planetary, ecosystem, and societal services on which the well-being of present and future generations depends. A holistic approach is thus needed to address the associated socio-ecological complexities, and to stabilize the global water system and associated material cycles.

A river basin approach is widely recognized to be well suited for water resource planning, demand management, and adaptation, and serve as a laboratory to study the interactions between basin-scale processes and global influences, such as climate change. The basinwide approach provides a good framework to address governance, resource management, and adaptation processes to global change at a scale.

In the face of global environmental change, institutional and governance challenges often stand as a predicament to respond to and mitigate scarcity and degradation. The paper by Mori Clement et al. explores the institutional rigidity of agricultural policy making in Uzbekistan, a country facing extreme water scarcity during the dry seasons. The paper explores how competition for water between crops can influence consumers' welfare, and how a flexible management system can enhance resilience while easing the hard economic trade-offs.

Adaptation may include choices of unconventional resource use. The paper by Karimov et al. illustrates the case study of the same river basin, and studies the strengths, weaknesses, and opportunities of reusing salt-enriched water resources of agricultural origin. Their analysis reveals the suitability of such unconventional resource use for the development of intensive aquaculture-agriculture systems.

Sharing the benefits of transboundary basin resources with those who reside within the basin, is a key principle of Integrated Water Resources Management (IWRM). The study of Hensengerth et al. addresses the concept of benefit sharing as a means of fostering the cooperative use of international rivers. It explores incentives that can encourage benefit sharing and identifies different benefit-sharing mechanisms pertaining to dam projects on the rivers Senegal, Columbia, Orange-Senqu, Nile, and Zambezi.

In the overall framework of IWRM, adaptive management is recognized to deliver an integrative, holistic, and learning-centered approach to mitigate anthropogenic changes. The paper by Villamayor-Tomas et al. aims to understand the ability of farmers in a large irrigation project to cooperate and adjust their water demands to cope with droughts. The paper formulates causal inferences on the basis of common pool resource (CPR) theory as well as qualitative and quantitative evidence, and highlights the necessity to adopt an integrating framework to understand water and land use dynamics and their possible evolution toward sustainable management.

Water supply on islands becomes a major challenge today due to anthropogenic pressure on water resources and global climate change. The paper by Hoff et al. addresses the responses of urban water systems to the Anthropocene, using the case study of a Mediterranean touristic island within the last decade. The authors explore key issues that need to be addressed by policy and practice in the field of sustainable freshwater management.

Following the green economy model, sustainable development in the Anthropocene requires an increase in green investment with focus on natural processes. The paper by Loučková et al. assesses the environmental aspects of current flood management strategies in the Czech Republic to understand the implementation of "green" measures at national and regional level. The study identifies governance gaps and the traditional public mentality that hinder wide-scale use of environment-friendly measures in flood risk management.

Studies on changes of cultural and socio-political structures that shape and are shaped by the environment are equally important to understand the river systems in the Anthropocene as rivers and their ecologies are not simply natural systems; they are human systems as well. The paper by Kelly et al provides a historical perspective to anthropogenic environmental changes, tracing back different dimensions of human–nature entanglements with case studies from the early nineteenth century until today.

The key question around the Anthropocene, whether humans have permanently changed the planet, is a question that is inherently interdisciplinary. As a historian, Scarpino calls for interdisciplinary approaches to the Anthropocene, recognizing that explaining the long-term human impact on earth systems is the domain of humanists and social scientists. When addressing the role of human agency, rivers in the Anthropocene can be valuable case studies for understanding the complex and changing connections between human culture and nature.

Chapter 7
The Role of Institutions and Water Variability in Food Security in Uzbekistan: The Case of Rice Markets in the Khorezm Region

Yadira Mori-Clement, Anik Bhaduri and Nodir Djanibekov

Abstract During the last three decades, water scarcity in Uzbekistan has seriously affected agricultural production and rural income. The allocation of water resources is heavily influenced by natural factors but also by the institutional rigidity of agricultural policy making in Uzbekistan, which represents a core pillar of the agricultural policy inherited from the former socialist system. The objective of this paper is to explore the effect of water competition between the state promoted cotton and farmer desired rice crops on the domestic price of rice, and evaluate how possible changes in these settings would influence the rice prices. Under more flexible institutional conditions, the price path of rice would be smoothed as the farmers can improve its production. Moreover, the impact of a more flexible state procurement for cotton will be more favorable not only for farmers, but also for consumers as it might protect the latter from extreme price fluctuations in markets as a result of changes in domestic rice supply.

Y. Mori-Clement (✉)
Agricultural and Food Economics (AFECO) Programme, University of Bonn, Nussallee 21, 53115 Bonn, Germany
e-mail: yadira.mori.clement@gmail.com

A. Bhaduri
Global Water System Project (GWSP), Center for Development Research (ZEF), University of Bonn, Walter-Flex-Str. 3, 53113 Bonn, Germany
e-mail: abhaduri@uni-bonn.de

N. Djanibekov
Leibniz Institute of Agricultural Development in Transition Economies (IAMO), Theodor-Lieser-Str. 2, 06120 Halle (Saale), Germany
e-mail: djanibekov@iamo.de

A. Bhaduri et al. (eds.), *The Global Water System in the Anthropocene*,
Springer Water, DOI: 10.1007/978-3-319-07548-8_7,

Introduction

During the Soviet Union period, Uzbekistan was a net food deficient country that relied largely on food imports from other Soviet republics including other Central Asian countries (Babu and Reidhead 2000). During that period, irrigated agriculture in Uzbekistan focused on cotton cultivation, while rainfed wheat production was far from satisfying the local needs. In the post-Soviet Union period with the collapse of regional trade and disruption of food supply, Uzbekistan has implemented import substitution and trade protectionism policies to promote its food self-sufficiency program as the sectoral goal for agriculture (Babu and Pinstrup-Andersen 2000). As a consequence of such policies towards food self-sufficiency, domestically produced wheat increased several folds (e.g. from 1 million tons in 1992 to 6.5 million tons in 2012), and thus resulting in a drastic decline in its imports. However, this is not the case for rice production, which is the second most important cereal crop after wheat in Uzbekistan and whose demand is projected to grow faster in future (Gupta et al. 2009). In addition, rice production remains not only a source of food, but also a source of generating incomes, particularly among rural households.

Paddy rice cultivation in Uzbekistan depends heavily on irrigation water, which is highly influenced by hydrological cycles and, hence, climate changes could lead to the exacerbation of the fluctuations of water supply in the Amudarya and Syrdarya rivers. Despite its high economic value, rice production in Uzbekistan has dropped almost threefold mainly during the last two decades primarily due to the decline in irrigation water supply (FAO 2011). As a consequence, per capita supply of rice for food consumption fell from 14 kg in 1992 to 8 kg in 2009 (FAO 2011), while the market prices followed an upward trend with periods of strong fluctuations (Mori-Clement et al. 2014). The peak in rice price was reached in 2008, which coincided with the worst drought on record that affected Uzbekistan and other countries in Central Asia.

In Uzbekistan, the competition for water resources between rice and cotton plays a major role in influencing the national production of rice and its domestic price behavior as the irrigation periods for cotton and rice production overlap (Aldaya et al. 2010). The allocation of water resource between these two crops is largely influenced by institutional rigidity of agricultural policymaking in Uzbekistan with respect to cotton production. Cotton production still represents a core pillar of the agricultural policy that was inherited from the former socialist planning system. Following the independence in 1991, the state policy of cotton production remained strongly linked to the interests of the national budget earnings via taxing cotton producing farmers (Pomfret 2008). The settings of the cotton policy have been thoroughly analyzed (Djanibekov et al. 2013) showing that according to the cotton production policy, the state determines the area of farmland that has to be cultivated under cotton. This is a *location-based setting* of the cotton procurement policy that imposes cotton to be grown in farm fields where potentially highest yields can be obtained. According to this setting, a farmer usually has to allocate around 50–60 % of his cropland for cotton cultivation, while in the remainder land he has the opportunity to grow other crops like rice. The *quantity-based setting* imposed by the state obliges a farmer to provide raw cotton yield at levels assigned according to soil-fertility status of his lands. Finally, the state purchases the entire cotton harvest from farmers at prices below border prices. In 2003–2009, the average price paid to Uzbek farmers for raw cotton was about USD 290 t^{-1}, lower than the ones observed in 2003 in Kazakhstan (USD 550 t^{-1}) and Kyrgyzstan (USD 450 t^{-1}), which abolished the cotton procurement policy in 1990s (Pomfret 2008). A crux of this policy is farmers' fulfillment of the production targets (Djanibekov et al. 2013).

Despite bringing export revenues in form of hard currency to meet governmental expenditures (CDPR 2008), the existent cotton policy has been causing economic losses to farms since it prevents them to grow more profitable crops and reduced farm incentives to produce more cotton beyond the production target. Furthermore, allocation of the vast areas of land for cotton cultivation, made this crop the most important in terms of water consumption (Aldaya et al. 2010). Given the maintenance of the large cotton cultivated areas by the national administration, any decline in water availability affects the production of rice both via decline in cultivated area and yield per hectare. Although there is a broad literature focused on water allocation in a context of water scarcity, there are few studies that analyze how policies might influence water allocation in agriculture and the consecutive impacts on crop price determination in markets. Therefore, the objective of this paper is therefore twofold. First, given the present institutional setting and recurrent variability of water supply, we explore the effect of water competition between rice and cotton on the domestic price of rice. Second, we evaluate how possible changes in the institutional settings would influence rice prices.

Methodology

The analyses are based on data for the Khorezm province, situated in the North-Western part of Uzbekistan, surrounded by the Karakum and Kizilkum desserts. Khorezm is located in the lower Amudarya River basin. This river is the most important water source for agriculture in the province. The province has around 275,000 ha of land made suitable for irrigated agriculture (FAO 2003). River water flow generally provides enough water to satisfy demands in Khorezm. Water shortage during the vegetation period has however been experienced in the region during the last 30 years affecting seriously agricultural production and rural incomes. Though cotton is considered the most important crop, Khorezm is also well-known for its rice production (Bobojonov 2009). In contrast to cotton, rice production in Khorezm has rather national than local significance since the province produces about one-third of the national rice production on 17 % of its total arable land (Djanibekov 2008). Although rice officially is cropped on around

10 % of the total irrigated land in Khorezm, it consumes about 20 % of the total crop irrigation water supply (Djanibekov et al. 2012).

In our study, we apply a scenario-based simulation approach based on mathematical programming, which is appropriate to understand the ex ante effects of the changes in cotton policy on rice prices in Uzbekistan. A Dynamic Partial Equilibrium Single-Market approach has been adopted to evaluate the impact of the cotton quota system on rice prices in domestic markets through farm decisions on water allocation between cotton and rice. This approach allows determining the equilibrium prices and quantities endogenously based on the assumptions about the behavior of economic agents and the policy context embodied into the model (Bellù and Pansini 2009). Moreover, this approach considers also the effects on prices and quantities originated by responses that are embodied into the model by means of supply and demand functions as well as it analyzes sub-sectoral policy measures and, in general, those policy changes which impacts on macro aggregates are more limited (Bellù and Pansini 2009).

An important feature of this model is the incorporation of water allocation decisions. We have introduced this component to identify how farmers would allocate water resources (i.e. what the optimal allocation pattern would be) subject to water supply and the design of institutional constraints.

We assume n farmers in Khorezm, where each farmer is denoted by i, with concave production and benefit functions. We consider that farmers cultivate j crops where j = cotton (c) or rice (r) over time (t). Each farmer is endowed with A_i amount of land, which can be used to cultivate both crops. A^c represents the area used to cultivate cotton and A^r represents the rice area. Thus, the total agricultural area per farmer is defined as:

$$A_t^r + A_t^c = A_t \tag{7.1}$$

In this model, we assume that the water supply variable (W^s) evolves according to the geometric Brownian process[1]:

$$dW^s = u \cdot W^s \cdot dt + \sigma \cdot W^s \cdot dz \tag{7.2}$$

where $u \cdot dt$ is the mean of the water variable, $\sigma \cdot dz$ is a random component that depicts the drift in water variable or uncertainty, and dz is defined as:

$$dz = \varepsilon(t) \cdot \sqrt{dt} \tag{7.3}$$

[1] The stochastic process of Geometric Brownian motion is best fitted with a log normal distribution. We have used annual water supply data of the Urgench district, Khorezm in Uzbekistan during period 1998–2009 to verify such distribution. Geometric Brownian motion has been used earlier in several studies in the context of water variability, to simulate water inflow over different time horizons (Fisher and Rubio 1997; Roseta-Palma and Xepapadeas 2004; Bhaduri et al. 2011).

The error term $\varepsilon(t)$ is normally distributed. Water variability can be defined as:

$$\frac{dW^s}{W^s} = u \cdot dt + \sigma \cdot dz \tag{7.4}$$

The water consumption (usage) is denoted by w^j for $j = c, r$. Hence, the water constraint can be described as follows:

$$w^r \cdot A_t^r + w^c \cdot A_t^c = W_t^d \leq W_t^s \tag{7.5}$$

We consider that the government sets a target yield and area for cotton, while rice is grown according to farmer's choice. The target cotton yield is denoted as $\hat{y}^c$, while the target cotton area is $\hat{A}^c$. Thus, total cotton production target for a given farm can be defined as $\hat{y}^c \cdot \hat{A}^c$.

A farmer can produce cotton at yields higher than $\hat{y}^c$ which again is sold at the state price. If $\hat{y}^c \geq y^c$, then the surplus production is represented by $[\hat{y}^c \geq y^c] \cdot \hat{A}^c$. If the farmer does not reach the cotton yield target, he has to pay a lump sum, tax or penalty for noncompliance. We consider the penalty as a fixed amount, X. Assuming inter-annual uncertainty in water supply and other factors, yields may vary over years. Farmers have knowledge about the probability distribution of yields. The cumulative distribution of cotton yield is defined by $F(\hat{y}^c) = P[y^c < \hat{y}^c]$. It is also assumed that yields y_t^j, are a function of water usage w^j. Therefore, we assume two distinct situations in which a farmer fulfills or fails the cotton production target:

$$y^c \geq \hat{y}^c, \quad where\ P[y^c \geq \hat{y}^c] = 1 - F(\hat{y}^c) \text{ and}$$
$$y^c < \hat{y}^c, \quad where\ P[y^c < \hat{y}^c] = F(\hat{y}^c)$$

Although farmers do not have to pay the penalty in reality, a monetary payment in the model works as an incentive to obligate farmers to achieve the cotton target production imposed by the government.

The main equations of the rice market model with inter-year storage (Helmberger and Chavas 1996) are represented as follows:

$$A_t^r = S(P_{t-1}^r, W_t^s) \tag{7.6}$$

$$H_t^r = A_t^r \cdot y_t^r \tag{7.7}$$

$$D_t^r = D(P_t^r) \tag{7.8}$$

$$H_t^r + I_{t-1}^r = D_t^r + I_t^r \tag{7.9}$$

$$P^r_{t+1} = P^r_t + \theta \cdot ED^r_t \tag{7.10}$$

where Eq. 7.6 describes the rice planted area, which depends on expected rice price, and water availability. Equation 7.7 describes the domestic rice production, H^r_t, which is defined in terms of planted area and yield per hectare. The domestic demand for rice, D^r_t, which we have assumed as non-stochastic, is defined in Eq. 7.8. A market-clearing condition is expressed in Eq. 7.9, where I^r_t is the rice stock. Equation 7.10 represents the price adjustment rule, where ED_t defines the excess of demand of the whole rice market (Heemeijer et al. 2009). In this case, the market price will increase or decrease depending on an excess of demand (supply).

In this way, the price equation helps to meet the excess demand and adjusts the market price for next period depending on the excess demand. We assume that the change in market price is determined by a continuous, monotonically increasing function of excess demand, with the nonnegative parameter θ measuring price adjustment flexibility. Parameter θ has a great effect on the price behavior. As it increases, the market price fluctuates more erratically, while its value is small, the price curve is almost in a steady state. As θ achieves some value, more irregular secondary fluctuations appear on the regular cycle, especially on its turning parts, which may describe the critical price behavior resulting from the difference between demand and supply (Li and Rosser 2001).

The expected Net Benefit function of the farmer is represented by:

$$\begin{aligned} E(NB) = {} & [1 - F(\hat{y}^c)] \cdot P^c \cdot [E(y^c) - \hat{y}^c]\hat{A}^c + F(\hat{y}^c) \cdot (-X) \\ & + (P^r - C^r) \cdot E(y^r) \cdot (A - \hat{A}^c) \end{aligned} \tag{7.11}$$

where P^j and C^j are the price and marginal cost of jth crop $j = c,\ r$.

The farmer's objective is to maximize his net benefit. For this goal, the choice variables are crop area, and the main constraints are water supply and the settings from cotton policy (target area and yield). Therefore, the maximization problem can be set up as follows, where the objective function is maximized over a certain number of years (t) and r is a discount rate: Max $\sum_0^T nNB^{-rt}$, subject to Eqs. 7.1 and 7.5. Finally, Eqs. 7.6–7.10 characterize domestic rice market.

The sector-level model was programmed in the General Algebraic Modeling System (GAMS), and solved as a non-linear optimization using the numerical solver CONOPT3.

Parameters, Assumptions and Scenarios

Table 7.1 provides the details on the parameter values and functional forms used in the model. We assume that the price of rice is mainly determined by domestic forces. This assumption is not completely unrealistic as the trade of rice is limited,

Table 7.1 Parameters and values of the model

	Parameter (units)	Value
r	Discount rate	0.05
P^c	Price of cotton (USD t^{-1})	250
$\hat{y}^c$	Target cotton yield (t ha^{-1})	3
A^c	Target cotton area (ha)	18
A^r	Rice area (ha)	12
A	Total area of average farm (ha)	30
w^r	Technically optimal water use rate for rice (1,000 m^3 ha^{-1})	26
w^c	Technically optimal water use rate for cotton (1,000 m^3 ha^{-1})	10
W^s	Water supply to the representative farm (1,000 m^3)	492
y^r *(mean)*	Yield of rice	
y^r	$= y^r\ (mean) + \mu^r$, where μ^r follows SND with mean 0 and SD σ^r	
y^c *(mean)*	Yield of cotton	
y^c	$= y^c\ (mean) + \mu^c$, where μ^c follows SND with mean 0 and SD σ^c	
P^r	Price of rice (USD t^{-1})	1,000
θ	Parameter for price sensitivity	0.010
D_t^r	Total demand of rice in Khorezm (t)	32,000
a	Annual increase in demand of rice	1.05
pop	Projected population in Khorezm (persons)	2,000,000
pcap	Per capita consumption of rice (t)	0.016
n	Total number of farmers	900
u.dt	The mean of water supply (1,000 m^3)	15
σ.dz	The variance of water supply (1,000 m^3)	8

and only imports are allowed and managed by the government during periods of scarcity (Robinson 2008). We consider the time horizon of 20 years as long term period during which the farmer maximizes his net benefit.

Quadratic yield-to-water response functions were parameterized for cotton and rice using the official recommendations on crop irrigation in Khorezm. Using a crop-water response functions, we take only costs related to irrigation as variable, while other production costs (fertilizers, seeds, labor and machinery) are assumed to be fixed. The irrigation costs depend on water application per hectare and the cultivation area of each crop.

The penalty component of the model implies that if farmers do not fulfill the production target, they will be penalized. This penalty consists of a monetary payment that represents the difference between actual cotton produced and the level assigned by the cotton production target multiplied by the local cotton price.

We formulate different policy scenarios to investigate the effect of different designs of the cotton policy on the price of rice. Each setting of the cotton policy is characterized by different policy measures as well as market status; for instance, the level of cotton production (determined by the target area imposed by government vs. farmer's free choice), the value of cotton price (set as fixed prices by the state vs. market-derived prices) as well as access to international cotton markets (close economy vs. open economy).

Table 7.2 Description of policy scenarios

	Scenarios				
	S1	S2	S3	S4	S5
Cotton target area	60 % of farmers total area	30 % of farmers total area	15 % of farmers total area	30 % of farmers total area	No target area
Cotton target yield	Yes	Yes	Yes	Yes	No target yield
Cotton price	Fixed by the government	Fixed by the government	Fixed by the government	Fixed by the government	Determined in international markets
Access to international cotton markets	No	No	No	Only if farmers produce more than target level	Yes

The description of the policy scenarios is provided in Table 7.2. *Scenario* 1 (S1) simulates a price path of rice under the current policy of cotton production, the so-called *cotton target system*, where the cotton area (60 % of farmer's land area), cotton yield as well as cotton prices are determined by the government. Under this scenario, if the farmers do not fulfill cotton production target, they have to pay a certain amount of penalty to the government. *Scenario 2* (S2) pursues to simulate a situation under a more flexible version of the cotton policy as it assumes a less restricted cotton target area (only 30 % of the farmer's land area), while *Scenario 3* (S3) simulates an even more flexible cotton policy (15 % of the farmer's land area). Other assumptions are kept similar to *Scenario 1* in both scenarios. *Scenario 4* (S4) simulates a similar situation explained under the context of *Scenario 2*, but the main difference is that farmers have the possibility of accessing the international cotton markets in case they produce more than the targeted production. In this case, they are allowed to sell the produced surplus at international prices. Finally, *Scenario 5* (S5) models a context without any cotton production target. Under this scenario, cotton prices are equal to the international market prices. We assume that in such situation Uzbek farmers would behave as price takers.

To analyze the effect of agricultural policy on rice price behavior, we simulate how the price path of rice evolves during a time horizon of 20 years, in these five selected scenarios. In each scenario, farmers have to face similar changes in water conditions under different settings of cotton policy. The connection between changes in water availability and rice prices is given mainly through total sown area and crop yields, which affect directly supply in domestic markets.

Results

Simulation results of selected scenarios are graphically depicted in Fig. 7.1. Five price paths of rice are displayed with a selected water supply pattern. Based on the Brownian motion process to simulate periods of normal water availability and extreme water conditions over 20 years (WS), *S1* depicts the highest price levels compared to the other scenarios, while the lowest paths were obtained in *S2* and *S3*.

An intermediate situation is observed under *S4*, a setting with limited access to international cotton markets, while *S5* depicts a setting under a free market condition. In *S1*, the scenario with the most restrictive settings of cotton policy and very close to current policy measures presently applied in Uzbekistan, the rice price followed an increasing trend during 20 years. In years of a pronounced scarcity in water availability, rice prices went up. Yet, this pattern is not significantly reverse in periods of more water availability.

According to the model results, it would be economically more optimal for farmers to prioritize the water allocation to cotton to ensure its production even in periods of normal water availability. This is mainly because of the imposed penalty component, which is a key pressure to allocate water to the state target crop cotton. However, this penalty is not the only variable that affects farmers' water allocation decisions. The water variability might reinforce the effect of the penalty in water allocation decisions. As periods of water shortage might generate uncertainty, farmers would be afraid of not producing enough to meet the cotton target. Therefore, they would cultivate mainly cotton to ensure its production, even in periods of improved water availability. As a consequence, the penalty might create disincentives to diversifying crop portfolio.

The findings of the scenarios that simulate price paths under more flexible cotton policies (*S2* and *S3*, with target areas of 30 and 15 %, respectively) show that a reduction in the imposed cotton target area would allow growing more rice, increasing production and thus lead to smoother price paths. As these settings of cotton policy comprise more flexible targets, farmers would have more freedom to make production decisions. Although farmers still must deliver a certain amount of cotton production, the governmental imposition of less restrictive conditions offers farmers more flexibility to allocate water resources to rice. As a consequence, rice supply might increase and thus smoothen the price path of rice, even in periods of water shortage.

It is important to point out that there is only a slight difference between the price paths under *S2* and *S3* (with target areas of 30 and 15 %, respectively). Although less restrictive production conditions reduce significantly price path levels (with respect to *S1*), these changes remain similar after achieving a certain level of target area. In other words, the effect of relaxing the cotton policy is more evident if the changes in current conditions are sufficiently large to serve as incentives to farmers. That is why after a certain level, reductions in the cotton target area would not have any additional effect on price paths of rice.

An intermediate situation is observed under *S4* (similar settings as under *S1*, but with limited access to international cotton markets if farmers produce above the targeted yield). In this scenario, the price path of rice follows a similar pattern as obtained under *S2* and *S3*, but after the tenth year, the price path diverges following an upward trend. This price behavior is caused mainly by the (limited)

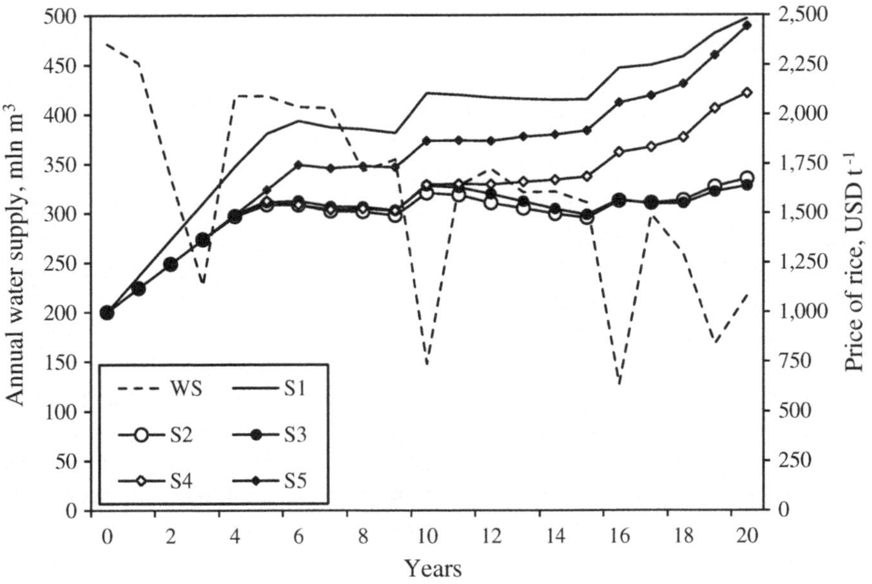

Fig. 7.1 Price path of rice under selected scenarios

access to the international cotton market. As higher prices are observed in international markets in comparison to the domestic cotton prices, farmers could see these as important economic incentives to improve cotton production and, thus, to achieve a cotton surplus to be sold externally at more attractive prices.

At the beginning of *S4*, farmers might not take advantage of having (although limited) access to international cotton market as they do not count with the experience/resources/technology to increase cotton yields enough to obtain a surplus. That is why the price path of rice follows a similar pattern to those observed under *S2* and *S3*. However, farmers might become more productive in producing cotton after a certain period and after accessing the international prices and finally sell the extra production at higher prices. At the point where farmers are able to improve cotton yields, they might opt to allocate more water resources in producing cotton instead of rice, reducing its supply and affecting price determination in domestic markets. This effect was obtained from the tenth year when a higher price path level was achieved.

Finally, *S5* simulates a situation in which the cotton market has been completely liberalized. There are no cotton targets, so farmers are free to decide. Under this setting, the price path of rice reaches the second highest level in comparison to other four simulated scenarios. Since cotton on the international markets has been showing in recent years more attractive prices than domestic rice markets, farmers will prefer to allocate more water resources to cotton production instead of rice, reducing rice supply and, consequently, pushing its price to follow an upward trend.

Discussion and Policy Implications

Results have highlighted the dynamic impact of water inflow on rice prices in the short term. This effect can be transmitted and smoothed through changes in rice stocks. However, if stocks have reached low levels, then prices become more sensitive to disturbances in supply, exacerbating the effect of any shock. In this context, it seems that rice price peaks in Uzbekistan during the last years have been a reflection of minimal stock levels. Although there is no available data about domestic capacities of rice stocks in Uzbekistan to corroborate it, theory supports this explanation.

As there is no specific policy concerning to rice stocks, the government might consider the implementation of storage schemes at regional level, which might help to smooth supply shocks as well as to improve food access in markets in the short term (Deaton and Laroque 1992). Thus, a regional store under the framework of a national food security program would emerge as an alternative. Although there is a state procurement for wheat, which aims at enhancing food security in Uzbekistan, this crop is cultivated during winter time and thus does not compete with cotton for water resources.

The competition for water resources, as a result of the imposing state production targets, showed to be a price determinant for domestic rice.

From farmer's side, the high share of arable land allocated to the two state target crops cotton and wheat prevents farmers to grow more rice and potentially to cultivate other crops that would increase income security and help farmers to cope with price and weather uncertainties (Bobojonov 2009). A less strict cotton policy, in other words a reduction in cotton target area, could make farmers consider more than at present market signs, in case becoming exposed to these and become better linked to markets, instead of political pressure in production decisions. Therefore, the findings underline that flexible policies might improve conditions for efficient risk-management in agriculture.

Concerning the cotton production, as observed in *S4*, limited access to external cotton markets (with more attractive prices) might work also as an incentive for farmers to improve cotton yields. This measure is to be considered however as a partial solution to confront yield stagnation, a problem faced by the cotton sector in Uzbekistan since many years (Abdullaev 2009). From the consumers' side, as cotton cultivation is mandatory to the detriment of food crops, consumers will have to face higher and more volatile rice prices as a result of insufficient supply in domestic markets. A more flexible cotton policy might also lead rice prices to develop a more stable path over time and protect consumers from severe fluctuations.

The side effects of imposed settings of the cotton policy in other crop markets can be considered for changing the design of this policy to improve rural food security. Policy instrument combinations might be an alternative to protect the most vulnerable consumers and improve crop accessibility. Another alternative would be to combine the latter with price stabilization schemes (Newbery and

Stiglitz 1981), which might smoothen the rice path in the long term, protecting not only consumers', but also producers' welfare.

Other policy measures may include the implementation of water policies, focused on improving, for instance, irrigation efficiency. The overall irrigation efficiency of approximately 36 % (Tischbein et al. 2012) and a field application efficiency of 45 % (Awan et al. 2011) in the study region show that large amounts of water do not reach crops directly although perhaps indirectly through groundwater contribution (Akhtar et al. 2013). The water policies should consider measures thus not only to enhance irrigation efficiency at system's level, but also at farmer's level. At farmer's level, the promotion of water efficient technologies and trainings about agricultural practices[2] might contribute in enhancing field water management.

Given that all these policy measures involve different levels of investment as well as time horizons to obtain results, short-run, intermediate-run and long-run strategies are required. The reduction of the cotton target areas might be considered as a short-run strategy; the implementation of storage schemes as an intermediate-run strategy, while improving the access to external markets as well as regional integration might be considered as long-run strategies.

Conclusions

The simulation analysis introduces the institutional factor to observe how a policy may affect farmer's decision making concerning water allocation and its consequences on price behavior of rice and consumer welfare in the long term. Consecutive water shortage episodes might put farmers under uncertain conditions, affecting their expectations. As crop storage responds to shocks and farmer's expectations, this response though storage has the ability to smooth out peaks.

The findings of the potential impact of different scenarios of cotton policy changes on rice markets in a context of water variability over a horizon of 20 years illustrated that under more flexible settings of cotton policy, the price path of rice would be smoother since farmers can meet more irrigation water for growing this crop. A more flexible state procurement policy is likely to be more favorable not only for farmers, but also for consumers. A relaxation of the present cotton policy might provide farmers higher incomes as well as it might protect consumers from rice price hikes during drought years. Under the current settings of the cotton policy, the price path of rice would be highest with hikes in years with extreme falls in water availability. Under conditions where farmers have greater control over production decision-making, higher and more stable incomes might be

[2] Many management practices have been developed to reduce cereal yield reductions in water-limited environments (Heisey and Morris 2006). Adequate land leveling, improved tillage methods are some examples of on-farm practices for rice, which could enhance water use in rice fields.

obtained. Thus, a more flexible cotton policy could be much desired by farmers in years when the probability of receiving enough water is expected to be below the levels observed in normal years.

Although the simulation results are not entirely realistic as they do not consider the impact of other variables, they illustrate the principles of methodological interest and some rough quantitative guidance. The model might serve as a useful tool for simulating each scenario, taking into account that some level of uncertainty in the results remains due to existing data limitations and assumptions.

References

Abdullaev I, de Fraiture C, Giordano M et al (2009) Agricultural water use and trade in Uzbekistan: situation and potential impacts of market liberalization. Int J Water Resour Dev 25(1):47–63

Akhtar F, Tischbein B, Awan U (2013) Optimizing deficit irrigation scheduling under shallow groundwater conditions in lower reaches of Amu Darya River Basin. Water Resour Manag 27(8):3165–3178

Aldaya M, Muñoz G, Hoekstra AY (2010) Water footprint of cotton, wheat and rice production in Central Asia. UNESCO-IHE Institute for Water Education, Value of Water Research Report Series No 41, Delft, p 31

Awan UK, Tischbein B, Conrad C et al (2011) Remote sensing and hydrological measurements for irrigation performance assessments in a water user association in the lower Amu Darya River Basin. Water Resour Manag 25(10):2467–2485

Babu S, Pinstrup-Andersen P (2000) Achieving food security in Central Asia: current challenges and policy research needs. Food Policy 25(6):629–635

Babu S, Reidhead W (2000) Poverty, food security, and nutrition in Central Asia: a case study of the Kyrgyz Republic. Food Policy 25(6):647–660

Bellù LG, Pansini RV (2009) Quantitative socio-economic policy impact analysis: A methodological introduction. Food and Agriculture Organization of the United Nations, Issue paper, EASYPol Module 068, Rome

Bhaduri A, Manna U, Barbier E et al (2011) Climate change and cooperation in transboundary water sharing: an application of stochastic Stackelberg differential games in Volta River Basin. Nat Resour Model 24(4):409–444

Bobojonov I (2009) Modeling crop and water allocation under uncertainty in irrigated agriculture: a case study on the Khorezm region, Uzbekistan. Dissertation, ZEF, University of Bonn

CDPR (2008) The resource curse. Centre for Development Policy and Research, Development Digest, 1. Available via SOAS. http://www.soas.ac.uk/cdpr/publications/dd. Accessed 11 Dec 2011

Deaton A, Laroque G (1992) On the behaviour of commodity prices. Rev Econ Stud 59(1):1–23

Djanibekov N (2008) A micro-economic analysis of farm restructuring in the Khorezm region, Uzbekistan. Dissertation, ZEF, University of Bonn

Djanibekov N, Bobojonov I, Djanibekov U (2012) Prospects of agricultural water service fees in the irrigated drylands, downstream of Amudarya. In: Martius C, Rudenko I, Lamers JPA, Vlek PLG (eds) Cotton, water, salts and soums: economic and ecological restructuring in Khorezm. New York, Springer, pp 389–411

Djanibekov N, Sommer R, Djanibekov U (2013) Evaluation of effects of cotton policy changes on land and water use in Uzbekistan: application of a bio-economic farm model at the level of a water users association. Agr Syst 118:1–13

Fisher A, Rubio S (1997) Adjusting to climate change: implications of increased variability and asymmetric adjustment costs for investment in water reserves. J Environ Econ Manag 34(3): 207–227

Food and Agriculture Organization (2003) Fertilizer use by crops in Uzbekistan. Food and Agriculture Organization of the United Nations, Rome

Food and Agriculture Organization (2011) FAO statistics division. http://faostat.fao.org. Accessed 11 Dec 2011

Gupta R, Kienzler K, Martius C et al (2009) Research prospectus: a vision for sustainable land management research in Central Asia. ICARDA Central Asia and Caucasus Program, Sustainable Agriculture in Central Asia and the Caucasus Series No.1. CGIAR-PFU, Tashkent

Heemeijer P, Hommes C, Sonnemans J et al (2009) Price stability and volatility in markets with positive and negative expectations feedback: an experimental investigation. J Econ Dyn Control 33(5):1052–1072

Heisey PW, Morris ML (2006) The economic impact of water-limited conditions on cereal grain production. In: Ribaut J-M (ed) Drought adaptation in cereals. The Haworth Press, New York, pp 17–48

Helmberger P, Chavas J-P (1996) The economics of agricultural prices. Prentice Hall, Upper Saddle River, p 356

Li H, Rosser JB (2001) Emergent volatility in asset markets with heterogeneous agents. Discrete Dyn Nat Soc 6(3):171–180

Mori-Clement Y, Bhaduri A, Djanibekov N (2014) Food price fluctuations in Uzbekistan: Evidences from local markets in 2002–2010. In: Lamers JPA, Khamzina A, Rudenko I, Vlek PLG (eds) Restructuring land and water use: technologies, policies and practices for the lower Amudarya region. V&R Unipress GmbH, Göttingen (in press)

Newbery D, Stiglitz J (1981) The theory of commodity price stabilization: a study in the economics of risk. Oxford University Press, Oxford, p 480

Pomfret R (2008) Tajikistan, Turkmenistan and Uzbekistan. In: Anderson K, Swinnen J (eds) Distortions to agricultural incentives in Europe's transition economies. World Bank, Washington D.C., pp 297–338

Robinson I (2008) Food markets and food insecurity in Tajikistan, Uzbekistan, Kyrgyzstan, Kazakhstan. World food programme, regional market survey for the central Asian region, Report of Mission

Roseta-Palma C, Xepapadeas A (2004) Robust control in water management. J Risk Uncertainty 29(1):21–34

Tischbein B, Awan UK, Abdullaev I et al (2012) Water management in Khorezm: current situation and options for improvement hydrological perspective. In: Martius C, Rudenko I, Lamers JPA, Vlek PLG (eds) Cotton, water, salts and soums: economic and ecological restructuring in Khorezm. New York, Springer, pp 69–92

Chapter 8
Dams on Shared Rivers: The Concept of Benefit Sharing

Waltina Scheumann, Ines Dombrowsky and Oliver Hensengerth

Abstract In recent years, the concept of benefit sharing has been proposed as a means of fostering the cooperative use of international rivers. Most of the relevant literature focuses on opportunities for the generation of net benefits from cooperation; however, little attention has so far been paid to specific mechanisms for benefit sharing applied to the specific case of dams on international rivers. This paper fills this gap and asks both what incentives can be offered to encourage benefit sharing and what benefit-sharing mechanisms can be identified. Based on a conceptual approach, dam projects on the rivers Senegal, Columbia, Orange-Senqu, Nile and Zambezi are reviewed in order to explore the benefit-sharing mechanisms used at international levels. The paper also finds that negative environmental impacts are largely neglected, while social costs are not fully accounted for. The paper advocates for linking interstate with domestic benefit-sharing mechanisms which might be the ultimate step towards a socially inclusive, sustainable dam development.

Introduction

Most parts of the global water system are composed of regional sub-systems—rivers and groundwater aquifers—which are shared between riparian states. Instead of utilizing these parts of the global water system unilaterally irrespective of the impacts on neighbours and the environment, states need to cooperate when

W. Scheumann (✉) · I. Dombrowsky
Deutsches Institut für Entwicklungspolitik, Bonn, German
e-mail: waltina.scheumann@die-gdi.de

I. Dombrowsky
e-mail: ines.dombrowsky@die-gdi.de

O. Hensengerth
Oliver Hensengerth, Northumbria University, Newcastle, UK
e-mail: oliver.hensengerth@northumbria.ac.uk

A. Bhaduri et al. (eds.), *The Global Water System in the Anthropocene*,
Springer Water, DOI: 10.1007/978-3-319-07548-8_8,

building single or multi-purpose dams on shared rivers in order to avoid conflicts. Solutions to avoid conflicts and to achieve fair outcomes are, however, not only confined to inter-state negotiations but must also include agreements with affected communities on the national level. The national interest of governments must be tied to the well-being of local communities in order to make large dam projects legitimate interventions into the natural and social environment.

The last decade has seen a debate on benefit sharing on international rivers. The main idea of the concept of benefit sharing is to move from sharing water to sharing the benefits the users gain from its use. In principle, dams (single and multi-purpose) may play an important role in benefit-sharing schemes; however, their role has yet to be systematically explored. This is a serious deficit not least because of a renewed interest in multi-purpose dams in the context of development and climate change. The reasons put forward in favour of dam construction are:

- the huge untapped hydropower potential in Asia and Africa (IEA 2011);
- the storage potential of dams have for supplying water households and industry;
- the yet undeveloped land that could be irrigated (World Bank 2007; Africa Recovery 2004) and
- the role dams play in strategies to mitigate (low carbon) and adapt (storage) to climate change (WSSD 2002; Bates et al. 2008).[1]

Given that most large rivers are shared by several countries, many of these new dams will be built on international rivers. Benefit sharing can be seen as a means of stimulating cooperation because it can prevent conflicts by focusing on the sharing of the benefits from an international river rather than the mere sharing of the water itself. Benefit sharing can be seen as the translation into practice of international water law, and specifically the principles of equitable and reasonable utilization, and of the absence of harm, which international and regional conventions emphasize.

However, there is no denying that single and multi-purpose dams are among the most controversial infrastructure projects because of their significant environmental and social impacts (WCD 2000a). They interrupt river flows, change a river's course, a river's sedimentation regime and water quality, to mention but a few of their impacts (McCartney 2009). They often entail the acquisition of land and, therefore, the physical relocation and displacement of people. If such impacts are not adequately addressed, some already vulnerable populations are likely to be further impoverished, which will undermine development objectives (Cernea and Mathur 2008).

The overall objective of this study is to analyse the essential elements and content of the benefit-sharing concept and its applicability to *dams* on shared rivers. This chapter will consider the following questions:

[1] This is not to say that single and multi-purpose dams are the only means; they are just one of several options—for water supply and for electricity generation.

- What are the peculiarities of applying benefit sharing to *dams* on shared rivers?
- Which types of benefit-sharing mechanisms are applicable to dams, how do they differ, and where can they be found in practice?

The study is based on a review of the conceptual literature on benefit sharing and of empirical studies on the rivers Senegal, Columbia, Orange-Senqu, Nile, Zambezi and Parana. We intend to develop the concept further by considering its applicability to single and multi-purpose dams. Since the majority of the cases reviewed show that environmental and social costs have yet not been adequately addressed and are not part of the economic calculation, we also intend looking for proposals which link interstate benefit-sharing arrangements with local arrangements targeting project affected people.

Conceptualizing Benefit Sharing on Dams on International Rivers

The Concept

The concept of benefit sharing in the context of shared rivers entails a change from the mere volumetric allocation of water to the allocation of the benefits gained from the use of the river (Sadoff and Grey 2002, 2005; Klaphake 2005; Phillips et al. 2006; Dombrowsky 2009). The prospect of potentially gaining higher benefits by cooperating rather than by maintaining the status quo or by taking unilateral action may encourage states to cooperate with each other in their use of shared rivers.

The concept suggests that countries can turn the perceived zero-sum game of water allocation, i.e. allocating more water to country A results in less water for country B, into a positive-sum game, i.e. a win-win situation in which all riparian countries are better off with cooperation than without (Biswas 1999, Giordano and Wolf 2003). This can be achieved by viewing the use of water from an economic perspective: rather than conceptualizing water use in quantitative terms, states should conceive of the river as a productive resource, and focus on the benefits they receive from its use. They should attempt to increase and ideally maximize the economic benefits from its use and to share them in a manner that all parties are better off than they were in the status quo ante.

The notion of benefit sharing in the use of shared rivers has been advanced by Sadoff and Grey (2002) among others. They define benefit sharing as ‘any action designed to change the allocation of costs and benefits associated with cooperation’ (Sadoff and Grey 2005). In doing so, they distinguish four categories of benefits through which cooperating states can produce win-win situations:

- benefits to the river: improve the ecological sustainability of the watershed;
- benefits from the river: water-related economic benefits by developing irrigation, generating hydropower, improving flood control or enhancing navigation;

Table 8.1 Externalities of upstream dams in a downstream country (authors' compilation)

Benefits to upstream state A from a single or multi-purpose dam	Externalities in downstream state B
Hydroelectricity	(−) changed flow and sedimentation regime
	(−) peak flows
	(−) seasonal imbalance
Flood control	(−) changed flow and sedimentation regime
	(−) peak flows
	(+) regularized flow
Irrigation /drinking water	(−) changed flow and sedimentation regime
	(−) peak flows
	(−) seasonal imbalance
	(−) high to low water extractions
	(+) regularized flow
Improved navigability	(+) increased trade

- benefits due to reduced costs because of the river: reduction in political conflict and associated costs of conflict, when countries shift the policy focus from dispute to cooperation and development;
- benefits beyond the river: improved regional infrastructure, markets and trade ultimately resulting from benefits derived because of the river (Sadoff and Grey 2002, p. 393, Table 8.1).

Similarly, Phillips et al. (2006) argue that benefits can be generated in the economic, the environmental or in the security arenas and that activities in these various spheres may have spill-over effects. They propose to identify security, economic and environmental drivers in international river basins and, on that basis, opportunities for development at various levels (household, sub-national, national, regional, global) within each of these spheres.

The benefit categories developed by Sadoff and Grey (2002) and Phillips et al. (2006) serve as a starting point for *benefit generation*. They can contribute to a deeper understanding of the range of sectors that can be included in generating benefits from cooperation and of the possible size of the "basket of benefits" (Phillips et al. 2006).

Going beyond benefit generation, Dombrowsky (2009, 2010) seeks to determine what incentives each of the riparian states involved has to negotiate and enlarge the basket of benefits and how riparians might distribute or *share* costs and benefits. To that end, she argues that it is useful to analyse the benefits and the potential negative and positive external effects of actual or planned water uses by individual states. The advantage of this approach is that it shows directly how cooperation alters the payoffs for each participating state compared to the status quo or unilateral action. This can then be used as a basis for identifying a benefit-sharing mechanism which so changes the allocation of costs and benefits that every state will be better off compared to the status quo.

In the following, this approach will be applied to dams; various kinds of benefits and external effects associated with dams will be analysed.

Benefits and External Effects Associated with Dams

Our starting point for thoughts on benefit sharing in the context of dams on *shared rivers* is the interest of individual basin states in developing their water resources for the benefit of their national economies. That interest may extend to energy production to meet energy security needs, the expansion of irrigated agriculture to meet food security needs, the mitigation of hazardous floods and droughts, and the improved navigability of rivers to enable trade (see Table 8.1).

However, the generation of benefits through the construction of a dam in one country may have external effects both on local populations and on other countries. Such "external effects" or "externalities" occur when the use of water by one agent directly affects the use of water by another, and when these effects are not "mediated by prices" (Mas-Colell et al. 1995, p. 352), i.e. when they are not reflected in the cost-benefit calculus of the agent causing them. In the case of transboundary externalities, an upstream dam may, for instance, produce negative externalities downstream by reducing downstream water flow for irrigation, navigation or drinking water supply, or by increasing peak floods. Conversely, the upstream dam may also produce positive externalities downstream when the upstream dam improves flood protection downstream. However, the construction of a dam downstream may also produce a negative externality upstream by extending the reservoir across the border into the upstream state, where it inundates land on its territory. Thus gaining benefits from a dam on the territory of one riparian country may have negative or positive external effects on other riparian countries. On a transboundary river (i.e. a river crossing an international border) these effects may occur downstream or upstream; on rivers forming state borders they tend to be more reciprocal (Dombrowsky 2007, 2010).

When conceptually focussing on the international dimension of benefit sharing, there is not automatism which leads to a fair, equitable allocation of benefits within a country. Even though the country may gain overall economically (because of benefits from hydropower), one section in the country may gain (say hydropower), while another section (say farmers) may experience some loss (see Duflo and Pande (2007) who analyzed the distributional effects of large dams built for irrigation).

Table 8.1 explores potential benefits of building a dam in upstream state A and potential associated positive and negative externalities in downstream state B in greater detail. Multi-purpose dams combine the externalities of single-purpose projects.

Hydro-Political Constellations and Incentive Structures Relevant for Negotiating Benefit Sharing of Dam Projects

It will be clear from the above that the opportunities and incentives for cooperation on dams on shared rivers and the content and applicability of benefit-sharing mechanisms depend on the following factors:

- The location of a dam on a shared river (hydro-political constellation). The defining element in the location of a dam is the spatial position of the river to the national border. Two types of river-border relationships can be distinguished: (i) a transboundary river crosses an international border from the upstream country to the downstream country, the dam being located in either the upstream or the downstream country or on the border itself; (ii) the river forms the international border, the dam being located in both riparian countries.
- The purpose(s) of a dam on a shared river with regard to achieving national (or sub-national) development objectives.
- Each state's specific interest in cooperation.

The hydro-political constellation and the purpose(s) determine a dam's benefit, cost and externality streams. In the following, the paper identifies typical constellations and, for each constellation, potential reasons and incentives for states to cooperate (for a more detailed explanation, see Hensengerth et al. 2012). In doing so, it initially conceptualizes states as unitary actors. This, however, is an abstraction for analytical reasons in order to understand the incentive structure at the international level. This does not imply that we suggest that states are unitary actors. To the contrary, international negotiations have to be conceptualized as two-level games (Putnam 1988): at the international level, state representatives negotiate international agreements; at the national and sub-national level, the political process determines whether a respective international agreement will be ratified (see also Fischhendler et al. 2004; Meijerink 1999). Depending on the opportunities of affected populations to voice their concerns, this national level process may—and we argue that it should—involve negotiations on the compensation of and sharing of benefits with those who are negatively affected by the international agreement. The liberalization of the energy sector which in most cases cover electricity *generation*, changes the scene further because private investors come into play and become a key concern of interstate as well as domestic negotiations.

Hydro-Political Constellation (1): A Dam on a Transboundary River—Externalities Downstream

The dam is located in upstream state A and produces positive and/or negative externalities in downstream state B (Fig. 8.1).

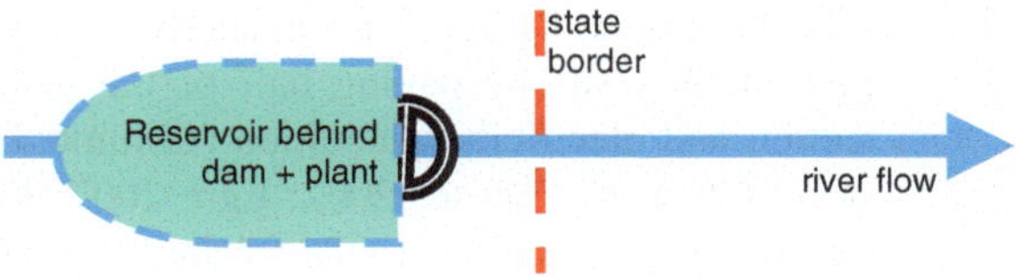

Fig. 8.1 Hydro-political constellation (1): A dam on a transboundary river in upstream state A with externalities in downstream state B

The possible reasons for cooperating in this constellation include:

(i) *Financial or economic constraints on unilateral action.* Where unilateral action is subject to financial constraints, a state lacks the financial resources and/or technical capacity needed to build the dam alone, although it would be economically viable, and therefore asks co-riparian countries to contribute towards its cost and offers to share the benefits generated by the dam.
This is the case for the River Senegal: the costs of the jointly owned infrastructure, the Manantali and Diama dams, were shared in proportion to expected irrigation, navigation and hydropower benefits (Yu and Winston 2008; AfDB 1988; Kirschke 2010; Alam et al. 2009; Kipping 2005). Where unilateral action is subject to economic constraints, the project does not make economic sense for the upstream state on its own. An example would be an upstream dam the cost of which would exceed benefits within upstream country A. If the dam generates positive externalities for the downstream state (by regulating flows, for example), the project may become economically viable if the downstream state contributes towards the cost of the project. In this case, the project is rational only collectively, not unilaterally, since it pays off only if the benefits to all riparians are considered in the cost-benefit analysis and if they all contribute to the project costs, e.g. in proportion to the benefits they will derive from the project (for a more detailed explanation, see Dombrowsky 2009). However, as Badhuri has shown, benefits may change with variability in the flow of water (Bhaduri et al. 2011).

(ii) *Altered dam design increases aggregate net benefits.* An altered, jointly agreed dam design that takes external effects into account increases overall aggregate net benefits at basin level. The downstream state participates in the establishment of an upstream project to increase the basin-wide benefits of the project compared with the upstream state's unilateral alternative. This is possible if the alteration of the project increases aggregate net benefits. However, the altered design typically leaves the upstream country worse off (otherwise, the upstream state would have pursued this alternative from the beginning). In that case, the downstream country will approach the upstream country and ask for an alteration of the dam design and compensate it for any consequent losses.
This is the case for the Columbia River: the alteration upstream increased aggregated net benefits: Canada built dams for downstream flood control

and additional hydropower generation; USA compensated Canada for investment costs by paying half of the value of downstream flood protection and electricity generation (Columbia River Treaty 1961; Muckleston (2003); Égré et al. 2002; Égré 2007; Krutilla 1967). The remaining benefits of cooperation are then shared (for a more detailed explanation see Dombrowsky 2009).

(iii) *The downstream country wants to build a dam on the territory of an upstream country.* This is the case, for example, if the dam site is more favourable for the achievement of the downstream country's national objectives than an alternative on its own territory, say flood control. Locating the dam in another state thus produces higher aggregate net benefits than a project alternative within national boundaries. The downstream country at least contributes to the cost of financing the investment and of operating the dam. The upstream country has an incentive to cooperate if it derives net benefits from the project.
One example is Iraq's 1946 agreement with Turkey to build a dam on Turkish territory (Kibaroglu et al. 2011, p. 391). The selection of the dam site promised to be more effective for flood control and higher rates of return on investments. The Lesotho Highlands Water Project (LHWP) is another example: South Africa pays investment and operation costs and external costs of storage and transfer of water from Lesotho; Lesotho receives in-kind hydropower benefits; net benefits of cooperation compared to unilateral action are shared (Yu and Winston 2008; Klaphake and Scheumann 2009; Égré 2002, 2007). While the upstream state is interested in building a dam in case (ii), it is the downstream state that is interested in doing so in case (iii).

Hydro-Political Constellation (2): A Dam on a Transboundary River—Externalities Upstream

The dam is located in the downstream state close to the international border, which causes externalities in the upstream state in that it inundates land in the upstream state (Fig. 8.2).

Possible reasons for cooperation in this constellation include:

(i) *The downstream country wishes to maintain good-neighbourly relations.* It therefore compensates the upstream state for damage and shares the benefits of the project with it, as happened in the case of the Aswan High Dam (Agreement between the Republic of the Sudan and the United Arab Republic for the Full Utilization of the Nile Waters 1959; Scudder 2003). The construction of the Aswan High Dam entailed cooperation under an international treaty that provided, among other things, for a single compensatory payment for flooding upstream. However, it is questionable whether this was sufficient to offset damage caused by the reservoir upstream (in which case it would not be reasonable to speak of benefit sharing).

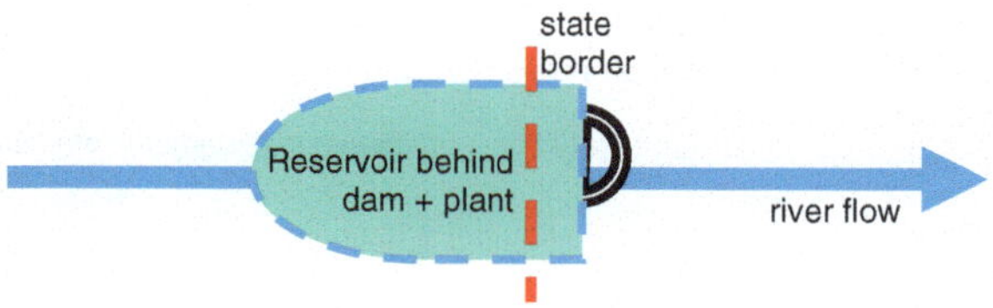

Fig. 8.2 Hydro-political constellation (2): A dam on a transboundary river in downstream state B with externalities in state A

(ii) *The downstream country wishes to avoid negative externalities* affecting the upstream country. This was the case for the Bui Dam (Fink 2005; Hensengerth 2011). An international dimension was entirely avoided in the Bui case when Ghana reduced the dam height during the planning process and thus avoided damage upstream. Hence, there was no benefit sharing in the end.

Hydro-Political Constellation (3): A Dam on a Border Formed by a Transboundary River

The dam is located where a river flows from state A to state B (Fig. 8.3).

The reason for cooperation in this constellation is:

- *Benefits can be gained only though cooperation, but externalities are asymmetrical*. An agreement is required to build a dam on the border between upstream and downstream states. The cost of building the dam is shared according to the allocation of benefits. In addition, the asymmetrical externalities can be compensated for with a side-payment by downstream state B to upstream state A or vice versa. Examples are the planned so-called 'friendship dams' on the Turkish-Syrian border[2] and on the Turkish-Bulgarian border which both have yet not materialized.

Whether and how state B compensates state A or vice versa, and what benefit-sharing mechanism is created, depends on the externalities and the incentive structure. This case represents an intermediate constellation between constellations 1 and 2, because territory of the upstream state is flooded and the downstream state may suffer the effects of upstream water storage and/or release. However, unlike the states in constellations 1 and 2, those in constellation 3 cannot act unilaterally.

[2] http://en.rian.ru/world/20110206/162478744.html, access 23 July 2013.

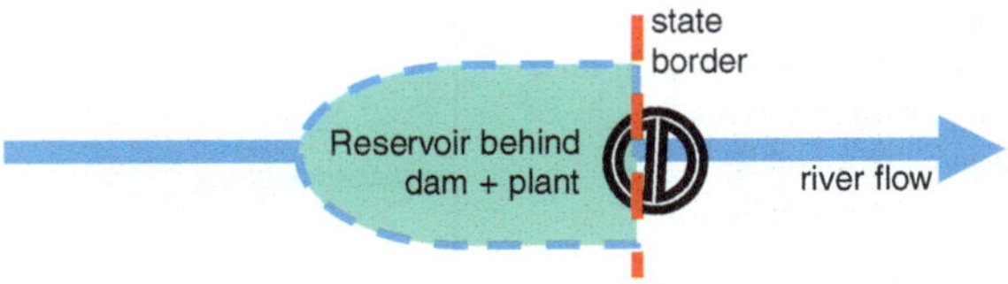

Fig. 8.3 Hydro-political constellation (3): A dam on a border crossed by a transboundary river

Fig. 8.4 Hydro-political constellation (4): A dam on a border river

Hydro-Political Constellation (4): A Dam on a Border River

The dam is located on a border-forming river (Fig. 8.4). Externalities will therefore affect both states.

The reason for cooperation in this constellation is:

- *Benefits can be exploited only through cooperation.* An agreement is required to build the dam on a river which forms the border between the riparian states. Due to sovereignty issues benefits can therefore be gained only through cooperation. Benefits are in principle symmetrical while externalities depend on the particular topographic circumstances. The costs of the joint investment are shared; the benefits are allocated according to investment shares. This was the mechanism for the Itaipu Dam shared by Paraguay and Brazil (Égré et al. 2002; Égré 2007; Scudder 2005), the Kariba Dam shared by Zimbabwe and Zambia (WCD 2000b; ZRA 1987; Tumbare 2002; Klaphake and Scheumann 2009), and is planned to be the case for Ruzizi III and Rusumo Falls (Dombrowsky et al. 2014).

Negotiating Benefit-Sharing Mechanisms

The states then enter into negotiations to "internalize" the external effects in a comprehensive cost-benefit calculus. In the process of internalizing the externalities, negative externalities are turned into costs and positive externalities into benefits. Ideally, all related costs (including capital, operating and maintenance, opportunity and external costs) and all related benefit streams (including direct and indirect use values, positive externalities and intrinsic values) are taken into account in the sharing of the benefits of dams (see, for example, Rogers et al. 2002).

The first step, therefore, is for the negotiating parties to determine whether cooperation will produce net benefits and whether it has the potential to make all the various parties better off compared with the status quo or unilateral action. This is the case if the net benefits of cooperation minus the net benefits of no or unilateral action are positive. Net benefits are the sum of all benefit categories minus the sum of all cost categories.

As their second step, the parties have to determine how cooperation will affect each party's cost-benefit calculus. Given that individual parties may be made worse off by cooperation compared with no or unilateral action, it may be helpful to use side-payments and issue-linkages during the negotiations (Klaphake 2005; Dinar 2006; Dombrowsky 2009, 2010). A side-payment is monetary compensation for losses incurred by one party as part of an overall deal vis-à-vis no or unilateral action made by those who gain from cooperation. Thereby, the negotiating partners will make sure that nobody is worse off after the deal than before it. Issues are linked when separate areas of negotiation are discussed simultaneously with a view to a joint settlement being reached (on, say, water and security or trade). They can be understood as side-payments in kind, the partners coming to a quid pro quo arrangement by linking issues, rather than making a monetary side-payment. Issue-linkage has the advantage of avoiding the loss of face that might be seen to be associated with a side-payment (see, for example, Mäler 1990).

However, as the definition of benefit sharing implies that all affected parties will be better off than they were before, not every side-payment or issue-linkage necessarily constitutes a form of benefit sharing per se. A side-payment that compensates only for losses, but does not make the party better off compared with the status quo would not be considered benefit sharing. For example, a one-time monetary payment by state A to state B for damages caused in B by a dam in A may not be considered benefit sharing if the payment does not make state B better off compared to the situation without the dam in place.

After compensating for potential losses incurred by individual negotiating partners, the parties can share the net benefits of cooperation. One way is to split the net benefits of cooperation equally, unless there are good reasons for unequal splitting.

Benefit sharing is easiest if all cost and benefit streams associated with the dam are quantified in monetary terms. However, environmental costs (i.e. loss of environmental services and costs for implementing environmental management plans) in particular are not usually expressed in market prices and their monetization can be methodologically complex (as well as time-consuming and costly) (see, for example, Garrod and Willis 1999). The results are often disputed and it is also disputed in what way negative effects can be adequately compensated. An alternative is to describe in qualitative terms those costs and benefit streams that cannot easily be quantified.

All the riparian states affected should perceive the project as beneficial—or at least acceptable—and as better than doing nothing or acting unilaterally.

Note that because the affected local population is typically not directly represented in international negotiations, international negotiations should be conceived as two-level games. Therefore, full compensation of negative social and environmental effects requires that governments negotiate with those affected and ensure that they are at least fully compensated or that they even enjoy a portion of the benefits of cooperation at the domestic level. All cases reviewed show that negative social and environmental effects are often not adequately addressed, and therefore not included in the cost-benefit calculation.

On the basis of the above description and based on an analysis of the dam cases reviewed, three types of benefit-sharing mechanisms can be distinguished:

(A) In the case of jointly owned infrastructure, costs are usually borne in proportion to benefits gained (Rivers Senegal, Zambezi and Paraná);
(B) When the design of a unilateral project upstream is changed to increase aggregate net benefits, the party altering its dam design is compensated for any losses incurred as a result of the alteration (in the form of monetary compensation, for instance), and the net benefits of cooperation compared to unilateral action by the upstream country are shared (in monetary terms or in kind) (River Columbia);
(C) When a downstream country finances a dam upstream in order to increase aggregate net benefits, it can be expected that investment and external costs are covered, with the upstream state participating in the benefits of the project and the two countries sharing the net benefits of the joint project compared to the downstream country's unilateral alternative (River Orange-Senqu).

In each case, the benefit-sharing mechanism consists of the combination of all the elements used to balance costs and benefits and may or may not include compensation.

Table 8.2 summarizes the cases reviewed, showing their hydro-political constellations, incentive structures and benefit-sharing mechanisms. As has already been said, with the exception of the Lesotho Highland Water Project (Yu and Winston 2008, pp. 61–63), the initial benefit-sharing mechanisms described in Table 8.2 did not take negative social and environmental effects adequately into account.

Full Accounting of Social and Environmental Costs, and Sharing Benefits from Dam Operation

One of the arguments advanced in the debate on dams is that, in too many cases, only a small segment of society benefits, while people living near the site of a dam have to endure its negative impacts: the flood control achieved with a dam may benefit those living downstream or in urban centres, while the assets of those living in or near the reservoir area disappear beneath the water; power generation benefits urban centres and industry, sometimes at some distance from the dam site, but not necessarily those living nearby; people living close to the reservoir may also have to endure such adverse environmental or health effects as eutrophication of a reservoir and waterborne diseases.

Table 8.2 The cases in brief (authors' compilation)

River /project	Hydro-political constellation and dam purposes	Incentive structure	Benefit-sharing mechanism
Manantali and Diama dams on River Senegal (Senegal, Mali, Mauritania, Guinea)	(1) and (2): upstream and downstream dams on transboundary river: hydropower, navigation, irrigation and flood control	(i) financial constraints on all riparians—project rational only collectively (ii) Senegal and Mauritania lack appropriate dam sites for hydropower	(A) sharing of cost of jointly owned infrastructure in proportion to expected irrigation, navigation and hydropower benefits; OMVS[1] attracts funding sources
Canadian dams on Columbia River (Canada, USA)	(1) upstream dams on transboundary river: hydropower and flood control	(ii) flood control benefits USA; electricity gain for Canada	(B) increase in aggregate net benefits through altered dam design upstream: Canada builds dams for downstream flood control and upstream hydropower generation; USA compensates Canada for investment costs by paying half of the value of downstream flood protection and electricity generation
LWHP on River Senqu-Orange (Lesotho, South Africa)	(1) upstream dams on transboundary river: hydropower and water supply	(iii) increased water supply for South Africa; electricity gain for Lesotho	(C) South Africa pays investment and operating costs and external costs of storage and transfer of water from Lesotho; Lesotho receives in-kind hydropower benefits; net benefits of cooperation compared to unilateral action are shared (royalties)
Aswan High Dam on Nile River (Egypt, Sudan)	(2) downstream dam on transboundary river: hydropower and irrigation	negative externality upstream	none, but compensation
Kariba Dam on River Zambezi (Zambia, Zimbabwe) Itaipu Dam on Río Paraná (Brazil, Paraguay)	(4) dam on border river: hydropower	symmetrical benefits, but not necessarily symmetrical externalities	(A) joint investment, benefit allocation according to investment shares

[1] Organisation pour la Mise en Valeur du Fleuve Senegal

Basic requirements for the construction of socially and environmentally acceptable dams are recommended, for example, by the World Commission on Dams (WCD) (2000a) and the Safeguard Policies of the World Bank (2001), which address involuntary resettlement practices and environmental planning and management. Both policies call not only for compensation for the loss of land and other property but also for the creation of new income-generating opportunities, for the restoration of the livelihoods of the populace affected (Pearce 1999) and for the costs of environmental management plans to be covered.

To prevent benefits from being shared between states at the expense of those who live near a dam, the mitigation of the adverse effects of a dam project on a transboundary or border river is a cost component which should be taken fully into account when net benefits are calculated (see Duflo and Pande 2007).

There are monetary and non-monetary mechanisms that do more than compensate for environmental, social and economic losses but share part of the benefits generated by dam operations (Égré et al. 2002, p. 2, 2008, pp. 317–357; Trembath 2008, pp. 375–393) with those affected—to varying degrees—by the project. Funds for local or regional development, for instance, are derived from the revenue stream of the project when in operation (Haas 2009).

The World Bank (Égré et al. 2002) and the United Nations Environment Programme Dams and Development Project (UNEP DDP) (Égré 2007) reviewed compensation options in dam projects on international, transboundary and national rivers and identified a number of monetary and non-monetary mechanisms:

- redistributing revenues from dam operation to local /regional authorities in the form of royalties tied to power generation or water charges;
- establishing development funds financed from power sales to provide, for example, seed money for economic development in the project-affected area;
- part or full ownership of the project by project-affected people who share profits and risks;
- levying property taxes on dam owners (e.g. hydropower corporations) or on a dam's property value (taxes are not related to revenues generated, but are a fixed charge) which are then transferred to local authorities, communities affected or river basin authorities;
- granting preferential electricity rates and subsidized irrigation and drinking water to local companies and project-affected populations;
- allocating fishing rights to resettlers in the newly created reservoir and hiring project-affected people for construction works (Égré et al. 2002, p. 3, 2008, p. 318ff).

The cases we reviewed showed that compensation payments were made in the early phases of projects. Re-negotiations at a later point in time comprised means to correct inadequate compensations, and, in some instances, the establishment of local benefit-sharing mechanisms, and funds for community development. (see Table 8.3).

If effectively implemented, such mechanisms have the potential not only to increase domestic acceptability of dam projects but to foster socially inclusive

Table 8.3 Social and environmental impacts in the cases reviewed (authors' compilation)

River /project	Compensation /benefit sharing Environment
Manantali and Diama dams on river Senegal (Senegal, Mali, Mauritania, Guinea)	Social and environmental effects not considered during planning and construction, no provision for compensation Set up of an Environmental Impact Mitigation and Monitoring Programme in 1998 to address environmental and social impacts
Canadian dams on Columbia river (Canada, USA)	No compensation for losses during planning and construction. Compensation and establishment of a local benefit-sharing mechanism as a result of re-negotiations (Columbia Basin Trust)
LWHP on River Senqu-Orange (Lesotho, South Africa)	Compensation for environmental and social losses A Fund for Community Development was established later on
Aswan high dam on Nile river (Egypt, Sudan)	Compensation and post-relocation development projects (better in Egypt than in Sudan)
Kariba dam on river Zambezi (Zambia, Zimbabwe)	Each government had the responsibility for managing resettlement (details are unknown)

Source Hensengerth et al. (2012), pp. 11–25

development. However, if these mechanisms are applied, they certainly affect the distribution of rents and require changes of decisions about who should be the beneficiaries of these rents (Égré et al. 2008; Trembath 2008).

Conclusions

This study has explored incentive structures and mechanisms for sharing the benefits of dams on shared rivers. It posits that the opportunities for deriving benefits from cooperation depend on the alignment of hydrological and political boundaries and the location of the dam in relation to them (i.e. hydro-political constellations), on the aims and external effects of each dam, and on the willingness of states to cooperate. It has been argued that it is rational for riparians to cooperate if each is able to generate higher aggregate net benefits than it would attain without cooperation. In doing so negative social and environmental outcomes and costs for their mitigation should be fully taken into account and compensated.

The cases reviewed in this study have the following four incentive structures for cooperation on dams on shared rivers:

1. cooperation enables economic or financial constraints on unilateral action to be overcome (River Senegal),
2. an altered dam design upstream increases net aggregate benefits (Columbia River),

3. locating a dam upstream increases aggregate net benefits (River Orange-Senqu),
4. a joint dam on a border river enables mutual benefits to be achieved (River Zambezi, Río Paraná).

The case studies also conclude on three different types of benefit-sharing mechanisms:

(A) costs are shared in relation to benefits in jointly owned dams (River Senegal, River Zambezi, Río Paraná);
(B) the party altering its unilateral dam design is compensated for losses incurred as a result of this alteration, and net benefits of cooperation are shared (Columbia River);
(C) the downstream state convinces the upstream state to build a dam, and covers the cost and shares the net benefits of the dam (River Orange-Senqu).

Benefit sharing seems to be particularly straightforward where the riparians decide to co-own the infrastructure involved from the outset. The costs are then shared in proportion to benefits (Type A) or vice versa. Interestingly, this happens not only on border rivers as in the cases of the Parana, but also on transboundary rivers for which the River Senegal is an example (Hensengerth et al. 2012, pp. 11–15).

While benefits have been shared in the cases mentioned, it is also evident that, in many cases, the environmental and social impacts on the population affected by dams were not taken into account from the outset, and that some projects had to be renegotiated at a later stage (see the creation of the Columbia Basin Trust arrangement, Table 8.3).

Furthermore, even if benefits of cooperation can be identified and cooperation therefore appears rational for all riparians, a number of factors[3] influence whether benefit-sharing schemes materialize, including foreign policies of and power relations between basin states (Frey 1993; LeMarquand 1977; Song and Whittington 2004; Zeitoun and Jägerskog 2011); a history of cooperation that states can build on (Giordano and Wolf 2003); national water policies and preferences (Waterbury 1997); third parties' involvement (Mostert 2005); regional initiatives and the degree of political and economic integration of regions (Durth 1996).

The study could not find any evidence of benefit sharing where negative externalities occur upstream. In the case of the Aswan High Dam on the Nile, the downstream riparian, Egypt, provided some compensation for resettlement in Sudan upstream, but it is questionable whether this can be described as benefit sharing.

From an academic point of view, the typology developed should be tested further in additional case studies (forthcoming Dombrowsky et al. on hydro-projects of the Ruzizi and Rusumo Falls in Africa's Great Lake Region). It is also worth taking a closer look under which conditions domestic arrangements are really and effectively implemented, and to study how international benefit-sharing

---

[3] For a discussion of factors see Hensengerth et al. 2012, pp. 26–30.

schemes on shared rivers are combined with domestic arrangements (Dombrowsky et al. 2014; Skinner et al. 2014).

Acknowledgement This study was co-funded by DIE and the GIZ Sector Project "Policy Advice for Sustainable Hydropower".

References

Africa Recovery (2004) vol 17, no 4. http://www.un.org/ecosocdev/geninfo/afrec/vol17no4/174ag.htm. Accessed 2 Aug 2011

African Development Bank (1988) Diama dam project: project performance evaluation report. African Development Bank, (no place)

Agreement between the Republic of the Sudan and the United Arab Republic for the Full Utilization of the Nile Waters of 1959 (1959) In: Scheumann W, Schiffler M (eds) (1998) Water in the Middle East: potential for conflicts and prospects for cooperation. Springer, Berlin, pp 186–190

Alam U, Dione O, Jeffrey P (2009) The benefit-sharing principle: implementing sovereignty bargains on water, in Political Geography, vol 28. pp 90–100

Bates BC, Kundzewicz ZW, Wu S, Palutikof JP (eds) (2008) Climate change and water. Technical paper of the intergovernmental panel on climate change. IPCC Secretariat, Geneva

Bhaduri A, Manna U, Barbier E, Liebe J (2011) climate change and cooperation in transboundary water sharing: an application of stochastic stackelberg differential games in volta River Basin. Nat Resour Model 24(4):409–444. Further information: onlinelibrary.wiley.com. doi:10.1111/j.1939-7445.2011.00097.x/abstract. Accessed 10 Jan 2014

Biswas AK (1999) Management of international waters: opportunities and constraints. Int J Water Resour Manage 15(4):429–441

Cernea MM, Mathur HM (2008) Can compensation prevent impoverishment? Reforming resettlement through investments and benefit-sharing. Oxford University Press, Oxford

Columbia River Treaty (1961). http://www.ccrh.org/comm/river/docs/cotreaty.htm. Accessed 2 Aug 2011

Dinar S (2006) Assessing side-payments and cost-sharing patterns in international water agreements: the geographic and economic connection. Political Geogr 25(4):412–437

Dombrowsky I (2007) Conflict, cooperation and Institutions in international water management. an economic analysis. Edward Elgar, USA

Dombrowsky I (2009) Revisiting the potential for benefit sharing in the management of trans-boundary rivers. Water Policy 11(2):125–140

Dombrowsky I (2010) The role of intra-water sector issue linkage in the resolution of transboundary water conflicts. Water Int 35(2):132–149

Dombrowsky I, Bastian J, Däschle D, Heisig S, Peters J, Vosseler C (2014) International and local benefit sharing in hydropower projects on shared rivers: the Ruzizi III and Rusumo Falls cases (submitted to Water Policy)

Duflo E, Pande R (2007) Dams. Q J Econ 122:601–646

Durth R (1996) Grenzüberschreitende Umweltprobleme Und Regionale Integration. Zur Politischen Ökonomie von Oberlauf-Unterlauf-Problemen an internationalen Flüssen. Nomos, Baden-Baden

Égré D (2007) UNEP dams and development project: compendium on relevant practices, 2nd stage. Revised final report, benefit sharing issue. UNEP, (no place)

Égré D, Roquet V, Durocher C (2002) Benefit sharing from dam projects. Phase 1: desk study. Final report. World Bank, Washington DC

Égré D, Roquet V, Durocher C (2008) Benefit sharing to supplement compensation in resource extractive activities: the case of dams. In: Cernea MM, Mathur HM (eds) Can compensation prevent impoverishment? Reforming Resettlement through Investments and Benefit-Sharing. Oxford University Press, Oxford, pp 317–357

Fink M (2005) Integrating the world commission on dams recommendations in large dam planning processes: the case of Bui, Ghana. Mimeo, Dortmund

Fischhendler I, Feitelson E, Eaton D (2004) The short-term and long-term ramifications of linkages involving natural resources: the US-Mexico transboundary water case. Environ Plann C: Gov Policy 22(5):633–650

Frey FF (1993) The political context of conflict and cooperation over international River Basins. Water Int 18:54–68

Garrod G, Willis K (1999) Economic valuation of the environment: methods and case studies. Edward Elgar, Cheltenham

Giordano MA, Wolf AT (2003) Transboundary freshwater treaties. In: Nakayama M (ed) International waters in Southern Africa. University Press, New York, pp 71–100

Haas L (2009) Introducing benefit sharing around large dams in West Africa: drawing on regional and international experience. Global Water Initiative and International Institute for Environment and Development, UK

Hensengerth O (2011) Interaction of chinese institutions with host governments in dam construction. The Bui Dam in Ghana. Discussion Paper 3. German Development Institute, Bonn

Hensengerth O, Dombrowsky I, Scheumann W (2012) Benefit sharing in dam projects on shared rivers. Discussion Paper 6. German Development Institute, Bonn

International Energy Agency (2011) World Energy Outlook 2011. OECD, IEA, Paris

Kibaroglu A, Scheuman W, Kramer A (eds) (2011) Turkey's water policy. National framework and international cooperation. Springer, Berlin

Kipping M (2005) Wasserkonflikte und Wasserkooperation am Senegalfluss. In: Kipping M, Lindemann S (eds) Konflikte Und Kooperation Um Wasser: Wasserpolitik Am Senegalfluss Und Internationales Flussmanagement Im Südlichen Afrika. LIT Verlag, Münster, pp 22–107

Kirschke S (2010) Grenzüberschreitendes Flussgebietsmanagement: Die Rolle Von Benefit Sharing-Mechanismen Im Rahmen der Senegalkooperation. Eschborn, GTZ

Klaphake A (2005) Kooperation an internationalen Flüssen aus ökonomischer Perspektive: das Konzept des Benefit Sharing. Discussion Paper 6. German Development Institute, Bonn

Klaphake A, Scheumann W (2009) Understanding transboundary water cooperation: evidence from Sub-Sahara Africa. In: Swatuk LA, Wirkus L (eds) Transboundary water governance in Southern Africa. Exploring underexplored dimensions. Nomos, Baden-Baden, pp 47–72

Krutilla JV (1967) The Columbia river treaty: the economics of an international River Basin development. Johns Hopkins University Press, Baltimore

LeMarquand DG (1977) International rivers. The politics of cooperation. Westwater Research Center, University of British Columbia, Vancouver

Mäler KG (1990) International environmental problems. Oxford Rev Econ Policy 6(1):80–107

Mas-Colell A, Whinston MD, Green JR (1995) Microeconomic theory. Oxford University Press, New York

McCartney M (2009) Living with dams: managing the environmental impacts. Water Policy 11(1):121–139

Meijerink SV (1999) Conflict and cooperation on the Scheldt River Basin: a case study of decision making on international Scheldt issues between 1967 and 1997. Kluwer Academic Publishers, Boston

Mostert E (2005) How can international donors promote transboundary water management? Discussion Paper 8. German Development Institute, Bonn

Muckleston KW (2003) International management in the Columbia River Basin. UNESCO, Washington DC

Pearce DW (1999) Methodological issues in the economic analysis for involuntary resettlement operations. In: Cernea MM (ed) The economic of involuntary resettlement. Questions and challenges. World Bank, Washington DC, pp 50–82

Phillips D, Daoudy M, McCaffrey S, Öjendal J, Turton A (2006) Transboundary water cooperation as a tool for conflict prevention and broader benefit-sharing. Swedish Ministry of Foreign Affairs, Stockholm

Putnam RD (1988) Diplomacy and domestic politics: the logic of two-level games. Int Org 42(3):427–460

Rogers P, de Silva R et al (2002) Water is an economic good: how to use prices to promote equity, efficiency, and sustainability. Water Policy 4:1–17

Sadoff CW, Grey D (2002) Beyond the river: the benefits of cooperation on international rivers. Water Policy 4(5):389–403

Sadoff CW, Grey D (2005) Cooperation on international rivers: a continuum for securing and sharing benefits. Water Int 30:420–427

Scudder T (2003) The Aswan high dam case. (Unpublished manuscript)

Scudder T (2005) The Kariba case study. Working paper 1227. California Institute of Technology, Pasadena

Skinner J, Krauss J, Newborne P (2014) Redistribution of revenues from hydropower dams: review of benefit-sharing mechanisms and local control (draft September 2013)

Song J, Whittington D (2004) Why have some countries on international rivers been successful negotiating treaties? A global perspective. Water Resour Res 40:W05S06. doi:10.1029/2003WR002536

Trembath BP (2008) Beyond compensation: sharing of rents arising from hydropower projects. In: Cernea MM, Mathur HM (eds) Can compensation prevent impoverishment? Reforming resettlement through investments and benefit-sharing. Oxford University Press, Oxford, pp 375–393

Tumbare MJ (2002) Co-operation in the Zambesi River Basin. In: Al Baz I et al (eds) Co-operation on transboundary rivers. Nomos, Baden-Baden, pp 101–112

Waterbury J (1997) Between unilateralism and comprehensive accords: modest steps toward cooperation in international River Basins. Int J Water Resour Dev 13(3):279–289

World Bank (2001) Operational policy 4.12: involuntary resettlement; operational policy 4.12, Annex A: involuntary resettlement instruments; bank procedure 4.12: involuntary resettlement. http://web.worldbank.org/WBSITE/EXTERNAL/PROJECTS/EXTPOLICIES/EXTSAFEPOL/0,,contentMDK:20543978~menuPK:1286647~pagePK:64168445~piPK:64168309~theSitePK:584435,00.html. Accessed 31 July 2011

World Bank (2007) Africa region: irrigation business plan. Revised Draft, Washington DC (October 23)

World Commission on Dams (2000a) Dams and development. A new framework for decision-making. Earthscan Publication Ltd, London

World Commission on Dams (2000b) Kariba dam Zambia and Zimbabwe. Final report, November. http://www.dams.org/docs/kbase/studies/cszzmain.pdf. Accessed 31 July 2011

World Summit on Sustainble Development Plan of Implementation (2002) World summit on sustainble development plan of implementation. http://www.johannesburgsummit.org/html/documents/summit_docs/2309_planfinal.htm. Accessed 14 Dec 2011

Yu T, Winston H (2008) Benefit sharing in international rivers: findings from the Senegal River Basin, the Columbia River Basin, and the Lesotho highlands water project. World Bank, Washington DC

Zambezi River Authority Agreement between the Republic of Zimbabwe and the Republic of Zambia concerning the utilization of the Zambezi River, 28 July 1987. http://www.law.co.zw/downloads/statutes/20/Zambezi%20River%20Authority%20Act.pdf. Accessed 3 Aug 2011

Zeitoun M, Jägerskog A (2011) Addressing power asymmetry: how transboundary water management may serve to reduce poverty. Report 29. SIWI, Stockholm

Chapter 9
Challenges and Solutions for Urban-Tourist Water Supply on Mediterranean Tourist Islands: The Case of Majorca, Spain

Angela Hof, Macià Blázquez-Salom, Margalida Comas Colom and Alfredo Barón Périz

Abstract Water supply on Mediterranean tourist islands becomes a major challenge due to anthropogenic pressure on water resources and global climate change. Asymmetries of water availability and demand are shown for the case of Majorca, focusing on the urban-tourist sector. Quality tourism development is increasing water demand, its unequal access and unsustainable use. Supply solutions rely on water supply enhancement through technological fixes, such as shipping freshwater from the mainland, or the desalination of brackish groundwater and seawater. Supply enhancement is entrenched in urban-tourist and demographic growth as major drivers and results of the Spanish economic development from 1995 to 2007, with urban-tourist growth being supported by technical water supply solutions. Instead of redirecting the discourse to water demand management, supply enhancement and technological, market-oriented solutions for accommodating rising water demand are favored. These supply solutions retard innovative and proactive water policies. On the other hand, successful public policies have constrained urban sprawl and golf courses development in order to enhance natural resources management; particularly in the Balearic Islands in comparison with the Spanish coastal areas in general (Rullan 2011). Palma and Calvià are studied in detail as two mature and representative island tourist destinations in the Mediterranean: Their urban-tourist water supplies increasingly rely on desalinated

A. Hof (✉)
Landscape Ecology/Biogeography, Geography Department, Ruhr University Bochum, Universitaetsstrasse 150, 44780 Bochum, Germany
e-mail: angela.hof@ruhr-uni-bochum.de

M. Blázquez-Salom
Research Group on Sustainability and Territory (GIST), Earth Sciences Department, University of the Balearic Islands, Carretera de Valldemossa, Km. 7.5, 07122 Palma (Majorca), Spain

M. C. Colom · A. B. Périz
Govern de Les Illes Balears, Conselleria D'Agricultura, Medi Ambient I Territori, Servei D'Estudis I Planificació, Direcció General de Recursos Hídrics, C/Gremi de Corredors, 10 (Polígon Son Rossinyol), 07009 Palma (Majorca), Spain

A. Bhaduri et al. (eds.), *The Global Water System in the Anthropocene*, Springer Water, DOI: 10.1007/978-3-319-07548-8_9,

water, provided through expensive infrastructure. Notwithstanding, the production of desalinated water dropped by 20 % since the beginning of the current financial and economic crisis. Reasons for this are the costs for fueling the desalination plants, and the opportunity to extend the overexploitation of the underground water tables that recovered to a better shape thanks to supplementation by desalination and thanks to a long wet period since 2008. The current institutional framework and the detailed characteristics of the regional regulatory framework of water supply is analyzed, showing that meanwhile the regional government should implement the European Union Water Framework Directive, current policies demonstrate a strong normative recoil. Water demand, supply and the reliance on non-conventional sources are spatially uneven, embodied in uneven sociospatial water supply and consumption. However, the current discourse reduces the associated environmental and societal problems to questions of supply enhancement and may retard innovative and proactive water policies dealing with global change pressures such as climate change.

Introduction

Mediterranean environments are facing specific challenges of water resources management that are inherent in natural and environmental conditions, with climate change aggravating the situation (Correia 1999; Iglesias et al. 2007). Annual and inter-annual asymmetries of water availability and water demand have necessitated water management since the beginning of agricultural land use in ancient societies (about 2,700 B.C.), which in turn was the basis for the first urban centers engaging in global trade and cultural exchange. Today, the spatially and temporarily uneven distribution of precipitation and runoff requires the construction of costly water reservoirs and high levels of wastewater treatment. Water scarcity in terms of average quantity per capita is often not the main problem. The main challenge of water management in Mediterranean environments is the high cost of making water available and guaranteeing sufficient supply to increasingly demanding and competing agricultural, urban and tourist sectors. The agricultural sector remains the main land use category with the highest water needs and the shift towards irrigated agriculture is ongoing in the Mediterranean. However, Mediterranean economies and societies are increasingly becoming service oriented and highly urbanized. The strongest development pressures are therefore coming from the agricultural sector on the one hand and the urban and tourist sector on the other hand. The water reservoirs are already under pressure and water supply increasingly relies on desalination and reuse of treated water.

The Mediterranean is the world's largest and oldest tourist destination, and in many countries, tourism is a crucial economic activity. Tourism must be considered as part of social and economic developments that exert pressures when identifying current and future water management challenges (Gössling et al. 2012).

Tourism is also a major driver of the urbanization process that concentrates water pressure along the Mediterranean coastline. Here, urbanization and tourist development overlap and a general observation is that water consumption by the tourism sector is not well documented by statistics at present. In fact, tourist water demand is included in the urban demand in most Mediterranean countries, concealing the fact that locally, tourist demand may exert the highest pressure on water resources and more than 50 % of municipal water uses can be related to tourism (Eurostat 2009; Gössling et al. 2012; Tamoh et al. 2008).

The Mediterranean is among those regions that are most affected by global climate change. Water supply in the region is likely to be impacted by reduced rainfall amounts, increased average temperatures, higher occurrence of droughts, and heat waves (Alcamo et al. 2007; Palutikof and Holt 2004). By 2050, a 30–50 % decline in freshwater resources in the Mediterranean basin is expected (Milano et al. 2013). Water management is becoming a major challenge for Mediterranean countries and the tourism sector (Eurostat 2009; Iglesias et al. 2007).

In this context, it is instructive to look at Mediterranean tourist islands where water challenges are advanced and environmental consequences crystallize locally. Many Mediterranean regions will face the problems presently beginning to be addressed in Mediterranean tourist islands.

Mediterranean Islands are particularly challenged by water resources management due to their geographical isolation and the impossibility to draw on more distant or diverse aquifers. Characteristic is their dependence on natural renewable resources (groundwater and surface water, including reservoirs) and increasingly, dependence on non-conventional water resources (sea and groundwater desalination, treated wastewater). The assessment of water management on the major Mediterranean Islands of Corsica, Crete, Cyprus, Majorca and Sicily pointed to the importance of a responsible awareness of the agricultural, domestic, and tourism sectors of their usage of water to ensure a sustainable water management in accordance with the European Union Water Framework Directive (Donta and Lange 2008). The Water Framework Directive already provides a legal framework for the policy objective that all waters (surface, underground and coastal) ought to attain a good and non-deteriorating status. The new approaches and responsible awareness needed must integrate the ecological, economic and social aspects of water policy at the level of river basins. Like for the whole Mediterranean, the main objectives are to protect and improve the islands' aquatic environment and to make a contribution to sustainable, balanced and equitable water use. However, the current situation on the islands is characterized by vulnerability to extreme weather events and the projected effects of climate change. The major challenges are to meet the growing water demands, to control environmental pollution and to improve sanitation infrastructure. The recommended policy and management options have been repeatedly laid out in detail and need not be repeated here (Candela et al. 2005; Donta and Lange 2008). Climate change adaptation measures have been described for Mediterranean agriculture (Candela et al. 2012) and the utilization of non-conventional water sources such as desalinated seawater, treated wastewater and brackish water are discussed as means of novel approaches to sustainable water resources management (Gikas and Angelakis 2009; Lazarova et al. 2001). Despite significant vulnerabilities to climate change, this challenge is hardly reflected in sustainable tourism policy and planning documents (compare Dodds and Kelman 2008, on Malta and Majorca). Water relevant policies still adhere to the hydraulic paradigm and focus mainly on water supply enhancement, rather than water demand management.

In the Spanish coastal areas and the Balearic Islands, the trend of urban and socio-economic growth is exerting a particularly strong pressure on water resources. Urban and demographic growth have been identified as the most significant expressions of the process of socio-economic tourist modernization in Spain (Mantecón 2010). In the urban realm, urban sprawl processes and the expansion of new urban landscapes have been extensively discussed with respect to increasing water consumption. The demand for water appears to be rising inexorably due to the combined effects of urban and tourist growth. More disperse, low density urban landscapes evolve that are characterized by a spacious residential matrix with large plots occupied by gardens and swimming pools (Domene et al. 2005; Domene and Saurí 2006; Parés et al. 2013; Vidal et al. 2010). As a result, urban water uses have a growing share of total water demand, which now is up to 10 % (2002) with the domestic sector consuming 70 % of urban water use. Per-capita water consumption has increased since 1996 at an annual average rate of 2 % and water prices increased by 4 % every year over the last decade. Recreational uses (tourism, golf courses, swimming pools, second homes) are the most rapidly growing water uses (Maestu and Gómez 2010). As Spanish water law prioritizes urban water supply, the projections for climatically induced and increasing competition for water between the urban, tourist and agricultural sectors is foreseeable. In this context it is worth mentioning that the revenue from water uses in tourism can be up to 60 times higher compared to agriculture (Gössling et al. 2012), and that the financial returns of golf courses are substantially higher per unit area than those of agriculture (Rodriguez Diaz et al. 2007).

As a consequence of these complex socio-economic developments, urban water demand and water supply have become the most dynamic sector of water resources management (Maestu and Gómez 2010; March and Saurí 2010; Masjuan et al. 2008). In most coastal areas urban water supplies are intrinsically linked to and influenced by the tourist sector, increasingly based on desalination and subject to privatization processes. While the privatization of traditionally public urban water management is taking place already (Saurí et al. 2007; Romero Renau 2006), public and academic debates seem to lag behind, despite the urgent need to discuss how different underlying objectives (e.g. social equity, economic efficiency and environmental conservation) can be combined within a coherent implementation framework (Cabrera et al. 2010).

This contribution describes the situation in Majorca as representative and major Mediterranean tourist island with advanced problems and elaborated solutions to manage the urban-tourist water supply. On closer inspection, this case study reveals the major responses of Mediterranean urban water systems to the

Anthropocene within the last decade. At the same time, it reveals important issues that need to be addressed by policy and practice in the field of sustainable freshwater management. In the following, the case of Majorca is presented to discuss if and how this island is approaching the shift in mind-set that is required to address the water challenges of tomorrow. With a focus on the urban and tourist sector, issues like variability in supply and increasing demands for water are outlined and the present water governance and related institutional and technological innovations are critically reflected.

Case Study Majorca

Majorca is a major Mediterranean tourist destination with a success story of sustained mass tourist flows since the 1960s and constant reinvention of itself. In the last five decades (1960–2011), Mallorca has been visited on average by 8 million tourists annually and has achieved average visitor growth rates of 6.6 %, underscoring the long-lasting success of its 'sea and sand' tourist model that is known as 'Balearic model' of package tourism (Aguiló et al. 2005; Buswell 2011). The pivotal role of water for tourism sustainability and the mismatch between water demand and water supply in Majorca has been discussed earlier (Essex et al. 2004; Garcia and Servera 2003; Kent et al. 2002). Like for other Mediterranean Islands, a range of policy and management options have been proposed, given the severity of overexploitation of water resources on the island and a per capita freshwater availability that indicates a condition of absolute water scarcity (Donta and Lange 2008). Given the high population density and high tourist demand, the water demand from the domestic, urban-tourist sector in Majorca is high. Because high amounts of water are also consumed for agriculture, the island has to deal with crucial water shortages and with water scarcity. Indicative for this is a per capita freshwater availability reaching values below 500 m^3/person/year. This widely used water stress indicator (Falkenmark et al. 1989) designates a condition of water scarcity if freshwater availability is between 500 and 1,000 m^3/person/year and a threshold level of 1,700 m^3/person/year which stands for an irregular and local water shortage.

Recent developments in the tourist sector that are marketed under the term 'quality tourism' since the mid-1990s (Schmitt and Blázquez 2003) have proliferated residential tourism and have accelerated urban growth. Residential tourism generally refers to the phenomenon of property ownership and short-term residence of foreign people in tourist areas (O'Reilly 2007). Golf tourism and residential tourism in particular serve as a highly profitable complement to the island's mass tourist sector. As a result, water uses in the domestic sector have increased and diversified, accelerating demand and requiring that water management objectives have to be balanced dynamically (Essex et al. 2004; Hof and Schmitt 2011). On this background, the following sections outline the state of water resources and describe the dynamics of urban-tourist water supply in Majorca.

Table 9.1 Water resources of the Balearic Islands and Majorca (Hm^3/year)

		Surface water (Hm^3/year)	Groundwater (Hm^3/year)	Total (Hm^3/year)
Balearic Islands	Natural resources	120	472.7	592.7
	Usable resources		303.5	
	Available resources	7.2	260.2	267.4
	% Available/ Natural	6	55	45
Majorca	Natural resources	120	374.1	494.1
	Usable resources		250.2	
	Available resources	7.2	220	227.2
	% Available/ Natural	6	58.8	45.9

Source own elaboration from Conselleria de Medi Ambient (2013)

Water Resources and Water Consumption

Groundwater is the main available water resource in Mallorca (volume: 220 Hm^3/year), and together with surface water collected in two dams constructed in the Tramuntana mountain range, constitutes 227.2 Hm^3/year (Table 9.1). This available water is the exploitable resource that could be technologically and economically utilized without undesirable effects to potential ground or surface water resources. Expressed as an exploitation index, this represents 45.9 % of the total natural water resources in Majorca (Conselleria de Medi Ambient 2002). Human activities in Majorca consume between 87.1 % (2001) and 90.7 % (2006) of these available water resources, which indicates that the island is severely water scarce (Conselleria de Medi Ambient 2002, 2013). Formerly (2001), the highest demand was coming from agriculture (62.1 %, calculated according to agrarian census data), and followed by domestic demand (37.3 %), golf courses (0.4 %) and industry (0.2 %). According to 2006 data, the highest water demand is coming from the domestic, urban-tourist sector (Conselleria de Medi Ambient 2013). The 157 Hm^3 of exploited underground water resources were shared by: urban-tourist demand (54.98 %), exurban domestic demand (18.03 %), irrigated agriculture (25.31 %, calculated according to remote sensing mapping), livestock husbandry (1.10 %), industry (1.14 %) and golf courses (0.19 %).

The person related water exploitability index, which considers the population and the exploitable water resources as m^3/person/year, shows for Majorca the anthropogenic pressures on the water resources. Taking into account the above-mentioned available water resources on the island and the permanent population on the one hand, and the population including mean seasonal population on the other hand, a comparison for the last decade (1999–2012) shows that the total available water resources for per capita use is 24.9 % (permanent population) and 18.2 % (population including mean seasonal population) lower in 2012 compared to 1999. The main reason is the demographic growth as a result of a population increase in Majorca of 33 % in that period (1999–2012). As a result of demographic changes and constantly high tourism intensity levels, in 2012, the total available water for use per capita is 259 m^3/person/year for the permanent population and 209 m^3/person/year for the population including mean seasonal population. Both values fall below the threshold level of 500 m^3/person/year indicating a main constraint to life. If the natural resources instead of the available resources in Table 9.1 are considered, the water for use per capita is 564 m^3/person/year for the permanent population and 455 m^3/person/year for the population including mean seasonal population. Given these constellations, it is not surprising that in addition to the use of treated wastewater, Majorca has since 1999 also turned to desalination to secure water supply. The current (2006) mix of sources for water supply on the island reflects the managerial solution to anthropogenic pressures on scarce water resources. Of the total 209.71 Hm^3 water consumed, 74.95 % are from groundwater bodies and only 2.95 % from water reservoirs. Treated wastewater contributes a 12.44 % proportion to water consumption and desalinated water 9.66 %. In the following, the origins of water supplied to and used in the island municipalities are discussed. Next, the water supply to the main urban-tourist zone of the island is described with a focus on the dynamics of water resource use over the last decade.

Water Consumption and Pricing, Tourist Infrastructure and Water Loss in the Distribution Network on the Island of Majorca

Majorca ranks among the Spanish Mediterranean tourist zones with the highest per capita water consumption rates (Saurí et al. 2011). Tourism is so much at the core of the Majorcan economy that its contribution to urban growth and water demand is seen as being part of the solution, not part of the problem. Relating available data on urban water consumption to inhabitants on the one hand and showing tourist bed capacity in relation to inhabitants on the other hand reveals the strong and positive correlation between per capita water consumption and tourism infrastructure (Fig. 9.1). In general, water consumption in the mass tourist segment is lower than in the luxury, quality tourist segment (Deyà Tortella and Tirado 2011; Hof and Schmitt 2011). In Majorca, water at consumer level is only paid by households, farms, enterprises, commercial bodies and/or industries connected to the municipal distribution networks. Owners of wells only pay for the expenses of building, maintenance of infrastructure and operation. For groundwater, sanitation taxes over half of the extraction concession is applied. Water for agricultural use is not subject to pricing as farmers obtain it from their individual well. The drinking water supply price usually includes a fixed service quota which includes infrastructure use and maintenance and a variable consumption quota (Candela et al. 2005). Substantial differences in municipal water prices exist that are not well

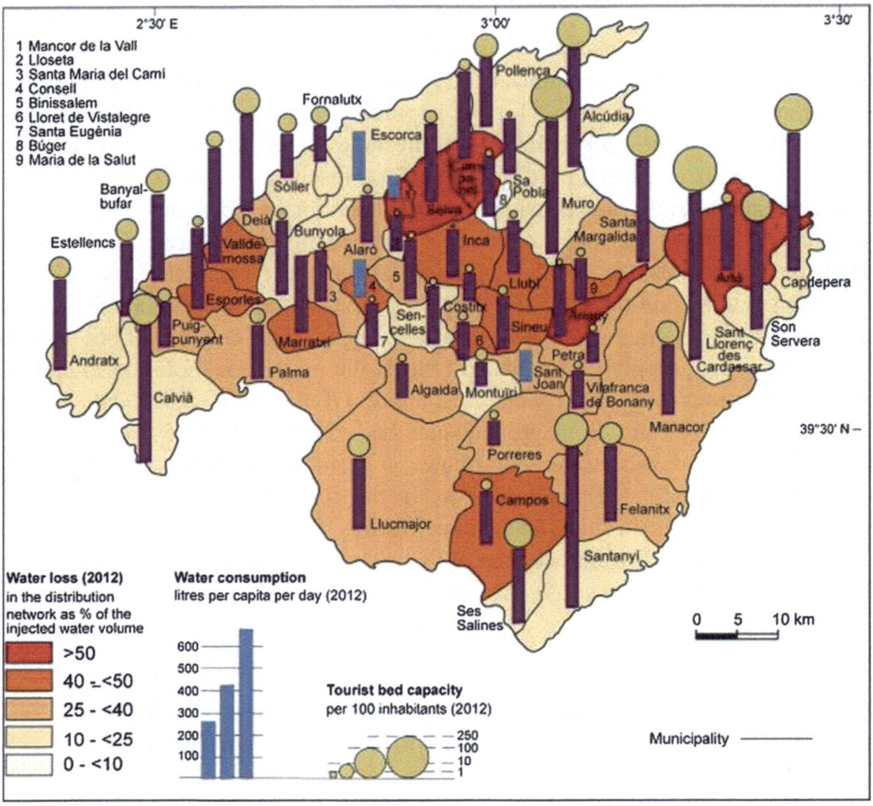

Fig. 9.1 Spatial pattern of tourist infrastructure, urban water consumption, and leakages in the urban water distribution network per municipality

explained by any of the relevant technical variables. With current water tariffs set on political grounds, the current prices favour a low economic level of leakage on the side of water supply companies and for the users, there are no real incentives to use water rationally. Like elsewhere in Spain, water supply in urban areas in Majorca is in many respects in need of reform, and particularly pricing policy is at the moment inconsistent with the manifold problems in the water sector (compare Cabrera et al. 2010).

The water distribution network on the island of Majorca is in no better shape than that in the rest of the country. In Spain, where water policy has focused almost exclusively on the management and supply of water resources, the modernization of the water distribution networks has never been a priority. In the whole of Spain, the volumetric efficiency (the quotient between the metered water and the injected water) is usually less than 70 % (Cabrera et al. 2010). In Majorca, it is 69.4 % on average and ranges from 97 to 38 % for the 53 municipalities (2012).This means that in the most extreme case (Selva), 62 % of the water supplied is lost through leakages in the urban distribution network (Fig. 9.1).

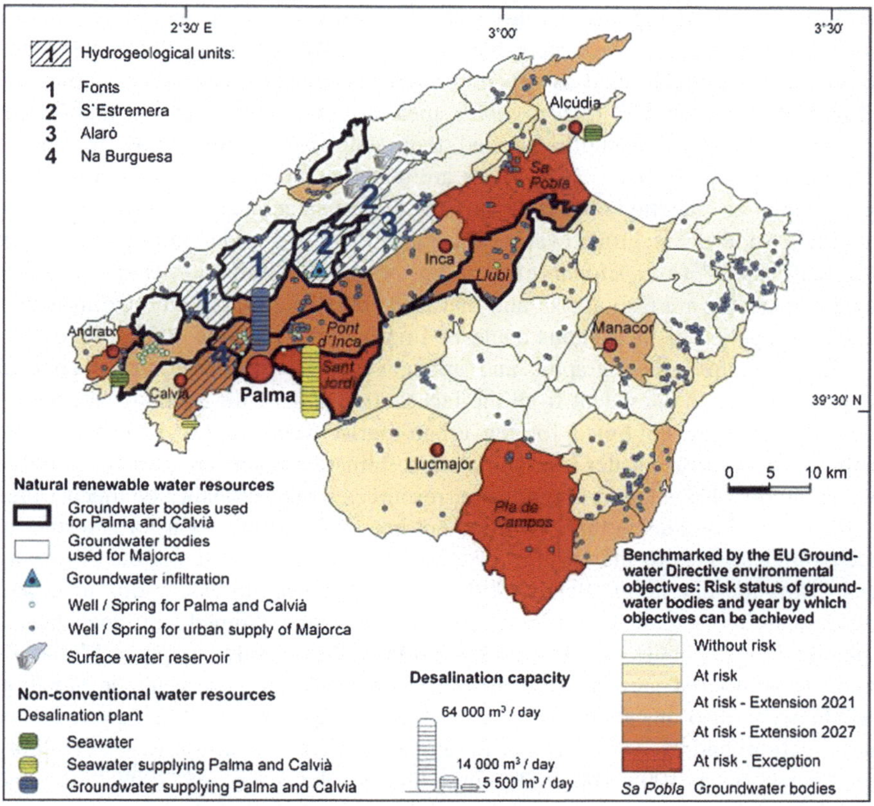

Fig. 9.2 Map of natural renewable and non-conventional water resources for urban-tourist water supply to Palma and Calvià and the status of exploited groundwater bodies in Majorca Island. In the 2001 Hydrological Plan of the Balearic Islands (PHIB), the hydrogeological unit was the management unit, in the 2013 PHIB, the groundwater body is the management unit

State of Aquifers and Groundwater Bodies and the Role of Desalination for Urban-Tourist Water Supply and Its Long-Term Dynamics

For long, groundwater was the main available water resource on the island and the state of aquifers and groundwater bodies is under close monitoring. If judged by the environmental objectives of the European Union's Groundwater Directive, half of the exploited groundwater bodies on the island are classified as in good quantitative and qualitative status (Fig. 9.2). Groundwater bodies classified as "without risk" have either high recharge rates due to precipitation or artificial recharge from wells, i.e. groundwater infiltration (Fig. 9.2). In terms of water quality, these groundwater bodies exhibit acceptable salinity levels. Groundwater bodies that are classified as "at risk" should be under managed extraction and salinity monitoring regimes because saltwater intrusion, and in some cases, nitrate pollution levels are above acceptable limits. Twenty percent of the underground water bodies are classified as severely overused. Seven groundwater bodies are classified as "extended" (*prorrogable*), meaning that they are at risk and that, if preventive measures would be applied, their status could improve by the year of the extension, in 2021 or 2027. These are tentative deadlines to reconsider these water units as 'healthy', but only in case that management measures would be applied. Another six groundwater bodies are classified under the description of "exceptionable" (*excepcionable*) or in an "exception" situation (Fig. 9.2). This means that they are also at risk and that it is unrealistic that preventive measures could be applied or their status could be improved by the year 2027.

The economic, demographic and urban weight of the urban-tourist zone of Palma and Calvià is evident from the fact that 15.4 % of the groundwater bodies on the island are exploited for the urban-tourist water supply to this area. In addition, two surface water reservoirs and desalination plants are providing for the water supply. The state of groundwater resources would probably be much worse if the water shortages in the 1990s had not been the starting point for resorting to technological fixes to the upsurge of urban-tourist water consumption. A turning point was the operation 'Barco' (1995–1997) that was necessary due to urgent water scarcity. A total of 16.6 Hm^3 of freshwater were shipped from the Spanish mainland to contribute on average 14.4 % to the annual water supply to Majorca's main urban-tourist area. In that period, groundwater and seawater desalination plants were also installed. They contribute on average 17.2 % (groundwater desalination) and 21.7 % (seawater desalination) to the urban-tourist water supply of Palma since 1995 (Fig. 9.3). In addition, the years of 2008–2009 and 2009–2010 were exceptionally wet, resulting in higher recharge rates of the groundwater bodies compared to the period from 2004 to 2008. Groundwater desalination is using water from the hydrogeological units of Na Burguesa (H.U. Na Burguesa, compare Figs. 9.2 and 9.3) and wells of Pont d'Inca aquifer, reducing the pressure on other groundwater bodies but worsening the salinity intrusion of these hydrogeological units. This is why it is obvious that the water supply of Majorca's capital city and most important tourist areas has become highly dependent on desalination technology. In the current crisis context, it is noteworthy that the volume of the produced desalinated seawater has dropped to a merely 20 % of the average volumes produced in the period of 1999–2010 (Fig. 9.3). This signifies not only the resort to groundwater bodies rendered possible by the only very recently improved recharge rates, but also that the desalination plants operate below capacity. The situation is also pointing to the large differences in resource costs, which are 0.823 €/m^3 for desalinated water (including amortization cost) and 0.179 €/m^3 for water from groundwater wells (excluding amortization cost, which is negligible). The water from desalination plants are sold at real costs, without including the investment cost paid by the Balearic Government. This has allowed municipal enterprises to make big profits by overestimating their predicted requirements of expensive desalinated water achieving a high water price, thus big revenues and stressing the cheap water

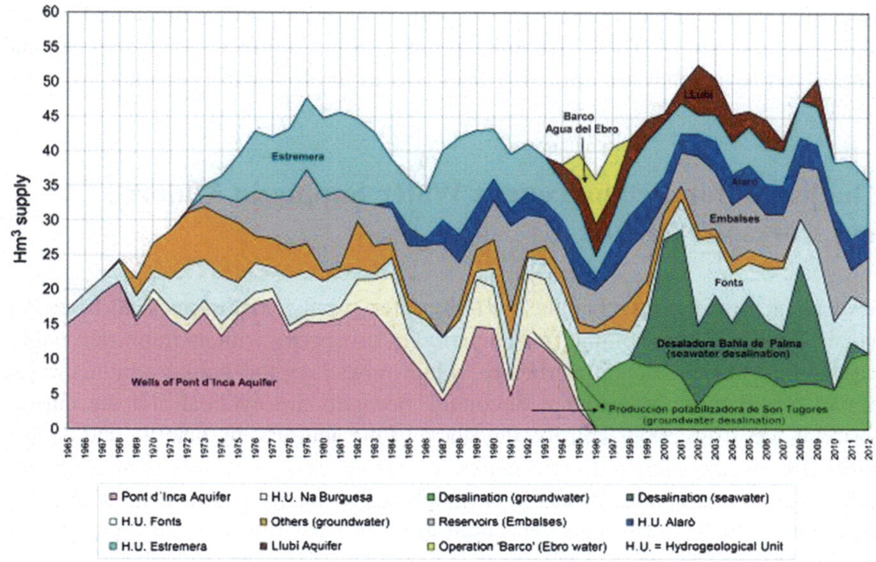

Fig. 9.3 Dynamics in the sources of water supply to the Palma metropolitan area and the Bay of Palma tourist zone

resources. The cost of the water transferred from wells within hydrogeological units is often negotiated and fixed over several years according to an agreement between a water distribution company and local or municipal water supply companies. No regulations exist regarding the price set by the enterprises. This has generated situations when a distribution company buys bulk water quantities at expensive rates, and may thus negotiate a higher price for consumers by adding high costs of production or purchase (Candela et al. 2005).

Given its relatively high cost, desalinated water is used in Majorca mainly within the urban-tourist sector. Available data for the years 2000–2008 show that on average, 92.6 % of the total amount of desalinated water produced was supplied to the urban-tourist areas of Palma and Calvià, with Palma alone having a share of 51.3 %. Considering the fact that 47 % of the Majorcan population live in the city, this share seems reasonable. However, as the main urban agglomeration, Palma and its tourist sector are drawing on the whole range of water resources (Fig. 9.3), while Andratx municipality is relying completely on desalinated water for urban-tourist supply and Calvià municipality had on average 63.2 % of water supplied for urban-tourist use originating from desalination plants. The most extreme years were 2007 and 2008, when 84 % of water supplied for urban-tourist use in Calvià originated from desalination plants. In Palma municipality in this period (2000–2008), 22.4 % of water supplied for urban-tourist use originated from desalination, with a peak in the years 2000 and 2001 (up to 36.1 %). Although it is difficult to prove a causal relationship between the current financial and economic crisis and the recent shift from desalination to groundwater exploitation, the dynamics in the urban-tourist water supply system are amply illustrated by the Majorcan case study. These dynamics point to some future water challenges and unresolved issues that are critically discussed in the following section.

Challenges for Urban-Tourist Water Supply in Majorca and Unresolved Issues

Several factors indicate that the current water supply conditions could become worse. Firstly, we are coming from a period of prosperity with an important public expenditure to develop infrastructure, while nowadays the public administration cannot afford more investment. Secondly, droughts are cyclical and the current period is wet from 2008. Thirdly, the shape of the aquifers, in quantity and quality terms (for instance, jeopardized due to seawater intrusion or nitrates contamination from intensive agriculture), has been much worse in the past and their ongoing exploitation could mean their deterioration in the near future. Lastly, energy prices, on which desalinization efficiency is dependent, are increasing due to the peak oil and with the depletion of fossil energy resources.

The European Union has already taken into consideration these scenarios and enacted the 2000/60/CE Water Framework Directive (WFD) that obligates the member states to apply precautionary measures to ensure the good state of the resources and the full cost recovery of their human use. The elaboration of the Hydrological Plan of the Balearic Islands (PHIB) should be devoted to enact the WFD principles, but the needed adoption of new limits to the water supply would threaten the speculation on the urban-tourist growth and the agrarian sector business' profitability, particularly based on the European Union granting scheme. Two versions of the PHIB have been approved initially, the first one in February 2011 (Conselleria de Medi Ambient 2002) and the second one in July 2013 (Conselleria de Medi Ambient 2013), due to a change of government. The 2013 PHIB promotes deregulation of the underground water table exploitation and eliminates the preventive measures that formed part of the 2011 initial approval. Some examples will illustrate the 2013 normative recoil: firstly, it allows any new well construction for private purposes with permission to extract up to 7,000 m^3 per year. Secondly, it also allows more new wells for salty water extraction (to be desalinized by particular consumers, such as hotels) within a 200 m buffer from the seafront. Thirdly, it permits to continue the exploitation of groundwater bodies that have been damaged due to historical overuse and that are therefore still at risk. Under these resource conditions, in the face of climate change and the projected decline of freshwater resources, this deregulation of groundwater exploitation runs counter sustainable water resources management. Fourthly, using desalinated water for golf course irrigation is also permitted or even promoted in order to make the current desalinization infrastructure more profitable. Fifthly, the limitation of water extraction beyond 80 % of the aquifer recharge estimations has been withdrawn. Sixthly, the autonomous sewage treatment systems have been deregulated; according to the last PHIB approval they wouldn't need administration authorization but only a responsible statement by the promoter. Seventhly, the current version of the Hydrological Plan weakens the regulation of sources of punctual pollution, such as the petrol stations. Lastly, only to mention important changes with wider environmental implications, some wetlands protection measures have been removed. Wetlands protection has been a controversial issue due to their seafront location, where more urban and tourism development pressure is exerted. Consequently, a detailed inventory of wetlands annexed to the 2011 PHIB has not been included as a protective normative in its successor version of 2013. Owing to these controversial issues, the PHIB 2013 administrative process of report adopted by the Spanish National Council of the Water, on the 29th of July 2013, was not unanimous because the environmentalist and scientific representatives voted against its approval.

Implementation of the WFD would imply full water cost recovery, with very important increases of water prices. This is because Spain has very low water prices, with urban rates in the lower segment of the OECD countries (Garrido and Calatrava 2010). There is a political and technical consensus that the Spanish water prices should be higher in order to internalize the full social, environmental and economic cost of its supply. Detailed regulations to implement measures of full costs of water supply are included in the PHIB 2013 (Conselleria de Medi Ambient 2013). In general, awareness tools, incentives and sanctions for water saving across the agricultural, urban and tourist sectors are promoted. The specific measures focus on water pricing for the urban and industrial sector. Dealing with the water rate structure, the following measures are proposed: rate adjustment so that charges are linked to volumes actually consumed, promotional contracts for representative sectors, surcharges on peak or high season (seasonal rates), and establishing a quota volume per household depending on the number of registered inhabitants. This would require a modification of the increasing block rate structure in all municipalities to provide price signals to customers which serve as incentives for them to use water efficiently, encouraging them to modify their behavior in particular directions. Prior to this, cost analysis and rate-structuring have to take place in order to ensure full cost recovery. Additional price signals would include bonuses and awards for proven responsible uses and decreased consumption, consumption penalties on outdoor applications, equilibrium prices for different sectors, proper application of canon sanitation and impact of all services of the integrated water cycle, and an annual update of the rate structure. Notwithstanding these detailed proposals, recently only an increase of the sanitation charges has been applied in 2013 (Resolution number 5708 2013). The current scenario is that the public authorities are promoting the privatization of the public water supply in order to delay and translate its prices' increases to the concessionary companies' decisions (Morán 2013; Trillas 2013). Two more factors promoting the concession of water supply to private corporations should be considered: Firstly, the revenues realized by the public administrations for their sale would palliate their public deficit, which is embarrassing Spain as participant of the EU common currency zone. Secondly, international financial liquidity is looking for new products to be commoditized, for instance through the privatization of welfare state services such as the water supply, sewage treatment and sanitation, in addition to the public health services or education. In this way, the current crisis allows the artificial creation of value out of water, treating the resource as if it were a commodity or financial product. According to the Spanish Environment Secretary of State, the public administrations have made an expenditure of 2,400 million Euros to construct desalination plants (mainly with EU financial support), that are now producing water under the 10 % of their total capacity (Morán 2013). Full water cost recovery should include the amortization of this huge investment through its pricing. As this public authority pronouncement explains, the case of the Balearic Islands where the desalination infrastructure is overdimensioned and underused is not an exception in Spain.

A New Water Culture is needed as an alternative policy and management approach to the water supply needs (Aguilera Klink 2008; Esteban 2008; Esteban and Naredo 2004). Among other measures, the New Water Culture promotes that priority should be given to managing the demand instead of managing the supply. This option would include water saving measures and matching the water quality with the water needs, for instance through a better and sound use of treated waste water. A New Water Culture aims at pricing consumption according to its use, charging the more lavish and immoderate uses higher than those that are only devoted to fulfill basic needs. The current block rate system is giving no incentive to save water and is in favor of high end consumers (Garrido and Calatrava 2010), particularly those developing quality tourism activities (such as golf) and luxury residential development. However, there are several obstacles for adoption of a New Water Culture. One obstacle is water pricing policy. No public debate has been initiated about internalization of full costs though water prices increase; this seems to be due to its unpopularity and the government's subordination to electorate opinion. Substantial differences in prices exist between the 53 island municipalities, reflecting that current water rates are set on political grounds, with pricing policy frequently subject to political decisions and electoral debate, rather than the real costs (Candela et al. 2005). Water prices are set by each individual municipality. When the service is offered by a private or public company the final price is negotiated between the company and the city council and has to be authorized by a price commission, which regulates the prices for public utilities being under monopoly. A severe shortcoming of the system is that new consumer-price negotiations never consider a revision of the last negotiation agreement in order to study the previous predictions and control the profits (Candela et al. 2005). With this heterogeneous water pricing policy, it is difficult to make assumptions about the willingness of the different actors to change their behavior regarding water demand or to pay higher water prices. Another obstacle for the adoption of a New Water Culture is the dependency on tourism as major pillar of the economy and a questionable willingness to introduce higher water prices for the tourism industry. The fate of a previous policy instrument exposed a hostile political climate for taxing for environmental purposes. The Autonomous Community of the Balearic Islands introduced an eco-tax on tourist nights spent to promote environmental protection, with extra resources for environmental policy, building clearance and the conservation of natural and cultural value. However, water conservation was not at all an objective of the eco-tax. The eco-tax was in effect from March 2002 to March 2003 and was withdrawn by the newly elected Autonomous Community government, not least because it was highly unpopular with stakeholders and it had met fierce opposition from local hoteliers and international tour operators.

Another important aspect is infrastructure improvement, repairing network leakages as a priority and managing natural water resources efficiently instead of relying on the supply by desalination technology which relies on fossil fuel consumption. Last but not least, urban and demographic growth should comply with the territorial socio-environmental carrying capacity. Instead of attending the problem from a segmented focus (for instance, of water supply detached from regional planning), a New Water Culture advocates integrative perspectives, regulating future frameworks towards sustainability. To illustrate this last proposal and dealing with our case study, Majorca has nowadays already 23 golf courses; how many more could it hold? The Balearic Islands have a long political tradition of enacting planning constraints, from natural areas and seashore buffers set aside for protection, to limits to urban development and limits to the expansion of tourist bed stock. These constraints have been based, among others, in the argument of this Mediterranean Island's water scarcity. Although technological approaches and solutions allow us to overcome these limitations, perhaps its unsustainability in terms of the tradeoffs between energy consumption of desalination technology and its important role for water supply augmentation should make us reconsider their priority. So far, the public institutions and the taxpayers bear the cost of increasing the water supply. A broader discussion about water (demand) management and the necessary contributions by the agricultural, urban, tourist and industrial sectors is still in its initial stage.

Conclusions

Spain has very low water prices, with urban rates in the lower segment of the OECD countries. So much so that there is a political and technical consensus that the Spanish water prices should be higher in order to internalize the full social, environmental and economic costs of its supply. The current institutional scenario, which has been shown for Majorca, is that the public administrations deregulate the water supply sector and promote its privatization. If this trend continues and regulative policies are minimized, this could lead to further overexploitation of the underground water resources and increasing dependence on desalination technology driven by fossil fuel consumption. Because desalination technology is energy-intensive, it is considered a 'maladaptation strategy' to water shortages induced by climate change (Saurí et al. 2011).

Alternative policy approaches, such as those known under the denomination of the New Water Culture, offer more sound and sustainable water management practices. The adoption of a New Water Culture is confronted with the forces of institutional inertia in the political arena. The initial approval of the Hydrological Plan of 2001 was in 2011, and the memorandum and other annexes are from the years before. In fact, even the PHIB approved in 2013 has most of its material based on memories approved many years ago. Given these inertia, it seems very optimistic to expect that water-related policies are timely framed so that society could be induced to the make the needed changes. The evidence that the new legislation doesn't choose preventive measures is reason for skepticism if the objective of more sustainable freshwater management has arrived on the political agenda yet. This contribution has turned the spotlight on the water challenges of tomorrow—the shift in mind-set that is required to address these in Majorca is yet to come.

References

Aguiló E, Alegre J, Sard M (2005) The persistence of the sun and sand tourism model. Tour Manag 26(2):219–231

Aguilera KF (2008) La nueva economía del agua. Madrid: los Libros de la Catarata

Alcamo J, Moreno J, Nováky B, Bindi M, Corobov R, Devoy RJN, Giannakopoulos C, Martin E, Olesen J, Shividenko A (2007) Europe. In: Parry ML (ed) Impacts, adaptation and vulnerability, contribution of working group II to the fourth assessment report of the intergovernmental panel on climate change. Cambridge University Press, Cambridge, p 541–580

Buswell RJ (2011) Mallorca and tourism. History, economy and environment. Channel View Publications, Bristol

Cabrera E, Cabrera Rochera E, Cobacho R (2010) Water supply in urban areas. In: Garrido A, Llamas MR (eds) Water policy in Spain. CRC Press, Boca Raton, pp 77–84

Candela L, Bejarano C, Igel W von, Alonso A. (2005) Report on Majorca. In: Donta AA, Lange MA, Herrmann A (eds) Water on Mediterranean Islands: current conditions and prospects for sustainable management. ZUFO-Reports vol 5. Centre for Environmental Research University of Muenster, Muenster, pp 11–95

Candela L, Elorza FJ, Jiménez-Martínez J, von Igel W (2012) Global change and agricultural management options for groundwater sustainability. Comput Electron Agric 86:120–130. doi:10.1016/j.compag.2011.12.012

Conselleria de Medi Ambient (ed) (2002) Plan Hidrológico de las Islas Baleares. Palma de Majorca

Conselleria de Medi Ambient (ed) (2013) Plan Hidrológico de les Illes Balears. Memoria. Palma de Majorca, p 474

Correia FN (1999) Water resources in the Mediterranean Region. Water Intl 24(1):22–30. doi:10.1080/02508069908692130

Deyà Tortella B, Tirado D (2011) Hotel water consumption at a seasonal mass tourist destination. The case of the island of Majorca. J Environ Manag 92(10):2568–2579. doi: 10.1016/j.jenvman.2011.05.024

Dodds R, Kelman I (2008) How climate change is considered in sustainable tourism policies: a case of the Mediterranean Islands of Malta and Mallorca. Tour Rev Intl 12(14):57–70

Domene E, Saurí D, Pares M (2005) Urbanization and sustainable resource use: the case of garden watering in the metropolitan region of Barcelona. Urban Geogr 26(6):520–535

Domene E, Saurí D (2006) Urbanisation and water consumption: influencing factors in the Metropolitan Region of Barcelona. Urban Stud 43(9):1605–1623

Donta AA, Lange MA (2008) Water management on Mediterranean Islands: pressure, recommended policy and management options. In: Koundouri P (ed) Coping with water deficiency, vol 48. Springer, Netherlands, pp 11–44

Essex S, Kent M, Newnham R (2004) Tourism development in Majorca: is water supply a constraint? J Sustain Tour 12(1):4–28

Esteban A (2008) Herencias Y Problemas de La Política Hidráulica Española. Bakeaz/FNCA, Bilbao

Esteban A, Naredo JM (2004) Ideas y propuestas para una nueva política del agua en España. Bilbao:Bakeaz/FNCA; 2004

Eurostat (2009) MEDSTAT II: 'water and tourism' pilot study. Methodologies and working papers, Luxembourg

Falkenmark M, Lundqvist J, Widstrand C (1989) Macro-scale water scarcity requires micro-scale approaches. Nat Resour Forum 13(4):258–267. doi:10.1111/j.1477-8947.1989.tb00348.x

Garcia C, Servera J (2003) Impacts of tourism development on water demand and beach degradation on the island of Majorca (Spain). Geografiska Annaler, Series A 85(3/4):287–301

Garrido A, Calatrava J (2010) Trends in water pricing and markets. In: Garrido A, Llamas MR (eds) Water policy in Spain. CRC Press, Boca Raton, pp 131–144

Gikas P, Angelakis AN (2009) Water resources management in Crete and in the Aegean Islands, with emphasis on the utilization of non-conventional water sources. Desalination 248(1–3):1049–1064. doi:10.1016/j.desal.2008.10.021

Gössling S, Peeters P, Hall CM, Ceron J, Dubois G, Lehmann LV, Scott D (2012) Tourism and water use: supply, demand, and security. An international review. Tour Manag 33(1):1–15. doi:10.1016/j.tourman.2011.03.015

Hof A, Schmitt T (2011) Urban and tourist land use patterns and water consumption: evidence from Mallorca, Balearic islands. Land Use Policy 28(4):792–804. doi:10.1016/j.landusepol.2011.01.007

Iglesias A, Garrote L, Flores F, Moneo M (2007) Challenges to manage the risk of water scarcity and climate change in the Mediterranean. Water Resour Manag 21:775–788. doi:10.1007/s11269-006-9111-6

Kent M, Newnham R, Essex S (2002) Tourism and sustainable water supply in Majorca: a geographical analysis. Appl Geogr 22(4):351–374

Lazarova V, Levine B, Sack J, Cirelli G, Jeffrey P, Muntau H, Salgot M, Brissaud F (2001) Role of water reuse for enhancing integrated water management in Europe and Mediterranean countries. Water Sci Technol 43(10):25–33

Maestu J, Gómez C (2010) Water uses in transition. In: Garrido A, Llamas MR (eds) Water policy in Spain. CRC Press, Boca Raton, pp 39–48

Mantecón A (2010) Tourist modernisation and social legitimation in Spain. Intl J Tour Res 12(5):617–626. doi:10.1002/jtr.784

March H, Saurí D (2010) The suburbanization of water scarcity in the Barcelona Metropolitan Region: sociodemographic and urban changes influencing domestic water consumption. Prof Geogr 62(1):32–45

Masjuan E, March H, Domene E, Saurí D (2008) Conflicts and struggles over urban water cycles: the case of barcelona 1880–2004. Tijdschrift voor Economische en Sociale Geografie 99(4):426–439

Milano M, Ruelland D, Fernandez S, Dezetter A, Fabre J, Servat E, Fritsch J, Ardoin-Bardin S, Thivet G (2013) Current state of Mediterranean water resources and future trends under climatic and anthropogenic changes. Hydrol Sci J 58(3):498–518. doi:10.1080/02626667.2013.774458

Morán C (2013) La reforma del agua dará más garantías al inversor privado. Entrevista a Federico Ramos, Secretario de Estado de Medio Ambiente". Expansión, p 17

O'Reilly K (2007) Emerging tourism futures: residential tourism and its implications. In: Geoffrey C, Sibley R (eds) Going abroad. Travel, tourism, and migration: cross-cultural perspectives on mobility. Cambridge Scholars Pub., Newcastle, pp 144–157

Palutikof J, Holt T (2004) Climate change and the occurrence of extremes: some implications for the Mediterranean Basin. In: Marquina Barrio A (ed) Environmental challenges in the Mediterranean 2000–2050. Kluwer Academic Publishers, Dordrecht, pp 61–73

Parés M, March H, Saurí D (2013) Atlantic Gardens in Mediterranean climates: understanding the production of suburban natures in Barcelona. Int J Urban Reg Res 37(1):328–347. doi:10.1111/j.1468-2427.2012.01118.x

Resolution number 5708 (2013) Resolución del vicepresidente económico, de Promoción Empresarial y de Empleo de 18 de marzo de 2013 por la que se publica la actualización de las bases, los tipos de gravamen y las cuotas tributarias de los tributos propios de la Comunidad Autónoma de las Illes Balears para el año 2013. Official Bulletin of the Balearic Islands 43:14448–14516. http://www.atib.es/normas/Resoluci%C3%B3%20actualitzaci%C3%B3%202013_s.pdf. Acessed 7 Jan 2014

Rodriguez Diaz JA, Knox JW, Weatherhead EK (2007) Competing demands for irrigation water: golf and agriculture in Spain. Irrig Drain 56(5):541–549. doi:10.1002/ird.317

Romero Renau Ld (2006) Privatització de la gestió de l'aigua i govern del territori a l'arc mediterrani espanyol. Documents d'anàlisi geogràfica (Doc. Anàl. Geogr.) 48:35–59

Rullan O (2011) La regulación del crecimiento urbanístico en el litoral mediterráneo español. Ciudad y Territorio: Estudios Territoriales 168:279–297

Saurí D, Olcina J, March H, Martin-Vide J, Vera F, Padilla E, Serra-Llobet A (2011) ESPON climate: climate change and territorial effects on regions and local economies. Final Report Annex 4: Case Study Mediterranean Coast of Spain

Saurí D, Olcina J, Rico A (2007) The March towards privatisation? Urban water supply and sanitation in Spain. J Compar Social Welf 23(2):131–139. doi:10.1080/17486830701494616

Schmitt T, Blázquez M (2003) Der dritte Tourismusboom auf Mallorca (1991–2000). Zukunftsweisender Trend oder überschrittener Zenit? Tourismus J 7(4):505–522

Tamoh K, von Igel W, Soler M, Candela L (2008) Analysis of water use patterns and conflicts in the Sa Pobla Plain and Alcudia Bay (Majorca, Spain). Options Mediterraneennes 83:131–142

Trillas A (2013) Rescate a cambio de gestión privada. Alternativas económicas, 5 (July/August), pp 8–11

Vidal M, Domene E, Saurí D (2010) Changing geographies of water-related consumption: residential swimming pools in suburban Barcelona. Area 43(1):67–75. doi:10.1111/j.1475-4762.2010.00961.x

Chapter 10
Unconventional Water Resources of Agricultural Origin and Their Re-utilization Potential for Development of Desert Land Aquaculture in the Aral Sea Basin

B. K. Karimov, M. Matthies and B. G. Kamilov

Abstract The average long-term annual volume of unconventional mineralized water resources of agricultural origin (MWRA) in Aral Sea basin (ASB), consisting of drainage (about 95 %) and wastewater from industry and municipal users (about 5 %) in 2000–2009 was around 30 km^3 with 24.3 km^3 generated on Uzbek territory. Only 3 % (0.7 km^3) of MWRA are used for irrigation again and for fish-farming. During the last decade, a group of German and Uzbek scientists has jointly investigated the suitability of the hydrological, hydrochemical and hydrobiological regimes of MWRA and their ecological sustainability for intensive desert/arid land aquaculture (DALA) development. A SWOT analysis of the strengths, weaknesses, opportunities of re-using MWRA for DALA, and threats of vulnerability has revealed suitability (both in terms of water quantity and quality) of MWRA for the development of intensive aquaculture-agriculture systems. The latter implies better management practices including combined production of fishes and other aquatic organisms, diversification of cultured species and usage of halophytes, where the water enriched with biogenous will be used for plant growth. At least 10 km^3 MWRA are generated and flowing annually into the natural depressions plus more than 15 large brackish water lakes (3–12 g salt/L) with 9,000 km^2 of total water surface area fed by drainage waters. This excess of unconventional water in the region will allow production of more than 300,000 t/y fish additionally in Uzbekistan, which will generate substantial employment and income for people in rural areas.Aral Sea basin · Mineralized waters · Desert/arid land aquaculture · SWOT analysis

---

B. K. Karimov (✉) · B. G. Kamilov
Institute of Gene Pool of Plants and Animals World of Uzbekistan Academy of Sciences, 232 Bagi-Shamal Street, Tashkent 100053, Uzbekistan
e-mail: Karimov@sarkor.uz

M. Matthies
Institute of Environmental Systems Research, University of Osnabrück, Barbarastrasse 12, 49069 Osnabrück, Germany

A. Bhaduri et al. (eds.), *The Global Water System in the Anthropocene*,
Springer Water, DOI: 10.1007/978-3-319-07548-8_10,

Introduction

Current statistics provided by the United Nations Development Program/United Nations Office to combat Desertification (UNDP/UNSO) indicate that about 13 % of the total world population is living in arid zones (Smith et al. 2008).

Former Central Asian Soviet countries are experiencing an increasing imbalance between availability and demand for land and water resources at the local and national levels. The size of areas reaching the limits of their production capacity is fast increasing, above all due to acute water scarcity growing as well as to salinization as a consequence of in-appropriate water and land management. While living standards, accessibility, and quality of health and educational services are steadily growing on average in the region because of the technological progress, transition and integration into the world market economy, the environmental situation demonstrates the reverse trend. The natural environment, especially water-related ecosystems, are worsening gradually, which is primarily related to local, regional and global environmental problems such as increasing water deficit due to irrational irrigation, climate change, air, water, and soil pollution deforestation, soil erosion and salinization etc. Today many regional rivers do not reach their natural end points and wetlands are disappearing.

The development of agriculture, including aquaculture and capture fisheries in arid lands in the Aral Sea basin (ASB), has one very common problem, namely—deficit of river freshwater because of irrational and inefficient use (huge percolation, seepage and evaporation losses) of water for irrigation. The distribution of water resources is extremely uneven and determined by different surface flow generation conditions, which are favorable in the mountainous areas and unfavorable in vast plain areas covered by deserts and semi-deserts. In Uzbekistan, a deeply landlocked country, almost 80 % of the surface is desert, dominated by the Qizilqum Desert of the north-central part of the country. The mountains of the far southeast and far northeast, which are foothills of the Tian-Shan Range, reach 4,500 m in elevation. Some 10.5 % of Uzbekistan's land, most of it in the Fergana Valley, is classified as arable and 0.8 % is planted to permanent crops. About 0.4 % is forested.

Much of the agriculture of this arid region developed in response to the agricultural plans of the former Soviet Union. For instance, the irrigated area in the Aral Sea basin was 2 million ha in 1900, 3.2 in 1913, 4.3 in 1933, and about 8 million in the 1990s (Fig. 10.1). In this period, the irrigation activities in ASB were directed mainly to the growing of cotton, whereas the production of other agricultural goods, especially the production of meat and fish has been widely neglected. For example, about 73.4 % of the irrigated land at the beginning of 1990s was set aside for cotton production, which was unprecedented in the world's agricultural practice. Today total area of cotton fields is decreased considerably, i.e. 46 % of total 4.2 million ha irrigated land in Uzbekistan (Ruecker et al. 2007). Water resources were overcommitted to agriculture and were transferred unsustainably from all river systems, which caused reduction of river flow into the Aral Sea of 5 km^3 in opposite to about 56 km^3 in early 1960s. This aggravated the

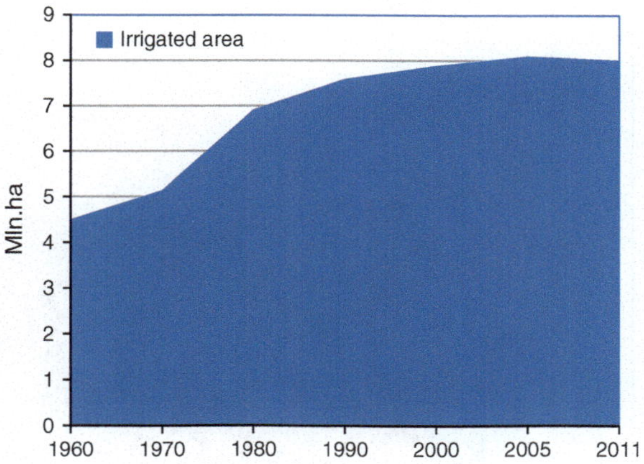

Fig. 10.1 The dynamics of irrigated land area in ASB in 1960–2011 (*data source* http://www.cawater-info.net/analysis/water/asb_dynamics_ru.pdf). Accessed Dec 16 2013

ceasing of the fishery potential of the Aral Sea itself because of desiccation and disappearance of wetlands and migration ways for spawning, as well as water quality problems in the Aral Sea, e.g. extremely high salt content (Kamilov et al. 2004). This has led to a situation, where the protein supply for the population could only be met through the import of meat and fish (Karimov 2003; Karimov et al. 2002, 2004, 2005).

The specific objective of this paper is to develop knowledge and understanding, necessary to assess the status of unconventional water resources of agricultural origin and their re-utilization potential for food production by developing desert and arid land aquaculture in Aral Sea basin. Furthermore, attention of decision makers, politicians and agriculturists, international river basin organizations and relevant non-governmental organizations should be drawn on the measures to be taken to re-integrate unconventional water resources in strategies following the concept of Integrated Water Resource Management (IWRM). The proposed ideas are an innovative approach, at least for countries situated within the Aral Sea basin for re-utilization of return waters of anthropogenic origin based on empirical ecological, hydrobiological and ecotoxicological investigations. Suggested ideas contribute to better management practices of creating substantial employment and new sources of income generation in rural areas.

Study Site and Data Compilation

The Aral Sea Basin

The Aral Sea basin is situated within Central Asia (CA) and covers an area of 2.2 million km^2 and is home to around 50 million people. It comprises the drainage area of two major rivers, the Amudarya and the Syrdarya and the Aral Sea itself

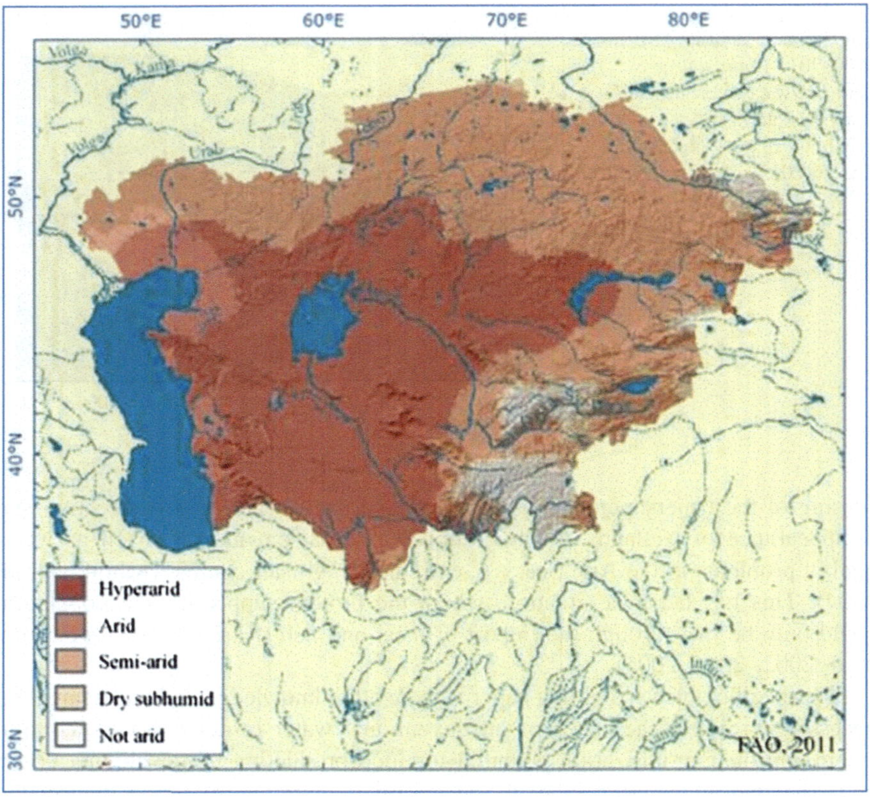

Fig. 10.2 Aridity in ASB (FAO 2011)

(Fig. 10.2). Around 2,034 million km^2 are under arid regime (FAO 2011a). The average annual precipitation is about 100–200 mm in the plains and 30–50 % of the total rain is in spring, 25–40 % in winter, 10–20 % in autumn and 1–6 % in summer (UNDP 2008).

Water bodies in the basin of the Aral Sea can be grouped as follows: natural water bodies; rivers, streams and lakes; primary artificial freshwater water bodies; irrigation canals, reservoirs and ponds; secondary artificial brackish water bodies; drainage canals, lakes for storage of the residual non-conventional mineralized water resources of agricultural origin (Karimov 1994).

In accordance with the international agreements and depending on annual water discharge, Uzbekistan is allocated from 45 km^3 (2001) to 62 km^3 2005) of river water per year. General water-consumption in Uzbekistan during mid 1990s was stabilized at the level of about 62–65 km^3 per year while the total freshwater resources of the ASB are about 115.6 km^3 per year (United Nations 2010). The total amount of the surface runoff from internal rivers formed within the territory of Uzbekistan is only 11.5 km^3 (UNDP 2007). Most of agricultural land in the

Table 10.1 Use of water resources in Uzbekistan (average data for 2002–2004 and 2007–2011 (National report 2005, 2013)

Used water resources	2002–2004, km^3/%	2007–2011, km^3/%
Total	55.1/100	49.8/100
Irrigated agriculture	49.7/90.2	45.2/90.8
Other Users		
Municipal and drinking water supply	3.4/6.2	2.4/4,8
Industry	1.2/2.2	1.6/3.2
Fisheries	0.8/1.4	0.6/1.2
Other users total	5.4/9.8	4.6/9.2

Republic of Uzbekistan belongs to arid zone creating thus the need for irrigation and making agriculture the main water consumer in the region. Therefore, this paper will focus mainly on this country.

During 2002–2004, about 90 % (49.7 km^3) water is used in agriculture, 2.2 % (1.2 km^3) in industry, 1.4 % (0.8 km^3) in fisheries and 6.2 % (3.4 km^3) in municipal and drinking water supply. During 2007–2011, general water consumption decreased (average 49.8 km^3), however, about 90 % of water still was used for irrigated agriculture (Table 10.1).

Unconventional Mineralized Water Resources of Agricultural Origin (MWRA)

The average long-term annual volume of drainage waters in ASB, consisting of drainage (about 95 %) and wastewater from industry and municipal users (about 5 %) has varied in 1990–1999 between 28.0 and 33.5 km^3. In 2000–2009 it was approximately 30 km^3 of which 24.3 km^3 were generated on Uzbek territory. About 13.5–15.5 km^3 per year is accumulated in the Syrdarya River basin and about 16–19 km^3 in the Amudarya River basin (Yakubov et al. 2011, http://www.cawater-info.net/index_e.htm). However, only a very small volume of these so-called unconventional water resources are re-used. During 2002–2011 about 3–7 % (0.7–1.7 km^3) of these unconventional waters were re-used for irrigation, especially during drought periods, e.g. in 2000–2002 (National Report 2005, 2013).

MWRA with mineralization above 2.5 g/L and elevated chloride concentration has very limited suitability for crop irrigation, especially in middle and lower reaches (Yakubov et al. 2011). More than 55 % returns to rivers and about 30 % end up in natural depressions, from which the water evaporates (http://www.cawater-info.net/index_e.htm). Taking into account the current situation in the water sector of ASB it is unlikely that in the near future MWRA will be desalinated for use by other sectors. The example of Israel demonstrates the potential for re-use of MWRA in the domestic and industrial sectors (Hulata and Simon 2011).

There is about 100,000 km of drainage canals (collectors) in Uzbekistan. The total length of the large main collectors is about 3,560 km (National report 2013).

Fig. 10.3 Transboundary Main Collector "Ozerniy" in Amudarya River lower reach, flowing through Qarakum sand desert to Sarikamish Lake in Turkmenistan (photo of B. Karimov)

Table 10.2 Hydrological parameters of some main collectors in ASB

Collector	Length, km	Discharge (max.), m^3/s	Annual discharge sum (max.), km^3
Central-Golodnostepskiy	86	90	2.1
Ozerniy	190	151	1.5
Devankul	100	63	0.2
Yujniy	96.2	100	0.7
Severniy-2	154	19.5	0.3
Parsankul	62	50	1.2
Parallel	55	40	
Achikul		49.5	1.6
KS -1		6.80	0.3

A number of them exceeds 50 km of length and reaches discharges of 40–100 m^3/s (Fig. 10.3). The annual discharge sum of some large collectors is comparable with that of rivers, e.g. Ozerniy 1.5 km^3 and Central-Golodnostepskiy 2.1 km^3 (Table 10.2).

About 50 water reservoirs with total surface area of 1,760 km^2 and total volume of 21 km^3 were build up and exploited at present in Uzbekistan. The majority of reservoirs in the region were constructed for either irrigation supply and/or energy needs, and the demands of these sectors dictate when and how much water is released. While reservoirs such as Todakul and Talimarjan in Uzbekistan, for example, offer significant potential for culture-based fisheries due to the relatively high volumes (0.325 and 0.125 km^3, respectively) of remaining water after exhaustive water abstraction for irrigation purposes (the 'dead' level), others e.g. such as Kattakurgan, Uchkurgan and Kuymazar have much lower dead levels (0.010, 0.16 and 0.047 km^3 respectively). It is not just the level of residual water, which impacts fisheries in the regional water bodies, but also water release during the summer (irrigation) or winter (hydro-power) can sweep native and restocked fish, released larvae and fingerlings, and food sources downstream.

Results

SWOT Analyses

Environmentalists, decision makers, scientists and farmers are concerned about human health aspects of fish as a food. During the last decade, a group of German and Uzbek scientists has jointly investigated the ecohydrological and ichthyological status within the framework of German Environmental Foundation (DBU) funded Project: "Sustainable Aquaculture in Recirculating Systems; Feasibility Study for the Catchment Area of the Aral Sea" (Wecker et al. 2007). The authors have revealed a great potential of MWRA for the development of intensive sustainable desert/arid land aquaculture in ASB. The investigations focused on the assessment the hydrological, hydrochemical and hydrobiological suitability of MWRA and the ecological sustainability of aquaculture development.

The subsequent SWOT analysis, conducted by authors using FAO methodology (FAO 2011b) of strengths, weaknesses, opportunities re-using MWRA for DALA (Desert/Arid Land Aquaculture), and threats of vulnerability provide a starting-point for developing measures on MWRA re-use for water managers and farmers, which are currently not practicing such opportunity. DALA based on MWRA re-use is not widespread in the region so that the results of the pilot adaptation projects can form a basis for popularizing these opportunity.

Strengths

1. Large volumes of MWRA water—about 20–25 km^3 in Uzbekistan.
2. MWRA are available all-round the year in main collectors and therefore less vulnerable to water scarcity.
3. Vast land resources around collector drainage canals and lakes for landless farmers.
4. Sufficient human resources.
5. State support programs to fisheries development in the country.

6. Availability of consumer markets.
7. No or low conflict potential with other common users of water body.

Weaknesses

1. Poorly developed infrastructure (roads, electricity, etc.) in the vicinity of MWRA sources.
2. Absence of technologies adapted to local DALA.
3. Absence of DALA pilot farm for training of interested aqua-farmers.
4. Education and research institutes poorly funded and DALA development is not well-organized.
5. Lack of availability of and access to high quality fish feeds.

Opportunities

1. Re-use of land and water for production of large quantities of fish, shrimp and other aquatic biota.
2. Substantial employment and additional income generation for rural population.
3. Improving food safety and population health situation in the country.
4. Biodiversity conservation improvement via declining anthropogenic impact (commercial fishery and hunting, unregistered, unregulated, illegal fish capture, poaching, etc.) to wild fish populations.

Threats

1. Diminishing water flow during years with low water or droughts.
2. Low water quality due to enhanced mineralization and use of prohibited pesticides.
3. Careless stocking could adversely impact the aquatic biodiversity of water bodies.

Concerning threat 1, Fig. 10.4 clearly shows the availability of sufficient quantities of water in main collectors all-round the year. Concerning threat 2, Fig. 10.5 shows that the lower and upper values of total water mineralization ranges from 2 to 6 g/L, and annual average values within 2–5 g/L. This is lower than the "critical salinity range" for the fertilization and survival of the eggs of most freshwater fish (Khlebovich 1974). Earlier we have demonstrated (Karimov and Keyser 1998) that Aral Sea water is less toxic than standard sea water due to specific salt composition. In salt composition of MWRA, sulfates are the leading anions followed by chlorides in percentage proportion of 48–15 %, which is considerably different than that in sea water (7.7–55.3 %). Sodium and magnesium are the most abundant cations followed by potassium (Fig. 10.6). Furthermore, MWRA contain more calcium and magnesium than sea water (6.1–4.2 vs. 1.2–3.7 %), which reduce the negative effects of more toxic monovalent ions such as sodium and chloride, and potassium and calcium act here as antagonists of sodium (Kanygina 1957; Doudoroff 1957). Indeed, drainage water from ASB region was less toxic for fertilization and embryonic development of carp (Cyprinus carpio L.) than sea water (Karimov and Keyser 1998). These findings

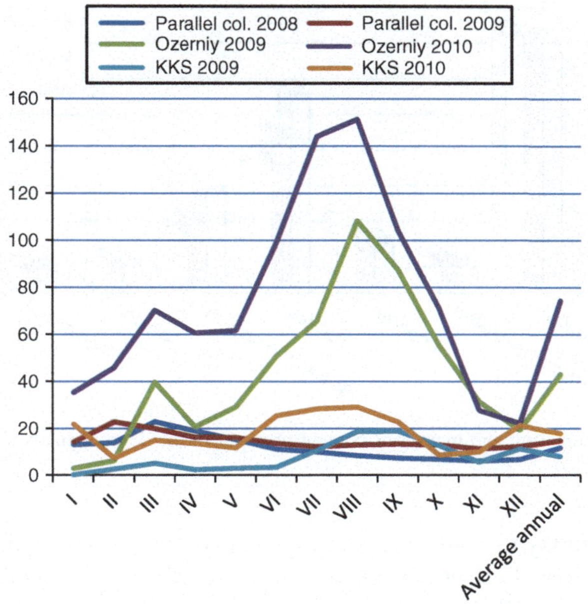

Fig. 10.4 The dynamics of discharge of some large main collectors in Amudarya River basin, m^3/s

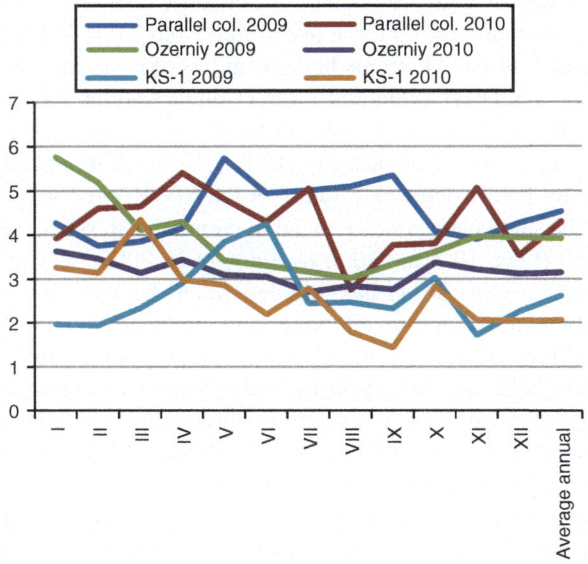

Fig. 10.5 Annual dynamics of mineralization in some main collectors in Amudarya River basin in 2009–2010, g/L

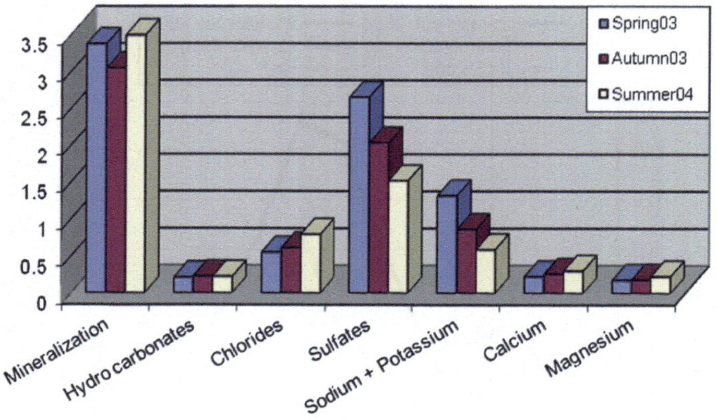

Fig. 10.6 Mineralization and main ions content of water in the delta of main collector Central-Golodnostepskiy, g/L

imply the possibility to conduct fertilization and swelling phases of fish eggs under artificial conditions. Further development of embryos and raising of food fish can then take place in MWRA at about 6 g/L or even higher mineralization directly depending on the age of fish.

During the Soviet era and in the early 1990s in the ASB high pollution levels of DDT and its metabolites DDE and DDD as well as Lindane (γ-HCH) and other various, ecologically dangerous pesticides have been detected almost in all aquatic ecosystems and their compartments (sediments, plants, fishes) (Karimov and Razakov 1990; Karimov et al. 1995). This was mainly due to the rapid increase of irrigated area used mainly for cotton monoculture from 2.0 to 7.2 million hectares between 1925 and 1980s and extreme high quantities of pesticides used per hectare of irrigated land—about 25 kg/ha and more (United Nations 2010). Recent eco-toxicological studies in 2002–2006 and later have revealed that the pollution levels of former commonly used pesticides, e.g. DDT, γ-HCH, DDE and DDD have now fallen significantly below detectable limits in the MWRA fed water bodies from 1993 to 2004. The same pattern was noticed in common fish species like carp, pike perch, roach, etc. (Figs. 10.7 and 10.8, Karimov et al. 2005; Wecker et al. 2007).

In 2006–2010 Crootof (2011) has sampled water, lake surface sediments and cores, fish, zooplankton, macroinvertebrates and algae, as well as deployed semi-permeable membrane devices (SPMDs) as passive pesticide sampler in Khorezm lakes, Northern Uzbekistan. Her results have also revealed that pesticide and metal contamination in all four lakes was relatively low. While γ-HCH concentrations were below water quality guidelines for aquaculture, DDT was above recommended water quality guidelines for aquaculture, but these pesticides did not appear to be accumulating or harming fish species in the lakes. Sediment samples contained low levels of DDT, γ-HCH and metals, which were all below consensus-based probable effect concentrations.

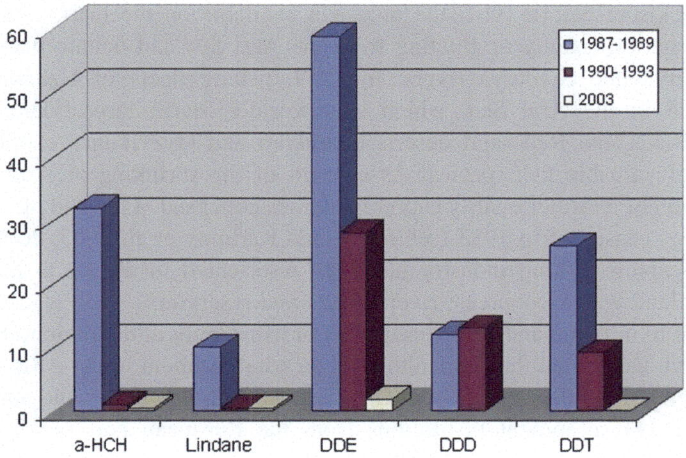

Fig. 10.7 Average bioaccumulation levels of chlorinated pesticides (DDT, DDD and DDE and sum of α- and γ-HCH) in various fish species (bream, carp, crucian carp, asp, roach, shemaya, pike, pike-perch, wels and snakehead) in lake Tuzkan, μg/kg wet weight

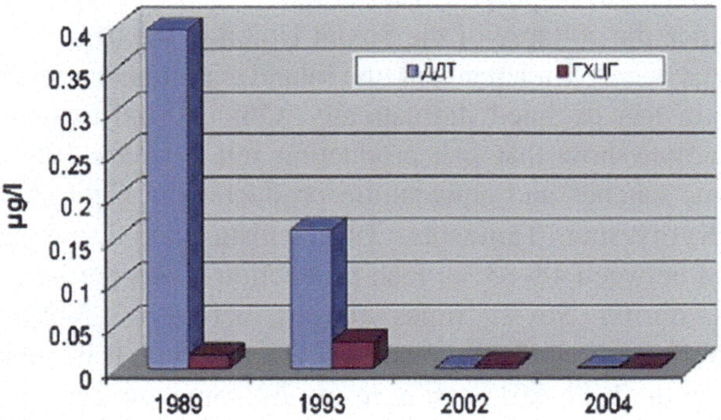

Fig. 10.8 Long-term dynamics of DDT, DDD and DDE and sum of α- and γ-HCH in water of Lake Muynak

Based on these studies it can be concluded that present hydrochemical and ecotoxicological indicators of MWRA are suitable to support healthy fish populations, which are safe for human consumption.

Use of MWRA for Fisheries

Desert/arid land aquaculture (DALA) is defined as "aquaculture activities in desert and/or arid lands which are characterized by low rain precipitation (<250 mm/year) and high evaporation rates" (FAO 2011a).

In Uzbekistan before 1961, the only fish available on the market came from capture fisheries, mainly originating from the Aral Sea and deltaic water bodies of Amudarya and Syrdarya rivers. In 1959, fishing fleets of Uzbekistan and Kazakhstan in the Aral Sea, which was world's fourth largest lake situated between Black and Red sand deserts (Qarakum and Qyzylkum), caught nearly 50,000 t of valuable fish species. As a result of the shrinking of Aral Sea and deterioration of its water quality capture volumes collapsed at the end of 1970s and sea fisheries ceased of in 1983 (see for details Karimov et al. 2005, 2009). Since then, the capture fishing industry has been re-oriented to all the available and suitable inland water resources: rivers, lakes and reservoirs.

In order to mitigate and compensate loss of fish supply in the region, the Soviet government had started the aquaculture sector establishment in the ASB countries since the end of 1960s, especially in Uzbekistan. Central fisheries and aquaculture research and development institutions from the European part of USSR were engaged in this process. Usually, unconventional land resources (with high soil salinity or sand deserts on the peripheries of cultivated lands) were used for construction of extensive and slightly semi intensive pond fish farms. As a result of these measures total fish production in the Republic of Uzbekistan in 1980s reached 25,000 t/yr. Thus, the loss of capture fishery volumes was compensated, mainly by development of pond fish farming.

However, since the collapse of the Soviet Union in the beginning of the 1990s, when these states became independent, the inland capture fisheries and aquaculture production again has declined dramatically. Official statistical figures for total fisheries production show that fish production fell between 1989 and 2010 significantly. Inland catches and aquaculture production in Central Asian republics (Kazakhstan, Kyrgyzstan, Tajikistan, Turkmenistan and Uzbekistan) decreased over that period between 48–85 %. Fish production levels per ha of pond systems in Uzbekistan during Soviet times ranged between 1.5–2 t/(ha·yr), which decreased until the early 2000s to about 0.6–0.7 t/(ha·yr). It increased again up to 0.8–1.2 t/(ha·yr) in 2011–2012 as a result of some government measures to develop this sector and improvement of fisheries statistics (Table 10.3).

Despite the fact that aquaculture is a water user rather than a consumer, today it consumes around 60 % of the 0.8 km^3 of annual water intake because large part of water will be lost through evaporation and percolation. As a result of this increase in capacity, the total water demands of the fishing industry will rise sharply and by 2025 will reach 2.4 km^3/yr (United Nations 2010).

In some large fish farms MWRA is already used for fish-culture by playing role of additional water source of fish ponds (Fig. 10.9). However, in most cases this requires to install heavy duty pumping aggregates and electricity, which can be afforded only by large fish farming enterprises. Moreover, the volume of MWRA use in these cases is unregistered and incurred losses in fish productivity are unknown.

Unfortunately, until today there is no attempt to develop aquaculture integrated with agriculture systems in ASB. Wecker et al. (2007) have for the first time analyzed the feasibility of this system for the ASB and suggested to introduce it in combination with recirculating aquaculture systems. These experiences can be

Table 10.3 Fish production in the Central Asian republics, t/yr (*Source* FAO and authors)

	1989	2008	2010	2012	2010 production (as a percentage of 1989 output)
Kazakhstan	89,508	55,902	46,827	40,000	52.3
Kyrgyzstan	1,447	100	390	297	27.0
Tajikistan	3,547	172	544		15.3
Turkmenistan	52,974	15,016	15,000		28.3
Uzbekistan	25,526	7,900	10,700	25,900	41.9
Total	173,000	58,130	69,634		40.1

Fig. 10.9 Collector Achisay is the main water source of Damachi fish farm and collects drainage waters from irrigated fields, which are flowing alongside of fattening ponds, Uzbekistan (Photo of Karimov)

implemented in all desert lands with sufficient MWRA water and infrastructure. However, there is a need to conduct further specific research to find-out salt-tolerant crops suitable for cultivation in ASB as well as to evaluate and transfer world experience in the field of integrated agriculture systems.

Worldwide intensive aquaculture farming systems have been a successful economic activity and still continue to expand. Various national polices and concepts have been implemented in many countries for the appropriate sustainable development of fish farming, although enforcement is still unsatisfactory in several parts of the world. However, in Uzbekistan and other countries situated in ASB still there are no officially recognized and accepted concepts (Karimov et al. 2009). We have revealed that there are different aquaculture development concepts feasible for Uzbekistan and other arid ASB. The most promising of them are the following:

- Cage culture systems in large collectors and lakes;
- Flow-through farms integrated with agriculture;
- Integrated/poly-culture pond farms;
- Fisheries enhancement.

The future development of these offered concepts desires an evaluation in relation to the aquaculture situation and fish market.

Discussion

MWRA and peripheral lakes fed by them being linked to aquifers are the main reserves for aquatic flora and fauna biodiversity conservation and are sources of valuable yields of fish, substantial employment and income generation for fishermen. In Uzbekistan almost 100 % of capture fishery based on fish yields from artificial brackish water lakes are fed by MWRA. The regulation of hydrological and water quality regime in favor of fisheries enhancement may result in sharp increase of economic indicators of MWRA re-use, which is further supported by restocking, introduction and acclimatization of new species and rehabilitation of extinct species as well as biosaline agriculture development (salt tolerant crops cultivation and agroforestry).

In the world's desert and arid lands despite of acute water shortage there are a number of successful examples of DALA development. Accessible, low-cost, subsurface, brackish geothermal water found in the Negev Desert in Israel with its moderate salinity (3–7 g/L), mineral composition, constant warm temperature (39–41 °C), purity and availability is highly suitable for aquaculture. This water provides opportunity to produce such edible fish species, as for example barramundi (Lates calcalifer), red drum (Sciaenops ocellatus), European seabass (Dicentrarchus labrax), North African Catfish (Clarias Gariepinus), Nile tilapia (Oreochromis niloticus) and white leg shrimp (Litopenaeus vannamei). In particular, the catfish density can reach 125 kg/m^3 and shrimp density 4 kg/m^3 (Appelbaum 2011).

Sadek (2011) has reported that today Egyptian desert aquaculture comprises more than 100 intensive tilapia rural farms and 20 commercial aquaculture farms with combined water surface area of about 893 ha and annual total fish production of 13,000 t. The reared edible fish species include those in Israel plus common carp (Cyprinus carpio), silver carp (Hypophthalmichthys molitrix), grass carp (Ctenopharyngodon idellus), etc. Production of tilapia (in densities of 20–30 kg/m^3 to market size of 250–400 g in 6–8 months) is possible due to suitable warm climate. The water source comes from underground water reserves and/or agricultural drainage. The latter has salinity ranging from 0.5 to 26 g/L and temperature from 22 to 26 °C.

Development of aquaculture integrated with agriculture systems in desert and arid lands reusing MWRA is also a very promising culture practice. This system uses water twice: first in aquaculture production units (basins, tanks and earthen ponds) and subsequently to produce crops (mostly vegetables and melons) in irrigated agriculture. By this way water enriched with organic wastes is re-used as a fertilizer for salt-tolerant crops. Aquaponic system, which combines intensive aquaculture with hydroponics, may be considered as a highly developed version of this system. It has become common to use irrigation reservoirs for fish culture in integrated farming systems in Israel (Hulata and Simon 2011). Aquaculture integrated with agriculture systems is since 2000s the most common farming system in Egypt and large numbers of desert land owners have established fish rearing facilities (Sadek 2011).

The expected increases in population in the ASB region over the next 10 years require more food, much of which is to come from intensification of agriculture which requires the rearranging of cropping patterns for the benefit of food crops. The demand on fish products will be also greater than at present. Governments have the task to provide an enabling environment for enhancing fish production in support of meeting the animal protein requirements of the populations' diet by ensuring sufficient water supplies for aquaculture, optimization of fish yield from aquafarms and development of intensive technologies including the diversification of types and objects of aquaculture.

It is unlikely that the present wasteful irrigation technologies in ASB region will essentially be improved in the nearest future. Therefore the generation of large volumes of MWRA is unavoidable, which requires the development of concepts of their utilization in national economies. Some authors even suggest increasing utilization level of MWRA in ASB up to 50–60 % from 15–17 % at present (Yakubov et al. 2011). Therefore, adaptation of agriculture to re-use MWRA in fisheries sector (aquaculture integrated with agriculture systems) to provide food-fish and salt-tolerant crop production for population provides unique solution for this task.

Conclusions

The most important threat to aquatic food production in Central Asian Republics is the lack of or low availability of river water, especially in middle and lower reaches of rivers. Optimization of fish yield from aquafarms and development of intensive technologies in these areas requires sufficient water supplies to ensure substantial employment and income generation for people in rural communities.

Our ecohydrological investigations have revealed suitability (both quantitative and qualitative) of MWRA for the development of intensive aquaculture-agriculture systems. The latter implies better management practices including the combined production of fish and other aquatic organisms and salt-tolerant crops, diversification of cultured species and halophytes, where the water enriched with biogenous fertilizers will be used for plant growth. There is an excess of unconventional water in the region. At least 10 km^3, which is approximately 50 %

from all generated MWRA, are flowing annually into the natural depressions. Additionally, more than 15 large brackish water residual lakes (3–12 g/L) with 9,000 km^3 of total water surface area are fed by MWRA waters in Uzbekistan. The re-use of MWRA by developing intensive fish farming in flow-through tanks and cage culture in lakes will allow increasing food fish production in Uzbekistan up to more than 300,000 t/yr in the nearest future.

Transfer of world experience of accessible, low-cost DALA systems of rearing edible fish species using brackish waters is most prospective promising way forward. These fish species could be indigenous: common carp (Cyprinus carpio), introduced: silver carp (Hypophthalmichthys molitrix) and grass carp (Ctenopharyngodon idellus), or exotic: North African Catfish (Clarias Gariepinus), Nile tilapia (Oreochromis niloticus) white leg shrimp (Litopenaeus vannamei).

Acknowledgements The present study was supported by grant of the Deutsche Bundesstiftung Umwelt (DBU, 2006–2007) and the State Grant of the Republic of Uzbekistan FA-A6-T233 (2012–2014).

References

Appelbaum S (2011) Aquaculture experiences in the Negev Desert in Israel. In: Crespi V, Lovatelli A (eds) Aquaculture in desert and arid lands: developmental constraints and opportunities. FAO technical workshop, 6–8 July 2010, Hermosillo. FAO fisheries and aquaculture proceedings, No. 20. Rome, p 113–118

Crootof AB (2011) Assessing water resources in Khorezm, Uzbekistan for the development of aquaculture. A Thesis submitted in partial fulfillment of the requirements for the degree of Master of Science in Hydrology. University of Nevada, Reno, 114 p

Doudoroff P (1957) Water quality requirements of fishes and effects of toxic substances. In: Brown ME (ed) The physiology of fishes, vol 2. Academic Press, New York, pp 403–430

FAO (2011a) Aquaculture in desert and arid lands: developmental constraints and opportunities. In: Crespi V, Lovatelli A (eds) FAO technical workshop, 6–8 July 2010, Hermosillo, Mexico. FAO fisheries and aquaculture proceedings, vol 20. Rome, 202 p

FAO (2011b) EAF planning and implementation tools. SWOT (strength, weaknesses, opportunities and threat) ANALYSIS. EAF Tool fact sheets (online). Rome. Updated 18 Oct 2011. http://www.fao.org/fishery/eaf-net/eaftool/eaf_tool_45/en. Accessed 12 Dec 2013

Hulata G, Simon Y (2011) An overview on desert and aquaculture development in Israel. In: Crespi V, Lovatelli A (eds) Aquaculture in desert and arid lands: developmental constraints and opportunities. FAO technical workshop, 6–8 July 2010, Hermosillo. FAO fisheries and aquaculture proceedings, No. 20. Rome, p 85–112

Kamilov B, Karimov B, Keyser D (2004) The modern state of fisheries in the Republic of Uzbekistan and its perspectives—World. Aquac Mag 35(1):8–14

Kanygina AV (1957) Effect of industrial sewage waters solvable mineral salts on the water organisms. Nauka, Moscow, 126 p (in Russian)

Karimov BK (1994) Ecological classification of hydroecosystems in the Aral Sea basin. Uzbek Bio J 3:35–37 (in Russian)

Karimov BK (1995) Ecological and toxicological problems of the state and of fishery use of hydroecosystems in the Aral Sea basin. Avtoreferat of Doctoral Dissertation, "Fan" Press, Tashkent, 51 p (in Russian)

Karimov BK (2003) Saving the Aral Sea. Humboldt-Cosmos 82:24–25

Karimov BK, Keyser D (1998) The effect of salt composition on the salinity tolerance of mirror carp (Cyprinus carpio L.) during early ontogeny. Archive Fish. Mar. Res. 46(3):225–239

Karimov BK, Razakov RM (1990) The evaluation of toxicological situation on example of Central Asian region. In: Lvov VA, Kuzin AK (eds) The fundamentals of water protection. Kharkov, pp 26–34 (in Russian)

Karimov B, Keyser D, Kurambaeva M (2002) Impact of agriculture on deltaic hydroecosystems of the Aral Sea basin: ecotoxicological studies. In.: Contributions to the 4th international symposium on "physiology and behaviour of wild and zoo animals". Berlin, p 183

Karimov B, Joldasova I, Blanchoud H, Kurambaeva M, Mullaboev N, Chevreuil M (2004) Problems of sustainable fishery and aquaculture in the southern Aral Sea region under anthropogenic impact. In: Abstract volume: the 14th Stockholm water symposium, Aug 16–20, pp 272–273

Karimov B, Lieth H, Kurambaeva M, Matsapaeva I (2005) The problems of fishermen in the Southern Aral Sea Region. J Mar Sci/Special Issue: Mitig Adapt Strat Glob Change 10:87–103 (Nihoul JCJ, Aladin N, Glantz M (Guest eds))

Karimov BK, Kamilov BG, Upare M, Van Anrooy R, Bueno P, Shohimardonov DR (2009) Inland capture fisheries and aquaculture in the republic of Uzbekistan: current status and planning. FAO Fish Aquac Circular Rome 1030(1):124 p

Khlebovich VV (1974) Critical salinity of biological processes. Nauka, Leningrad, 236 p (in Russian)

National Report on the state of nature protection and natural resources utilization in the Republic of Uzbekistan in 2002–2004 (2005) State Committee on Nature Protection of Tashkent, 131 p (in Russian)

National Report on the state of nature protection and natural resources utilization in the Republic of Uzbekistan in 2008–2011 (2013) State Committee on Nature Protection of Tashkent, 254 p (in Russian)

Ruecker GR, Shi Z, Mueller M, Conrad C, Ibragimov N, Lamers JPA, Martius C, Strunz G, Dech SW (2007) Cotton yield estimation in Uzbekistan integrating modis, landsat etm + and field data. ISPRS Archives XXXVI-8/W48 Workshop proceedings: Remote sensing support to crop yield forecast and area estimates, pp 123–128. http://www.isprs.org/proceedings/XXXVI/8-W48/123_XXXVI-8-W48.pdf. Accessed 23 Aug 2013

Sadek S (2011) An overview on desert aquaculture in Egypt. In: Crespi V, Lovatelli A (eds) Aquaculture in desert and arid lands: developmental constraints and opportunities. FAO technical workshop, 6–8 July 2010, Hermosillo. FAO fisheries and aquaculture proceedings, No. 20. Rome, p 141–158

Smith M, Veth P, Hiscock P, Wallis LA (2008) Global desert in perspective. In: Smith M, Veth P, Hiscock P (eds) Desert peoples: archaeological perspectives. Blackwell Publishing, pp. 1–14

United Nations (UN) (2010) Uzbekistan second review. Environ Perform Rev Ser 29:201

United Nations Development Program (UNDP) (2007) Water critical resource for Uzbekistan's future) Tashkent, 128 p

United Nations Development Program (UNDP) (2008) Environmental profile of Uzbekistan based on indicators. 88 p

Wecker B, Karimov B, Kamilov B, Matthies M, Lieth H (2007) Sustainable aquaculture in recirculating systems; Feasibility study for the catchment area of the Aral Sea. Contribution no. 40 of the Institute of Environmental Systems Research, University of Osnabrueck. (Ed. Prof. Dr. Michael Matthies). ISSN Nr. 1433-3805. http://www.usf.uos.de/projects/usf/literatur/beitraege/texte/040-weckeretal.pdf. Accessed 23 Aug 2013

Yakubov XE, Yakubov MA, Yakubov SX (2011) Collector-drainage waters in Central Asia and estimation of their use for irrigation. IWP AS Uzbekistan, SIC ICWC, Tashkent, 189 p

Chapter 11
A Historian's Perspective on Rivers of the Anthropocene

Philip V. Scarpino

Abstract The assertion by leading scientists that the Anthropocene has replaced the Holocene as the most recent geological epoch represents an important opportunity to promote interdisciplinary communication. Determining the existence of a new geological epoch and naming it the Anthropocene is largely the domain of science. Analyzing and explaining the long-term human impact on earth systems that produced and sustained that epoch is the domain of humanists and social scientists, such as historians and archaeologists, who address the complex and changing connections between culture and nature. Historians look at the evolving relationship between people and their environment. History adds both the longue durée and context to present-day understanding of the interplay between human and natural systems. Writing in Smithsonian Magazine (January 2013), Joseph Stromberg, journalist and science writer, posed an essential question: "Have human beings permanently changed the planet?" His question is simultaneously historical and interdisciplinary; it highlights the role of human agency driven by an evolving mosaic of human culture. Rivers offer a metaphor for understanding the human environmental experience. As such they present an opportunity for the real and sustained interdisciplinary study, communication, and collaboration that could yield a credible and effective answer to Stromberg's question.

In 1949, Aldo Leopold posthumously published A Sand County Almanac and Sketches Here and There. Leopold was a forester trained at Yale University, a contributor to the development of ecological science, creator of the discipline of wildlife management in the United States, conservationist, wilderness advocate, professor, and one of the leading environmental thinkers of the 20th century (Flader 1974 and Meine 1988). One of the essays in A Sand County Almanac is titled "Song of the Gavilan." (The Gavilan is a river in the Southwestern United

P. V. Scarpino (✉)
Department of History, IUPUI, Indiana University-Purdue University Indianapolis, Indianapolis, IN, USA

A. Bhaduri et al. (eds.), *The Global Water System in the Anthropocene*, Springer Water, DOI: 10.1007/978-3-319-07548-8_11,

States). In that essay, Leopold examined the relationship between the process of scientific research and creation of knowledge as he understood those things during his professional lifetime.

> There are men charged with the duty of examining the construction of the plants, animals, and soils which are the instruments of the great orchestra. These men are called professors. Each selects one instrument and spends his life taking it apart and describing its strings and sounding boards. This process of dismemberment is called research. The place for dismemberment is called a university.
> A professor may pluck the strings of his own instrument, but never that of another, and if he listens for music he must never admit it to his fellows or to his students. For all are restrained by an ironbound taboo which decrees that the construction of instruments is the domain of science, while the detection of harmony is the domain of poets. (Leopold 1949)

Leopold was pointing out in 1949 a set of problems that continues to plague our understanding of the environment in general and rivers in particular: Extreme specialization; the lack of communication across disciplines; and a corresponding inability to see and understand the big, interdisciplinary picture. As a distinguished scientist, Leopold also recognized that the interaction between human and natural history was an important variable in understanding the modern environment:

> That man is, in fact, only a member of a biotic team is shown by an ecological interpretation of history. Many historical events, hitherto explained solely in terms of human enterprise were actually biotic interactions between people and land. The characteristics of the land determined the facts quite as potently as the characteristics of the men who lived on it. (Leopold 1949)

In quarter century after World War II, ecology had a major impact on reshaping the way that many people in North America and Western Europe understood their relationship with the environment. Writers like Rachel Carson, author of Silent Spring, popularized a version of ecosystem theory that provided the intellectual and philosophical underpinning of the modern environmental movement. (Carson 1962) The impact of Silent Spring extended across North America and Western Europe. Silent Spring was an international best seller, published in fifteen countries and translated into numerous European languages; it also impacted public policies on pesticides in several nations in North America and Western Europe.[1] (McCormick 1989 and Lear 1997) Basic ecological principles like interconnectedness and interdependence, reinforced by international concerns over radioactive fallout and Strontium 90, pointed to the conclusion that people who make war on

[1] The information on the European languages comes from http://www.environmentandsociety.org/exhibitions/silent-spring/silent-spring-international-best-seller. The web site's author Dr. Mark Stoll, an Associate Professor of History at Texas Tech University, worked with the Rachel Carson Center for Environment and Society in developing the web site. Stoll's list of the languages into which Silent Spring was translated includes: "German in 1962; French, Swedish, Danish, Dutch, Finnish, and Italian in 1963; Spanish, Portuguese, and Japanese in 1964; as well as in Icelandic in 1965, Norwegian in 1966, Slovenian in 1972, Chinese in 1979, Thai in 1982, Korean in 1995, and Turkish in 2004." Stoll also notes that abridged versions appeared in several popular European periodicals.

nature ultimately make war on themselves—a realization that energized the environmental movement. Preceded by Aldo Leopold and followed by other writers such as Barry Commoner (1971) and Rene Dubos and Barbara Ward (1972) Rachel Carson called for a fundamental reform in western society eschewing traditional attitudes in favor of a relationship with nature organized around ecology-based principles. Carson ended Silent Spring with a warning that continued emphasis on controlling nature when combined with powerful new products of modern science, such as synthetic, organic pesticides, presented the potential for significant unintended consequences.

> The 'control of nature' is a phrase conceived in arrogance, born of the Neanderthal age of biology and philosophy, when it was supposed that nature exists for the convenience of man. The concepts and practices of applied entomology for the most part date from that Stone Age of science. It is our alarming misfortune that so primitive a science has armed itself with the most modern and terrible weapons, and that in turning them against the insects it has also turned them against the earth. (Carson 1962)

The ecology-based environmental movement in the 1960s and early 1970s created both a constituency and the political will for dramatic shifts in public policy in North America and Western Europe. (Wooster 1994 and McCormick 1989) It is no accident that the UK, Canada, and the United States all passed national water pollution control legislation within a few years of each other (Scarpino 2010 and Hassan 1998):

1. US, Water Quality Improvement Act, 1970 and Federal Water Pollution Control Act Amendments, 1972
2. Canada Water Act, 1970; Canada/Ontario Act, 1971
3. Canada/U.S., Great Lakes Water Quality Act, 1972
4. UK, Water Act of 1973, established integrated management of water in England and Wales

In 1969, human ecologist, Paul Shepard, addressed the transformative promise of popularized ecology in his introduction to Subversive Science.[2] In company with Aldo Leopold and Rachel Carson, Shepard understood the potential for an ecological perspective to bring about a fundamental shift in the bed rock values and attitudes that shaped the relationship between people and nature. Shepard argued that "the ideological status of ecology is that of a resistance movement. Its Rachel Carsons and Aldo Leopolds are subversive." (Shepard 1969) Despite Shepard's assessment (and the grassroots passion of the environmental movement

---

[2] Paul Shepard developed a tremendous reputation as a human ecologist and environmental philosopher. His career was markedly interdisciplinary. He earned an A.B. in English and Wildlife Conservation from the University of Missouri in 1949; his MS in conservation from Yale University in 1952, writing a thesis on art and ecology in New England; and, his PhD from Yale in 1954, an interdisciplinary degree in Conservation, Landscape Architecture, and the History of Art. From 1973–1994, he was the Avery Professor of Natural Philosophy and Human Ecology, Pitzer College and the Claremont Graduate School, Claremont, CA. See: http://paulhoweshepard.wordpress.com/bio/.

in North America and Western Europe), ecology never fulfilled its subversive potential. Since the 1970s, the science of ecology and the popular idea of ecology have followed different developmental trends. The environmental movement lost its grass-roots power and became dominated by professionals who fought for political consensus and policy victories. Increasingly specialized ecological science added tremendously to knowledge, but it did so within the confines of narrower and deeper intellectual silos. By the end of the 20th century, advances in ecological science no longer had the public impact and influence that ecosystem theory exhibited during the height of the popular environmental movement. Modern environmental science looked much like the critique of his peers that Aldo Leopold leveled in 1949.

It is in that context that the proposed designation of the Anthropocene as the most recent geological epoch represents an important opportunity to promote interdisciplinary communication. In 2000, Paul J. Crutzen, a Dutch atmospheric chemist, coined the term "Anthropocene." Crutzen was a co-winner of the Nobel Prize in 1995 for his work on the ozone layer. (Crutzen 2002; Pearce 2007; Crutzen and Schwagerl 2011) In 2007, Libby Robin and Will Steffen described the Anthropocene in the following terms: "A new geological epoch. . . reflecting the scientific fact that anthropogenic change is now shaping planetary systems. Describing changes to the Earth system over time demands understanding of the history of the biophysical factors, the human factors and their integration." (Robin and Steffen 2007) By focusing on human agency, the concept of the Anthropocene holds out the potential for recovering the unrealized promise of ecology: Transforming popular attitudes towards nature and integrating the study, understanding, and management of human and natural systems. Paul Crutzen originally dated the start of the Anthropocene "immediately following the invention of the steam engine in 1784." (Crutzen and Steffen 2003; Crutzen 2002; Stromberg 2013). Discussion continues among scientists over when the Holocene transitioned to the Anthropocene, in addition to the broader debate about designating the Anthropocene itself (Zalasiewicz et al. 2008).

A historian would argue that significant human impact on the earth began several centuries before the rapid rise in the use of fossil fuels that followed invention of the steam engine in 1784. One example may be found in Alfred W. Crosby's brilliant book, Ecological Imperialism: The Biological Expansion of Europe, 90 0-1900 (1986). Crosby documents the tremendous influence that Europeans had on facilitating the moving and mixing of species around the surface of the earth—a process that was well underway before the age of steam transportation. Indeed, the human-facilitated, world-wide moving and mixing of species remains one of the key issues associated with modern rivers and one of many variables that points to a transition from the Holocene to the Anthropocene.

Determining the existence of a new geological epoch and naming it the Anthropocene is largely the domain of science. Analyzing and explaining the long-term human impact on earth systems that produced and sustained that epoch is the domain of humanists and social scientists, such as historians and archaeologists, who address the complex and changing connections between human culture and nature. Historians look at the evolving relationship between people and their environment. History adds both the *longue durée* and context to our present-day understanding of the interplay between human and natural systems.

Historically, it is useful to examine the interaction between people and their environment, including rivers, through the lens of a definition of material culture coined by a historical archaeologist named James Deetz. Until his death in 2000 Deetz was one of the World's leading historical archaeologists. Professor Deetz defined material culture as "that portion of man's physical environment purposely transformed by him according to culturally dictated plans." (Schlereth 1989) Deetz's deceptively simple definition opens the door to a complex and sophisticated examination of the relationship between people and their surroundings. Human beings, like most other living organisms, modify their surroundings to better suit their needs. People undertake these modifications based upon the attitudes and values embedded in their cultures. By the early 21st century, there were very few places on the surface of the Earth that have not been so modified as to meet Deetz's definition of material culture. Debate over designating the Anthropocene as a new geological epoch further elevates the role of human agency and creates the possibility for a new picture of the intertwined evolutionary course of human and natural history; understanding the Anthropocene calls for collaboration across disciplines to bring that picture into focus.

In the past two centuries, the environmental history of rivers in Western Europe and North America followed similar developmental trajectories. The actions of people transformed these rivers to fit James Deetz's definition of material culture. The scale of the impacts and the rate of change accelerated significantly between the middle of the 19th century and the late 20th century. Historically, cities and towns in Western Europe and North American placed multiple and conflicting demands on rivers to meet a variety of clean-water needs and simultaneously to serve as a sink for untreated wastes. As cities grew and industries expanded, rivers became convenient, nearby, and relatively inexpensive repositories for untreated domestic and manufacturing waste. Dumping waste into rivers solved an immediate, local problem, but vastly complicated the challenges associated with river pollution, as contaminated water crossed multiple political boundaries and precipitated prolonged conflicts over which level of government bore the responsibility for paying for pollution control.

The witch's brew of pollutants dumped into rivers damaged habitat and caused a major public health crisis. Communities faced a rising epidemic of water-borne illness such as cholera and typhoid. The addition of chlorine to drinking water and the development of the rapid sand filter made it possible by the early 20th century to treat polluted water and make it potable. (Hassan 1998) These relatively inexpensive technical fixes solved the immediate public health crisis and took the pressure off of governments to do something about water pollution. As a result, river habitat in Europe and North America continued to decline in many cases into the 1970s.

In addition to untreated wastes, a variety of other uses significantly modified river environments. Rivers provided a continuing transportation option, which frequently involved substantial improvements such as dredging and construction of locks and dams to deepen and stabilize navigation channels. Dams also generated water power and later electricity. Drainage of wetlands altered habitat and facilitated agricultural expansion, which in turn accelerated run off and washed silt, nutrients, and eventually agricultural chemicals into rivers. Demands for flood control resulted in construction of reservoirs, levees, and flood walls. In many cases, the relatively level land of flood plains attracted commercial and residential development. Parallel modes of transportation in the form of railroads and highways cut off rivers and limited access between land and water. Many communities literally and figuratively turned their backs on their badly contaminated rivers (Scarpino 1997).

The Tyne River, which flows through Newcastle into the North Sea in the Northeastern UK, offers a useful, microcosmic look at human impact on a single river between the middle of the 19th century and the latter part of the 20th century. The entire River Tyne is about 100 km (62 miles) in length, with the final 23 km (14 miles) a tidal estuary. Newcastle is located about 16 km (10 miles) from the mouth. Mining for Lead and Zinc took place in the drainage of the Tyne for centuries, but peaked in the late 19th century and ended in 1917. Northeast England became a center of coal mining and the Tyne served as the major coal port. Port development led to significant dredging of the Tyne between 1860 and 1888. Port improvements attracted more industry, so that by the late 19th century tanneries, breweries, gas works, alkali and coke producers dumped burgeoning levels of industrial waste into the River.[3] The arrival in the early 20th century of treated water and indoor water closets generated a rising volume of domestic sewage, which mixed with the industrial effluents pouring into the Tyne (Milner et al. 2004).

In 1950, the Tyne River Authority issued a celebratory booklet describing 100 years of River improvement carried out under its tutelage, which not only documented dramatic change in the river but also revealed attitudes toward those changes. The booklet contrasts historic and "nostalgic" views of the Tyne with the working river of the 1950s by juxtaposing engravings and paintings by artist, J. W. Carmichael (undated but likely completed between the mid-1820s and the late 1850s), with photographs from the middle of the 20th century.[4] In the Chairman's "Forward," W. A. Souter explained "Carmichael's pictures produce a nostalgic feeling at the loss of the scenic beauties of the lower reaches of the Tyne.

[3] A useful summary of industrial development and river improvements may be found in Henry E. Armstrong, "Port Sanitary Administration on the Tyne: A 7 Years' Retrospective (1881–1887), in Proceedings of the Society of Medical Officers of Health and reprinted from Public Health (June 1888): 4–10. Armstrong was the Medical Officer of Health for Newcastle and the associated port. Accession, 604/1207, Tyne and Wear Archives, Newcastle, UK. Hereafter, Tyne and Wear Archives.

[4] John Wilson Carmichael, 1799–1868, lived and worked in Newcastle until he moved to London in 1846; known as a marine painter, See: Marshall Hall, The Artists of Northumbria, p. 71. The booklet also uses the work of others but features Carmichael's art.

Nevertheless we are a mercantile nation and commerce, which is our life blood, takes some toll of natural beauty." Souter stated that Newcastle had become the most important coal-shipping port in the United Kingdom; the U.K's. second leading ship building center; and the largest ship-repairing port in the world. He added that the "dredged depths make the river available to the largest vessels afloat." In 1850, he noted that at low tide it was possible to ford the river at the bar near the mouth and at Newcastle. By 1950, average navigation depths at low tide ranged from 25 to 30 feet.[5]

Between the late 19th century and the mid-1960s, the City of Newcastle steadily expanded its sewer system, intended to convey untreated domestic sewage directly to the Tyne River. Between 1861 and 1896, the population of Newcastle grew from 109,108 individuals living in 13,979 houses to 212,223, residing in 29,600 households. In 1896, the City Engineer reported just over 162 miles of sewers serving Newcastle.[6] While population growth slowed in the 20th century, the volume of sewage grew, as did the size of the sewer system emptying directly into the Tyne, driven in part by the conversion to indoor water closets. In 1912, investigators recorded a dissolved oxygen level of zero in the Tyne, dockside at Newcastle. (Milner et al. 2004) By the middle of 1929, Newcastle's expanding sewer system served an estimated the population of 283,400 living in 66,529 houses. The City's Medical Officer reported "there are 319 miles of sewers discharging directly into the Tyne, which is tidal, at various points along the seven miles of river frontage."[7]

An important salmon fishery disappeared from the River as a direct result of habitat degradation largely a consequence of pollution. (Champion 1990 and Marshal 1992) When authorities banned net fishing in the Tyne in 1934, the fishery was no longer viable due to plummeting catches. The Fishery Board for the Fishery District of the River Tyne reported a total of 7,243 salmon taken by net in 1934 from the River Tyne and the North Sea. In any event, consumers had grown reluctant to buy fish that tasted like tar.[8] (Milner et al. 2004) The Annual Report for 1945 of the Fishery Board for the River Tyne pointed out that the "pollution of the Estuary of the River Tyne still continues to become an increasing menace to

---

[5] Tyne Improvement Commission: Centenary, 1850–1950, "Foreward by the Chairman," p. 3, Accession 604/1230, Tyne and Wear Archives.

[6] Newcastle City Library, Proceedings of the Council of the City and County of Newcastle upon Tyne for 1895-96; being the 61st Year After the Passing of the Municipal Reform Act, reference to 1891, pp. 110–111; "Report of the City Engineer to the Town Improvement and Sanitary Committees, for the Year Ending March 25, 1896," Table IV, "Lengths and Descriptions of Sewers Known to Exist in the City," p. 20; Table XIII "Census of the City from the Year 1801–1896," p. 33. Hereafter, City Library, Proceedings.

[7] City Library, Proceedings for 1930–1931, Part II "Annual Report of the Medical Officer of Health on the Sanitary Condition of the City during the Year 1930," pp. 45, 49.

[8] The Fishery Board for the Fishery District of the River Tyne, Annual Report for the Year 1943 and Yearbook for 1944, Appendix B, p. 28, Tyne and Wear Archives, Accession 3983/9, Newcastle, UK, Fishery Board for Fishery District of the River Tyne Annual Reports 1943–1947. Hereafter, Fishery Board, Annual Report.

the whole river." Despite some ups and downs, the number of salmon taken by net in the river and nearby sea had fallen to 617 in 1945.[9]

Throughout the 1950s, pollution persisted as a serious problem in the River Tyne, accompanied by a reluctance of the City Council to pay for ameliorative action. In 1954, a report issued by the Newcastle city council treated addressing pollution of the Tyne as a luxury that could be reasonably postponed:

> There is no doubt that with the general improvement in the standard of living, people are now becoming more sensitive to the condition of the river. . . It is felt that the correct attitude. . . is that the removal of sewage from the River Tyne is a problem which should be tackled, and that this should be done as soon as funds reasonably become available, in other words it is something of a luxury to be afforded when other necessary improvements in the standard of living of the people of Tyneside have been completed. (Hassan 1998, p. 120)

Five year later in 1959, Newcastle had entered a long period of deindustrialization. A population of 271,000 living in 87,993 households employed "463.05 miles of sewers in the City, discharging directly into the River Tyne at various points along the 8½ miles of river frontage."[10] Even though the population had declined, the number of houses and the size of the sewer system disgorging untreated waste into the Tyne River had both expanded. The condition of the River was so bad in 1969 that a public meeting about pollution of the Tyne taking place in Newcastle's Moot Hall a few hundred meters up the hill from the River had to be adjourned because the stench from the nearby waterway became overwhelming (Hassan 1998).

During the last third of the 20th century—driven by a shift in values prompted by the ecology-based environmental movement—communities across Europe and North America rediscovered and reconsidered the value of their rivers. Since the 1970s, Europeans and North Americans did an increasingly effective job of cleaning up point sources of pollution so that visually, aesthetically, and biologically rivers saw dramatic improvement. At the same time, non-point sources of pollution presented increasingly serious and difficult challenges. Again, the Tyne River offers a case in point. By the late 1980s, thanks to a combination of sewage treatment and rapid industrial decline the River Tyne witnessed impressive change for the better, including the return of Atlantic salmon and sea trout. (Milner et al. 2004) In 2013, the River is a visual and aesthetic centerpiece for the city. Very little physical evidence remains of the ship building and repair facilities or of the factories and quays that once lined the Tyne from Newcastle to the sea.[11]

By the middle of the 20th century, navigable rivers and their tributaries in Western Europe and North American had become parts of an interconnected,

---

[9] Fishery Board, Annual Report 1945, Tyne and Wear Archives, pp. 20, 36.

[10] City Library, Proceedings for 1960–1961, "Annual Report of the Medical Officer of Health for the Year 1959," pp. 26, 27, 28.

[11] Observations on the present-day "look" of the Tyne River and its surroundings are based on field work by the author in 2011 and 2013.

world-wide, maritime transportation system that offered significant environmental implications and challenges—among them the accelerated movement of species. The Great Lakes and associated St. Lawrence, Mississippi and Ohio Rivers in the United States and Canada illustrate that point. A series of navigational improvements between the mid-1820s and the late 1950s connected formerly isolated bodies of water via deep-water navigation channels to the Atlantic Ocean and the Gulf of Mexico.[12]

These improvements, which successfully met their primary, single-purpose goal of establishing continuous, deepwater navigation, also produced a number of unintended and unanticipated consequences. Among the most significant, was facilitating the migration of invasive species—a process illustrated by the story of Zebra Mussels. Zebra Mussels originated in the Baltic Sea and made the trans-Atlantic crossing to the Great Lakes by hitching a ride in the ballast water of an unknown freighter. First sighted near Detroit in 1986, they spread rapidly through the Great Lakes. Between 1990 and 1995, Zebra Mussels colonized the St. Lawrence, Mississippi, and Ohio River systems in the United States and Canada transported by commercial and recreational vessels plying the navigation channels.[13] A map of their distribution pattern in 1993 almost perfectly aligns with the inland transportation system. (See Map 1) Their introduction and rapid proliferation through Canadian and U.S. waters offers a reminder of the continuing importance of "ecological imperialism" (Crosby 1986) and one more bit of evidence of the degree to which the intended, unintended, and unanticipated consequences of human actions have profoundly altered rivers and other waterways, such as the Great Lakes.

In 1995, The Toronto Globe and Mail printed a remarkable insightful environmental analysis of Lake Ontario, which also readily applies to the rivers and related waterways of North America and Western Europe:

> The road leading to the Glenora fisheries research station on the north shore of Lake Ontario carries a lesson on its shoulder. First there is a front-yard display of plastic cows. Then, as the road curves to meet the lake, there is a house sided with panels of fake stonework.

---

[12] The State of New York completed the Erie Canal in 1825 connecting Lake Erie to the Hudson River, which was navigable to New York City. The former Erie Canal remains open to recreational use. The St. Lawrence Seaway opened in 1959, creating a deepwater channel from Lake Ontario to the Atlantic Ocean. For an excellent overview of navigational improvements on the Great Lakes, See: "A Chronology of Great Lakes Navigation," prepared at Northern Michigan University, http://www.nmu.edu/upperpeninsulastudies/node/63. For a useful overview of the construction of the Chicago Sanitary and Ship Canal connecting Lake Michigan to the Mississippi River via the Illinois River, See: "Constructing the Sanitary and Ship Canal," http://www.encyclopedia.chicagohistory.org/pages/300018.html. Helpful on navigational improvements to the Ohio River, U.S. Army Corps of Engineers, "History of Navigational Development on the Ohio River, http://www.lrd.usace.army.mil/Missions/CivilWorks/Navigation/OhioRiverNavigation/History.aspx.

[13] http://nationalatlas.gov/dynamic/an_zm.html provides a sequence of maps that illustrate the spread of zebra mussels from 1986 through 2010.

> And then there is the lake, seemingly a refuge from all that is artificial. On the land side is an environment shaped by human hands, while in the depths, human impact is minimal. But that assumption is deceiving. Just like the false stonework and fake cows, Lake Ontario's ecosystem is contrived by humans. . . .[14]

The Great Lakes, along with the St. Lawrence, Mississippi, and Ohio Rivers have become classic illustrations of human material culture (Schlereth 1989); heavily humanized, cyborg-like systems composed of natural and artificial parts. Placed into the larger matrix of an integrated, world-wide maritime transportation system, which includes the River Tyne in the UK, they reinforce the argument in favor of the Anthropocene.

In 1949, Aldo Leopold invited his colleagues and the readers of A Sand County Almanac to understand the natural world as an interdependent system. He criticized fellow scientists who were "charged with the duty of examining the construction of the plants, animals, and soils which are the instruments of the great orchestra." Instead of hearing "the great orchestra," his colleagues narrowed the focus of their research, specializing in a manner that he described as "dismemberment." While popularized ecology drove the post-WWII environmental movement and the science of ecology added greatly to what we know, neither the popular understanding of ecology nor the science of ecology became agents of cultural transformation—never fulfilling the "subversive" potential predicted by Paul Shepard and others (Shepard 1969).

The concept of the Athropocene invites us to figuratively hear the "great orchestra." It challenges us not to spend our professional lives in bottoms of disciplinary silos dismembering the individual "instruments" that together make up a fascinatingly complex and thoroughly humanized system. Writing in Smithsonian Magazine in January 2013, Joseph Stromberg, journalist and science writer, posed an essential question: "Have human beings permanently changed the planet?" His question is simultaneously historical and interdisciplinary; it highlights the role of human agency driven by the evolving mosaic of human culture. In many ways, rivers offer a metaphor for understanding the human environmental experience. As such they present an opportunity for the real and sustained interdisciplinary study, communication, and collaboration that could yield a credible and effective answer to Stromberg's question.

References

Carson Rachel (1962) Silent spring. Houghton Mifflin Company, Boston

Champion AS (1990) Managing a recovering salmon river—the river tyne. In: Mills D (ed) Strategies for the rehabilitation of salmon rivers: proceedings of the linnean society joint conference (November 1990), pp 63–72

Commoner Barry (1971) The closing circle: nature, man, and technology. Knopf, New York

[14] Toronto Globe and Mail, 7/15/95, D8.

Crosby Alfred (1986) Ecological imperialism: the biological expansion of Europe, 900–1900. Cambridge University Press, Cambridge
Crutzen PJ (2002) Geology of mankind. Nature 415:423
Crutzen PJ, Steffen W (2003) Editorial comment: how long have we been in the Anthropocene era? Clim Change 61(3):251–257
Crutzen PJ, Schwagerl C (2011) Living in the Anthropocene: toward a new global ethos, environment 360. http://e360.yale.edu/feature/living_in_the_anthropocene_toward_a_new_global_ethos/2363/. Accessed 24 Jan 2011
Flader SL (1974) Thinking like a mountain: Aldo Leopold and the evolution of an ecological attitude toward deer, wolves, and forests (Columbia. University of Missouri Press, Columbia
Hassan John (1998) A history of water in modern England and Wales. Manchester University Press, Manchester, pp 123–147
Lear Linda (1997) Rachel Carson: witness for nature. Henry Holt and Company, New York
Leopold Aldo (1949) A sand county almanac and sketches here and there. Oxford University Press, New York
McCormick John (1989) Reclaiming paradise: the global environmental movement. Indiana University Press, Bloomington
Marshall MW (1992) Tyne waters a river and its salmon. HF and G Witherby, Ltd, London
Meine Curt (1988) Aldo leopold: his life and work. University of Wisconsin Press, Madison
Milner NJ et al (2004) Environment agency: the role of stocking recovery of the river tyne salmon Fisheries," Environment Agency, National Fisheries Technical Team, Fisheries Technical Report No. 2004/1 (2004): 6. http://www.wyeuskfoundation.org/problems/downloads/Tyne%20Hatchery%20Report.pdf
Pearce F (2007) With speed and violence: why scientists fear tipping points in climate change. Beacon Press, Boston, p 44, http://www.gci.org.uk/Documents/wsav.pdf
Robin L, Steffen WW (2007) History for the Anthropocene. Hist compass 5(5):1694–1719. http://onlinelibrary.wiley.com/doi/10.1111/j.1478-0542.2007.00459.x/full
Scarpino PV (1997) Large floodplain rivers as human artifacts: a historical perspective on ecological integrity. U.S. Geological Survey, Special Refereed Report
Scarpino PV (2010) Addressing cross-border pollution of the Great Lakes. In: Behiels M, Stuart R (eds) Transnationalism in Canada-United States history into the twenty first century. McGill-Queen's University Press, Montreal pp 146–167
Schlereth TJ (1989) History museums and material culture. In: Leon W and Rosenzweig R (eds) History museums in the United States. University of Illinois Press, Champaign (On page 294 Schlereth quotes Deetz's definition of material culture)
Shepard Paul (1969) Ecology and man—a viewpoint. In: Shepard Paul, McKinley Daniel (eds) The subversive science: essays toward an ecology of man. Houghton Mifflin Harcourt, Boston
Steffen W et al (2003) Global change and the earth system: a planet under pressure. Springer, Berlin
Stromberg J (2013) What is the Anthropocene and are we in It?" Smithsonian Magazine (January 2013). http://www.smithsonianmag.com/science-nature/What-is-the-Anthropocene-and-Are-We-in-It-183828201.html
Ward B, Dubos, R (1972) Only one earth: the care and maintenance of a small planet. Norton & Company, New York (An unofficial report commissioned by the secretary-general of the United Nations Conference on the human environment. Prepared with the assistance of a 152-member committee of corresponding consultants in 58 countries)
Wooster D (1994) Nature's economy: a history of ecological ideas, 2nd edn. Cambridge University Press, Cambridge (first published by Sierra Club Books, 1977)
Zalasiewicz J et al (2008) Are we now living in the Athropocene?" GSA today (February 2008): 7, http://www.geosociety.org/gsatoday/archive/18/2/pdf/i1052-5173-18-2-4.pdf

Chapter 12
Transdisciplinarity, Human-Nature Entanglements, and Transboundary Water Systems in the Anthropocene

Jason M. Kelly

This essay introduces some major ideas and concepts relating to transdisciplinary approaches to Anthropocene river systems. The first section is a historical case study of George Catlin, an artist who traveled throughout the Great Plains in the 1830s. Catlin's writing embodies popular 19th-century ideas about "pristine" natural states which have continued to shape environmental thought into the 21st century. By exploring Catlin's descriptions of the Hidatsa, a community at the confluence of the Knife and Missouri Rivers, it shows the importance of integrating human-nature entanglements into studies of anthropogenic environmental change. Key to understanding these changes are cultural and socio-political structures that shape and are shaped by the environment.

The second part of this essay uses the example of the Cochabamba *Guerra del Agua* in 2000 to examine human-nature entanglements in transboundary water systems. Specifically, it considers the ways that human systems at multiple scales define boundaries, shape policy, and transform environments. Rather than simply focus on transboundary water systems as physical presences, it argues for the importance of expanding the definition of "transboundary" to include non-physical systems, including sociocultural structures and practices.

The two case studies are not meant to be exhaustive. Rather, they are surveys meant to exemplify the complex intersections of cultural, economic, and political interests around water resources in the Americas. They demonstrate current scholarly thinking related to human-nature entanglements and transboundary water systems. For more information on each of the case studies, the reader is directed to the bibliography.

Rivers and their ecologies are not simply natural systems; they are human systems as well. They shape human societies even as they are shaped by them. In fact, separating human and environmental systems is an artificial division which is

---

J. M. Kelly (✉)
IUPUI Arts and Humanities Institute and Department of History in the IU School of Liberal Arts at IUPI, Indiana University-Purdue University, Indiana, Indianapolis
e-mail: jaskelly@iupui.edu

A. Bhaduri et al. (eds.), *The Global Water System in the Anthropocene*, Springer Water, DOI: 10.1007/978-3-319-07548-8_12,

a product of historically constituted epistemological categories. (Castree and Braun 2005; Cronon 1996) Riverine ecologies emerge from a series of complex interactions between the atmosphere, hydrosphere, lithosphere, and biosphere. In the age of the Anthropocene, these biophysical systems have been dramatically affected by one species: humans. As a result, the study of anthropogenic biophysical processes has become a key area of focus for scientific research. However, the majority of studies focus on the consequences of human action, not on the systems driving human actions. In other words, researchers are increasingly gathering information about the effects of human-induced environmental change, but they are at the earliest stages of linking effects to causes.

Understanding the causes of anthropogenic environmental change requires the techniques developed in the social sciences and humanities. Anthropogenic changes result from patterns of human behavior, which are shaped and reshaped by sociocultural systems. For example, cultural beliefs can limit responses to ecological crises and therefore contribute to environmental feedback loops.

A complex network of disparate systems and objects are linked at the human-riverine interface, connected by a chain of historical material and sociocultural structures. This interrelation is what we might call an entanglement. (Hodder 2011, 2012; Latour 2005; Malafouris and Renfrew 2010; Demarrais et al. 2004) In the Anthropocene, when humans have had an increasingly powerful effect on global riverine environments, these entanglements are ever more apparent. As Matt Edgeworth has recently argued, “If the river ever was entirely a natural entity, it has long since been at least partially honed to fit human projects. If it ever was wholly wild and untamed, it has long since been at least partially domesticated. And if it ever was merely an environmental entity, it has long since become part of the cultural landscape” (Edgeworth 2011, 14).

Transdisciplinary approaches, which connect the strengths of the sciences, social sciences, and humanities, are more likely to result in a comprehensive understanding and sustainable solutions to human-induced environmental change, because they can better understand the entanglements of environmental systems and human systems. This perspective has been articulated by several research groups including IHOPE, the Integrated History and Future of People on Earth which is a project of the International Geosphere and Biosphere Programme (IGBP), and UNESCO’s International Hydrological Program (Costanza et al. 2007, 2012; Davies and M’Mbogori 2013; Hassan 2011; Hibbard et al. 2010; Hornborg and Crumley 2006; Mosley 2006; Sörlin 2012). In 2012, the European Science Foundation, Strasbourg and European Cooperation in Science and Technology, Brussels commissioned a report, titled *RESCUE*, which argued for more conceptual and methodological disciplinary integration (Jäger et al. 2012). *RESCUE* argued that the social sciences and humanities have been auxiliary to scientific environmental research—despite long decades of work in the environmental social sciences and humanities. And, while some social science research, including human population patterns, economies, and governance frameworks, has been integrated into scientific research, valuable insight from ethnography, social and cultural history, environmental ethics, and postcolonial literary criticism has remained peripheral. Bridging this divide will require projects that address the "two culture" problem directly, a task targeted by such collaborations as the Rivers of the Anthropocene project (rivers.iupui.edu).

In 1830, George Catlin, a Pennsylvania-born artist, arrived in St. Louis looking for fame and fortune. There, he sought the patronage of General William Clark, the U.S. Superintendant of Indian Affairs. A connection to General Clark was no mean feat. Clark was the famed explorer who—with Meriwether Lewis, Sacagawea and the "Corps of Discovery"—made a three-year transcontinental trek to the Pacific Ocean in 1804. In addition to connecting Catlin to networks of other patrons, Clark eventually helped him gain passage aboard the steamship *Yellow Stone* in 1832 (Dippie 1990). The voyage was meant to prove the "practicability of steam navigation" on the Missouri River, and it traveled as far north as Fort Union, 2,000 miles upriver from St. Louis (Buckingham and Buckingham 1832). Its success hastened the expansion of U.S. economic and imperial endeavors into the upper portions of the Louisiana Purchase. For his part on the voyage, Catlin took the opportunity to record ethnographic observations, topographical landscapes, and portraits of the indigenous peoples whom he encountered, eventually translating them into a series of books. Often capturing the mood of his contemporaries—philosophers, artists, and scientists alike—his publications saw the peoples and places of the American West through a lens of romantic sentimentality:

> And what a splendid contemplation too, when one (who has travelled these realms, and can duly appreciate them) imagines them as the *might* in the future be seen (by some great protecting policy of government) preserved in their pristine beauty and wildness, in a *magnificent park*, where the world could see for ages to come, the native Indian in his classic attire, galloping his wild horse, with sinewy bow, and shield and lance, amid the fleeting herds of elks and buffaloes. (Catlin 1850).

His descriptions are visions of unspoiled nature, which in its wildness was beautiful and worthy of preservation. In many ways, Catlin was a product of his time, but his statements demonstrate ideas that remain central to the construction of environmental discourse into the 21st century. (Anderson 2002).

In discussing the environment, Catlin followed the classical European distinction between the human and the natural worlds, which became increasingly important in the writings of seventeenth and eighteenth-century natural philosophers. "Nature" was that realm which existed beyond the artifice and invention of human societies. To Catlin and his contemporaries, nature could exist in a pure state. Untouched by humans, it could be "pristine". The value of preserving and protecting pristine nature was an important philosophical stance among 19th-century Romantics, who recoiled from the excesses of urbanization and industrialization. A corollary to this notion of the purity of untouched nature was the idea that some humans—groups designated by the Europeans as innocent, unchanging, and uncorrupted by society, such as the Tahitians and James Fenimore Cooper's Mohicans—could live harmoniously in a state of nature. As with the pristine state of flora and fauna, writers such as Cooper and Catlin, argued that "natural man" was similarly threatened by imperialism, Indian removal policies, and assimilation. Catlin went so far as to muse about a "*magnificent park*," which would bound and protect them.

This bifurcated approach to nature, in which humans either lived outside of it or in harmony with it, simplified the complex historical relations that humans have with their environments. Catlin was just one of many who ignored evidence of the substantial human-induced ecological impacts—even those by non-urbanized and unindustrialized communities. For example, Catlin's description of the Missouri River, the riparian ecology, and the prairie was one of nature in balance, full of rich alluvial soils and dotted with "luxuriant forest timber" (Catlin 1850). Settlements, such as those of the Hidatsa at the confluence of the Knife River and Missouri River, were communities in physical and spiritual balance with their surroundings.

Below this veneer however were the centuries-long processes of anthropogenic biophysical forces. The Hidatsa, for example, were one of many groups which practiced swidden agriculture, burning tracts of land to clear their fields. They also used controlled fires to burn grasses which would attract buffalo to new growth (Anderson 2002; Courtwright 2011). The "natural" grasslands were, in fact, environments created by humans. The change in flora altered the fauna as well. Archaeological evidence suggests that there was an influx of Deer Mice (Peromyscus) into the Hidatsa settlements after the 17th century. Part of this is attributable to climate change, but Hidatsa land use and grain storage also played a significant role (Ahler et al. 1993). As a "fire positive" species—meaning that it thrives in fire-prone ecologies—the Deer Mouse population was aided by human induced prairie fires (Kaufman and Kaufman 1997).

The Hidatsa leaders explained to Catlin that the Deer Mice were out of control by the 1820s, destroying their clothes and infesting their stores. However, the arrival of a new species—likely the long-tailed weasel—to the Hidatsa villages introduced a new predator into the local ecology. According to Catlin, the leaders saw no need to control the weasels, which quickly reduced the Deer Mouse population, as it seemed a blessing from the "Great Spirit." However, since the weasel thrived in riverine environments, their numbers skyrocketed. The Hidatsa soon found the food in their *caches*, or storage cellars, eaten. The floors of their houses even began collapsing from the burrows. (Catlin 1850) According to an official from the American Fur Company, their traders had been responsible for the introducing the weasels when the animals had escaped the company's keel boats.

The example of the Knife River Hidatsa community demonstrates the ways that human systems and natural systems are entangled. The prairie itself was maintained over millennia by indigenous peoples who used fire to shape their ecologies (Abrams and Nowacki 2008). The alluvial soils at the river confluence made the location an ideal one for agriculture, encouraging human settlement. And, a combination of changing climate and human land use patterns reshaped the fauna. With the arrival of Europeans with interests in fur trading and imperial expansion, the river became a central artery for ecological exchange.

The choice of the Hidatsa for their settlement and their continued mix of sedentary agriculture and hunting patterns were not simply environmentally predetermined either; they were the result of sociopolitical forces at the regional and international levels. The Knife River confluence put them at the center of a vast trading network in which they had become important agricultural exporters and traders by the mid eighteenth-century. Their emphasis on agriculture meant that they remained sedentary even as other groups in the Plains region became nomadic due to the social transformations effected by horses and guns. The expansion of European and American empires and the instability caused by Old World technologies, fauna, and disease not only reduced their population, but stifled their expansion and movement. Likewise, competition for the control of European trade, with groups such as the American Fur Company and Hudson's Bay Company, heightened rivalries among regional groups. By the 1830s, the Hidatsa were hemmed in on the west and south by the Lakota (Calloway 1982; Hanson 1986; Martin and Szuter 1999). But, their position on the Missouri River system meant that the American Fur Company established an important post nearby (Wishart 1992). And, with the trade came the potential for invasive species, such as the weasel.

Catlin's writing represents how powerful the trope of "pristine" natural states was in nineteenth-century—a trope that continues to permeate popular environmental discourse in the 21st century. Catlin argued for the preservation of a "natural" state even as his descriptions revealed significant anthropogenic effects on the Missouri and Knife River ecologies. And, in fact, these ecologies were part of a broader, international network of human activities that ranged across cultures and continents. Competition between rival groups—the Lakota, the British, the Hidatsa, and the Americans—for control of territory and trade meant that this section of the Missouri River would both shape their activities and in turn be reshaped by them.

Examining rivers as human-nature entanglements necessitates that we see them as complex, dynamic structures that include both material and non-material systems. In effect, they are geomorphological, biophysical, *and* sociocultural systems. This observation has implications that go to the heart of research questions and methodologies as well as environmental policy. Examining a river as a natural system means studying it as a complex ecology of interdependent non-living and living components. A river is not simply determined by flows, drainage systems, or basins; it is also a biological entity that is deeply entwined with regional flora and fauna, including humans. As part of this riverine ecology, humans make multiple demands on it. These demands emerge from technological, economic, social, or cultural needs and include sanitation, flood control, transportation, recreational space, sacred space, aquaculture, and wastewater runoff channels. As humans use rivers, they impose sociocultural frameworks on river systems, which include political, scientific, religious, economic, and ideological components. These sociocultural frameworks help determine human activities and can facilitate or limit actions, and consequently are key to understanding rivers as human-nature entanglements.

Take for example the notion of a transboundary water system. A transboundary system denotes a geomorphological structure that cuts across boundaries—typically those of a nation-state. However, in the age of the Anthropocene, a host of environmental factors, including non-point pollution and climate change, mean that even water basins bounded within a single nation-state can be affected directly by extraterritorial policies and actions. Furthermore, sociocultural frameworks can have both local and transnational aspects. For example, geopolitical agendas in one part of the world might determine (or at least direct) the water policies of a nation in another part of the world. Because of this, it is useful to designate two forms of transboundary water systems: contiguous and non-contiguous. A contiguous transboundary water system is one shared by two contiguous states. An example would be the Colorado River, which is shared by the United States of America and Mexico. A non-contiguous transboundary water system is one which is physically bounded within a state but which is embedded in a transnational sociocultural system. An example would be the Yangtze River in China, which is at the center of international debates over the environmental, human, and cultural consequences of damming rivers (Lee 2013).

Bolivia's *Guerro del Agua* of 2000 highlights a number of ways in which transnational sociocultural systems are entwined with local water systems and are central to understanding both human-nature entanglements and addressing policy issues. In 2000, Cochabamba, Bolivia was a city of approximately 500,000 people and growing. Situated in a valley in the Andes, it was the 3rd largest city in the country. As an industrial hub and center for migration, its recent history had been one of continuous demographic growth. However, the city's infrastructure did not keep pace with growth, leading to unequal access to public resources such as water. And, despite the fact that the name Cochabamba is derived from "Kucha Pampa"—the Quechua word for swampland—deforestation, drought, and an overburdened water table meant that water was a precious commodity. (Shultz 2008).

To obtain water, residents relied primarily on three means. About 50 % of the residents were supplied by the Servicio Municipal de Agua Potable y Alcantarillado (SEMAPA), which administered the public reservoirs, wells, and sanitation system. SEMAPA's system provided better service to wealthier districts, which left poorer neighborhoods to rely on water delivery trucks and rain barrels. Outside of the urban core, many residents obtained water from cooperative wells and water delivery systems. Farmers had even challenged SEMAPA over where it drilled wells because it lowered the water table and threatened their access to freshwater. In 1997, they organized the Federación Departamental Cochabambina de Regantes (FEDECOR), which aimed to protect customary and communal water rights—*usos y costumbres*. (Perreault 2008) In effect, Cochabamba was an amalgam of local water management regimes, which would soon be caught up in transnational economic policies.

In the 1980s, Bolivia's national debt and inflation—the product of military rule, financial mismanagement, and subsequent instability—led the Bolivian government to pursue a regime of neoliberal economic reforms, including the privatization of the national infrastructure (Kohl 2006). Set into motion by the Decreto Supremo 21060[1], the policies helped the Bolivian government delay loan repayments and guarantee new loans from the IMF and the World Bank. A letter of intent to the IMF in 1998 promised to privatize "all remaining public enterprises" (Müller and Morales 1998). The following year, the World Bank explicitly targeted the public water system in Cochabamba, noting that it should be privatized. As part of the restructuring, water fees would go up to pay for building the Misicuni Multipurpose Project, which was meant to provide electricity and freshwater through damming the Misicuni River. The World Bank report stated that no public subsidies should be used "to ameliorate the increase in water tariffs in Cochabamba. (World Bank 1999).

Compliance with the World Bank recommendations promised economic aid for the government. Within months the Bolivian government granted a single-bid contract to Aguas del Tunari, which was owned by the Bechtel Corporation subsidiary International Water Limited as well as several Bolivian companies with links to the Bolivian government. (Nickson and Vargas 2002 fn 10). The deal gave Aguas del Tunari a 40 year contract to run the water system with a promise of an average 16 % annual return on investment. These conditions meant that the average consumer saw an increase of 35 % in their water bills, but some saw increases as high as 300 %. For many of the poorest, these increases were nearly impossible to pay. Meanwhile the government passed legislation—(Ley 2029[2]: Ley de Servicios de Agua Potable y Alcantarillado Sanitario)—which commercialized the water supply and threatened the *usos y costumbres* of local cooperatives. (Assies 2003).

The reaction to these moves was quick and potent. Roads were blockaded by FEDECOR in protest, and this was soon followed by an alliance with the Federación de Trabajadores Fabriles de Cochabamba—an engagement that led to the creation of the Coordinadora de Defensa del Agua y de la Vida. Confrontations with the government centered on several concerns: the cost of water, the *usos y costumbres*, and anti-neoliberal sentiment. Government repression and violence was met with mass protest and resistance, eventually leading the government to end the contract with Aguas del Tunari and integrating the Coordinadora de Defensa del Agua y de la Vida into the management of SEMAPA.

Understanding the water war in Cochabamba requires an understanding of both the local environmental conditions as well as the transnational sociocultural frameworks in which it exists. While representatives on all sides were aware of the need for addressing the water supply issue, ideological positions helped determine solutions. There were three primary ideological frameworks for addressing the water system in Cochabamba: neoliberal market mechanisms, social democracy,

---

[1] Decreto Supremo 21060, 29 August 1985. http://www.gacetaoficialdebolivia.gob.bo/normas/buscar/21060. Accessed July 11, 2014.

[2] Ley 2029, 29 October 1999. http://www.lexivox.org/norms/BO-L-2029.xhtml. Accessed July 11, 2014.

and the moral economy of the *usos y costumbres*. Each was concerned, in its own way, with ameliorating the problems facing the city's residents—from poor sanitation to the absolute scarcity of water. The World Bank in alliance with the government, sought to address the problem by imposing an economic model that they imagined would boost economic growth through the market economy. Social democrats rejected privatization out-of-hand, preferring a state-based solution that would answer to the people. Those arguing for the *usos y costumbres* preferred a locally based program tied to community needs. Consequently, the local material conditions, which had been created by environmental and anthropogenic processes, were tied to transnational ideological debates, power struggles, and global economics. As such, the water system in Cochabamba was an transnational entanglement of environmental, social, political, technological, and cultural expectations and practices.

The short case studies described above only hint at the complex sociocultural-environmental entanglements that are central to the Cochabamba and the Missouri water systems. Only a transdisciplinary analysis of these entanglements—one that blends scientific knowledge of earth systems with social scientific and humanistic knowledge of human systems—is likely to get at the complex material and non-material interrelations and feedback mechanisms inherent to the system. This means that new transdisciplinary conceptual and methodological frameworks are necessary for studying anthropocene environments and developing policies to mitigate anthropogenic environmental impacts. New approaches will need to pay close attention to the role of human agency and the construction of sociocultural systems at multiple scales in order to understand the mechanisms by which sociocultural systems converge with environmental systems. Doing so will allow scholars to understand not only the effects of anthropogenic environmental change, but the processes that drive human behavior and action.

References

Abrams MD, Nowacki GJ (2008) Native Americans as active and passive promoters of mast and fruit trees in the eastern USA. Holocene 18(7):1123–1137

Ahler SA, Snyder LM, Falk CR, Semken HA Jr (1993) KNRI and Upper Knife-Heart Region unmodified faunal remains. In: The phase I archeological research program for the Knife River Indian Villages National Historic Site. U.S. Department of the Interior, National Park Service, Midwest Archeological Center 263–272

Anderson MK (2002) An ecological critique. In: Stewart OC (ed) Forgotten fires: Native Americans and the transient wilderness. University of Oklahoma Press, Oklahoma City, pp 37–64

Assies W (2003) David versus Goliath in Cochabamba: water rights, neoliberalism, and the revival of social protest in Bolivia. Latin Am Perspec 30(3):14–36

Castree N, Braun B (eds) (2005) Social nature: theory, practice, and politics. Blackwell, Malden

Buckingham JT, Buckingham E (eds) (1832) Trade in the West. In: The New England Magazine 3:262

Calloway CG (1982) The inter-tribal balance of power on the Great Plains, 1760-1850. J Am Stud 16(1):25–47

Catlin G (1850) Illustrations of the manners, customs, and condition of the North American Indians. 9th edn. vol. 1. Henry G. Bohn, London
Courtwright J (2011) Prairie fire: a great plains history. University Press of Kansas, Lawrence
Cronon W (1996) Uncommon ground: rethinking the human place in nature. W. W. Norton & Company, New York
Costanza R et al (2012) Developing an integrated history and future of people on Earth (IHOPE). Curr Opin Environ Sustain 4(1):106–114
Costanza R, Graumlich L, Steffen W (eds) (2007) Sustainability or collapse?: an integrated history and future of people on Earth. The MIT Press, Cambridge
Davies MIJ, M'Mbogori FN (eds) (2013) Humans and the environment: new archaeological perspectives for the twenty-first century. Oxford University Press, Oxford
DeMarrais E, Gosden C, Renfrew C (eds) (2004) Rethinking materiality: the engagement of mind with the material world. McDonald Institute for Archaeological Research. Oxbow Books and David Brown, Cambridge
Dippie BW (1990) Catlin and his contemporaries: the politics of patronage. University of Nebraska Press, Lincoln
Edgeworth M (2011) Fluid pasts: archaeology of flow. Bristol Classical, Bristol
Hanson JR (1986) Adjustment and adaptation on the northern plains: the case of equestrianism among the Hidatsa. Plains Anthropologist 31(112):93–107
Hassan F (2011) Water history for our times, IHP essays on water history, vol 2. UNESCO, Paris
Hibbard KA et al (2010) Developing an Integrated History and Future of People on Earth (IHOPE): Research Plan. IGBP Secretariat. http://www.uvm.edu/giee/pubpdfs/Costanza_2012_Current_Opinion_in_Environmental_Sustainability.pdf. Accessed August 15, 2013
Hodder I (2011) Human-thing entanglement: towards an integrated archaeological perspective. J Roy Anthropol Inst 17(1):154–177
Hodder I (2012) Entangled: an archaeology of the relationships between humans and things. Wiley, London
Hornborg A, Crumley CL (eds) (2006) The world system and the Earth system: global socioenvironmental change and sustainability since the Neolithic. Left Coast Press, Walnut Creek
Jäger J et al (2012) Responses to environmental and societal challenges for our unstable earth (RESCUE), ESF Forward Look—ESF-COST "Frontier of Science" Joint Initiative. European Science Foundation, Strasbourg and European Cooperation in Science and Technology, Brussels. http://www.esf.org/fileadmin/Public_documents/Publications/rescue.pdf. Accessed August 15, 2013
Kaufman GA, Kaufman DW (1997) Ecology of small mammals in prairie landscapes. In: Knopf FL, Samson FB (eds) Ecology and conservation of great plains Vertebrates. pp 207–43. Springer, New York
Kohl B (2006) Challenges to neoliberal hegemony in Bolivia. Antipode 38(2):304–326
Latour B (2005) Reassembling the social: an introduction to actor-network-theory. Oxford University Press, Oxford
Lee YB (2013) Global capital, national development and transnational environmental activism: conflict and the Three Gorges Dam. Journal of Contemporary Asia 43(1):102–126
Malafouris L, Renfrew C (2010) The cognitive life of things: archaeology, material engagement and the extended mind. In: Malafouris L, Renfrew C (eds) The cognitive life of things: recasting the boundaries of the mind. McDonald Institute for Archaeological Research and Oxbow Books, Cambridge, pp 1–12
Martin PS, Szuter CR (1999) War zones and game sinks in Lewis and Clark's West. Conserv Biol 13(1):36–45
Mosley S (2006) Common ground: integrating social and environmental history. J Soc Hist 39(3):915–933
Müller H, Morales, JA (1998) Bolivian letter of intent to the IMF for economic reforms. http://www.imf.org/external/np/loi/081498.HTM. Accessed October 4, 2013
Nickson A, Vargas C (2002) The limitations of water regulation: the failure of the Cochabamba concession in Bolivia. Bull Latin Am Res 21(1):99–120

Perreault T (2008) Custom and contradiction: rural water governance and the politics of usos y costumbres in Bolivia's irrigators' movement. Ann Assoc Am Geogr 98(4):834–854
Shultz J (2008) The Cochabamba water revolt and its aftermath. In: Draper M, ShultzJ (eds) Dignity and defiance: stories from Bolivia's challenge to globalization. University of California Press, Berkeley, pp 9–42
Sörlin S (2012) Environmental humanities: why should biologists interested in the environment take the humanities seriously? Bioscience 62(9):788–789
Wishart DJ (1992) The fur trade of the American West: 1807–1840: a geographical synthesis. University of Nebraska Press, Lincoln
World Bank (1999) Bolivia Public Expenditure Review. Report number 19232-BO

Chapter 13
Eastern European Perspective on the Environmental Aspects in Current Flood Risk Management: The Example of the Czech Republic

Blanka Loučková

Abstract New approaches to flood risk management strategies, moving away from large-scale technical solutions towards a greater involvement of natural processes, have recently been observed in some European countries. The primary purpose of this paper is to analyze this trend in the Czech Republic, an eastern European country in which engineered approaches have been heavily employed in the recent past. To assess the environmental aspects of current flood management strategies, the degree of implementation of "green" measures in relevant documents was evaluated at two levels: national legislative level and the regional implementation level. While the shift towards natural solutions in flood protection is well reflected at national level, traditional engineered approaches favouring "grey" infrastructure still prevail at regional level. The study discusses possible reasons for this gap between levels of governance and looks at the obstacles that hinder the promotion of natural measures in flood risk management.

Introduction

The increasing frequency and severity of extreme flood events in recent decades in Europe, and in many countries worldwide, have involved huge levels of damage and great economic losses, not to mention loss of human life. Projections of warmer climate show a growth in the number of extreme precipitation events and the risk of flooding is expected to rise in many areas (IPCC 2007; EEA 2012). Moreover, human settlement in flood-prone areas has increased the vulnerability of

B. Loučková (✉)
Department of Human Dimensions of Global Change, CzechGlobe—Global Change Research Center, Academy of Sciences of the Czech Republic, Prague, Czech Republic
e-mail: louckova.b@czechglobe.cz

A. Bhaduri et al. (eds.), *The Global Water System in the Anthropocene*, Springer Water, DOI: 10.1007/978-3-319-07548-8_13,

communities to flooding (Barredo 2009; Schlüter et al. 2010; Feyen et al. 2012; Ballesteros-Cánovas et al. 2013). Management of flood risks thus poses substantial challenges for water managers and policymakers.

In the past, flood management strategies in Eastern Europe were based mainly on "prediction and control" approach, with centralized governance and massive technical infrastructure (dams, dikes, levees, embankments, etc.), all intended to play key roles in coping with hydrological extremes (Kundzewicz 2001; Pahl-Wostl 2007; Krysanova et al. 2008). In the last decade there has been a shift from such attempts at direct flood control to management of flood risk (Merz et al. 2010), encapsulated in the concept of integrated flood management (IFM), which aims to maximize the benefits of floodplains and minimize losses from flooding by using a comprehensive range of possible flood protection measures (WMO 2006). The key elements of IFM are multidisciplinary approaches to flood management at river basin level, combined with stakeholder involvement and preserving ecosystems and their benefits—"natural services" (WMO 2006).

This trend corresponds with the tasks prioritized by European Union water policy in recent years, including (i) improving the sustainable use of water resources and (ii) protection and maintenance of the resilience of water ecosystems (EEA 2012). Many recent EU policy documents highlight the importance of resilient ecosystems and their crucial role in hazard mitigation and climate change adaptation. The "Green Infrastructure", a planned EU network of natural and semi-natural areas and features delivering a wide range of benefits and services has recently been promoted (EC 2011). Key elements of the Green Infrastructure are natural water retention measures, adaptive measures that improve the retention potential of watersheds by restoring ecosystems and natural eco-hydrological processes (EEA 2012). Using the capacity of natural processes, these measures enhance the resilience of society to water-related natural hazards through regulating ecosystem services, simultaneously delivering a variety of ecosystem services from which people may benefit, such as water and material provision, water purification, air quality and climate regulation, recreation, aesthetic and cultural value (EC 2011).

A shift away from large-scale engineered solutions towards a "working with nature" approach in flood management has already taken place in some European countries (van der Grijp et al. 2006; Loftus 2011; Ford et al. 2011).

This paper describes a shift towards "green" solutions in flood strategies in the Czech Republic, a former communist country in which centralised command and control were absolute, and such engineered measures were vigorously imposed. These took up and reinforced an already very strong tradition in flood management. The main goal of this paper is to capture a recent change in attitude towards environmental solutions in flood management, as reflected in legislation and policy documents at both national and regional levels. The question addressed is: how are the natural (or "green") measures implemented in national policy documents and how are they incorporated into river basin management plans in the Czech Republic? The study also discusses obstacles that may play significant roles in thwarting the implementation of natural measures in flood management.

In this paper, natural or "green" measures are considered as action taken in watersheds that complements technical ("grey") infrastructure and that are based on the principle of "working with nature" (EEA 2012): restoration of rivers and floodplains, re-establishment of natural flow regimes in watercourses, increasing natural water retention in watersheds, afforestation, land cover and land use changes, sustainable forestry and agricultural practices, improvement of the permeability of surfaces in urban areas (EEA 2012).

The paper is organized as follows: Sect. "The History of Czech Flood Risk Management" describes the development of Czech flood risk management from totalitarian socialist times to the present, pointing out milestones—in particular, the end of the communist system, a series of catastrophic floods and finally integration into the European Union. In Sect. "Incorporation of Natural Flood Protection Measures into Legislation and Policy Documents" the degree of incorporation of natural measures into Czech flood management documents at national level is evaluated (3.1) and the actual state at regional level is investigated (3.2). Section "Main Obstacles to the Promotion of Natural Flood Protection Measures" discusses obstacles that might be hindering the promotion of natural measures in flood risk management. Section "Conclusions" concludes the paper and identifies further directions for research in this field.

The History of Czech Flood Risk Management

Flood Risk Management in Totalitarian Times (1948–1989)

During the "communist" period (a term used hereafter to summarise several decades of totalitarian government), flood risk management in the Czech Republic (at that time known as Czechoslovakia) was characterized by large-scale technical intervention, predominantly involving river training, building large reservoirs and the construction of embankments. Such a "prediction and control" approach gave priority to structural measures that had been centrally planned and widely implemented. Floodplains were drastically reduced, wetlands were drained and all large rivers were "trained" to meet the increasing demands of intensive collective agriculture. These massive interventions led to the disappearance of many characteristic river corridor habitats and left behind a landscape with reduced water storage capacity and high vulnerability to extreme events, particularly to floods (Hrádek and Lacina 2003). Natural ecosystem services were replaced by "grey" infrastructure and, far from important elements of flood protection, they were considered undesirable elements of landscape. These country-wide regulations led to a decrease in the retention capacity in watersheds, resulting in floods that are all the more severe and rapid in modern times (Kundzewicz 2001; Petz et al. 2012).

Table 13.1 Recent floods in the Czech republic (*source* the Ministry of the Environment and the Ministry of Agriculture of the Czech Republic, 2010)

Flood event	Damage (million EUR)	Death toll
1997 (summer type)	2 407.2	60
1998 (summer type)	70	10
2000 (spring type)	146.1	2
2001 (spring type)	38.5	0
2002 (summer type)	2 884.1	16
2006 (spring type)	238.4	9
2009 (summer type)	326.9	15
2013 (summer type)	*	10
Total 1997–2013	6 111.2	122

Two types of floods are typical: summer floods resulting from heavy storm rainfall and spring floods caused by rapid snowmelt and ice movement on rivers

*Not yet quantified at the time of manuscript preparation

Early 1990s–Present

The collapse of the communist system in 1989 triggered profound social, political, institutional and cultural changes. The state of water and flood management was recognized as unfavourable and was perceived as an urgent issue to be addressed. To ensure the restoration and improvement of the water regime in watersheds, the "Revitalization of River Systems" programme was undertaken by the Ministry of the Environment in 1992. This programme (finished in 2008) provided financial means for developing natural measures in the landscape that would slow run-off and increase biodiversity.

Integration of the country with the European Union in 2004 accelerated changes in attitudes towards environmental solutions in flood management. Joining the EU required the "approximation" of new legislation, including the Water Framework Directive (EC 2000) and the Flood Directive (EC 2007). To comply with EU legislation, several targets related to sustainable development and the protection of the environment had to be met. The "Environment" Operational Programme, with priority tasks such as "reducing the risk of flooding" and "optimization of the water regime in landscape" was approved by Czech Government in 2006.

The development of flood management in the Czech Republic was lent impetus by the occurrence of eight severe floods that hit the country in the last 16 years, leaving behind total damage to the tune of 6.1 billion Euros and 122 dead (Table 13.1). In particular, the disasters in 1997 and 2002 may be considered important landmarks in contemporary Czech flood risk management. Not only have early-warning and emergency systems been improved enormously since 1997, but the events also initiated a profound shift in the perception of floods per se and served to trigger changes in approaches to flood protection. An increasing number of experts have started to point out that solely technical solutions cannot protect society against flooding and that a change is needed in the existing approach towards flood control. This shift becomes quite clearly observable in legislation and is described in the following section.

Incorporation of Natural Flood Protection Measures into Legislation and Policy Documents

The documents that provide the underlying framework for Czech flood risk management at national level translated as: Strategy for Protection against Floods; Plan for Main River Basins; Spatial Development Policy; and Plan for Flood Protection using Technical and Natural Measures; were evaluated to investigate the degree of incorporation of natural measures and the type of measures proposed (Table 13.2). To establish the actual situation at regional level (at district level for the river basins involved), all eight regional river basin management plans were assessed.

The National Level

The current Czech flood risk management is based on the Strategy for Protection against Floods (further, "the Strategy") approved by the government in 2000 (updated in 2006). The Strategy acknowledges that effective flood protection is "a combination of measures in watersheds that promote natural retention of water and technical measures that attenuate flood flows" (MACR 2000, p. 2). It recognizes a suitable combination of natural and technical measures as a necessary tool for effective flood protection. The main natural measures proposed are land use and land cover changes, restoration of riparian vegetation, creation of natural inundation and infiltration structures and changes in the landscape in order to retain water and decelerate run-off.

The Plan for the Main River Basins (further, "the Plan") is an important strategic document for water planning. Issued by the Ministry of Agriculture in 2007, it includes flood protection among its three main topics. The document places great emphasis on preventive protection and recognizes that a suitable combination of measures in the landscape that increase natural water retardation and technical measures addressing flood run-off are necessary for effective flood protection (MACR 2007, p. 20). According to this document, an appropriate combination of natural water retention measures and technical measures should be favoured as a strategy in protection against floods. The natural measures proposed in the Plan are similar to those in the Strategy: creation of natural inundation areas and infiltration structures in watersheds, restoration of wetlands and regulated river channels, improvement of structure and species composition of forest ecosystems, and environment-friendly agricultural activities. Apart from these, the document also suggests financial incentives to convert arable land in flood-prone areas into permanent grasslands.

Table 13.2 Incorporation of natural flood protection measures in flood management documents at national level (documents arranged according to year of issue)

Document	Issued by	Aims of the document	Incorporation of natural measures	Proposed types of natural measure
Strategy for Protection against Floods (2000, updated 2006)	Ministry of agriculture	Reduction in flood risks and flood damage.	Yes (In combination with technical measures)	Enhancing water retention in watersheds through natural inundation spaces and infiltration structures in watershed. Land cover and land use changes. Restoration of riparian vegetation.
Plan for the main river basins (2007)	Ministry of agriculture	Protection and sustainable management of water resources, flood protection.	Yes (In combination with technical measures)	Enhancing water retention in watersheds through landscape restoration and natural inundation spaces. Land use and land cover changes. Restoration of riparian vegetation
Spatial development policy (2008)	Ministry for regional development	Minimization of flood damage.	Not specified	Not specified
Plan of flood protection using technical and natural measures (2010)	Ministry of agriculture and ministry of the environment	Reduction in flood damage. Change in attitudes towards floods: maximum water retention in landscape through enhancement of natural means.	Yes (In combination with technical measures)	Land use and land cover changes. Restoration of river channels and floodplains. Afforestation (fast-growing trees). Creation of natural inundation spaces.

The Spatial Development Policy is a strategic document for coordination of spatial planning. One of its priorities is "to create conditions for preventive protection of an area against potential risks and natural disasters within it (floods, landslides, erosion etc.) in order to minimize any damage" (MRDCR 2008, p. 18). This means provision of territorial protection for areas designated for flood protection measures, as well as the establishment of areas in floodplains that may be built upon, albeit only in exceptional cases. The policy devotes only one sentence (MRDCR 2008, p. 18) to the issue of natural retention of rainwater as an alternative to artificial accumulation of water; it does not go any further and nor does it specify any detail at all.

The Plan of Flood Protection using Technical and Natural Measures (MACR and MECR 2010) issued in 2010 may be considered as the "state-of-the-art" document in terms of promoting natural measures in Czech flood management. Considering natural measures as means of equal value to the engineered solutions currently predominant, it emphasizes the need for a shift in attitudes towards natural measures. Moreover, it calls for developing appropriate adaptation measures to climate change and highlights the interdependence of flood control measures and such an adaptation process. The massively preferred technical infrastructure is viewed here as only a partial solution that cannot prevent floods, inferring that environmental matters need to be taken into account. Priorities include changing landscape patches with high erosion rates into permanent grasslands, improving the hydromorphology of river corridors through restoration of rivers channels and floodplains, afforestation with fast-growing woody species and the creation of natural inundation areas in watersheds.

The Regional Level

The Czech Republic is divided into eight river basin districts; each of them has its own management plan for the years 2010–2015 (approved in 2009).[1] Based on the Plan for the Main River Basins and prepared by river basin authorities, these plans are the legally binding policy documents for the six-year period in the field of water management at the level of river basin district. Flood protection is one of the three primary goals, along with protection of water bodies and sustainable management of water resources. Each plan contains a summary of the proposed flood control measures ("grey", "green", "soft") planned in the district for the years 2010–2015, including the identification number, name, type of measure with a short description and anticipated implementation costs.

As such plans reflect the situation in flood protection at regional level, each of them was inspected as part of this study to establish the extent to which natural

[1] Czech version only (*Plány oblasti povodí*) available at http://eagri.cz/public/web/mze/voda/planovani-v-oblasti-vod/plany-povodi-pro-1-obdobi/plany-oblasti-povodi/.

Table 13.3 Planned flood protection measures in river basin districts (natural measures to all measures) (*Source* river basin management plans)

River basin district	Planned flood protection measures—natural/all	Recognition of natural measures
Ohře and Dolní Labe	1/30	Low
Horní and Střední Labe	2/97	Low
Berounka	0/77	None
Horní Vltava	1/116	Low
Dolní Vltava	0/42	None
Dyje	4/68	Low
Morava	3/72	Low
Odra	5/104	Low

measures are involved at this level in each particular river basin. According to the number of natural measures in the river basin management plans, four categories of plans were defined: plans with (1) "no recognition" of natural measures (zero natural measures in plan); (2) "low recognition" (proportion of natural measures 1–20 % of all measures); (3) "intermediate recognition" (proportion of natural measures 21–40 % of all measures); and (4) "significant recognition" (proportion of natural measures higher than 40 % of all measures). This classification was based on findings that to be effective, Green Infrastructure should encompass a sufficiently large area—relevant documents suggest a minimum of 40 % of total land area (including private gardens and green roofs) (DCLG 2009).

The results show (Table 13.3) that two of river basin management plans are classified as "no recognition" of natural measures (Berounka and Dolní Vltava), while six others may be classified as "low recognition". Table 13.3 also shows the total number of "green" measures as given in river basin management plans compared to all types of measures in each district planned for years 2010–2015. In general, "green" solutions are substantially less represented in river basin management plans than "grey" measures.

Main Obstacles to the Promotion of Natural Flood Protection Measures

The experience of a series of disastrous floods in a relatively short period of time (eight extreme flood events in the last 16 years) in the Czech Republic has stimulated a shift in attitudes to flood protection. This is reflected in flood management documents at the national level. As the evaluation of key national documents showed, incorporation of "green" measures at this level is generally satisfactory. Natural measures in combination with technical infrastructure are taken into consideration in three of four documents; the most often-proposed measures are creation of natural inundation spaces in watersheds, restoration of riparian vegetation, together with land use and land cover changes (Table 13.2).

The Plan of Flood Protection using Technical and Natural Measures may be considered the most advanced of these documents, where "green" measures are given top priority for effective flood risk management. However, a gap remains between the national level and its implementation at regional level in river basin management plans, as "green" measures represent a relatively tiny number of all planned flood protection measures. Although the environmental aspect is incorporated in national legislation and the fact that solely technical solutions in flood risk management can be counterproductive has been recognized, technical approaches are still favoured in Czech flood management at regional level.

There are many possible explanations for this inconsistency between national legislation and regional implementation, and any approach to a clearer picture will need further research. Some preliminary hypotheses are offered in the following section, based on the available literature and local experience. They might serve as a starting point for further discussion and research on this topic.

Governance Disunity and Fragmentation Between Policy Sectors

The responsibility for flood protection measures at national governance level is divided between two governmental bodies—the Ministry of Agriculture and the Ministry of the Environment. The implementation of technical measures is provided primarily through flood prevention programmes under the auspices of the Ministry of Agriculture, which financially supports the construction of water reservoirs and construction of embankments. While the Ministry of Agriculture advocates largely technical solutions, the Ministry of the Environment and various non-governmental organizations, along with local initiatives, are the main promoters of "green" flood protection measures. Hence governance responsibilities are highly fragmented between these two policy sectors, which results in highly inefficient performance. Discrepancies in sectoral coordination that hamper successful action have already been recognized in some research papers, particularly in terms of adaptation to climate change as a process (Peltonen et al. 2010, Shelfaut et al. 2011). Coordinated cross-sectoral interplay is crucial for effective execution of flood risk management and enhancing the resilience of society to extreme climate-based events.

Deep-Rooted and Long-Term Practices Among Water Managers

Established routines among water managers are very difficult to change, since they have developed over the course of some time (Pahl-Wostl 2007; Merz et al. 2010). This is relevant to the situation in the Czech Republic, where technical infrastructure played the most important role in flood protection during communist times and was favoured at the expense of all other solutions ("green" and "soft") Moreover, most of senior staff in today's water management authorities, with important decision-making rights, were educated when the water management profession was lacking in any environmental aspect whatsoever. The promoters of environmental solutions may be perceived by certain water managers as unqualified and unprofessional. Enhanced communication between these actors, supporting a change from conservative and narrowly focused approaches to more comprehensive and holistic ideas thus represents a major challenge.

Strongly Rooted Trust in Large-Scale Technical Infrastructure on the Part of the General Public

Most of Czech public feel safe behind large dams and do not trust "solutions that cannot be precisely calculated", as the natural measures are frequently perceived. Environmental principles in flood protection have not yet gained much public support and they are often seen as something "alternative" and not very effective. These attitudes may well have historical roots in the communist regime, in which much of the country's status was invested in visible and extensive technical infrastructure e.g. dams and reservoirs were promoted, while watercourses were regulated into artificial channels and riparian vegetation was systematically removed. Raising public awareness of the value of benefits from natural systems, along with expressing their effects in monetary and measurable terms, seem to be intrinsic to achieving any success for natural measures.

Comprehensive Land Consolidation Still in Process

Spatial planning is an influential tool for reducing flood impacts (Shelfaut et al. 2011). The process of comprehensive land consolidation, a powerful instrument with a multi-purpose objective (dealing with changes in land ownership, land conservation and flood control) has been taking place in the Czech Republic at the level of cadastral areas since 1990s (the cadastral unit is basic to Czech land law). According to policymakers and municipality authorities, plot fragmentation and problems of land acquisition from certain private owners are the main barriers to successful realization of the process.

Stakeholder Participation

Public participation, strongly advocated by many authorities in the field (Tierney et al. 2001; Dixit 2003; Wisner et al. 2004; Schelfaut et al. 2011) plays a crucial role in raising societal resilience and building adaptive capacity. Flood management at the regional level is missing the greater participation of local administrations in the decision-making process; flood protection measures are often organized by the state administration without wider public involvement.

Conclusions

A series of severe floods in the last two decades in the Czech Republic has triggered a legislative response and stimulated discussion about sustainable flood management and protection. The shift in approach towards the environmental dimension of flood management during the last decade is reflected in relevant legislation at national level. However, a gap still remains between the national and regional level. Czech regional planning authorities have not fully appreciated the importance of natural ecosystems in disaster risk management. Enhanced communication between actors at national and regional levels and monitoring the incorporation of "green" measures in regional plans and their conformity to national legislation—and even EU guidelines and legislation—might help to reduce the gap between levels of governance.

Further research is needed to answer the following questions: what are the main factors influencing the actors in the decision-making process in flood management? What are the attitudes of the relevant stakeholders (policymakers, water managers, regional spatial planners and others) towards natural measures? A research that would explain public preferences towards technical solutions in flood management is needed as well. Further, more examples that combine "grey" and "green" measures, demonstrating that these solutions can complement one another in real, measurable terms, are acutely needed in the Czech Republic, as they can help change public perception and the long-established practices of certain water managers. The implementation of "green" measures into flood risk management strategies poses a substantial challenge for policymakers. To facilitate this task, further research should also address the questions of the precise quantitative effects of wetlands in buffering floods, as well as the issue of determination of the area, character and distribution of natural measures in watersheds.

Acknowledgements The paper was prepared with the financial support of the Ministry of the Interior of the Czech Republic, grant no. VG20122015091 Integrated assessment of global change impacts on environmental security of the Czech Republic. The study was also supported by project CzechGlobe–Centre for Global Climate Change Impacts Studies, Reg. No. CZ.1.05/1.1.00/02.0073. The author would like to thank Mr. Tony Long for English style corrections.

References

Ballesteros-Cánovas J, Sanchez-Silva M, Bodoque JM, Díez-Herrero A (2013) An integrated approach to flood risk management: a case study of Navaluenga (Central Spain). Water Resour Manage 27:3051–3069. doi:10.1007/s11269-013-0332-1

Barredo JI (2009) Normalised flood losses in Europe: 1970–2006. Nat Hazards Earth Syst Sci 9:97–104

DCLG (2009) Planning policy statement: Eco-towns. A supplement to planning policy statement 1. Department for Communities and Local Government, London. https://www.gov.uk/government/uploads/system/uploads/attachment_data/file/7773/pps-ecotowns.pdf. Accessed 21 Aug 2013

Dixit A (2003) Floods and vulnerability: need to rethink flood management. Nat Hazards 28:155–179

EC (2000) Directive of the European parliament and of the council 2000/60/EC on establishing a framework for community action in the field of water policy. European Commission, Brussels

EC (2007) Directive of the European parliament and of the council 2000/60/EC on the assessment and management of flood risks. European commission, Brussels

EC (2011) Our life insurance, our natural capital: an EU biodiversity strategy to 2020. European Commission, Brussels

EEA (2012) European waters—current status and future challenges. EEA Report 9/2012. Copenhagen, p 52

Feyen L, Dankers R, Bódis K et al (2012) Flood hazard in Europe in present and future climates. Clim Change 112:47–62. doi:10.1029/2008JD011523

Ford JD, Berrang-Ford L, Paterson J (2011) A systematic review of observed climate change adaptation in developed nations. Clim Change 106:327–336. doi:10.1007/s10584-011-0045-5

van der Grijp NM, Olsthoorn AA, Tol RSJ, van der Werff PE (2006) Flood risk institutions and climate change in the Netherlands. In: Ruth M (ed) Smart growth and climate change: regional development, infrastructure and adaptation, 1st edn. Edward Elgar, Northampton, pp 243–268

Hrádek M, Lacina J (2003) Destructional landforms arised from extreme events in the Desná River valley and their vegetation. Moravian Geogr Rep 11(1):2–19

IPCC (2007) Climate change 2007: the physical science basis. Summary for policymakers. Contribution of working group I to the fourth assessment report of the intergovernmental panel on climate change. Cambridge University Press, Cambridge

Krysanova V, Buiteveld H, Haase D et al (2008) Practices and lessons learned in coping with climatic hazards at the river-basin scale: floods and droughts. Ecol Soc. 13(2):1–32

Kundzewicz ZW (2001) Water problems of central and eastern Europe—a region in transition. Hydrol Sci J 46(6):883–896

Loftus AC (2011) Lodz, Poland—river restoration as a catalyst for sustainable urban development. Case study of the SWITCH project. ICLEI (Local Governments for Sustainability), Freiburg. http://www.switchtraining.eu/fileadmin/template/projects/switch_training/files/Case_studies/Case_study_Lodz_preview.pdf. Accessed 28 Aug 2013

Merz B, Hall J, Disse M, Schumann A (2010) Fluvial flood risk management in a changing world. Nat Hazards Earth Syst Sci 10:509–527. doi:10.5194/nhess-10-509-2010

MRDCR (2008) Spatial development policy (Politika územního rozvoje). Ministry for Regional Development of the Czech Republic, Prague. http://www.uur.cz/images/publikace/metodickeprirucky/plnezneni/pur-cr-2008-2009/pur-cr-2008-en.pdf. Accessed 10 May 2013

MACR (2000) Strategy for protection against floods (Strategie ochrany před povodněmi). Ministry of Agriculture of the Czech Republic, Prague. http://www.mzp.cz/C1257458002F0DC7/cz/strategie_ochrany_povodne/$FILE/OOV_strategie_povodne_20000419.pdf. Accessed 10 May 2013

MACR (2007) Plan for the main river basins (Plán hlavních povodí). Ministry of Agriculture of the Czech Republic, Prague. http://eagri.cz/public/web/file/32725/PHP_anglicky_web.pdf. Accessed 10 May 2013

MACR, MECR (2010). Plan of Flood Protection using Technical and natural measures (Koncepce řešení problematiky ochrany před povodněmi v ČR s využitím technických a přírodě blízkých opatření). Ministry of Agriculture of the Czech Republic and Ministry of the Environment of the Czech Republic, Prague. http://eagri.cz/public/web/mze/ministerstvo-zemedelstvi/koncepce-a-strategie/koncepce-reseni-problematiky-ochrany.html. Accessed 10 May 2013

Pahl-Wostl C (2007) Transitions towards adaptive management of water facing climate and global change. Water Resour Manage 21:49–62. doi:10.1007/s11269-006-9040-4

Peltonen L, Juhola S, Schuster P (2010) Governance of climate change adaptation: policy review. Climate change: Impacts, costs and adaptation in the Baltic Sea region, project report. Available via BaltCICA. http://www.baltcica.org/results/events/baltcicanews_10.html. Accessed 30 June 2013

Petz K, Minca EL, Werners SE, Leemans R (2012) Managing the current and future supply of ecosystem services in the Hungarian and Romanian Tisza river basin. Reg Environ Change 12:689–700. doi:10.1007/s10113-012-0284-7

Schelfaut K, Pannemans B, van der Craats I et al (2011) Bringing flood resilience into practice: the FREEMAN project. Environ Sci Policy 14:825–833. doi:10.1016/j.envsci.2011.02.009

Schlüter M, Hirsch D, Pahl-Wostl C (2010) Coping with change: responses of the Uzbek water management regime to socio-economic transition and global change. Environ Sci Policy 13:620–636. doi:10.1016/j.envsci.2010.09.001

Tierney KJ, Lindell MK, Perry RW (2001) Facing the unexpected disaster preparedness and response in the United States. Joseph Henry Press, Washington, DC

Wisner B, Blaikie P, Cannon T, Davis I (2004) At risk. Routledge, London

WMO (2006) Environmental aspects of integrated flood management. Associated programme on flood management, flood management policy series, Technical Document No.3, WMO, Geneva, p 86

Chapter 14
Adaptive Irrigation Management in Drought Contexts: Institutional Robustness and Cooperation in the Riegos del Alto Aragon Project (Spain)

Sergio Villamayor-Tomas

Abstract This chapter aims to understand the ability of more than 10,000 farmers in a large irrigation project to cooperate and adjust their water demands to cope with droughts. Causal inferences are formulated with the aid of common pool resource (CPR) theory as well as qualitative and quantitative evidence. According to the analysis, a series of robust water management institutions as well as additional land use factors contribute to the collective adaptation of farmers in drought conditions. Water management institutions include a flexible common property regime, effective environmental and social monitoring mechanisms, and decentralized administrative leadership. Land use factors include the existence of a moderate heterogeneity of farmers in their dependence from irrigated agriculture, the relatively substitutability of high and low water demand crops and a strong mechanism of government-sponsored income support subsidies. Overall, the analysis illustrates the interest of understanding adaptation from the perspective of CPR theory, as well as the usefulness of integrating the study of water and land use dynamics to understand sustainable management in the irrigation sector.

Introduction

The increased global exposition to climate change disturbances such as droughts and floods has generated a new interest in understanding the manner by which communities in specific productive sectors at different scales cope with those threats (UN/ISDR 2004). This chapter aims to contribute to fill that gap by offering some explanations to the ability of more than 10,000 farmers in a large irrigation project in Spain to cope with droughts.

S. Villamayor-Tomas (✉)
Division of Resource Economics, Department of Agricultural Economics, Humboldt University, Berlin, Germany

A. Bhaduri et al. (eds.), *The Global Water System in the Anthropocene*, Springer Water, DOI: 10.1007/978-3-319-07548-8_14,

There is a long history of policy and research efforts focused on explaining the performance of large irrigation projects as a source of wealth and food security in developing and developed countries (Ostrom 1992a; Subramanian et al. 1997). Scholarship aiming to understand the robustness of those projects to disturbances is much less developed. Spain has a century-long history of such type of projects (Melgarejo Moreno 2000), many of which have successfully evolved to adapt to a variety of threats over time. In the last 20 years, however, a series of severe droughts, as well as other threats, have raised renewed concern about such adaptive capacity (Lopez Galvez and Naredo 1997).

Natural resource management scholarship in general and common pool resource (CPR) theory in particular can be very productive starting points to understand adaptation and robustness. From the perspective of CPR theory, the success of common property regimes like those embodied by Spanish irrigation associations can be judged on the basis of their ability to promote cooperation among their members and guarantee that the water needs of every farmer are satisfied on time (Lam 1998; Araral 2005). In this chapter, such ability is explained both with regard to water management institutions and land use factors.

The analysis consists on a case study of the *Riegos del Alto Aragon* (RAA) irrigation project. Data to assess the performance of the project was collected from public records and included over time meteorological, hydrological and crop data, as well as and spatial data. Secondary documents about the history of irrigation in the area, formal regulations related to water management, meeting minutes, registers of water rights and organizational charts were also used as a source of information. A total of 61 interviews were also conducted with cadres of the irrigation and water organizations at different governance levels. The sampling method was purposive and aimed at having representative understanding of management in the RAA project as a whole.

Case Background: The RAA Project

The RAA project is located in the inter-basin of the Gallego and Cinca rivers. The Gallego and the Cinca are two snow-melt dependent rivers that flow from the Pyrenees Mountains to the Ebro river valley, from the North to the South of the Spanish region of Aragon (see Fig. 14.1). The local climate is semi-arid Mediterranean continental, with a mean annual temperature of 14.5 °C, an annual precipitation of around 400 mm and an annual reference evapotranspiration (Hargreaves and Samani 1985, cited in Lecina et al. 2010) of around 1,100 mm (Lecina et al. 2010). A series of reservoirs and canals store and divert the water from the rivers to the 50 irrigation systems and more than 10,000 farmers who depend on the RAA project. The project encompasses more than 100,000 irrigable hectares and an average demand of around 750 million m^3 per year (according to 1970–2010 series). The reservoirs serve the RAA systems as well as other systems outside the project for a total average demand of around 1,500 million m^3.

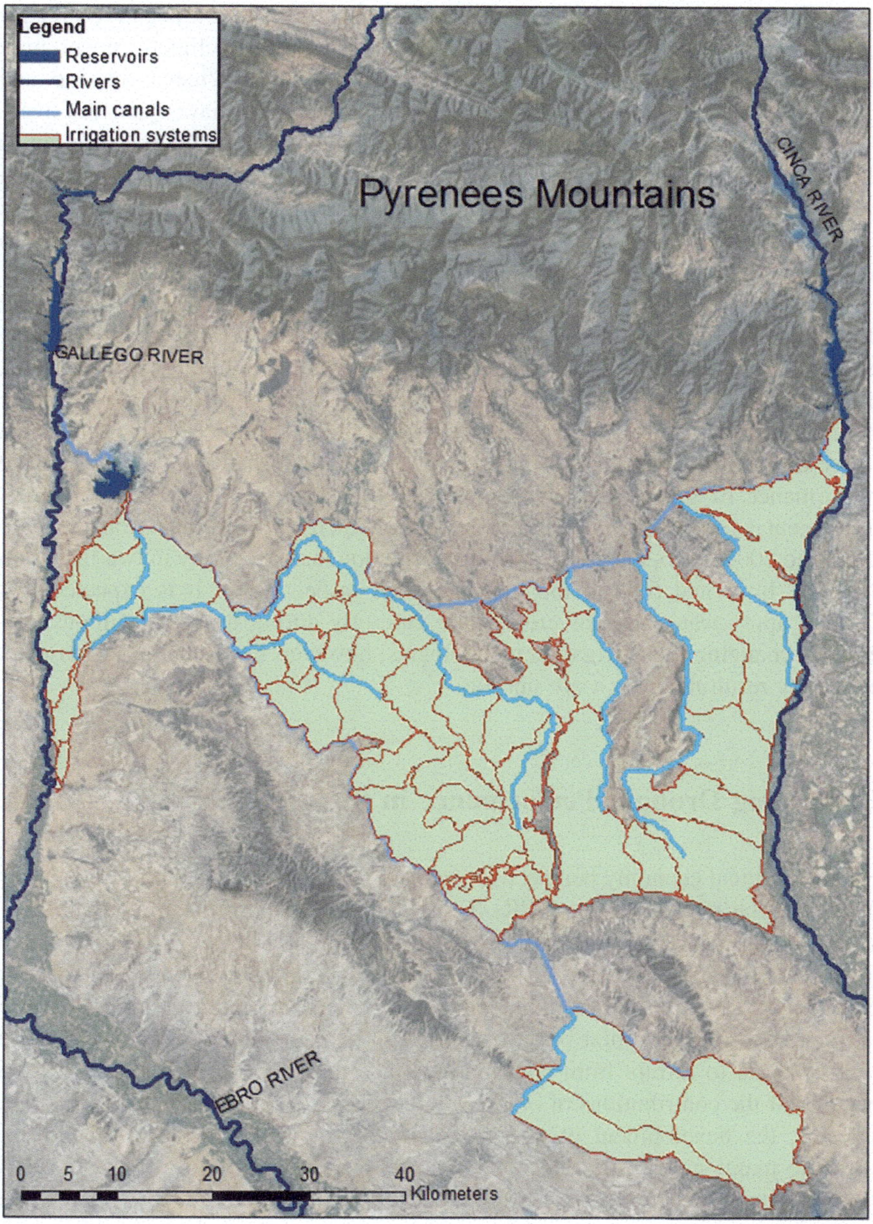

Fig. 14.1 The RAA irrigation project. *Source* Data obtained from GCRAA and Regional Government of Aragon

In the last 40 years the Ebro valley has indeed witnessed a negative precipitation trend (López-Moreno et al. 2010) and an increased climatic uncertainty caused by rapid changes between wet and dry periods (Vicente-Serrano and Cuadrat-Prats 2007). As illustrated in Fig. 14.2, droughts have been relatively frequent during at least the last 40 years. The drought of 2005–2006 stands out as the severest of the period. The reservoir inflows in 2005 decreased by more than 60 % of the inflows in 2004 and slightly less than 60 % of the average inflows from 1971 to 2003. Inflows in 2006 were also significantly lower than the series average. By 2007 inflows had recovered to normal levels.

As a consequence of the drought, the water effectively supplied to the project decreased by 45 %, from 815 to 451 hm^3 (see Fig. 14.3).The irrigation performance of the RAA project, however, only decreased a bit more than 20 %, meaning that the RAA project was able to mitigate an important portion of the drought's impact. This can be explained by the ability of the ensemble of farmers in the project to reduce their crop water needs by more than 20 % as well as the performance of the project's water allocation institutions. The project did not implement any infrastructure improvements to increase efficiency from 2004 to 2005 and 2006. Also, the use of sprinkler irrigation increased only in two irrigation systems (by 6 and 5 % of the irrigable area, respectively). It is expected that farmers apply water to their crops more carefully during droughts without necessarily changing their irrigation technologies; however, the potential increases in efficiency resulting from it are unclear.

Explaining Drought Performance in the RAA Project

From a political economy perspective, water in an irrigation system is an example of a common pool resource (CPR), i.e., is difficult to partition for private consumption and can be depleted (Ostrom and Ostrom 1977). In CPRs, sustainable management is usually tied to the resolution of cooperation problems, which are in many cases the result of social dilemmas. A social dilemma emerges because individuals can obtain joint benefits as a result of their joint actions but they are each tempted to refrain from contributing since they may receive part or all the benefits of the contributions of others whether they contribute or not. In irrigation systems, the development of and compliance with water allocation and infrastructure maintenance rules are good examples of the ability of farmers to overcome cooperation problems (Ostrom et al. 1994).

Droughts can threaten the ability of farmers to cooperate vis a vis water allocation in at least two interrelated ways. First, severe drops in water availability can increase uncertainty among farmers about the performance of the water allocation rules and thus augment the risk of water allocation problems. Promoting the robustness of those rules thus constitutes a first condition for an irrigation system to cope with droughts. Second, and most important, there is the challenge of adjusting water demand to the decreased water availability. No matter how well

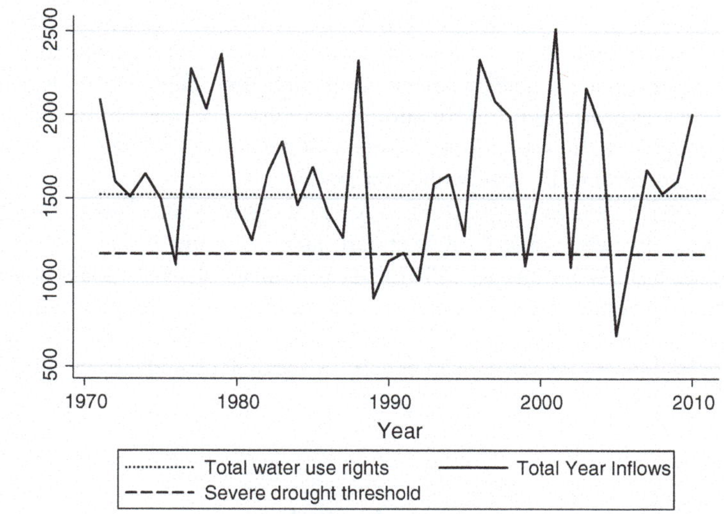

Fig. 14.2 Series of total inflows in the RAA reservoirs (million m^3). *Source* Data obtained from Ebro water agency. *Note* Series calculated from October to September of each year. Drought threshold: one standard deviation below the series mean (~1,200 million m^3) (Hisdal and Tallaksen 2000) [1,200 million m^3 is also close to the average consumption of water by the RAA project and the other irrigation systems that are served by the reservoirs (total demand of ~1,500 million m^3)]

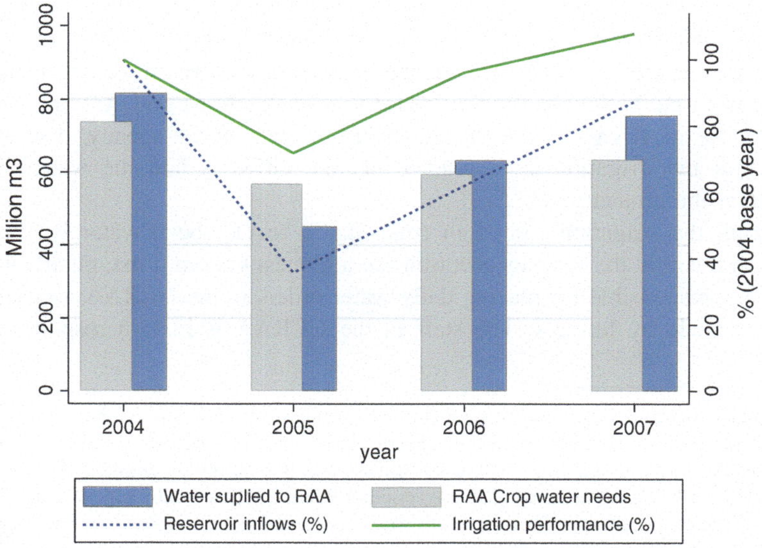

Fig. 14.3 Percentage change of reservoir inflows and irrigation performance in the area of study (2004–2007). *Source* Data obtained from Ebro water agency. *Note* All measures but the "Total water available" are calculated by aggregating irrigation system data ($n = 38$). The base year is 2004. *Note 2* Irrigation performance is calculated as the ratio between water withdrawn in an irrigation system and the system's water needs as estimated from the crops that were planted (Salvador et al. 2011)

water is allocated, if water supply and demand are not balanced, some fields will receive less water than they should and risk crop loses. Water demand strongly depends on the quantity and types of crops that are cultivated. Thus, given appropriate water availability information, the ability of farmers to cooperate and collectively adjust their cropping patterns constitutes a second important condition for an irrigation system to cope with droughts.[1]

CPR theory (Poteete et al. 2010) emerged in the 80s as an effort to understand whether, how and why some CPR users are able to cooperate and self-regulate their resource use. According to CPR theory a number of institutional and social factors can contribute to the emergence and endurance of cooperation in CPR management regimes (Poteete et al. 2010). Some of those factors can help to understand the ability of farmers in the RAA project to collectively cope with droughts.

Performing Water Management Institutions

Flexible Property Rights: Enhancing a Common Property Regime Through Temporary Quotas

Scholars contributing to CPR theory have traditionally focused on common property regimes. Water use and management in the RAA project is articulated through one such common property regime. All farmers across the systems share an equal right to use the water and then coordinate through a series of rules to allocate the resource. The water allocation process involves three organizational actors, from the bottom to the top: water user associations (WUAs), the General Community of RAA (GCRAA) and the Ebro river water agency. The WUAs operate at the irrigation system level and the GCRAA and the water agency operate at the project level.

During the irrigation campaign (mid-March to October), water is allocated across and within the systems according to a request system. First, guards in each WUA are responsible for placing daily water orders to the GCRAA according to requests made by farmers. The staff in the GCRAA is in turn responsible for

[1] Generally speaking, larger quantities of cultivated land as well higher water demand crops tend to yield higher returns. Thus, everything being equal, farmers would tend to resist reducing cultivated land or switching from high to low water demand crops during droughts. This would be aggravated by the existence of a collective action problem, as the costs of adjusting one's water needs are private but the benefits in terms of water conservation are shared. In irrigation systems individual farmers may not have the right to exclude other regime members from the benefits of water conservation efforts, unless there are specific rules about it. In that scenario, farmers who do not bear the water conservation costs may still receive enough water and enjoy similar production yields to those who do bear the costs. This would discourage farmers from making any water conservation efforts. Individual investments in irrigation efficiency via new technologies or practices would face a similar problem.

compiling the water orders from all WUAs and placing a unified order to the water agency. The water agency officials are then in charge of regulating the reservoirs that are connected to the project according to the water request made by the GCRAA, as well as serving the water to the irrigation systems. Once the water gets to the irrigation systems, WUA guards or the farmers are in charge of guaranteeing that the water gets to the plots as requested.

During droughts the water use right that is normally shared among all farmers in the project is "privatized" across systems according to a quota institution. Whenever water reserves at the beginning of the campaign and estimations about snow pack and snow melt are below a security threshold, the quota institution is implemented. According to the institution, the water that is available in the reservoirs at the beginning of the campaign is allocated among the systems on a per hectare basis. Quotas are non-transferable among WUAs by default, meaning that if a WUA does not use up its quota the water cannot be used by other WUAs. That said, farmers who own land in different systems can request a transfer of the quota that theoretically would correspond to their land from one of the systems to the other.

Probably the most persuasive argument for the use of private property rights is that owners, whether individual or collective, have an incentive to make efficient investments in resource conservation because they can be assured that only they will receive the benefits of such efforts (Copes and Anthony 2004; Acheson 2006). Common pool quotas, like those shared by farmers within the systems under study, have been also praised in other resource sectors as an effective way to balance resource use efficiency and risk under uncertainty conditions. When there is uncertainty about resource availability, pooled quotas allow users to share the risk of financial losses if the resource is more scarce than expected. In the irrigation sector, the mechanism would theoretically allow using the water conserved by farmers with lower dependence on irrigated agriculture to serve the needs of those that are more dependent on irrigation and tend to incur in riskier cropping plans during droughts (Holland 2010). As further illustrated in the sections below, this mechanism seems to be at play in the RAA project.

Transferability of rights can also facilitate rationalization and risk reduction by enabling the concentration of rights into uses that are more efficient or necessary (Copes 1986). This can be particularly beneficial in the irrigation sector during droughts, as water use rights can be transferred from areas where the costs of reducing acreage or switching crops are higher to areas where the costs are lower (Chong and Sunding 2006; Garrido 2007). Indeed, as reported by farmers and illustrated in Table 14.1, landowners in the RAA use the above mentioned quota transfer mechanism to concentrate the water in the systems where they can use sprinkler irrigation and where property is less fragmented. Everything being equal, sprinkler irrigation tends to be more water efficient than furrow irrigation (Lecina et al. 2010). Similarly, larger farms enable scale economies and reduce the transaction costs of participating in the water allocation process. The correlations between in-flow transfers and sprinkler irrigation and average farm size are significant; however, the strength of the relationship is only moderate. As mentioned, only landowners with land in two or more systems can request water transfers.

Table 14.1 Correlations between water, land and technology variables across RAA systems (2005)

	In-flow transfers	Average farm size	% sprinkler technology	% hydric soils
In-flow transfers	1			
Average farm size	0.2443*	1		
% sprinkler irrigation	0.4009*	0.2923*	1	
% hydric soils	−0.05	0.1	0.15	1

$n = 50$
*10 % significance
Note The quota transfers are measured in hectares, i.e. the number of hectares that would stop being irrigated in the system of origin, and would be in turn irrigated in the receiving system. Here the variable is computed as a percentage of the size of the receiving irrigation system
Source Data obtained from GCRAA and fieldwork

Table 14.2 Correlations between crop and land variables across RAA systems (2005 drought)

	% high water-demand crops	% low water-demand crops	Farm size heterogeneity	% number of small farms (<30 has.)	Average farm size
% high water-demand crops	1				
% low water-demand crops	−0.8*	1			
Farm size heterogeneity	0.367*	−0.522*	1		
% number of small farms	−0.337*	0.484*	−0.993*	1	
Average farm size	0.44*	−0.567*	0.939*	−0.939*	1

$n = 50$
*Significant at 5 %
Note Farm size Heterogeneity is measured as a fractionalization index. The fractionalization index measures the chances that two random hectares in an irrigation system belong to a small farm (<30 ha) and to a big farm (>30 ha) respectively
Source Data obtained from Regional Government of Aragon

This would be limiting the influence of technological improvements on transfers. Also, as shown in Table 14.2 (Sect. "Beyond Control and Water Institutions" there is a high correlation between average farm size and farm size heterogeneity. The coexistence of small number of large farms with large numbers of small farms would be moderating the impact of the former on water transfers. Finally, other factors like irrigation dependence, or the distance between the systems may affect the willingness of farmers to request water transfers. This would further limit the influence of both the technological and the farm size variable on the transfers.

Clear Physical Boundaries and Decentralized Management

The possibility to partition the collective water use right into pooled quotas effectively is enabled by a particular structure of physical and social boundaries. The irrigation project consist of two main canals that branch into a series of minor canals that allocate the water across the systems. The intersections of the infrastructure and the topography of the terrain result in a series of hydraulic sectors with clear physical boundaries.[2] The clarity of the boundaries facilitates a common understanding about which plots belong to which irrigation system and thus contribute to the enforcement of water use rights (Ostrom 1990; Cox et al. 2010).

Although the right to use the water in the RAA project is common to the ensemble of farmers, the water management right (Schlager and Ostrom 1992) is decentralized across systems, i.e. across WUAs. Decentralized management has been pointed as a factor of sustainability by CPR scholars because it permits decreasing the number of individuals involved in resolving collective-action problems (Coward Jr 1977; Ostrom 1990; Cox 2010). And, everything being equal, individuals in smaller and relatively autonomous groups can more easily come to and monitor collective action agreements than otherwise (Ostrom et al. 1994; Agrawal 2001).

Environmental and Social Monitoring

Finally, the robustness of the water allocation institutions in the RRAA project is enhanced though the effective monitoring of resource conditions and resource use. Environmental monitoring contributes to reduce uncertainty about resource availability and thus helps collective choice and institutional compliance (Cox et al. 2010). A good indicator of the environmental monitoring capacity in the RAA project is the amount of data generated and shared among the water agency, GCRAA and WUAs about reservoir levels and water use. Much of these data are used by leaders of those organizations to decide whether to activate the quota regime and to monitor its performance during the irrigation campaign. Monitoring of resource use makes those who do not comply with rules visible to the community, which facilitates the effectiveness of rule enforcement mechanisms and the performance of CPR management regimes (Cox et al. 2010). A good indicator of the strength of social monitoring in the RAA project is the notable decrease in the number of rule violation cases brought to the GCRAA executive board during the 2005 drought (see Fig. 14.4). As reported by officials from the GCRAA, that

[2] Both main and minor canals follow the contour lines of the terrain so water can be transported and then distributed to plots by gravity. Similarly, the drainage system is located at lower elevation than the conveyance canals but still at higher elevation than the hydrological system so runoffs can flow by gravity from the plots to the drainage system and then to the hydrological system.

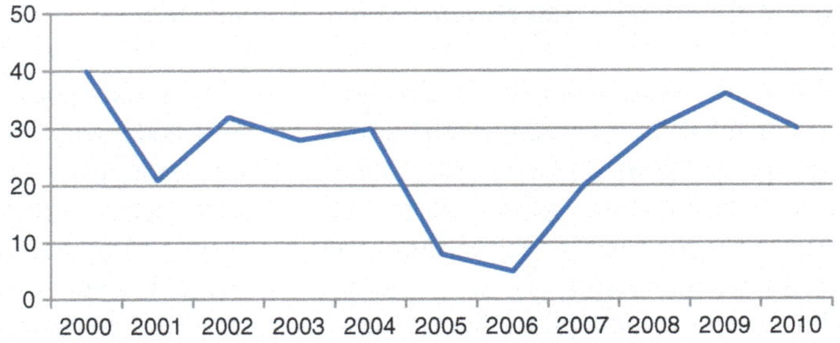

Fig. 14.4 Number of rule infraction cases brought to the GCRAA court per year (2000–2010). *Source* Data obtained from GCRAA

responds to the high visibility that rule infractions have during droughts, as well as the existence strong social norms about the need to cooperate particularly during periods of crisis.

(Decentralized) Administrative Leadership

Monitoring entails costs (Ostrom et al. 1994). Decisions about who should carry the monitoring and at which cost are thus a crucial aspect of monitoring effectiveness. In many long-lasting CPR regimes such monitoring as well as more general coordination roles are to a great extent carried by leaders (Ostrom 1992b; Agrawal 2001; Meinzen-Dick et al. 2002). In the RAA project, leadership is decentralized at different governance levels. That helps reducing the monitoring and coordination costs. As mentioned in the section above, the water agency, GCRAA and WUA representatives are all crucial in the activation of the quota regime during droughts. Additionally, secretaries from the WUAs are responsible for generating records of water requests and deliveries within the systems and use them to double-check that the systems do not go over their quotas. Finally, there is the monitoring carried by the water agency and WUA guards who provide first-hand information about incidences in the water delivery as well as non-forecasted changes in water availability.

Beyond Control and Water Institutions

The ultimate performance of the RAA project during droughts need also to be understood with regard the capacity of farmers to cooperate and collectively adjust their crop water needs to the decreased water availability (see Fig. 14.5). At least three factors contribute to that capacity in the RAA project.

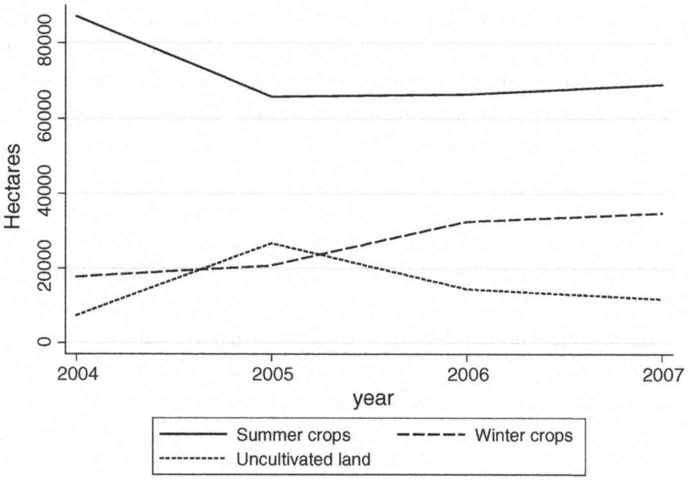

Fig. 14.5 Evolution of land use (hectares) in the RAA project during the 2005 drought. *Source* Data obtained from Regional Government of Aragon

First, there is the relatively high substitutability of high and low water demand crops like corn and alfalfa and barley and wheat, respectively. The production and market price of high water-demand crops is usually higher[3] and can yield notable returns; however, those crops also require more agricultural and labor inputs (i.e. herbicides, fertilizers, and water) and thus are more costly and risky to grow than low water demand crops what becomes especially relevant in drought periods.

The second factor is the existence of strong income support subsidies sponsored by the European Community Agricultural Policy (CAP). Currently those subsidies amount to 26 % of net farm incomes (Lecina et al. 2010). Indeed, much of the agricultural activities in the area would not exist without such subsidies (Arrojo and Bernal 1997). Most importantly, since the last reform of the CAP in 2003 the subsidies have been decoupled from production, meaning that farmers receive a fixed lump sum every year regardless of the crops cultivated and yields (Moreddu et al. 2004).

The third factor is the existence of heterogeneity in the dependence from irrigated agriculture by farmers. In the last decades, a series of migration waves from rural to urban areas in the region have resulted in a progressive lack of labor factor in the agricultural sector (CESA 2002). Also, the increasing price of agricultural inputs and the internationalization of agricultural markets have put the Spanish

[3] From 1990 to 2010, the average prices of corn and alfalfa (high water-demand crops) have been ~15.8 and ~10.8 E/100 kg respectively; and that of barley and wheat (low water-demand crops) ~13.5 and ~16.7 E/100 kg, respectively. That being, the average production of corn and alfalfa for years 1998, 2000, 2002, and 2007 was around 9,000 and 11,000 kg/ha respectively; and that of barley and wheat as around 4,000 kg/ha (Elaborated from data from Regional Government of Aragon).

agriculture under the need to increase its productivity (Mas Ivars 2012). In the RAA project, those phenomena have resulted in the progressive concentration of land in the hands of a small number of big landowners who aim to increase productivity by investing in economies of scale. This group coexists with an increasing number of part-time, small landowners who have found in the industrial and service sectors their main source of income. While big landowners are increasingly dependent on high-water demand (and more profitable) crops, smaller landowners enjoy more flexibility to combine high water-demand crops with lower water-demand crops if necessary. As illustrated in Table 14.2, small numbers of big farms tend to coexist with big numbers of small farms and with higher percentages of high water demand crops. As confirmed by farmers, during drought years, small landowners would be more willing to modify their cropping patterns and reduce the amount of high water-demand crops they grow; this in turn would allow big landowners to grow a higher percentage of such crops than otherwise.

The RAA Case in Global Perspective

The RAA case illustrates a specific combination of factors contributing to successful resource management under drought conditions. A number of those factors echo findings from similar studies in other countries and irrigation contexts.

Some of the findings of this study about water management institutions are congruent with the literature on irrigation policy and governance. Much of this literature developed from the 1970s to the 1990s with regard to the implementation and reform of state-promoted irrigation projects in developing countries (Uphoff et al. 1985; Cernea and Meizein-Dick 1992; Ostrom 1992a; Subramanian et al. 1997; Knox and Meinzen-Dick 2000). Factors contributing to successful management in those contexts include the multilevel organization of management tasks, monitoring and leadership, moderate water dependence, social cohesion among farmers, financial viability and agricultural policies that allow crop choice and provide adequate returns to irrigated production (Tang 1992; Subramanian et al. 1997). As illustrated in this study, a good number of these features, can also help understand the ability of irrigation projects to cope with external disturbances like droughts.

The results about the role of water management institutions are also convergent with the insights from drought robustness studies in other irrigation contexts. Cox and Ross (2011) and Cox (2014) assess the robustness of more than 70 traditional irrigation communities in the Taos valley, New Mexico. Similarly to the RAA project case, irrigation communities in Taos rely on decentralized common property regimes, the duties and initiative of strong leaders, and effective monitoring mechanisms based on both third party and mutual surveillance. Although coordination across the Taos communities is not institutionalized through a central organization like in the RAA project, the leaders of the communities do meet to coordinate whenever is necessary. Contrary to the RAA case, water allocation institutions appear to be sufficient to cope with water allocation uncertainties during both normal water availability and drought conditions. This is partially aided by the widespread access of the communities to ground water, which is a feature that is absent in the RAA case. Lam (2006) reviews the case of irrigation institutions in Taiwan. Similarly to the RAA case, the multilevel organization of leadership and tasks like monitoring contribute to performance during droughts. According to Lam (2006), the possibility that leaders and organizations at different governance levels and scales complement each other in activities like monitoring or conflict solving is an important factor of performance under disturbance conditions. Also like in the RAA case, Taiwanese irrigation systems use quotas; however, the quotas are used both during normal and drought conditions and are granted directly to farmers. During droughts, efforts at different scales are made to coordinate the allocation of the quotas depending on the severity of the drought.

Finally, the interest of this study on land use factors vis a vis drought adaptations resonates also with findings from drought studies in the agricultural sector in semi-arid countries. Liverman (1990, 1999), reports findings from studying drought management in Mexico. Like in the RAA project, irrigation in Mexico benefits from notable price support mechanisms and agricultural subsidies (Liverman 1999). As pointed by the author and others (Naylor and Falcon 2012), those subsidies have an important role to mitigate the economic impact of droughts in the short term but can also crowd out learning and innovation in the long term. Additionally, Liverman (1990) highlights the contribution of fertilizer use and improved seeds to reduced crop loses during droughts; as well as the advantages of private land ownership as compared to communally ownership. Mert et al. (2009) and Deressa et al. (2009) synthesize findings from agricultural adaptations to climate change in the Sahel, Africa. Like in this study, the authors highlight a correlation between reduced cropping efforts and drought periods. Additionally, the authors point to the widespread use of short cycle crop varieties, shifts in farming location, early and late planting strategies, and soil conservation practices as measures that contribute to reduced water needs during drought periods.

Discussion and Conclusions

As illustrated above, the RAA project has been able to mitigate to a great extent the impact of severe droughts like that of 2005. This can be understood with regard to (1) institutional robustness factors, such as the flexibility of the common property right regime, the strength of monitoring institutions and leadership; and (2) factors contributing to water demand adaptability such as crop substitutability, income support subsidies and heterogeneity of farmers in their dependence on water. While the former group of factors is under the relative control of farmers and public authorities in the area, the latter group is not. This constitutes a source of vulnerability in the RAA project.

The findings of this study are specific to the RAA case; however, they also resonate with findings from similar studies in other countries and irrigation contexts. Relevant factors highlighted both in this study and other studies include the decentralization of water management tasks and leadership, moderate water dependence, crop substitutability and agricultural subsidies. Further research might explore the impact of technological improvements and the intensification of water transfers on cooperation and robustness; whether the factors identified in this study are relevant to cope with disturbances other than droughts and in other productive sectors; and the implications of assessing irrigation performance both over time and space through a diversity of indicators.

References

Acheson JM (2006) Institutional Failure in Resource Management. Annu Rev Anthropol 35:117–134

Agrawal A (2001) Common property institutions and sustainable governance of resources. World Dev 29:1649–1672

Araral E (2005) Bureaucratic incentives, path dependence, and foreign aid: an empirical institutional analysis of irrigation in the Philippines. Policy Sci 38:131–157

Arrojo P, Bernal E (1997) El regadio en el valle del Ebro. In: López Galvez J, Naredo JM (eds), La Gestion del Agua de Riego. Fundacion Argentaria, Madrid, pp 139–181

Cernea M, Meizein-Dick R (1992) Design for water user associations: organizational characteristics. In: Developing and improving irrigation and drainage systems. Technical paper no. 178. World Bank, Washington, DC, pp 45–59

CESA (2002) Informe Socioeconomico de la decada 1991-2000 en Aragon. Consejo Economico y Social de Aragon, Zaragoza

Chong H, Sunding D (2006) Water markets and trading. Annu Rev Environ Resour 31:239–264

Copes P (1986) A critical review of the individual quota as a device in fisheries management. Land Econ 62:278–291

Copes P, Anthony C (2004) Socioeconomics of individual transferable quotas and community-based fishery management. Agric Resour Econ Rev 33:171–181

Coward EW Jr (1977) Irrigation management alternatives: themes from indigenous irrigation systems. Agric Adm 4:223–237

Cox M (2010) Exploring the dynamics of social-ecological systems: the case of the Taos valley acequias [thesis]. Type, Indiana University, Bloomington

Cox M (2014) Applying a social-ecological system framework to the study of the Taos Valley irrigation system. Hum Ecol 42(2):311–324

Cox M, Ross JM (2011) Robustness and vulnerability of community irrigation systems: the case of the Taos valley acequias. J Environ Econ Manag 61:254–266

Cox M, Arnold G, Villamayor Tomás S (2010) A review of design principles for community-based natural resource management. Ecol Soc 15

Deressa TT, Hassan RM, Ringler C, Alemu T, Yesuf M (2009) Determinants of farmers' choice of adaptation methods to climate change in the Nile Basin of Ethiopia. Glob Environ Change 19:248–255

Garrido A (2007) Water markets design and evidence from experimental economics. Environ Resource Econ 38:311–330

Hisdal H, Tallaksen LM (2000) Drought event definition. In: Assessment of the regional impact of droughts in Europe. Department of Geophysics, University of Oslo, Oslo

Holland DS (2010) Markets, pooling and insurance for managing by catch in fisheries. Ecol Econ 70:121–133
Knox A, Meinzen-Dick R (2000) Collective action, property rights, and devolution of natural resource management: exchange of knowledge and implications for policy. In: CAPRI working papers. CGIAR, Washington, DC
Lam W (1998) Governing irrigation systems in Nepal. ICS Press, Oakland, CA
Lam WF (2006) Foundations of a robust social-ecological system: irrigation institutions in Taiwan. J Inst Econ 2:203–226
Lecina S, Isidoro D, Playán E, Aragüés R (2010) Irrigation modernization and water conservation in Spain: the case of Riegos del Alto Aragón. Agric Water Manag 97(10):1663–1675
Liverman D (1990) Drought impacts in Mexico: climate, agriculture, technology, and land tenure in Sonora and Puebla. Ann Assoc Am Geogr 80:49–72
Liverman D (1999) Vulnerability and adaptation to drought in Mexico. Nat Resour J 39:99–115
Lopez Galvez J, Naredo JM (1997) La Gestion del Agua de Riego. In Economia y Naturaleza. Fundacion Argentaria, Madrid
López-Moreno JI, Vicente-Serrano SM, Angulo-Martínez M, Beguería S, Kenawy A (2010) Trends in daily precipitation on the northeastern Iberian Peninsula, 1955–2006. Int J Climatol 30:1026–1041
Mas Ivars M (2012) Analisis para la Mejora de la Productividad en Aragon. CESA, Zaragoza
Meinzen-Dick R, Raju KV, Gulati A (2002) What affects organization and collective action for managing resources? Evidence from canal irrigation systems in India. World Dev 30:649–666
Melgarejo Moreno J (2000) De la política hidráulica a la planificación hidrológica, un siglo de intervención del Estado. In: Melgarejo Moreno J, Barciela Lopez C (eds) El Agua en la historia de España. Publicaciones Universidad de Alicante, Alicante, pp 295–324
Mertz O, Mbow C, Reenberg A, Diouf A (2009) Farmers' perceptions of climate change and agricultural adaptation strategies in rural Sahel. Environ Manag 43:804–816
Moreddu C, Anton J, von Lampe M, Martini R, Tallard G, Vavra P (2004) Analysis of the 2003 CAP reform. OECD, Paris
Naylor R, Falcon W (2012) The Yaqui Valley's agricultural transition to a more open economy. In: Matson P (ed) Seeds of sustainability: lessons from the birthplace of the green revolution in agriculture. Island Press, Washington, DC, pp 107–38
Ostrom E (1990) Governing the commons. Cambridge University Press, New York
Ostrom E (1992a) Crafting institutions for sel-governing irrigation systems, vol pp. ICS Press, San Francisco, CA
Ostrom E (1992b) The rudiments of a theory of the origins, survival, and performance of common-property institutions. In: Bromley DW (ed) Making the commons work: theory practice and policy. ICS Press, San Francisco, CA, pp 293–319
Ostrom V, Ostrom E (1977) A theory of institutional analysis of common pool problems. In: Hardin G, Baden JA (eds) Managing the commons. Freeman, San Francisco, pp 157–172
Ostrom E, Gardner R, Walker J (1994) Rules, games and common pool resources. Michigan University Press, Michigan
Poteete AR, Ostrom E, Janssen M (2010) Working together. Princeton University Press, Princeton, NY
Salvador R, Martínez-Cob A, Cavero J, Playán E (2011) Seasonal on-farm irrigation performance in the Ebro basin (Spain): crops and irrigation systems. Agric Water Manag 98:577–587
Schlager E, Ostrom E (1992) Property-rights regimes and natural resources: a conceptual analysis. Land Econ 68:249–262
Subramanian A, Jagannathan NV, Meinzen-Dick R (1997) User organizations for sustainable water services. The World Bank, Washington DC, vol 354
Tang SY (1992) Institutions and collective action: self-governance in irrigation. Institution for Contemporary Studies, San Francisco, CA
UN/ISDR (2004) Living with risk: a global review of the international strategy for disaster reduction. United Nations Office for Disaster Risk Reduction (UNISDR), Geneva

Uphoff N, Meinzen-Dick R, Julien NS (1985) Improving policies and programs for farmer organization and participation in irrigation water management. Department of Government, Rural Sociology, and Regional Planning, Cornell University, Ithaca, NY

Vicente-Serrano SM, Cuadrat-Prats JM (2007) Trends in drought intensity and variability in the middle Ebro valley (NE of the Iberian peninsula) during the second half of the twentieth century. Theoret Appl Climatol 88:247–258

Part III
Ecosystem Perspectives in Water Resources Management

Human water security, the provision of water services at a high level of reliability has so far been achieved at the expense of the environment. High population growth, accelerating economic activities, and land use alterations are increasing the pressures on the quality and quantity of global water resources. The human aspiration to manage water primarily to fulfill humanity's growing needs has resulted in a violation of the basic requirements to sustain freshwater species and ecosystems with profound ecological consequences. This is threatening the sustainable use of freshwater ecosystems on which our planet depends.

Given the essential role of water in terms of its sufficient quality and quantity for both human life and biodiversity, there is an urgent need to find ways of mitigating the negative impacts of human activities on the water cycle. It requires more attention on how we can guarantee uninterrupted supply of ecosystem services to humans and the environment under different and changing conditions. There is a need to understand trade-offs between competing water demands (both quantity and quality) in order to achieve sustainable solutions, and to build capacity for holistic approaches enhancing and sustaining water security and the resilience of social-ecological systems. The combined concept of environmental flows and ecosystem services can serve as a framework in this regard. It allows for better understanding of water quality and quantity attributes and their interdependence. Further, tipping points and threshold values can be identified and thereby help to determine the potential of freshwater bodies and wetlands to provide essential ecosystem services sustainably.

The objective of this theme is to bring together different perspectives of ecologically sustainable water management drawing on different case studies, and focusing on how to improve the sustainable management of water resources and aquatic ecosystems. Further, the theme focuses on governance systems to facilitate the implementation of ecologically sustainable water management. It addresses how societal learning and decision-making processes can be supported to promote change toward enhancing water security and the resilience of social-ecological systems.

In the paper by Barbosa et al., two Brazilian case studies are presented that require the application of new practices to achieve improved environmental quality and conservation. The paper focuses on different practices that show how humans can live in harmony with environmental dynamics so that the environment can retain its biodiversity while human can still derive ecological services.

Bekchanov et al. in their paper show how to integrate both economic efficiency and environmental sustainability, which is essential for designing policies for a sustainable development. The paper illustrates the case study of water-based economy in Uzbekistan, and uses water productivity as a proxy indicator for the environmental fragility of the ecosystem, which is vital for defining development strategies. The paper derives a composite indicator with backward and forward linkage indices by using the multi-criteria analysis method-TOPSIS, which allows for direct ranking of economic sectors, and to formulate sectoral transformation measures guided by sustainable growth objectives in Uzbekistan.

Maintenance of freshwater biodiversity is a key element for the sustainability of freshwater ecosystems. In their paper, Garcia Moreno et al. address the need for careful thinking among landscape managers and policy makers about strategic adaptive management of freshwater systems in order to both effectively conserve natural ecosystems and continue to supply human populations with the freshwater they need. The management of human and environmental water needs is therefore challenging and calls for an integrative view on ecosystem services.

The paper by Knüppe and Pahl-Wostl addresses the management of human and environmental water needs. The paper analyzes different water governance and management systems, and highlights different characteristics that are assumed to be crucial for adaptive and integrated management with a focus on ecosystem services. The paper argues for a significant shift of current water management objectives that is required to ensure water security for current and future generations.

The paper by Patterson et al. explores the management problem of nonpoint source pollution through theory-informed empirical research, involving an in-depth case study in South-East Queensland, Australia. This paper focuses on the need to better understand how to manage nonpoint source pollution in practice, and argues how management efforts can help to enable different knowledge and institutional capacities that support practical action within complex, dynamic, and changing situations. The paper by Siew et al. explores the potential role of transdisciplinary research to support the implementation of the ecosystem services concept in land and water management. The paper presents experiences with the implementation of a transdisciplinary research approach to support the integration of ecosystem services into land and water management under climate change in the arid Tarim River Basin, Northwest China.

Chapter 15
The Missing Piece in the Conservation Puzzle: Cohesion Among Environmental, Economic and Social Dimensions

Francisco A. R. Barbosa, Paulina M. Maia-Barbosa, Diego G. F. Pujoni and Lorena T. Oporto

Abstract Managing water resources efficiently is a difficult and complex task. This task will continue to challenge those who prioritise sustainable development over economic growth. In this report, we discuss two Brazilian case studies of national relevance that urgently require the application of new practices to achieve improved environmental quality and conservation. The first case study involves the middle Rio Doce Lake System (RDLS), which is composed of approximately 300 water bodies with distinct morphometric, physical and chemical features and a range of different land use types. Eighteen lakes (8 within Rio Doce State Park) have been studied since 2000 after the implementation of the Long Term Ecological Research Programme (Brazil-LTER site #4). The aquatic communities studied were highly diverse, with 481 algae species, 346 zooplankton species, 58 families of benthic organisms and approximately 30 fish species (7 of which are exotic). Furthermore, the results of this programme confirm that community dynamics and ecological processes, such as life-history strategies, primary production and decomposition, are determined primarily by the water mixing pattern observed in most of the lakes. The introduction of exotic fishes was responsible for the local extinction of 7 species of native fauna, with cascading events affecting lower trophic levels, resulting in a modified aquatic community structure and diminished water quality. Interviews with local fishermen demonstrated that they understand the environmental impact of

F. A. R. Barbosa (✉) · P. M. Maia-Barbosa · D. G. F. Pujoni · L. T. Oporto
Laboratory of Limnology Ecotoxicology and Aquatic Ecology (LIMNEA),
Universidade Federal de Minas Gerais (UFMG), Av. Antonio Carlos,
Belo Horizonte, MG 6627 – 31270-910, Brazil
e-mail: barbosa@icb.ufmg.br

P. M. Maia-Barbosa
e-mail: maia@icb.ufmg.br

D. G. F. Pujoni
e-mail: diegopujoni@gmail.com

L. T. Oporto
e-mail: lorenatoporto@gmail.com

A. Bhaduri et al. (eds.), *The Global Water System in the Anthropocene*,
Springer Water, DOI: 10.1007/978-3-319-07548-8_15,

exotic species and are willing to participate in management actions, thus promoting cohesion between social and environmental dimensions. Currently, our main challenge is to partner with local fishermen to test new management methods such as selective overfishing that are potentially capable of decreasing or even halting the impacts of exotic species introductions. To disseminate scientific knowledge to the fishing communities, an environmental education (EE) programme has been developed with different social groups (school teachers, students, prefectures' technical staff and local residents) using participative techniques. This programme has the potential to improve residents' understanding of regional environmental problems and to help them to develop the critical thinking skills necessary to change opinions, concepts, habits and practices. The second case study involves the use of water in mining activities in Brazil. These mining activities have led to an intense conflict of interest among different environmental, economic and social goals. Assuming that water in itself is a concept with multiple dimensions, it is important to develop cohesion among different goals to achieve sustainability. In short, conservation must evolve from its original "isolate to preserve" approach to the modern "use sustainably to conserve" approach, in which people are the major drivers and users of the resource management process.

Introduction

A synthesis of the onset of conservation was published by Meine (2010), who identified the late 1930s as the period when an array of environmental problems, such as soil erosion, watershed degradation, urban pollution, deforestation and the depletion of fisheries and wildlife populations, brought together academic ecologists and resource managers interested in increasing biological conservation awareness, particularly with respect to biological diversity. According to Meine, it took no less than 50 years for conservation biology to become an organised field.

To the broader public, however, concern regarding the conservation of biological resources arose after the publication of Silent Spring by Rachel Carson (1962), in which the author drew attention to the denial of the negative environmental impacts of pesticides that, in turn, reduced humans' quality of life. At that time, economic considerations were perceived as more relevant than social and environmental considerations, and the development of a country was measured only in terms of its gross domestic product (GDP) (Costanza et al. 2009).

This book initiated worldwide concern regarding the conservation of natural resources, demonstrating that the emphasis on rapid economic progress might end in disaster if we neglect the interconnectedness between nature and humans. Currently, more than 50 years of theoretical and applied studies in various disciplines have addressed the major challenge of providing a sound theoretical basis for conserving species and maintaining ecological processes (e.g., primary production, decomposition) and ecosystem services.

We use Carson's book as a benchmark because its inspirational message brought together scientists and international non-governmental organisations (e.g., WWF, TNC, IUCN) to work towards a common goal. As a result of this teamwork, scientists have produced a theoretical basis for conserving species and maintaining ecological processes, and international organisations have advocated for the establishment and management of conservation areas worldwide (Claus et al. 2010).

Given current technology and the fast-growing population, we are depleting natural resources faster than regulatory agencies can respond (Berkes et al. 2006), resulting in an apparent trade-off between economic development and environmental protection. In fact, the term "trade-off" is misleading because predatory economic growth is not sustainable in the long run without environmental protection, and both can exist in a sustainable and synergistic way (Feiock and Stream 2001). It is true that "win-win" synergy examples are rare (Wunder 2001), but they are persuasive for those who prioritise sustainable development over economic growth.

The majority of conservation actions have been based on genetic principles such as genetic drift and inbreeding depression and on ecological theories such meta-population dynamics (Levins 1969) and island biogeography (MacArthur and Wilson 1963, 1967). Based on these principles and theories, biological reserves have been created and effectively managed across more than 6,000 conservation units worldwide. Other theories of conservation have also been applied, and the Theory of Pleistocene Refuges (Haffer 1969), based originally on the distribution of endemic birds within the Amazon Basin, has been particularly relevant in the Americas.

Thus, theory has made valuable contributions to conservation practices in both aquatic and terrestrial ecosystems (although the theoretical background is entirely based on the distribution of terrestrial plants and animals). However, the majority of conservation practices have practically excluded humans by isolating them from the areas to be conserved, in some cases even using fences to prevent human access (i.e., ecological sanctuaries; Sarkar 1999). Humans should strive to establish more of these unmanaged areas, as their value is undeniable particularly for the conservation of the gene pool (Noss 1991). However, this approach is becoming more difficult to implement. Furthermore, there are areas destined for exploitation where knowledge of conservation would help mitigate the inevitable impacts. Such situations encourage us to seek a new paradigm that would involve an optimal balance between the exploitation of natural resources and the conservation of biological diversity. This paradigm would recognise that sustainably managed areas are impacted and modified ecosystems but they can still conserve species and maintain ecological processes.

It is clear that it is impossible to conserve that which is not recognised as important. Therefore, the present study intends to illustrate that conservation must consider the multiple dimensions of water resources and develop cohesion among distinct goals. We consider the participation of local communities to be a key part of the conservation process in which local and non-scientific knowledge inform resource management and sustainable practices are reinforced. As stated by Schaller (1992) *apud* Jacobson and McDuff (1998), conservation problems go beyond the biological dimension because they are equally socially and economically important issues. In summary, we propose that conservation must evolve from its original "isolate to preserve" approach to a modern "use sustainably to conserve" approach.

Old Versus New Conservation Approaches

In the beginning, conservation relied mostly on the capacity to provide safe and controlled areas for wildlife, largely within settled areas. The majority of the created parks are likely remnants of natural areas not completely occupied by humans, either because they were large enough or because someone identified them as particularly deserving of preservation due to their uniqueness or because they harboured threatened flora and fauna. The "old" basic conservation approach can be summarised as "isolate to preserve": wildlife is protected by limiting human presence (Sarkar 1999).

We accept that such an approach has been effective for conservation at certain times and to a certain degree, depending mostly on the area and the remaining wildlife to be protected. However, new aspects and interests have arisen and converted the conservation of natural areas into a disputed issue, particularly in relation to areas that are valuable for food production and industrial activities. This conflict has led to an unfair balance between biodiversity conservation and the production of goods and services that are valuable in modern society. These new challenges have added complexity to the process of creating conservation units. Further views on conservation biology, including its historical foundations, growth as an interdisciplinary field and scope, can be found in Meine et al. (2006) and Meine (2010).

One particular conservation perspective has been gaining appreciation among conservationists: "use sustainably to conserve". In contrast to the perspective that a human presence is always harmful to the environment, this approach focuses on the sustainable use of resources. An extreme point of view, the "myth of the ecologically noble savage", states that indigenous people live in harmony with their environment because they have an extensive body of knowledge about it ("traditional ecological knowledge") and that conservationists should adopt this practice (Redford 1991). Regardless of whether these local populations threaten the environment, conservationists' efforts should concentrate on mitigating harmful practices and promoting beneficial ones, in particular by increasing contributions from social and humanities studies to develop sound conservation practices. Moreover, it seems plausible that "local resource users are also conservation agents", as noted by Claus et al. (2010). However, these authors also call attention to the fact that "not all indigenous or local people have developed or retained a conservation knowledge, but where this knowledge does exist it can be critical to, and in effect be, the conservation effort most needed".

Biological Invasions in the Rio Doce Lakes: A Case Study

The Vale do Aço Metropolitan Region is the largest steelmaking park in Brazil. It is located in the middle part of the Rio Doce Basin and has a lake system composed of approximately 300 natural water bodies with distinct water quality and land use characteristics. In the centre of this region lies the Rio Doce State Park (RDSP), the largest continuous Conservation Unit of Atlantic Forest in the state of Minas Gerais (35,970 ha), recently incorporated as a Ramsar site (RAMSAR 2010). Despite the fact that the lakes within the RDSP have been protected since 1944, several environmental impacts have been recently identified (e.g., illegal hunting and fishing, intentional and/or accidental introductions of exotic species and tourism activities on the shores of Lake Dom Helvécio; Gontijo and Britto 1997). Moreover, the extensive *Eucalyptus* plantations, the pasturelands and the fast-growing, unplanned urbanisation in the surrounding areas of the park deserve attention.

This area has been studied since the 1970s (e.g., Tundisi and Saijo 1997) and, since the implementation of Brazil's Long-Term Ecological Research Programme (Brazil-LTER- site # 4) in November 1999, has been sampled systematically. The region has great aquatic diversity: 481 phytoplankton species (Barros et al. 2013), 346 zooplankton taxa (Maia-Barbosa et al. 2014), 58 families of benthic organisms (Marques and Barbosa 2001) and 24 fish species (Godinho 1996). The climate is tropical semi-humid with dry winters (May–August) and rainy summers (September–April) determining a water column mixing in the winter and a thermal and chemical stratification in the summer. Micro-stratifications between day and night periods (atelomixis) have also been recorded (Barbosa and Padisák 2002). Both daily and annual circulation patterns are the main driving forces underlying plankton community structure (Barros et al. 2006; Souza et al. 2008; Barbosa et al. 2011), although human impacts (e.g., the introduction of exotic fish species) also have negative long-term effects on water quality and biodiversity (Pinto-Coelho et al. 2008; Maia-Barbosa et al. 2010). For instance, the introduction of exotic fish species is responsible for up to a 50 % reduction in the number of native fish, in particular as a result of their predation on small and mid-sized species (Godinho 1996; Latini and Petrere 2004).

The first introductions occurred in the early 1970s, starting with the intentional introduction of the peacock bass (*Cichla kelberi*, Kullander and Ferreira 2006) and the red piranha (*Pygocentrus nattereri*, Kner 1858) by local fishermen (Godinho et al. 1994). At the same time, the oscar (*Astronotus cf. ocellatus*, Agassiz 1831) was introduced, possibly by aquarists. The African catfish (*Clarias gariepinus* Burchell 1822) and tilapia species (*Tilapia rendalli*, Boulenger 1897 and *Oreochromis cf. niloticus*, Linnaeus 1758) were only recorded after 1999, most likely as a result of having escaped from culture tanks. The tamboatá (*Hoplosternum littorale*, Hancock 1828) was the latest introduction and most likely originated from the bait used by fishermen (Latini et al. 2004).

According to the model proposed by Blackburn et al. (2011), the majority of these exotic species are currently in the last invasive stage (Sunaga and Verani

1991; Godinho 1996; Latini et al. 2004; Oporto 2013). Besides In addition to decreasing the number of native species, the invasives have had a major impact by changing the feeding habits of the native species (Pompeu and Godinho 2001) and the structure of the fish assembly, resulting in a dominance of exotic species, fewer total species and a modified zooplankton community (Pinto-Coelho et al. 2008; Maia-Barbosa et al. 2010; Oporto 2013). Moreover, it is likely that the extent of the impact on ecosystem services has not yet been accurately measured.

The high risk of new species introductions deserves attention considering both the existence of sport fishing clubs outside the RDSP and public policies that encourage the cultivation of exotic species using net-tanks (Pelicice et al. 2013). Management plans for invasive species must consider both the prevention of new introductions and the control of existing populations (Blackburn et al. 2011). Furthermore, despite the facts that Brazil is the country with the greatest megadiversity, that Brazil is the first country to sign the Convention on Biodiversity and therefore is capable of impeding species invasions and that Brazil has passed specific laws to improve invasive species control (Oliveira and Pereira 2010), mitigation and prevention actions are scarce, except for those related to agricultural pest control (Speziale et al. 2012).

Sustainable Interventions: Experimental Management of Exotic Fishes

Humans' perceptions of their environment, along with their individual and collective vision (Maryn et al. 2003), influence the ecological, economic and social valuation of biodiversity (Castillo et al. 2005; Fischer and Young 2007). Considering the history of local fish introduction in the Rio Doce Lake System and considering that intentional introductions are one of the major problems in the management of invasive species around the world (Gozlan et al. 2010), it is essential to understand the relationship society has with these species to propose mitigation plans (García-Llorente et al. 2008) and prevent new invasions (Simberloff et al. 2013).

To incorporate social and scientific knowledge of invasive fish management and to promote cohesion among social and environmental dimensions, we conducted a study aiming to evaluate how local fishermen perceive the invasion process and its impacts on the native fish fauna (Oporto 2013). Semi-structured interviews were conducted with 56 fishermen in a recreational fishing club; the club lies in the surrounding areas of the RDSP and Lake Dom Helvécio, which is located within RDSP, where fishing for non-native species is allowed for visitors. Differences in the perceptions of fishermen related to age, fishing experience, the frequented lake or the lake's location (inside and outside the park) were assessed using the Chi-square test ($\alpha = 0.05$). The results revealed that fishermen are aware of the presence of introduced species and perceive some environmental impacts on the lakes. However, they do not necessarily perceive a direct correlation between these impacts and the presence of exotic species. Some of the fishermen even show interest in increasing the number of fish species, regardless of their origins. Interestingly, fishermen believe they can take part in the management practices concerning introduced species.

Incorporating local knowledge into scientific practices has been a promising approach for invasive species management and the prevention of new invasions (Carey et al. 2012). Considering the major role of the RDSP in conserving local diversity, it is essential to include local fishermen and other social groups (school teachers, students, prefectures technical staff and settlers) as partners in the original RDSP management plan and to allow these groups to take part in the decision-making process.

An experimental management study that integrates the knowledge of species and fisheries techniques from both local fishermen and scientific research has recently been proposed. This study involves the selective and intensive fishing of exotic species using specialised fishing gear (seine, rod and reel, gill and cast nets). The fishing effort is standardised and conducted by researchers, fishermen and members of the park administration. To evaluate the success of the management initiative, data on species richness, diversity and patterns of species dominance and abundance obtained during the project will be compared with data recorded during the period 2006–2010. Furthermore, we have planned environmental education activities within the park and at selected locations in its surroundings that aim to prevent new introductions and the dispersion of exotic species by local fishermen. These efforts are examples of the inclusion of local human populations in conservation activities, without whom it would be practically impossible to control biological invasions in the area.

The Environmental Education Programme in the Middle Rio Doce: Major Lessons

More than a half million people, divided across nine municipalities, live near the RDSP. The municipalities have different degrees of urbanisation (HDI 2010 from 0.62 to 0.77) and predominant economic activities, some of which are related to the inadequate use of natural resources (e.g., wood extraction, hunting, fishing). To develop a close link between these communities and the RDSP, it is necessary to understand how the communities see, feel and use the environment. Through discussions with local teachers in the areas surrounding the RDSP, it became evident that the knowledge of the ecosystem services provided by the park, including its biodiversity, is limited. This finding was not surprising considering that many Brazilians have received limited environmental education. This observation reinforces the fact that the creation of conservation units alone does not guarantee that the objectives for which they have been created will be fulfilled.

Through implementation of the project "educate for environmental action" (2000–2008), approximately 1,000 schoolteachers from 162 schools in 13 municipalities were chosen, based on their answers to a questionnaire, to take part in several activities, including short courses on selected environmental themes. Participative techniques were used to exchange experiences, enhance understanding of the region's environment reality and develop critical thinking skills that can help alter opinions, concepts, habits and practices. These efforts were expected to enhance the local population's acceptance of the conservation unit by increasing their participation in the decision-making process (Reed et al. 2006).

The major results of several research projects developed within the park were translated and shared with the participants. Several didactic-pedagogical materials were produced (booklets, games, biological collections) and donated to the partner schools. The teachers were encouraged to produce their own materials using adaptations based on the realities of their schools. Field excursions were organised to discuss local problems and possible solutions that consider social, economic and cultural aspects. The local media were mobilised to explain the programme, thereby establishing an effective channel through which to educate and promote the exchange of experiences, methods and values.

A particularly important challenge for the environmental education programme was to find means of interaction with an area occupied by members of the agrarian reform movement established in 2002, named "Settlement Chico Mendes II". This settlement occupies the largest forest fragment outside the RDSP (340 ha; 48 families) and has become an important threat to the local ecosystems, mainly because it has exacerbated deforestation and hunting pressures. The objectives of the programme were to encourage the conservation of local biodiversity and improve the quality of life of the settlers. Participative diagnoses were conducted to identify the major problems in the area, and six workshops were organised that addressed health conditions related to cultivation and soil, soil and water conservation, the flora and fauna of the Atlantic Forest, green manure and garbage. A local fair for exchanging seeds and taking guided tours of the RDSP was also organised.

During 2009 and 2010, activities were conducted with technical staff from local administrations that addressed the control, harvesting and disposal of the exotic species *Achatina fulica* (Giant African Snail). A total of 1,221 students attended the two events organised by the programme.

Our results show that interactions with local communities motivated changes in behaviour, although such motivation cannot be considered evidence of success because environmental education is a slow process. The groups that were involved had distinct experiences depending on age, culture and socio-cultural level. Furthermore, a change in habits and attitudes occurred across generations. The results showed that, to achieve the expected long-term results, it is necessary to make periodic contact with local communities and perform frequent evaluations of programmes and strategies so that corrections and adjustments can be made. The success of the project can only be confirmed when a change in understanding has been demonstrated not only at the individual level but also at the level of the social network. Once this change has occurred, we can conclude that "social learning" has taken place (Reed et al. 2010).

Re-Thinking Mining Activities: Scale and Territorial Impact and Management

Another important environmental impact in the Rio Doce Basin is the effluents that result from mining activities. The region is located in the Quadrilátero Ferrífero (Iron Quadrangle), one of the main iron ore deposits in the world, which makes Brazil's mining market quite robust. Together with the metallurgy sector, mining represents more than 50 % of the country's GDP. The mining sector brought the country US$ 28 billion in 2008, US$ 24 billion in 2009, US$ 39 billion in 2010 and a growth projection of US$ 50 billion for 2011, exhibiting tremendous growth (550 %) during the period 2001–2011 (IBRAM 2011). In 2012, mining production was estimated to grow 2–5 %, bringing approximately US$ 51 billion to Brazil (IBRAM 2012).

Impact assessment (IA) has become the standard policy tool used by governments to identify, analyse, predict and prevent impacts of human activities. If used correctly, IA can help make decisions more sustainable. According to Camagni (1998), there are four policy dimensions of sustainability: technological, behavioural, diplomatic and territorial. Territory, in this sense, is a complex concept that encompasses a physical space, its administrative or political level (national, regional, municipal), its type (urban, rural) and its functionality (watershed basin, agricultural areas, service areas).

The territory is a fundamental part of sustainability because many services of general interest (SGI), such as transport, electricity and wastewater, are constrained by natural geographic boundaries and should be analysed and managed as such (Balalia and Rauhut 2012). Nevertheless, most environmental policies have political and administrative boundaries, creating a chasm between spatial planning and natural geographic constraints that may lead to inefficient management strategies. There is a need to develop cohesion among policies to manage natural geographic boundaries and achieve more efficient results. The European Union was the first organization to use the expression territorial cohesion as a component that is required to achieve policy cohesion among Member States, which leads to improvements in accessibility, competitiveness, diversity and sustainability (European Commission 2010).

The need to include the territorial context in IA practices in order to assess the impact of policies and programmes led to the creation of the concept Territorial Impact Assessment (TIA). The TIA requires that we seek cooperation in both horizontal (among policies) and vertical (actors/stakeholders) terms and analyse the territory impact as a whole, thus avoiding the constraints imposed by political or administrative boundaries. The TIA also requires the implementation of management strategies to take into account that the degree of impact, the relevance of each criterion for IA and the intensity and vulnerability of the policies' application may differ from one territory to the next (Camagni 2009; Golobic and Marot 2011).

Water: Mining and Ecological Views

Natural resources are a good example of an SGI that has an explicit spatial dimension constrained only by natural geographic boundaries. However, the way natural resources are exploited has strong political and administrative components that differ among countries. Even within the same country, interests concerning environmental conservation and natural resource exploitation may diverge, creating tension between actors. There is increasing pressure for local water resources generated by a combination of urban expansion, large-scale agribusiness and mining operations. This increasing pressure has resulted in considerable changes in paradigms such as scale and territorial impact as well as seeing water through an economic lens (especially with respect to its role in mining) and from an ecological perspective. Despite several improvements that have been made regarding water economy and recycling water, the economic view considers water to be an indispensable and valuable product. The ecological view promotes the concept of water as an environment that is formed by a physical and chemical matrix in which basic processes (e.g., production, decomposition) maintain a little understood biota. There is a real need to promote cohesion between economic and ecological uses of water and territory to achieve sustainable natural resource use.

The new paradigm of territorial management offers a new vision of mining that focuses on the anticipation of impacts rather than their remediation, the integration of water resources management with research and development, the conciliation of mining with the conservation of natural resources and the adoption of integrated water resources management.

Humans: An Integral Part of Modern Conservation

As noted by Claus et al. (2010) (quoting Aldo Leopold 1935), one of the main challenges for modern ecology is combining the findings of human ecology (sociology, economy and history) and biological ecology (plant and animal community studies), fields that developed without close contact or collaboration. There is a clear and distinct valuation of these two sciences, and social sciences have clearly been undervalued by the ecological sciences with respect to conservation. Some authors explain this fact by the greater time consumed by social research, funding constraints and the current predominance of natural scientists among the conservation sciences and their pretence of understanding human behaviour. However, increasing the participation of social scientists in conservation projects is expected to be successful because it is currently recognised that conservation measures will fail if they do not consider humans as an integral part of modern conservation (Robinson 2006; Alves et al. 2012; Cooke et al. 2012).

The ecological "isolate to preserve" point of view is becoming obsolete within current conservation biology practices. Instead of trying to keep humans outside preserved areas, we can focus on teaching them how to live in harmony with environmental dynamics. The "use sustainably to conserve" point of view is more suitable for our socio-economic time because we live in an information society in which information is widely available. As shown for the Rio Doce basin, to achieve sustainability, we must promote cohesion among different points of view and set one unique goal for the territory that is being exploited. Companies, not only the mining ones, should change their view of water as simply a product and begin to think of it as an environment, as well. If they adopt this view, we can together develop sustainable ways to explore water usage, and both parties will benefit. The environment will retain its biodiversity, and companies will benefit from the ecological services that the conserved environment will provide.

In conclusion, we suggest three central points for consideration by modern society: (i) conservation initiatives must implement the "use sustainably to conserve" philosophy rather than the previous "isolate to preserve" philosophy; (ii) humans are an integral part of modern conservation, and therefore, social and economic aspects are key elements for any conservation plan; and (iii) water is, among other things, an environment of its own with biodiversity that should be assessed and protected.

References

Alves RRN, Vieira KS, Santana GG et al (2012) A review on human attitudes towards reptiles in Brazil. Environ Monit Assess 184:6877–6901

Balalia A, Rauhut D (2012) Assessing Territorial Impact Assessment: The Case of Services of General Interest. Paper presented at the 52nd Annual Congress of the European Regional Science Association, Bratislava, August

Barbosa FAR, Padisák J (2002) The forgotten lake stratification pattern: atelomixis, and its ecological importance. Verh Int Ver Theor Angew Limnol 28:1385–1395

Barbosa LG, Barbosa PMM, Barbosa FAR (2011) Vertical distribution of phytoplankton functional groups in a tropical shallow lake: driving forces on a diel scale. Acta Limnol Bras 23:63–73. doi:10.4322/actalb.2011.020

Barros C, Santos A, Barbosa F (2013) Phytoplankton diversity in the middle Rio Doce lake system of Southeastern Brazil. Acta Bot Bras 27:327–346

Barros C, Souza M, Barbosa F (2006) Seasonal mechanisms driving phytoplankton size structure in a tropical deep lake (Dom Helvécio Lake, South-East Brazil). Acta Limnol Bras 18:55–66

Berkes F, Hughes TP, Steneck RS et al (2006) Ecology, globalization, roving bandits, and marine resources. Science 311:1557–1558. doi:10.1126/science.1122804

Blackburn TM, Pyšek P, Bacher S et al (2011) A proposed unified framework for biological invasions. Trends Ecol Evol 26:333–339. doi: 10.1016/j.tree.2011.03.023

Camagni R (1998) Sustainable urban development: definition and reasons for a research programme. Int J Environ Pollut 1:6–26

Camagni R (2009) Territorial impact assessment for European regions: a methodological proposal and an application to EU transport policy. Eval Program Plan 32:342–350. doi:10.1016/j.evalprogplan.2009.06.014

Carey MP, Sanderson BL, Barnas KA et al (2012) Native invaders—challenges for science, management, policy, and society. Front Ecol Environ 10:373–381. doi:10.1890/110060

Carson R (1962) Silent spring. Houghton Mifflin, Boston

Castillo A, Magaña A, Pujadas A et al (2005) Understanding the interaction of rural people with ecosystems: a case study in a tropical dry forest of Mexico. Ecosystem 8:630–643. doi:10.1007/s10021-005-0127-1
Claus CA, Chan KMA, Satterfield T (2010) The roles of people in conservation. In: Sodhi NS, Erlich PR (eds) Conservation biology for all. Oxford University Press, Oxford, pp 262–283
Cooke S, Paukert C, Hogan Z (2012) Endangered river fish: factors hindering conservation and restoration. Endang Species Res 17:179–191
Costanza R, Hart M, Posner S et al (2009) Beyond GDP: the need for new measures of progress. Boston University, Boston
Commission European (2010) Investing in Europe's future—fifth report on economic, social and territorial cohesion. Publications Office of the European Union, Luxembourg
Feiock RC, Stream C (2001) Environmental Protection Versus Economic Development: A False Trade-Off? Public. Adm Rev 61:313–321. doi:10.1111/0033-3352.00032
Fischer A, Young JC (2007) Understanding mental constructs of biodiversity: implications for biodiversity management and conservation. Biol Conserv 136:271–282. doi:10.1016/j.biocon.2006.11.024
García-Llorente M, Martín-López B, González JA et al (2008) Social perceptions of the impacts and benefits of invasive alien species: implications for management. Biol Conserv 141:2969–2983
Godinho AL (1996) Peixes Do Parque Estadual Do Rio Doce. Instituto Estadual de Florestas / Universidade Federal de Minas Gerais, Belo Horizonte
Godinho AL, Fonseca MT, Araújo LM (1994) The Ecology of predator fish introductions: the case of Rio Doce valley lakes. In: Pinto-Coelho RM, Giani A, Von Sperling M (eds) Ecology and human impacton lakes and reservoirs in minas gerais with special reference to future development and management strategies. SEGRAC, Belo Horizonte, pp 77–83
Golobic M, Marot N (2011) Territorial impact assessment: integrating territorial aspects in sectoral policies. Eval Program Plan, 34:163–73. doi:10.1016/j.evalprogplan.2011.02.009
Gontijo BM, Britto CQ (1997) Identificação e classificação dos impactos ambientais no Parque Florestal Estadual do Rio Doce—MG. Geonomos 5:43–48
Gozlan RE, Britton JR, Cowx I et al (2010) Current knowledge on non-native freshwater fish introductions. J Fish Biol 76:751–786. doi:10.1111/j.1095-8649.2010.02566.x
Haffer J (1969) Speciation in Amazonian forest birds. Science 165:131–137
IBRAM (2011) Information and analysis of the Brazilan mineral economy 6th edition. http://www.ibram.org.br. Accessed 27 Aug 2013
IBRAM (2012) Information and analysis of the Brazilian mineral economy 7th edition. http://www.ibram.org.br. Accessed 27 Aug 2013
Jacobson SK, McDuff M (1998) Training idiot savants: the lack of human dimensions in conservation biology. Conserv Biol 12(2):263–267
Latini AO, Lima-Junior DP, Giacomini HC et al (2004) Alien fishes in lakes of the Doce river basin (Brazil): range, new occurrences and conservation of native communities. Lundiana 5:135–142
Latini AO, Petrere Júnior M (2004) Reduction of a native fish fauna by alien species: an example from Brazilian freshwater tropical lakes. Fish Manag Ecol 11:71–79. doi:10.1046/j.1365-2400.2003.00372.x
Levins R (1969) Some demographic and genetic consequences of environmental heterogeneity for biological control. Bull Entomol Soc Am 15:237–240
MacArthur RH, Wilson EO (1963) An equilibrium theory of insular zoogeography. Evolution 17:373–387
MacArthur RH, Wilson EO (1967) The theory of island biogeography. Princeton University Press, New Jersey
Maia-Barbosa P, Barbosa L, Brito S et al (2010) Limnological changes in Dom Helvécio Lake (South-East Brazil): natural and anthropogenic causes. Braz J Biol 70:795–802
Maia-Barbosa PM, Menendez RM, Pujoni DGF et al (2014) Checklist of zooplankton from natural lakes of Middle Rio Doce basin, Minas Gerais, Brazil. Biota Neotropica (in press)

Marques M, Barbosa F (2001) Biological quality of waters from an impacted tropical watershed (middle Rio Doce basin, southeast Brazil) using benthic macroinvertebrate communities as an indicator. Hydrobiologia 457:69–76

Maryn AA, Oliveira HT, Comar V (2003) A educação ambiental num contexto de complexidade do campo teórico da percepção. Interciencia 28:616–619

Meine C, Soulé M, Noss RF (2006) A mission-driven discipline: the growth of conservation biology. Conserv Biol 20:631–651

Meine C (2010) Conservation biology: past and present. In: Sodhi NS, Erlich PR (eds) Conservation biology for all. Oxford University Press, Oxford, pp 7–26

Noss R (1991) Sustainability and wilderness. Conserv Biol 5:120–122

Oliveira AES, Pereira DG (2010) Eradication of invasive exotic species: multiple views of the Brazilian reality. Desenv Meio Ambiente 21:173–181

Oporto LT (2013) Aspectos ecológicos e sociais da invasão de peixes em lagos naturais do médio rio Doce, MG. Ph.D. Thesis, Universidade Federal de Minas Gerais

Pelicice FM, Vitule JRS, Lima Junior, DP et al (2013) A serious new threat to Brazilian freshwater ecosystems: the naturalization of nonnative fish by decree. Conserv Lett. doi: 10.1111/conl.12029

Pinto-Coelho RM, Bezerra-Neto JF, Miranda F et al (2008) The inverted trophic cascade in tropical plankton communities: impacts of exotic fish in the Middle Rio Doce lake district, Minas Gerais, Brazil. Braz J Biol 68:1025–1037

Pompeu PS, Godinho AL (2001) Mudança na dieta da traíra Hoplias malabaricus (Bloch)(Erythrinidae, Characiformes) em lagoas da bacia do rio Doce devido à introdução de peixes. Rev Bras Zool 18:1219–1225

RAMSAR (2010) The RAMSAR list of wetlands of international importance. http://www.ramsar.org/pdf/sitelist_order.pdf. Accessed 27 Aug 2013

Redford K (1991) The ecologically noble savage. Cult Surv Q 15:46–48

Reed MS, Fraser EDG, Dougill AJ (2006) An adaptive learning process for developing and applying sustainability indicators with local communities. Ecol Econ 59(4):406–418. doi:10.1016/j.ecolecon.2005.11.008

Reed MS, Evely AC, Cundill G et al (2010) What is social learning? Ecol Soc 15(4):r1. http://www.ecologyandsociety.org/vol15/iss4/resp1/

Robinson JG (2006) Conservation biology and real-world conservation. Conserv Biol 20:658–669

Sarkar S (1999) Wilderness preservation and biodiversity conservation: keeping divergent goals distinct. Bioscience 49:405–412

Schaller GB (1992) Field of dreams. Wildlife Conservation September/October: 44–47

Simberloff D, Martin JL, Genovesi P et al (2013) Impacts of biological invasions: what's what and the way forward. Trends Ecol Evol 28:58–66

Souza MBG, Barros CFA, Barbosa FAR (2008) Role of atelomixis in replacement of phytoplankton assemblages in Dom Helvécio Lake, South-East Brazil. Hydrobiologia 607:211–224. doi:10.1007/s10750-008-9392-2

Speziale KL, Lambertucci SA, Carrete M et al (2012) Dealing with non-native species: what makes the difference in South America? Biol Invasions 14:1609–1621

Sunaga T, Verani JR (1991) The fish communities of the lakes in Rio Doce Valley, Northeast Brazil. Verh Int Ver Theor Angew Limnol 24:2563–2566

Tundisi JG, Saijo Y (eds) (1997) Limnological studies on the Rio Doce Valley Lakes. Brazilian Academy of Sciences, São Carlos, Brazil

Wunder S (2001) Poverty alleviation and tropical forests—what scope for synergies? World Dev 29:1817–1833

Chapter 16
Integrating Input-Output Modeling with Multi-criteria Analysis to Assess Options for Sustainable Economic Transformation: The Case of Uzbekistan

Maksud Bekchanov, Anik Bhaduri, Manfred Lenzen and John P. A. Lamers

Abstract Integrating economic efficiency and environmental sustainability indicators is essential for designing policies for a sustainable development. Given the growing pressure on water resources, efficient water use becomes an essential environmental criteria for formulating adjustment reforms. Despite the wide use of backward and forward linkages as well as direct and indirect resource (energy, water, etc.) uses based on environmentally extended input-output models for assessing the performance of economic sectors, the common practice of presenting different indicators separately obstructed a straightforward policy interpretation of results. To derive a composite indicator that allows to direct ranking of sectors, we combined therefore a direct and indirect water use intensities with backward and forward linkage indexes by using the multi-criteria analysis method-TOPSIS (Technique for order preference by similarity to ideal solution). The model was implemented to formulate sectoral transformation measures guided by sustainable growth objectives in Uzbekistan, Central Asia, which is a representative of an area with growing water scarcity. The results showed that the presently promoted crops under the state order system—cotton and wheat—and crop preferred by farmers—rice—are the least effective production options for reaching such a sustainable growth. It is argued therefore that unbiased support for all crops through adaption of the current state order system for cotton and wheat cultivation is needed to

M. Bekchanov (✉)
International Water Management Institute (IWMI), 127 Sunil Mawatha, Pelawatte, Battaramulla, Sri Lanka
e-mail: m.bekchanov@cgiar.org

A. Bhaduri
Global Water System Project, Walter-Flex-Str. 3, Bonn, Germany
e-mail: abhaduri@uni-bonn.de

J. P. A. Lamers
Center for Development Research (ZEF), University of Bonn, Walter-Flex-Str. 3, Bonn, Germany

M. Lenzen
School of Physics, A28, The University of Sydney, Sydney, NSW 2006, Australia

A. Bhaduri et al. (eds.), *The Global Water System in the Anthropocene*,
Springer Water, DOI: 10.1007/978-3-319-07548-8_16,

achieve a more diversified crop portfolio with an increased share of fruits and vegetables. A further development of agro-processing industries and livestock sector bears more potential for sustainable economic development than a further promotion of producing raw agricultural commodities. Investing in industrial sectors illustrated more potential than in agriculture related sectors when aiming at economic effectiveness and increased water use efficiency. It is concluded that, with a relevant sectoral transformation, Uzbekistan has high opportunities to cope with reduced water availability.

Introduction

Limited water resources for increasing needs by many economic sectors necessitate increasing water use efficiencies. Achieving the Millennium Development Goals (UN 2000) of alleviating poverty, eradicating hunger and providing basic sanitation also depends on how sustainably and efficiently the scarce water resources are managed (von Braun et al. 2003). Although to many, an efficient water use is a synonym to the adoption of water conservation technologies, this view exclude strategies of relocating water from its lower to higher valued uses through prioritizing production activities not only in accordance to their economic effectiveness but also according to water productivity. Relocation of water to its higher use is particularly challenging in dryland regions (Rosegrant et al. 2002) where irrigated agriculture is at the core of the livelihoods and water scarcity is a primary constraint for economic development. At present, the dryland regions host about 40 % of the global irrigated area and almost one third of the world's population (Millennium Ecosystem Assessment 2005:1).

To determine options of sustainable sectoral adjustments in arid regions, different measures of water use intensities have been considered. The virtual water use for instance, measured as the amount of water required per unit of output, became popular following the work by Allan (1997) who argued for importing the commodities with high water use intensity rather than producing them domestically in water scarce Middle Eastern countries. Several follow-up studies estimated virtual water contents of agricultural and food commodities in different dryland regions of the world to deduct favorable production, consumption and trade pattern changes (Wichelns 2001; Oki and Kanae 2004; Hoekstra and Hung 2005). Even though the implications of trade pattern changes based on physical water use intensities (e.g., water use per unit of physical output) of a particular activity in different regions are valid, this indicator has shortcomings in comparing virtual water contents of different activities. For example, when aiming for the formulation of adjustment policies, a comparison of the virtual water content of one kg meat to one kg wheat is inadequate since these two commodities have different economic and food/nutrition values.

Virtual water use expressed in economic terms, which is expressed as water use per unit of economic output rather than per unit of physical output, is an alternative to cope with the limitations of physical water content (Lenzen 2009). Estimations of water use per economic output have been common in input-output model based studies that aimed at comparing water use intensities by economic sectors (Lenzen 2009; Zhao et al. 2009; Lenzen 2003; Velasquez 2006). The unique structure of input-output models also allow to estimating indirect water uses embedded in intermediate inputs (Lenzen 2009). Environmentally extended backward and forward linkages based on input-output models opened thus a way for a comprehensive analysis of direct and indirect input uses by activities (Lenzen 2003). Yet, a straightforward ranking of economic sectors based on different linkage-based indicators is still lacking obstructing respective policy implications. This study introduced therefore a way of elaborating composite indicator based on economic and environmental performance indexes for ranking economic sectors. TOPSIS (Hwang and Yoon 1981), a multi-criteria assessment method, was used to integrate economic and environmental indicators. The model was applied to the case of Uzbekistan, a dryland country where water is critical to national development and individual livelihoods. Since the bulk of the water used in the country stems from outside its boundaries, increasing water scarcity poses challenges to downstream Uzbekistan searching for sectoral adjustments which could be guided by the development of less water-intensive sectors.

Case Study Region

Uzbekistan is one of the former Soviet Union countries that started its transition from a command-based to a market-oriented economy since 1991. The cotton self-sufficiency policy introduced during the Soviet period drove irrigation expansion in entire Central Asia (Fig. 16.1) and turned these five countries into the cotton belt of the Soviet economy (Glantz 1999). Between 1960 and 2000, the irrigated areas in Uzbekistan have expanded from 1.8 to more than 4 Million ha (Mha) (FAO 2000) whilst annual irrigation water withdrawals tripled to more than 56 km^3 (Orlovsky et al. 2000) resulting in heavy dependence of the national economy on irrigated agriculture. Although the share of agriculture in GDP reduced from 40 % in the early 1990s to 20 % after 2005, irrigated agriculture remained the backbone of the regional economies with an output share of more than 50 % in many districts of Uzbekistan (Bekchanov and Bhaduri 2013).

Due to an excessive water use and huge water wastage, waterlogging and land degradation have become a grave concern in some parts of the Amu and Syr Darya Basins whereas water availability to the more-downstream regions and environmental systems gradually decreased (Glantz 1999). The desiccation of the Aral Sea, as a consequence of a decade long excessive diversion of river flows for irrigated crop production, has been coined as one of the worst environmental disasters in the world (UN 2010). Despite attempts to reducing the size of water intensive sectors and consequently improving the environmental balance after 1991, cotton production is still the dominant agricultural activity occupying a

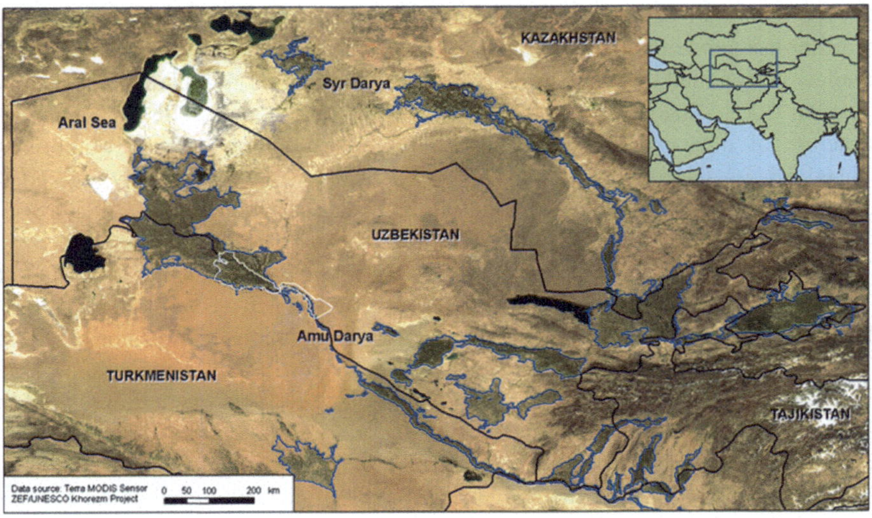

Fig. 16.1 Irrigated areas of Central Asia and the case study country Uzbekistan (ZEF/UNESCO Urgench Project 2013)

substantial share of croplands (Rudenko et al. 2009). The post-Soviet expansion of cereals production due to the declared grain self-sufficiency policy has increased the pressure on water resources further. Re-evaluating economic restructuring policies by integrating economic growth targets with environmental needs is of utmost importance for maneuvering the development path towards sustainability.

Methods and Data Sources

Leontief Model

The intersectoral financial flows in the economy were calculated using the input-output system according to Leontief (1951):

$$\mathbf{x} = \mathbf{A}\mathbf{x} + \mathbf{y} \tag{16.1}$$

where $\mathbf{x}$ is a $n \times 1$ vector of total production volume for each sector, $\mathbf{y}$ is a $n \times 1$ vector of final demand including private and government consumption, investment expenditures, changes in inventories, and exports. $\mathbf{A}$ is a $n \times n$ matrix of technical production coefficients.

In this model, with simple transformations, final demand is treated as an exogenous variable that determines the level of total production as:

$$\mathbf{x} = (\mathbf{I} - \mathbf{A})^{-1}\mathbf{y} = \mathbf{Ly} \tag{16.2}$$

where **I** is an $n \times n$ identity matrix and **L** is the $n \times n$ Leontief inverse matrix.

The element l_{ij} of the Leontief inverse **L** reflects the total requirements from sector i to provide a unit of the final demand for the commodities of sector j.

Ghosh Model

The Ghosh (1958) model was used to estimate the intersectoral allocation of primary and intermediate inputs:

$$\mathbf{x}' = \mathbf{x}'\mathbf{B} + \mathbf{v}' \tag{16.3}$$

where **B** is a $n \times n$ matrix of allocation coefficients that indicates a ratio of intersectoral intermediate inputs to the total inputs (raw sums of input-output table) and $\mathbf{v}'$ is a $1 \times n$ vector of primary factors, which includes capital, labor, and imports. The prime symbol (′) denotes matrix transposition.

Similar to (Eq. 16.2), with simple transformations, the relationship between the primary factors and the level of total production is obtained as:

$$\mathbf{x}' = \mathbf{v}'(\mathbf{I} - \mathbf{B})^{-1} = \mathbf{v}\prime\mathbf{G} \tag{16.4}$$

where **G** is an $n \times n$ Ghosh inverse matrix. The element g_{ij} of the Ghosh matrix **G** reflects the total required outputs from sector j to absorb a unit of the primary factors of sector i.

The Backward and Forward Linkage Indices

The Leontief inverse matrix (Eq. 16.2) allows measuring direct and indirect effects of a change in the final demand over production as well as to calculate the backward linkages (BLs). The BL of sector j that indicates an increase in total output of all sectors as a response to a unit increase in the final demand of sector j which was calculated following Rasmussen (1956) and Hirschman (1958):

$$BL_j = L_{*j} = \sum_i l_{ij} \tag{16.5}$$

where L_{*j} is the associated column sum of elements of the matrix **L** for sector j. BL is an indicator for the level of influence by sector j on the output of all sectors through its purchases.

For maintaining comparability, BLs were normalized around their average value to calculate BL indices (BLIs):

$$BLI_j = \left(L_{*j}/n\right)/L \tag{16.6}$$

where L^* is the average of BLs or the mean of all elements of the Leontief inverse **L** (Eq. 16.2).

Considering the relevance of the Ghosh model (Eq. 16.4) than the Leontief model for calculating forward linkages (FLs) (Beyers 1976; Jones 1976), the FL was elaborated as:

$$FL_i = G_{i*} = \sum_j g_{ij} \tag{16.7}$$

where G_{i*} is the associated raw sum of elements of the matrix **G** for sector i.

FL of the sector i indicates how much sector i influences on the output of all sectors through its sales (output supplies) and is interpreted as the total output required to absorb a unit of primary inputs.

FL indices (FLIs) were estimated by normalizing FLs around their average value:

$$FLI_i = (G_{i*}/n)/G^* \tag{16.8}$$

where G^* is the average of FLs or the average of all elements of the Ghosh matrix **G** (Eq. 16.4).

BLI and FLI of a sector reflect its influence and dependence on the remaining sectors of the economy, respectively. $BLI_j > 1$ indicates strong backward linkages of sector j which means that a unit increase in the final demand of sector j would result in a greater than average increase in total economic output (Lenzen 2003). In parallel, $FLI_i > 1$ shows strong forward linkages of sector i which means that a unit increase in primary inputs of sector i would require a greater than average increase in total economic output. A value of 10 for the respective linkage index, for instance, indicates that this linkage is 10 times stronger than the average of all sectors. If both conditions, $BLI_j > 1$ and $FLI_i > 1$, are fulfilled for any sector, then this sector is considered as a key sector which exhibits both greater than average influence and dependence on other economic sectors.

Direct and Indirect Water Uses

The integration of the virtual water content of commodities with BLIs and FLIs allows for a better-informed decision-making on economic restructuring. To estimate virtual water contents, direct water input coefficients (dw_j) were estimated initially as the ratio of total water use (W_j) to the total production volume of a given sector $j(Q_j)$:

$$dw_j = W_j / Q_j \tag{16.9}$$

Based on these direct water use coefficients and the Leontief inverse matrix elements, backward linkage-based full (direct plus indirect) water contents (vw_j) were calculated as:

$$vw_j = \sum_i dwi_j l_{ij} \tag{16.10}$$

Full water use content based on the Leontief model indicates the total (both direct and indirect) amount of virtual water that is required to produce a unit of final demand in sector j.

In parallel, a forward linkage-based full water use indicates the total (both direct and indirect) amount of virtual water that is required to absorb a unit of primary factors in sector i and which can be calculated as:

$$vw_i^G \sum_j dw_{ij} g_{ij} \tag{16.11}$$

Multi-criteria Ranking

Since the ranks of the sectors according to individual criteria differ from each other, multi-criteria decision analysis (MCDA) tools are used for ranking and selecting the most efficient options (sectors) in terms of economic and environmental efficiency (Wang et al. 2009). Several multi-criteria ranking methods exist, including the elementary methods such as the weighted sum method (CLG 2009; Simonovich 2009), unique synthesizing criteria methods such as AHP[1] (Saaty 1980) and TOPSIS (Hwang and Yoon 1981), and outranking methods such as ELECTRE[2] (Roy 1991) and PROMETEE[3] (Brans 1984). Since there is no clear evidence on preference of one method over the other, the TOPSIS method was chosen due to its easiness to handle and used to combine the BLIs, FLIs, and BL and FL-based virtual water use contents into a single composite indicator.

TOPSIS is based on the concept of ranking the options according to their closeness to the ideal option (the best alternative) and farness from the negative-ideal option (the worst alternative) (CLG 2009). The method involves several steps to evaluate the ranks of the sectors. At first, four separate indicators under

---

[1] Analytical Hierarchical Processes.

[2] Elimination et choice translating reality.

[3] Preference ranking organization method for enrichment evaluation.

consideration have been normalized to obtain the attributive values (d_{ik} (k stands for the attribute/criteria)) of the options in a $n \times m$ decision matrix (**D**):

$$d_{i1} = \frac{BLI_i}{\sqrt{\sum_i BLI_i^2}},\ d_{i2} = \frac{FLI_i}{\sqrt{\sum_i FLI_i^2}},\ d_{i3} = \frac{vw_i}{\sqrt{\sum_i vw_i^2}},\ \text{and}\ d_{i4} = \frac{vw_i^G}{\sqrt{\sum_i vw_i^{G2}}} \tag{16.12}$$

Based on the elements of this matrix and weights for each criterion (β_{ik}), the weighted normalized values (w_{ik}) of a $n \times m$ decision matrix (**W**) were evaluated as follows:

$$w_{ik} = \beta_{ik} d_{ik} \tag{16.13}$$

In this equation, β_{ik} depend on a subjectively chosen weight for environmental sustainability (α) which varies between 0 and 1:

$$\begin{aligned} \beta_{i1} = \beta_{i2} &= \frac{1}{2}(1-\alpha) \\ \beta_{i3} = \beta_{i4} &= \frac{1}{2}\alpha \end{aligned} \tag{16.14}$$

Attributive values of ideal option (p_k) were estimated as:

$$\mathbf{p} = \{p_k\} = \{\max_i w_{i1}, \max_i w_{i2}, \min_i w_{i3}, \min_i w_{i4}\} \tag{16.15}$$

where **p** is a $1 \times m$ dimensional vector of ideal values.

Meantime, attributive values of non-ideal option (q_k) were calculated as follows:

$$\mathbf{q} = \{q_k\} = \{\min_i w_{i1}, \min_i w_{i2}, \max_i w_{i3}, \max_i w_{i4}\} \tag{16.16}$$

where **q** is a $1 \times m$ dimensional vector of negative-ideal values.

The composite indicator was estimated based on the relative closeness of the option to the ideal option:

$$c_i = \frac{\sqrt{\sum_k (w_{ik} - q_k)^2}}{\sqrt{\sum_k (w_{ik} - q_k)^2} + \sqrt{\sum_k (w_{ik} - p_k)^2}} \tag{16.17}$$

The value of this indicator varies between 0 and 1 and equals to 1 for ideal option.

Data

Multi-criteria ranking and intersectoral linkage analyses were based on a national input-output table (IOT) which is extended with water use accounts. During the Soviet era, national statistical organizations were entrusted with the development of national and regional IOTs for Uzbekistan. After independence in 1991, IOTs have not been reported by the statistical agency anymore. Müller (2006) developed a national social accounting matrix (SAM) with twenty sectors for 2001 (hereafter referred to as SAM-2001). More recent IOT of Uzbekistan that includes thirteen sectors was developed for 2005 (hereafter referred to as IOT-2005) by researchers at the Center for Efficient Economic Policy (CEEP), the Center for Economic Research (CER), the Ministry of Economy (MoE) and Colorado University (UNDP 2006). Because IOT-2005 represents the most recent complete database, it was used for estimating the IOT with disaggregated agricultural accounts (hereafter referred to as AGRIOT-2005). Elements of disaggregated accounts for agricultural and agro-processing sectors in the AGRIOT-2005 were estimated considering the proportional shares of intermediate inputs in the SAM-2001. Production values, GDP, value added, exports, imports, and consumption levels across the sectors of the economy were based on the databases of the Asian Development Bank (ADB 2008), the National Statistical Committee of Uzbekistan (UzStat 2006, 2008) and CEEP (2006). The maximum entropy approach (Golan et al. 1996) was used for integrating data from multiple sources and eventually evaluating a balanced IOT. The calculations based on the entropy approach were conducted in GAMS (Brooke et al. 2006). The average exchange rate of 1 USD equaled to 1,128 UZS (as of 2005) was considered in the calculations.

Aggregated water use data (UNDP 2007) and existing water use requirements per head of livestock, per hectare of cropland, or per unit of production output were used to estimate water uses by subsectors of the agricultural and industrial sectors. For instance, total water consumption requirement for livestock production was estimated by multiplying the heads of each type of livestock (cattle, sheep, goats, pigs, horses, and poultry) (UzStat 2008) with annual water consumption requirement per head of livestock (CRIIWRM 1980). The recommended water use for each agricultural sub-sector was estimated by multiplying the cultivated land area of the respective crop (UzStat 2008) with the recommended per ha water uses (Müller 2006). Then, the relative shares of each agricultural sub-sector in the total agricultural water requirement were multiplied with actual total agricultural water use to derive actual sub-sectoral water uses.

Similarly, water use requirements by industrial sub-sectors were estimated as a multiplication of physical production volumes of industrial products (UzStat 2006) with water use requirements per unit of industrial output (State Construction Office 1978). The shares of required water uses for each industrial subsector in the total recommended industrial water use were multiplied with actual total industrial water use for estimating actual sub-sectoral water uses.

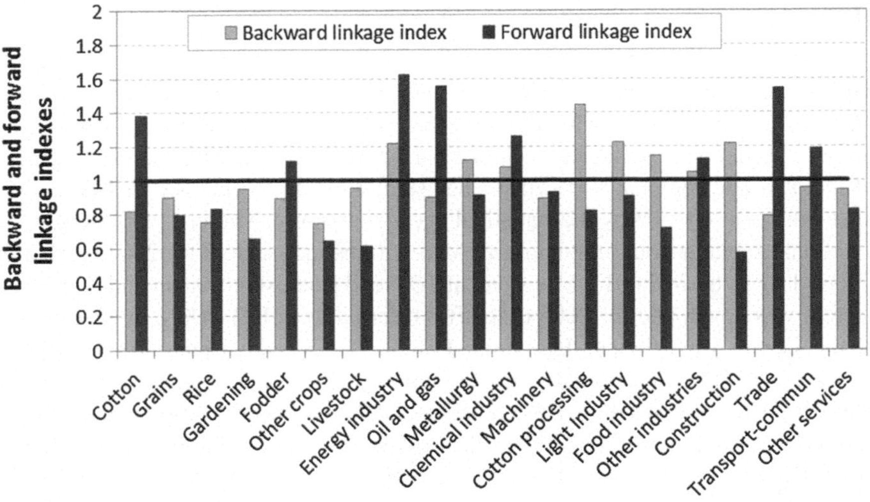

Fig. 16.2 Backward and forward linkage indexes by sectors of the economy in Uzbekistan for 2005

Results

Backward and Forward Linkages in the Economy

The industrial sub-sectors had generally higher BLIs, meaning had a higher influence on the economic sectors by demanding their outputs, than the agricultural sub-sectors. The BLIs for agriculture varied between 0.7 and 1.0 while BLIs in the industrial sector varied between 0.9 and 1.5 (Fig. 16.2). Thus, a unit increase in the final demand of any agricultural sub-sector would result in a lower than average increase in total economic output. In contrast, the BL of most of the industrial sectors was stronger than the average BL of all sectors. The fruits and vegetables sector (gardening) had the highest BLI among all agrarian sub-sectors (1.0) while the highest BLI among all sectors was observed for cotton processing (1.5).

Similar to BLIs, FLIs of the industrial sub-sectors were generally higher than those of the agricultural sub-sectors (Fig. 16.2) indicating thus a higher importance of the industrial sub-sectors than the agricultural sub-sectors as suppliers of intermediate inputs to the economic sectors. The FLIs for all agricultural sub-sectors varied between 0.6 and 1.4 while the FLIs for industrial sub-sectors varied between 0.7 and 1.6. The FLI for the raw cotton production sector was the highest among all agricultural sub-sectors confirming the massive volumes of raw cotton delivered to the cotton ginneries that virtually exist in each settlement across the country. The highest FLI valued of 1.6 was for the energy sector showing the importance of this sector as a supplier of electricity power to any sector. The FLIs for trading, transport and communication valued at 1.5 and 1.2 and were higher than the FLIs of most of the agricultural and industrial sectors.

Energy industry, chemical industry, and other industries (production of glass, furniture, construction materials, etc.) had higher than average BLIs and FLIs (Fig. 16.2). Thus, these sectors are key sectors of the economy and important both as demanders (consumers) of outputs of the economic sectors and as suppliers of intermediate inputs to the economic sectors.

Direct and Indirect Water Uses by Sectors

The analysis of the virtual water contents by sectors showed that in general agricultural sub-sectors required substantially higher amounts of water per unit of economic output than the other sectors (Fig. 16.3). Direct water use per unit of economic output was the highest for rice and valued at 44 m^3 USD^{-1}. Cotton and winter-wheat production used about 23 m^3 water per USD. Since per ha water uses of winter wheat was comparatively lower than that for the other crops examined, its high direct water use per economic output is understandable when realizing an undervaluation of wheat production outputs because of the low state prices for grain. In contrast, no government production targets and procurement prices exist for the production of fruits and vegetables (gardening). The direct water uses per economic output of this sector valued at 11 m^3 USD^{-1} was substantially lower than the water use intensities of rice, cotton and wheat. Among the industrial sub-sectors, the highest direct water consumption per economic output, valued at 3.4 m^3 USD^{-1}, was for the energy industry. Direct water use per unit of the economic output for the remaining industrial and service sectors was negligible since the non-agricultural sectors produced about 75 % of GDP in 2005 while using less than 10 % of total available water.

Full (direct plus indirect) water uses to produce one unit of the final demand were generally higher in crop production than any other sector considered. However, there was no significant difference between direct and full water uses of the crop production sub-sectors indicating that the water embedded in their intermediate input uses were low. In contrast, full water uses were considerably higher than direct water uses for the livestock sector and agro-processing industries since substantial amounts of intermediate inputs were delivered to these activities from the water intensive agricultural activities.

Forward linkage based virtual water contents were also generally higher for the crop production sub-sectors than the industrial sub-sectors. Considerable differences between the forward linkage-based virtual water contents and direct water

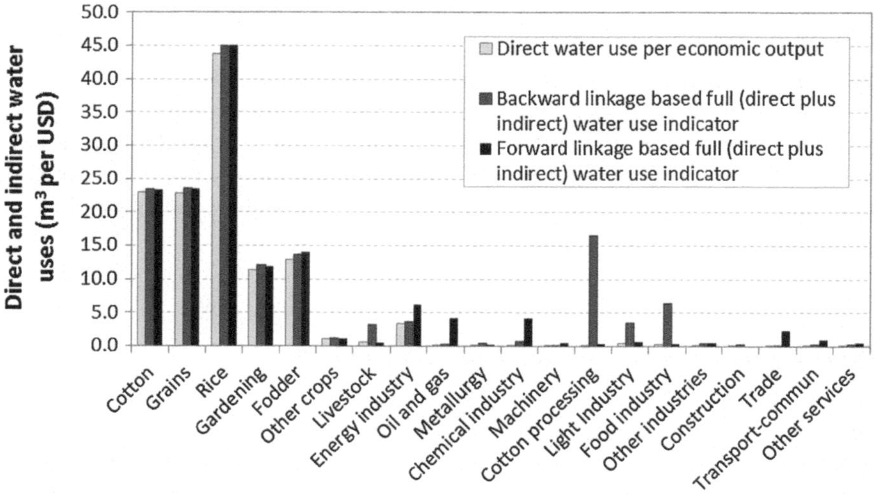

Fig. 16.3 Direct and indirect water use contents by sectors of the economy in Uzbekistan

uses were found for the energy sector, oil and gas industry, chemical industry and trade showing that a considerable amount of virtual water was embedded in intermediate input deliveries from these sectors to other sectors.

Multi-criteria Ranking of the Economy Sectors

Ranking the sectors according to their backward and forward linkages and full water use contents resulted in different rankings under each single criterion (Table 16.1). Integrating the separate attributive values of the options while considering the different weights for the environmental criteria, e.g. water use factor, showed that rice and wheat production were the least favorable activities. Rice production was unfavorable due to its huge direct and indirect water use requirements (ranked 20) and very low backward linkage indexes (ranked 19). Cotton production was more favorable than the other agricultural sectors except the production of fodder crops when the environmental factor was considered less important in choice-making. But when considering the weight to the environmental factor at least equal to the weight of the economic factor, cotton lost its prime position becoming less favorable than the production of fruit-vegetables, fodder crops and other crops including sweet beet and sunflower. The livestock sector generally had higher ranks than the crop production activities. Agro-processing industries such as cotton processing, light industry, and food processing were more favorable than any other agricultural sub-sectors. Agricultural related sub-sectors (e.g. agricultural and agro-processing) generally had lower ranks than the other sectors.

Table 16.1 Single and multi-criteria rankings of the economic sectors

Sectors	Ranking the sectors due to single criterion				Multi-criteria ranking based on TOPSIS		
	BLI	FLI	BLVW	FLVW	0.2	0.5	0.8
Cotton	17	4	18	18	15	18	18
Grains	13	15	19	19	19	19	19
Rice	19	12	20	20	20	20	20
Gardening	11	17	15	16	18	16	16
Fodder	15	8	16	17	17	17	17
Other crops	20	18	10	11	17	14	9
Livestock	9	19	11	8	16	13	11
Energy industry	3	1	13	15	1	3	14
Oil and gas	14	2	3	14	2	5	12
Metallurgy	6	10	7	2	8	7	3
Chemical industry	7	5	9	13	3	2	10
Machinery	16	9	1	7	10	9	4
Cotton processing	1	14	17	3	9	15	15
Light Industry	2	11	12	9	7	8	8
Food industry	5	16	14	4	12	12	13
Other industries	8	7	8	5	5	1	1
Construction	4	20	6	1	13	11	7
Trade	18	3	2	12	4	6	6
Transport and communication	10	6	4	10	6	4	2
Other services	12	13	5	6	11	10	5

BLI Backward linkage indexes; *FLI* Forward linkage indexes; *BLVW* Backward linkage based virtual water use; *FLVW* Forward linkage based virtual water use. Weights of 0.2, 0.5, and 0.8 were considered for environmental factor (α). Ranks are in ascending order, e.g. rank 1 stands for the most favorable option

Discussion

Crops such as cotton, wheat and rice that presently dominate the agricultural crop production portfolio in Uzbekistan require substantially higher water use per economic output compared to the other economic sectors. However, because of its crucial role in hard-cash income generation, cotton production is practiced on at least 40 % of the total irrigated cropland under the strict state scrutiny (Rudenko et al. 2009). Therefore, human capital and market infrastructure has also been aligned to keep cotton production and export as a major national activity. Although the irrigation expansion for cotton production since the 1960s has been acknowledged for the improved employment opportunities in rural areas of Central Asia (Rudenko et al. 2009), the past cotton production practices have also contributed to the chain of environmental disasters including water overuse, soil salinization, water logging, and the desiccation of the Aral Sea (WBGU 1998). To maintain environmental sustainability, present practice of differential crop support in Uzbekistan either should be phased out or equal importance should be given to other crops. Lowered government intervention and improved pricing system for agricultural production are likely to create additional incentives for farmers to maintain crop diversification and improve their crop rotations which in turn could result in increased water use efficiency and income generation (Bobojonov et al. 2012). The liberalization of the present state procurement system may also motivate producers a wider adoption of water saving technologies to earn higher incomes (Bekchanov et al. 2010). In contrast, continuing the heavy reliance on scarce water resources for obtaining risky revenues from raw commodity exports (Rudenko et al. 2009) without any technological improvements may lead to irreversible environmental deteriorations in the long-run.

When aiming at increasing overall water productivity, promoting agro-processing industries seems to be more promising than solely concentrating on the agricultural sector. The present findings together with these of Rudenko et al. (2009) suggest that upgrading the cotton value-chain through expanding the production of cotton-made commodities such as clothes, which have higher value-added, bear more prospects for producers to gain higher incomes with concurrently lower water use rates. However, when promoting the development of agro-processing sectors further, stakeholders have to cope with the present lack of investment capacities, up-to-date technologies, and processing expertise. Furthermore, this kind of sectoral transformations require improvements in the legislation and institutional environment, the reallocation of labor resources, and development of new trading strategies. A success of sectoral adjustments through enhancing agro-processing industries strongly depends also on establishing a competitive environment for domestic producers through limiting the state monopoly in the cotton sector. Since substantial efforts are required at all fronts to prepare a necessary enabling environment for such transformation reforms, the risks are high indeed, but so will be the gains.

Promoting the non-agro-processing industries and services sectors further also seems to bear more potential than the present pathway and certainly shows to be more promising than that of any agricultural or agro-processing industries when prevailing water productivity and economic growth potential are the main decision-criteria. Particularly, the energy industry, chemical industry, and construction materials production sectors turned out to be the key sectors of the economy to support further as evidenced by their high BLIs and FLIs. A further development of these sectors is also recommendable when water use aspects of these activities are considered in the multi-criteria choice analysis. Even though water requirements in the industrial sub-sectors are at present much lower than those in the agricultural sub-sectors, waste water from industrial processes is known to be much more hazardous than the return water flows in agriculture (Chapagain and Hoekstra 2004). Thus, the negative influence of the return flows on natural ecosystems should not be forgotten in promoting strategies for the development of the agro-processing and other industrial sectors.

Hazardous atmospheric emissions from the industrial sectors often are much more harmful to the environment than the emissions from the agriculture-related sectors. Since our analyses excluded these environmental factors, the inclusion of more environmental impact indicators would improve certainly the results which in turn would enable to make more robust conclusions on the sustainable development potential of the industrial sectors in Uzbekistan. Additional investigations on the potential magnitude of change in sectoral outputs due to structural adjustment policies and related changes in labor, capital and water demands would be needed and this could be considered by using a computable general equilibrium model.

Conclusions

Sustainable economic development necessitates an integration of economic and environmental impact indicators to lay the basis for better-informed policy decisions. For a water-based economy like in Uzbekistan, water productivity is a useful proxy indicator for the environmental fragility of the ecosystems that is vital for defining development strategies. To arrest and perhaps even inverse the on-going environmental degradation in the long run as well as cope with the present and expected water scarcity, Uzbekistan must seek ways to restructure its domestic production practices and give emphasis on the sectors and commodities with both higher value-added and less water requirements. A restructuring of land and water use while upgrading the agricultural value chains and the transformation of the economy towards industrial sectors would result in a more efficient use of the limited natural resources.

Acknowledgments This paper was prepared based on the results of a PhD study supported by the Bundesministeriums für Bildung und Forschung (BMBF) through the IPSWaT scholarship Program. This work was also partly funded by the BMBF and the Australian Academy of Science under their joint Australia-Germany Researcher Mobility Call 2010-2011. The authors thank Dr. Arnim Kuhn (ILRI, Bonn University), Dr. Claudia Ringler (IFPRI), Prof. Dr. Joachim von Braun (ZEF, Bonn University) and two anonymous reviewers for their constructive comments on earlier versions of this manuscript.

References

Allan JA (1997) 'Virtual water': a long term solution for water short. Middle Eastern Economies?, In: Paper presented at the 1997 British Association Festival of Science, Roger Stevens Lecture Theatre, University of Leeds, Water and Development Session, TUE.51, 14.45

Asian Development Bank (ADB) (2008) Key economic indicators (Online database). http://www.adb.org. Accessed 20 Aug 2008

Bekchanov M, Bhaduri A (2013) National tendencies and regional differences in industrialization and small business development in Uzbekistan. In: Paper presented at IAMO forum "Rural areas in transition: services of general interest, entrepreneurship and quality of life", IAMO/Halle (Saale), June 19–21, 2013

Bekchanov M, Lamers JPA, Martius C (2010) Pros and cons of adopting water-wise approaches in the lower reaches of the Amu Darya: a socio-economic view. Water 2:200–216. doi:10.3390/w2020200

Beyers WB (1976) Empirical identification of key sectors: some further evidence. Environ Plan A 8:231–236. doi:10.1068/a080231

Bobojonov I, Lamers JPA, Djanibekov N et al (2012) Crop diversification in support of sustainable agriculture in Khorezm. In: Martius C, Rudenko I, Lamers JPA, Vlek PLG (eds) Cotton, water, salts and soums: economic and ecological restructuring in Khorezm, Uzbekistan. Springer, Dordrecht, pp 219–233

Brans JP (ed) (1984) Operational research'84. North-Holland, Amsterdam

Brooke A, Kendrick D, Meeruas A, Raman R (2006) GAMS-user's guide. GAMS Development Corporation, Washington DC

Center for Efficient Economic Policy (CEEP) (2006) Economy of Uzbekistan (Statisticial Analytical Review). CEEP, Tashkent

Chapagain AK, Hoekstra AY (2004) Water footprints of nations. Value of water research report Series No.16, vol 1, UNESCO-IHE, Delft

Communities and local government (CLG) (2009) Multi-criteria analysis: a manual. https://www.gov.uk/government/uploads/system/uploads/attachment_data/file/7612/1132618.pdf. Accessed 03 Feb 2013

CRIIWRM (Central Research Institute of Integrated Water Resources Management—Sentralniy Nauchno-Issledovatelskiy Institut Kompleksnogo Ispolzovaniya Vodnih Resursov) (1980) Aggregated norms of water use and discharge in animal husbandry and in rural settlements with centralized water sewage system. CRIIWRM, Minsk

FAO (2000) Carbon sequestration options under the Clean Development Mechanism to address land degradation. World Soil Resources Reports No. 92, Food and Agriculture Organization of the United Nations, Rome

Ghosh A (1958) Input-output approach in an allocation system. Economica 25:58–64

Glantz MH (ed) (1999) Creeping environmental problems and sustainable development in the Aral Sea basin. Cambridge University Press, Cambridge

Golan A, Judge G, Miller D (1996) Maximum entropy econometrics: robust estimation with limited data. John Wiley & Sons, Chichester

Hirschman AO (1958) The strategy of economic development. Yale University Press, New Haven

Hoekstra AY, Hung PQ (2005) Globalisation of water resources: international virtual water flows in relation to crop trade. Global Environ Change Part A 15(1):45–56

Hwang CL, K Yoon (1981) Multiple attribute decision making. Lecture Notes in Economics and Mathematical Systems, vol 186. Springer, Berlin

Jones LP (1976) The measurement of Hirschman linkages. Quart J Econ 90(2):323–333

Lenzen M (2003) Environmentally important paths, linkages and key sectors in the Australian economy. Struct Change Econ Dynam 14(1):1–34

Lenzen M (2009) Understanding virtual water flows: A multiregion input-output case study of Victoria. Water Resour Res 45:W09416. doi:10.1029/2008wr007649

Leontief W (1951) The structure of the American economy, 2nd edn. University Press, New York

Millenium Ecosystem Assessment (2005) Ecosystems and human well-being: desertification synthesis. World Resources Institute, Washington

Müller M (2006) A general equilibrium modeling approach to water and land use in Uzbekistan. Dissertation, Bonn University

National Statistical Committee (UzStat) (2006) Industry in Uzbekistan. Statistical bulletin. UzStat, Tashkent, Uzbekistan

National Statistical Committee (UzStat) (2008) Statistical yearbook, 1991–2007. Statistical bulletin. UzStat, Tashkent, Uzbekistan

Oki T, Kanae S (2004) Virtual water trade and world water resources. Water Sci Technol 49:203–209

Orlovsky N, Glanz M, Orlovsky L (2000) Irrigation and Land degradation in the Aral Sea Basin. In: Breckle SW, Vesle M, Wuecherer W (eds) Sustainable land use in deserts. Springer, Heidelberg, pp 115–125

Rasmussen PN (1956) Studies in intersectorial relations. North-Holland, Amsterdam

Rosegrant M, Cai X, Cline SA (2002) World water and food to 2025: dealing with scarcity. IFPRI, Washington

Roy B (1991) The outranking approach and the foundations of ELECTRE methods. Theor Decis 31(1):49–73

Rudenko I, Lamers JPA, Grote U (2009) Could Uzbek farmers get more for their cotton? Europ J Devel Res 21(2):283–296

Saaty TL (1980) Analytic hierarchy process. McGraw-Hill, New York

Simonovich S (2009) Managing water resources: methods and tools for a systems approach. UNESCO, Paris

State Construction Bureau (SEV and VNI VODGEO Gosstroya SSSR) (1978) Aggregated norms of water consumption and sewage in various sectors of industry, Stroyizdat, Moscow

UN (2010) Shrinking Aral Sea underscores need for urgent action on environment: Ban. News report on visit of UN Secretary-General Ban Ki-moon to the Aral Sea. http://www.un.org/apps/news/story.asp?NewsID=34276. Accessed 12 Jun 2012

United Nations Development Program (UNDP) (2006) Modeling the impact of changes in tax rates on key macroeconomic parameters. UNDP, Tashkent

United Nations Development Program (UNDP) (2007) Water—critical resource for Uzbekistan's future. UNDP Energy and Environment Unit, Tashkent

United Nations General Assembly (UN), 2000. Resolution 2, Session 55, United Nations Millenium Declaration, on 18 Sept 2000

Velazquez E (2006) An input–output model of water consumption: analysing intersectoral water relationships in Andalusia. Ecolog Econ 56:226–240

von Braun J, Swaminathan MS, Rosegrant MW (2003) Agriculture, food security, nutrition and the millennium development goals. International Food Policy Research Institute (IFPRI), New York

Wang JJ, Jing YY, Zhang CF, Zhao JH (2009) Review on multi-criteria decision analysis aid in sustainable energy decision-making. Renew Sustain Energy Rev 13(9):2263–2278

WBGU (German Advisory Council on Global Change) (1998) Worlds in transition: Ways towards sustainable management of fresh water resources. Springer, Berlin

Wichelns D (2001) The role of 'virtual water' in efforts to achieve food security and other national goals, with an example from Egypt. Agric Water Manage 49(2):131–151

ZEF/UNESCO Urgench Project (2013). GIS database, Urgench, Uzbekistan. http://www.khorezm.zef.de. Accessed 04 Dec 2013

Zhao X, Chen B, Yang ZF (2009) National water footprint in an input-output framework: a case study of China 2002. Ecol Model 220(2):245–253

Chapter 17
Sustaining Freshwater Biodiversity in the Anthropocene

Jaime Garcia-Moreno, Ian J. Harrison, D. Dudgeon, V. Clausnitzer, W. Darwall, T. Farrell, C. Savy, K. Tockner and N. Tubbs

Abstract Globally, fresh water is a limited resource, covering only about 0.8 % of the world's surface area. With over 126,000 species living in its ecosystems, freshwater harbours a disproportionate share of the planet's biodiversity; it is essential for life, and central to satisfying human development needs. However, as we enter the Anthropocene, multiple threats are affecting freshwater systems at a global scale. The combined challenges of an increasing need for water from a growing and wealthier human population, and the uncertainty of how to adapt to definite but unpredictable climate change, significantly add to this stress. It is imperative that landscape managers and policy-makers think carefully about strategic adaptive management of freshwater systems in order to both effectively conserve natural ecosystems, and the plants and animals that live within, and continue to supply human populations with the freshwater benefits they need. Maintaining freshwater biodiversity is necessary to ensure the functioning of

J. Garcia-Moreno (✉)
Amphibian Survival Alliance, PO Box 20164, 1000 HD Amsterdam, The Netherlands
e-mail: jaime.garciamoreno@iucn.org

J. Garcia-Moreno
Het Haam 16, 6846 KW Arnhem, The Netherlands

I. J. Harrison (✉)
Conservation International, Center for Environment and Peace, 2011 Crystal Drive, Suite 500, Arlington, VA 22202, USA
e-mail: iharrison@conservation.org

I. J. Harrison
6180 E Camden Rd, Flagstaff, AZ 86007, USA

D. Dudgeon
School of Biological Sciences, The University of Hong Kong, Pok Fu Lam Road, Hong Kong SAR, China

V. Clausnitzer
Senckenberg Museum of Natural History Görlitz, PF 300154, 02806 Görlitz, Germany

A. Bhaduri et al. (eds.), *The Global Water System in the Anthropocene*, Springer Water, DOI: 10.1007/978-3-319-07548-8_17,

freshwater ecosystems and thereby secure the benefits they can provide for people. Thus freshwater biodiversity is also an important element of viable economic alternatives for the sustainable use of the freshwater ecosystems natural capital. In order to achieve this we need to do a better job at monitoring our freshwater biodiversity, understanding how the ecosystems function, and evaluating what that means in terms of service delivery.

The Global Freshwater Crisis

Fresh water is essential for life, and thus its provision for agriculture, sanitation, and domestic use is central to meeting many of the Millennium Development Goals and the more-recently proposed sustainable development goals (Griggs et al. 2013; Pahl-Wostl et al. 2013a). However, from a global perspective it is an absolutely limited resource, representing no more than 0.008 % of the volume of water on Earth and covering only about 0.8 % of the global surface area (Mittermeier et al. 2010; see Fig. 17.1).

Fresh water is also a highly threatened resource. A characteristic of the Anthropocene world is a 'pandemic array' of human transformations of the global water cycle (Alcamo et al. 2008), including changes in physical, biogeochemical and biological processes. Water scarcity and quality degradation already impact more than 2.5 billion people on Earth, and by 2030 human demand for water is expected to exceed reliable freshwater supply by 40 % (Addams et al. 2009). There is, and will be, every attempt to close this water gap in order to support

W. Darwall
Freshwater Biodiversity Unit, IUCN Global Species Programme,
219c Huntingdon Road, Cambridge CB3 0DL, UK

T. Farrell
Conservation International, Greater Mekong Programme,
PO Box 1356, Phnom Penh, Cambodia

C. Savy
International Finance Corporation, 2121 Pennsylvania Ave. NW,
Washington, DC 20433, USA

K. Tockner
Leibniz-Institute of Freshwater Ecology and Inland Fisheries,
Müggelseedamm 310, 12587 Berlin, Germany

N. Tubbs
The Royal Society for the Protection of Birds, UK Headquarters,
The Lodge, Sandy, Bedfordshire SG19 2DL, UK

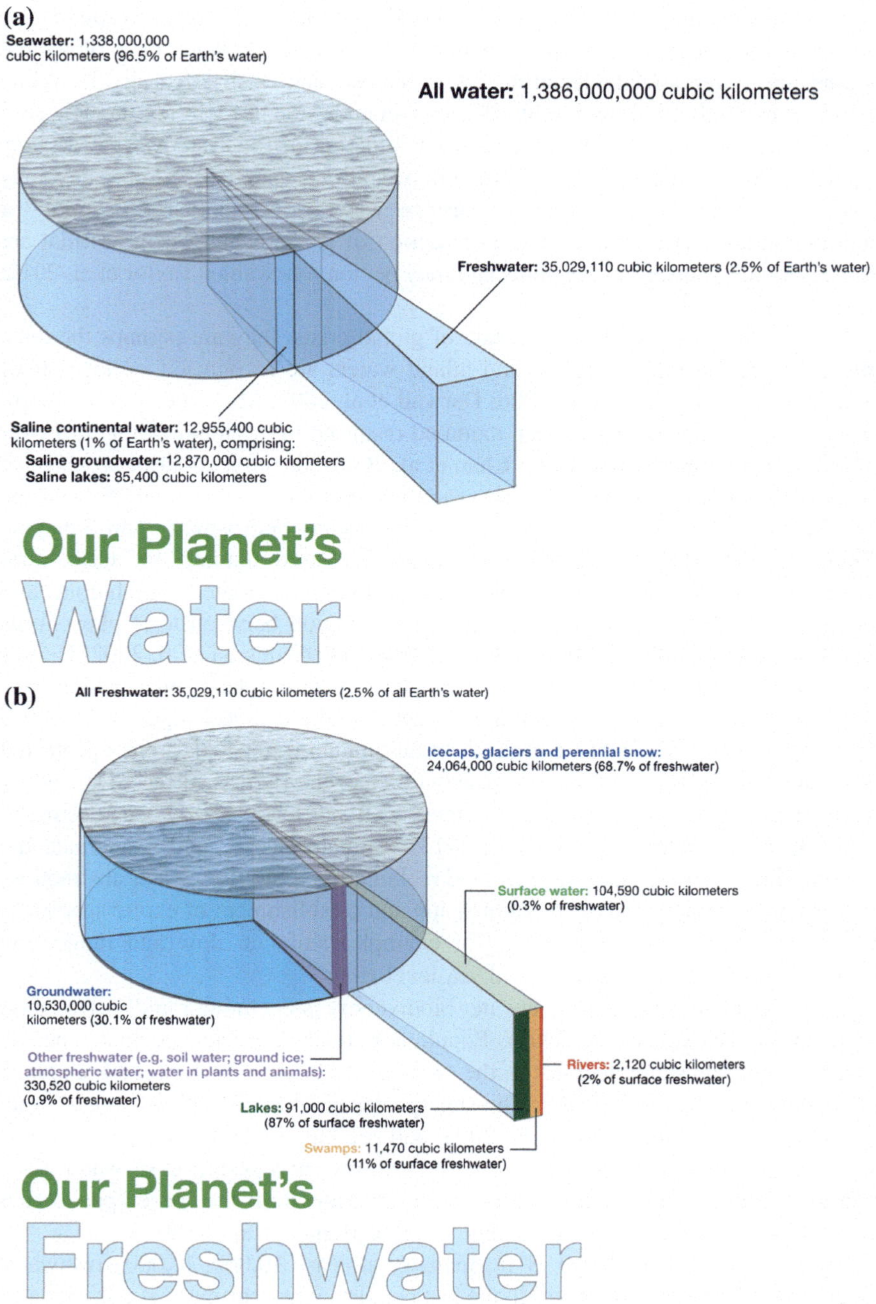

Fig. 17.1 **a** Approximate quantity and proportionate amounts of all water on earth; **b** approximate quantity and proportionate amounts of fresh water on earth. Illustration prepared by Stephen Nash

social and economic growth around the world. Nations have already responded to the threats to human water security by massive investment in water technology and engineered systems (Zehnder et al. 2003; Vörösmarty et al. 2010, 2013). While these engineered solutions might address human water needs, they are not concerned with the biodiversity and ecological function of the systems. Instead they often add to existing threats to biodiversity and ecosystem function. They may involve increased appropriation of surface water flows that are essential for environmental needs, and increased extraction of groundwater resources that are also essential to surface ecosystems and may be non renewable (Taylor et al. 2012; Foster et al. 2013).

Fresh waters are therefore in a state of global crisis; they are perhaps the most imperilled ecosystems on Earth, and inland waters are recognised as hotspots of endangerment (Dudgeon et al. 2006; Darwall et al. 2009; Mittermeier et al. 2010). Nearly every major river has been dammed resulting in the impoundment of over 10,000 km^3 of water (Chao 1995; Chao et al. 2008), the equivalent of around five times the volume of the Earth's rivers, and reservoirs trap more than 25 % of the total sediment load that formerly reached the oceans (Vörösmarty and Sahagian 2000). Around 70 % of available surface water is used annually for agricultural purposes alone (Wallace et al. 2003). Nutrient runoff has created algal blooms and anoxic dead zones. There is a very strong correlation between total phosphorus inputs and phytoplankton production in freshwaters (Anderson et al. 2002; Heisler et al. 2008), and runoff aggravates the formation of coastal dead zones, which have now been reported to affect a total area larger than the United Kingdom (Diaz and Rosenberg, 2008). More than two thirds of our upland watersheds are not protected (Thieme et al. 2010). Wetlands cover about 6 % of the Earth's surface. Depending on the region, between 30 and 90 % of these wetlands have already been destroyed or are heavily modified (Junk et al. 2013). Climate change will exacerbate the existing threats on wetlands such as land reclamation, pollution, water abstraction, overuse of resources, and facilitate invasion and establishment of exotic species as habitat conditions alter, reflecting (for example) shifts in flow and inundation patterns, increasing temperature and sea level rise.

There are clear signs that freshwater biodiversity is declining rapidly (Dudgeon et al. 2006; Darwall et al. 2009). Population trend data indicate that whereas terrestrial species show declines in the order of 25 % (95 % CL: 13–34 %) since 1970, the equivalent value for freshwater species is 37 % (21–49 %)—nearly one and a half times as high (Loh et al. 2005; and see Fig. 17.2). It should be stressed that these population trend data are based entirely on a selection of water-associated vertebrates, and lack adequate representation from the more species-rich invertebrates (Cardoso et al. 2011; but see also Balian et al. 2008).

While existing knowledge is inadequate, at least 10,000–20,000 freshwater species have become extinct within the last century or are currently at risk globally (Strayer 2006; Strayer and Dudgeon 2010). The IUCN Red List of Threatened Species currently only gives partial coverage to the world's freshwater species, currently listing 23,291, or 18.5 % of all known freshwater species. Accepting that the data may be biased towards inclusion of threatened species present in a region

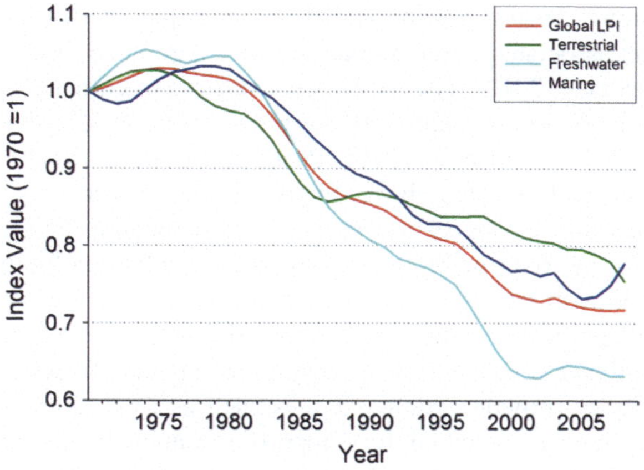

Fig. 17.2 The Living Planet Index (LPI) tracks the fate of populations of thousands of vertebrate species, just like a stock market index tracks the price of a basket of shares. The global LPI (*red line*) has declined by 28 % between 1970 and 2008. The global LPI can be split into its components by realm: terrestrial (*green line*), freshwater (*light blue line*), and marine (*dark blue line*). While all components have declined, freshwater has done so much more (37 %) than the marine (22 %) and the terrestrial (25 %) ones

(rather than the more recent trend to provide a comprehensive coverage of all species regardless of the threat; see Darwall et al. 2009, Carrizo et al. 2013), the trends are nevertheless disturbing: 30.1 % of all freshwater species that have been assessed by IUCN are classified as threatened (i.e., 'Critically Endangered'. 'Endangered', or 'Vulnerable' according to Red List criteria (IUCN 2013). Amphibians, a primarily freshwater taxon, are the second most threatened group of organisms (after cycads) that have been assessed globally (IUCN 2013; see Text Box 1); but, in intensively-developed regions, over one third of the species in other freshwater taxa are threatened also (e.g. Kottelat and Freyhof 2007; Jelks et al. 2008; Cuttelod et al. 2011; Collen et al. 2014). Although knowledge of freshwater biodiversity is improving (Clausnitzer et al. 2009, 2012; Darwall et al. 2009; Tisseuil et al. 2012; see Text Box 2), information gaps in the tropics (Balian et al. 2008) mean that the overall threat extent may be even greater than currently estimated. The possible extinction of the Yangtze River dolphin, *Lipotes vexillifer* (Turvey et al. 2007; Smith et al. 2008), which would be the first human-caused extinction of any cetacean, is not only emblematic of the perilous state of freshwater biodiversity, but indicative of our reluctance to effectively address conservation needs. It is a matter of great concern that freshwater biodiversity is largely neglected or insufficiently addressed in almost all water-development projects (Pahl-Wostl, pers. comm.; Vörösmarty et al. 2013); for example, the Bonn declaration that resulted from the Global Water System Project, which gave rise to this volume, mentions biodiversity only implicitly.

The increasing stress on water resources that is associated with increasing population and economic growth of the Anthropocene will likely commit us to further extinctions. To this can be added a substantial (perhaps unquantifiable) extinction debt associated with human actions that have been taken already (Strayer and Dudgeon 2010). The likely consequences of climate change for water availability in rivers do not augur well for biodiversity, at least for some regions (Ngcobo et al. 2013; Reid et al. 2013; Pearce-Kelly et al. 2013; Tedesco et al. 2013). Moreover, and as noted above, likely adaptation measures to be taken by humans to adjust to a warmer world may also be damaging (Palmer et al. 2008), and scenarios for the riverine biota in areas where the human footprint is already pervasive (see Vörösmarty et al. 2010) are especially bleak. Biodiversity loss has been shown to significantly affect the ecological function of ecosystems (Hooper et al. 2012). In the case of freshwater ecosystems this may mean that they have a reduced capacity to provide certain services such as food, nutrient cycling, and water filtration that are essential for supporting human livelihoods and health, beyond the supply of water itself (Horowitz and Finlayson 2011; de Groot et al. 2012; and see below).

Importance of Freshwater Biodiversity

There are at least 126,000 species of freshwater animals and vascular plants; this is estimated as perhaps up to 12 % of all known species on earth, and includes one-third (>18,000 species) of vertebrates, which is far more than would be expected from the limited extent of inland waters (Abramovitz 1996; Dudgeon et al. 2006; Balian et al. 2008, 2010). This total number of species is certainly an underestimate (Balian et al. 2010) since it omits several taxonomic groups that are likely to be rich in freshwater species (e.g., fungi, algae, several 'protozoan' taxa). It also does not account for the fact that many new species are being described annually, even in the case of the better known groups such as freshwater fishes and amphibians (for example, since 2005 amphibians are being described at a rate of one new species every 2–3 days; Frost et al. 2006; Reid et al. 2013). Nor does it account for recent losses of species that became extinct before they could be described by scientists. An almost unknown ecosystem type is the vast groundwater body. An estimated 50,000–100,000 stygobiont species, i.e. species that finish their entire life cycle in the subterranean freshwater realm, occur globally (Culver and Holsinger 1992). However, less than 10 % of these species are described up to now (Stoch and Galassi 2010). Ground waters are characterized by a very high proportion of endemic and cryptic species, although there is a major lack of information on their ecology and their functional performance.

Freshwater organisms and their ecosystems are valuable in their own right, but are also vital for providing people with many different goods and services (de Groot et al. 2012; Russi et al. 2013). Russi et al. (2013) have noted that the biodiversity of wetland ecosystems are at the core of the nexus between water, food and energy. However, while biodiversity loss does affect ecosystem function (Hooper et al. 2012; see above), there is limited understanding of this relationship for many ecosystems. It is not known how much biodiversity could be lost without seriously jeopardizing ecosystem functions and services, which makes it very difficult to accurately predict the management needs of freshwater systems under changing environmental pressures (Dudgeon 2010; Stuart and Collen 2013). While much research has yet to be conducted, there is evidence that biodiversity improves water quality (Cardinale 2011) and that the loss of biodiversity impacts human livelihood and well-being (Cardinale et al. 2012). To some extent it may seem obvious that we should expect some relationship between biodiversity and ecosystem functioning as, for example, conservation of fish biodiversity is necessary to maintain a productive fishery (Reid et al. 2013). One possible relationship is that ecosystem function may be enhanced in a near-linear fashion as species richness increases. Alternatively, the loss of species may have no effect on function until some critical threshold, or tipping point, is reached whereupon the remaining species can no longer compensate for loss of the others and complete failure may occur. A third possibility is that functioning may be unaffected by the loss of certain species, but greatly impacted by the loss of others, or even by the order in which they are lost. This last 'idiosyncratic hypothesis' holds that the identity of species lost may be more crucial than the number remaining, and there is some evidence that this relationship applies in freshwater ecosystems (e.g. McIntyre et al. 2007; Gessner et al. 2010; Capps and Flecker 2013). Recent findings (e.g., Cardinale 2011; Cardinale et al. 2012), and uncertainty over the form of the relationships between biodiversity and ecosystem functioning (see Dudgeon 2011; Tomimatsu et al. 2013), strongly suggest that it would be prudent to adopt the precautionary principle and minimize further species declines or losses. By the same token, the introduction of non-native species may have marked effects on ecosystem functioning (reviewed by Strayer 2010; see also Capps and Flecker 2013), and should be avoided.

Valuing Freshwater Biodiversity and Ecosystems

Appreciation of the need to protect species and nature for their own sake is taken as axiomatic by many scientists, but is often put aside when it comes to addressing the pressing demands of growing human populations and their need for water security and other necessities (Vörösmarty et al. 2013). One good rationale for halting the degradation and destruction of freshwater systems is that of enlightened self-interest; people rely on rivers lakes and wetlands—not only for water, but the other goods and services that they provide that are of immense value, far beyond the mere economic value of water (Costanza et al. 1997; Russi et al. 2013).

Economic values of inland wetland ecosystem services are typically higher than those of many terrestrial ecosystems. For example, the total economic value of inland wetlands (exclusive of lakes and rivers) was estimated at 25,682 Int.$/ha/year, compared to 5,264 Int.$/ha/year for tropical forests (where 'Int' refers to a translation of the original values into US$ values on the basis of Purchasing Power

Parity; see de Groot et al. 2012). The non-market services of freshwater ecosystems (e.g., regulating, habitat, and cultural services) represents 94 % of the overall economic value of inland wetlands, and 55 % of the overall economic value of rivers and lakes, according to the data provided by de Groot et al. (2012) (and see Text Box 3 for discussion of a specific example of non-market services). There is now a growing appreciation that sustainable use of all types of wetlands is usually economically more beneficial than conversion to alternative uses if all or most services are taken into account (de Groot et al. 2012). Jenkins et al. (2010) showed that restoration of wetlands in the Mississippi Alluvial Valley can provide a high return on the public investment for the restoration.

This potential economic return from careful management of the natural capital of freshwater ecosystems is important for both regional and global economies. Currently up to 0.75 trillion dollars (750 billion USD) is spent per year to maintain the infrastructure and operating costs of water management around the world, and two-thirds of this expenditure is in America and Europe (Zehnder et al. 2003; Addams et al. 2009; Vörösmarty et al. 2013; Boccaletti, pers. comm). These costs are likely to increase as middle and low income countries start to become more affluent and develop their own infrastructure. Hence, it is important to look beyond the traditional reliance on hard-path infrastructure and to work with nature, and use the natural capital it provides (Palmer 2010; Vörösmarty et al. 2013). The objective of such an approach should be to meet the requirements of regional and global economies while also reducing the intensity of threats to the biodiversity supported by these ecosystems (Totten et al. 2010).

Conservation Gaps (Protected Areas and Their Management)

Despite its ecological, economic, and cultural importance, freshwater biodiversity is evidently not adequately protected by existing conservation actions. Darwall et al. (2011b) compared the distribution of threatened freshwater species (crabs, fishes, molluscs, and odonates) with the distribution of protected areas in Africa. Their results showed that while 84–100 % of the studied species had some part of their range in protected areas, only 50 % or fewer of the species had at least 70 % of their range (mapped to river catchments) contained within a protected area (see red boxes in Table 17.1). Given the high degree of connectivity within freshwater ecosystems, such that impacts can spread rapidly and from areas far outside of the protected part of a species range, this lack of protection leaves freshwater species highly vulnerable.

It has also been shown that freshwater ecosystems are not adequately included in the global network of protected areas (e.g., Allan et al. 2010; Herbert et al. 2010). Globally, almost 70 % of rivers have no protected areas in their upstream catchment (Lehner et al. in prep), and yet upper catchment protection is important because this affects the delivery of water in adequate quantity or quality to

Table 17.1 Percentage of species within existing protected area networks in Africa

	(a) Intersect PA [n = 2,725]		(b) 70 % catchment in PA [n = 619]		(c) Catchment contains a designated Ramsar site [n = 190]	
	Total taxa (%)	Threatened taxa (%)	Total taxa (%)	Threatened area (%)	Total taxa (%)	Threatened taxa (%)
Amphibia	95.7	99.4	70.8	49.2	62.2	45.3
Birds	99.1	96.2	95.9	74.2	91.7	61.4
Mammals	97.6	98.4	88.4	98.4	80.1	62.5
Crabs	92.5	88	50	36	44.3	24
Fishes	87.4	93.9	48.5	31.4	46.6	33
Molluscs	80.8	84.1	21.7	33.1	54.8	35.2
Odonata	86.4	100	73.7	50	82.4	39.7

Percentage of species from major taxonomic groups (a) captured within protected areas based on overlap of any point of occurrence in the species range with a protected area; (b) based on the overlap of 70 % of the species range (mapped to river catchments) with a protected area; and (c) based on presence of the species within catchments that also contain a Ramsar site. The lower four groups (crabs, fishes, molluscs, and odonates) are the freshwater groups assessed as part of IUCN's Global Freshwater biodiversity Assessment; the top three groups are other higher vertebrates that have been previously assessed, for comparison. Adapted from Darwall et al. (2011a)

downstream habitats. There is, therefore, an important need for careful consideration of optimum placement of protected areas to secure freshwater biodiversity under rapidly environmental alterations.

Holland et al. (2012) describe a methodology for identifying priorities for freshwater protected areas via the development of freshwater Key Biodiversity Areas (KBAs), which has also been used by institutions and funding organisation for planning frameworks (e.g. the Critical Ecosystem Partnership Fund). Freshwater KBAs are defined on presence of threatened and endemic species or ecologically unique assemblages of species (Table 17.2), and are mapped using HydroBASINS (Lehner 2012) which is the best available digital hydrology resource for mapping connectivity within catchments, incorporating river basin boundaries, lakes, and river networks.

The application of these methods to Africa and several parts of Asia (Allen et al. 2010, 2012; Darwall et al. 2011b; Molur et al. 2011) has identified a large number of potential KBAs which may be compared to protected areas to identify gaps in both spatial coverage and management focus. Once these gaps have been identified it is then possible to start developing management plans to address those gaps. However, equally as important as identifying the sites where protected areas should be implemented, is identifying the proper management plans for these locations. Abell et al. (2007) described an integrated approach to selecting and managing freshwater protected areas that first identifies focal sites or habitats that are important for species or communities, then defines critical management zones that would support the integrity of these areas, and subsequently embeds these zones within a wider catchment management scheme that integrates multiple user needs (Fig. 17.3). Such focal sites and crucial management zones would be

Table 17.2 Criteria and thresholds for defining freshwater KBAs, based on Holland et al. (2012)

Criteria	Threshold
1. Globally threatened species or other species of conservation concern	One or more CR, EN, or VU species
2. Species (or infraspecific taxa as appropriate) of restricted range)	20,000 km^2 for crabs, fish and molluscs and 50,000 km^2 for odonates
3. Group of species that are confined to an appropriate biogeographic unit or units	At least 25 % of the total species from a specific taxonomic group occurring within a subcatchment must be restricted to the ecoregion (Abell et al. 2008) in which the subcatchment is located

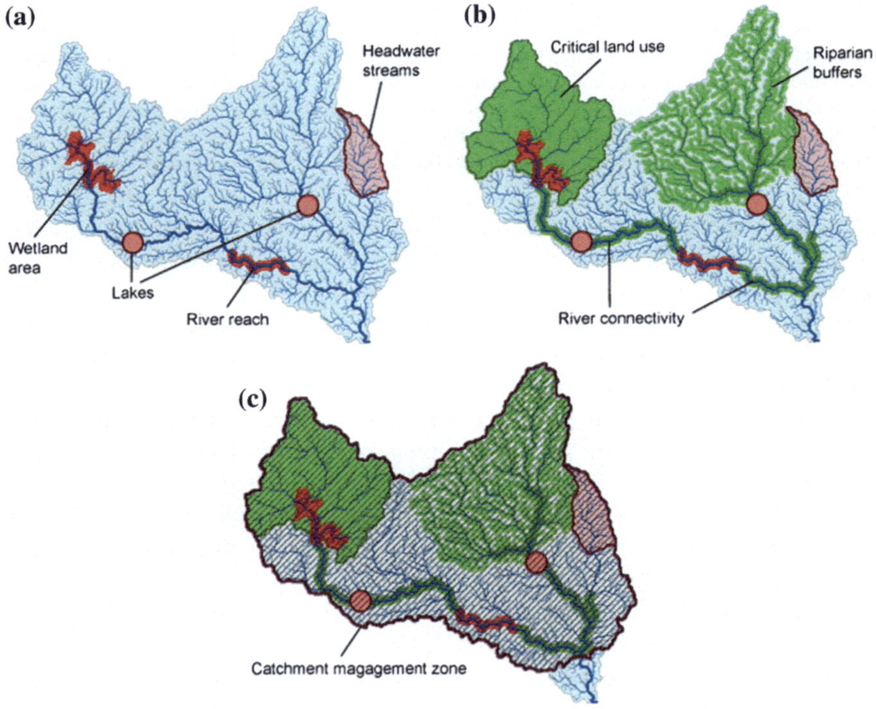

Fig. 17.3 Schematics of proposed freshwater protected area zones as proposed by Abell et al. (2007). **a** Freshwater focal areas, such as particular river reaches, lakes, headwater streams, or wetlands supporting focal species, populations, or communities. **b** Critical management zones, like river reaches connecting key habitats or upstream riparian areas, whose integrity will be essential to the function of freshwater focal areas. **c** A catchment management zone, covering the entire catchment upstream of the most downstream freshwater focal area or critical management zone, and within which integrated catchment management principles would be applied. (Reprinted from Abell et al. (2007). Copyright (2007), with permission from Elsevier)

represented as part of the management approach within a freshwater KBA. The objective is to move beyond protection directed just to the actual sites holding target species, towards protective management of the wider associated catchment.

Freshwater Management Plans

The importance of well-thought out management structures has been highlighted by several studies (e.g., Broadmedow and Nisbet 2004; Dudgeon et al. 2006; Ramsar Convention Secretariat 2010), and simple, single-factor, 'rules of thumb' approaches to management are often unsuccessful. For example, Pittock et al. (2010) outlined the status of five wetlands sites in the Murray Basin, each of which is recognised as an "icon site" for the restoration of ecological health in the basin by the Australian government. Despite such recognition, all of these sites have experienced declines in ecological character. Despite this deterioration, there was limited implementation of any conservation or mitigation measures, and degraded habitat was not compensated nor had it been restored in any way. The most recent government initiatives have been to change flow patterns, but apparently not in a carefully thought-out way, with the result that more stress is placed on some areas in favour of others (Pittock et al. 2010). In addition, a single focus on flows, important as they are, is not a sufficient management response to the array of threats these wetlands face, and a series of multiple-factor initiatives integrated across all five sites would have been more likely to result in conservation gains.

Protected area managers often tend to underestimate the stress on freshwaters in protected areas (Thieme et al. 2012). In addition, even in developed countries, resources are limited: a third of the protected areas in the southeastern United States surveyed by Thieme et al. (2012) lacked any budget for freshwater management or protection, and over half had no staff time allocated to freshwater management activities. At the European level, almost 70 % of rivers fail to achieve "good ecological status" according to the EU Water Framework Directive, and most likely will not meet this status until 2015 or later unless there is significant extra allocation of resources to river protection.

There are a number of specific challenges that face those attempting to manage fresh waters with the aim of conserving biodiversity, while meeting human needs for water. While terrestrial conservation strategies tend to emphasize areas of high habitat quality that can be bounded and protected, this 'fortress conservation' approach is not suitable for river segments or lakes embedded in unprotected drainage basins unless the boundaries can be drawn at a catchment scale (see, for example, Dunn 2003). This is hardly ever possible, and the shortcomings inherent in fortress conservation are particularly acute for freshwater biodiversity because protection of a particular component of the biota or habitat, for example in rivers, requires control over the upstream drainage network, the surrounding land and riparian zone, and—in the case of anadromous species and the risk of invasive species—downstream reaches as well. It is a major challenge to reconcile the need for a catchment-scale approach to conservation of freshwater biodiversity when this requires that large areas of land need to be managed in order to protect relatively small water bodies.

Thus all the necessary elements for freshwater management and the conservation of its biodiversity need to be included in water policies. Management of water resources must take account of aquatic biodiversity in and of itself, as well as its contribution to ecosystem functions and the goods and services used by humans, while also establishing monitoring schemes that can underpin adaptive management. Planning conservation initiatives or the activities needed to support them—for example, establishing protected areas and conducting biological inventories (Gaston et al. 2008; BioFresh, 2013)—requires high-quality spatial data on patterns of biodiversity and threat. Unfortunately, prioritization of conservation activities has been largely directed at terrestrial habitats, focusing on primarily terrestrial vertebrates as target species (e.g. Rodrigues et al. 2004). Identification of areas that support particularly high freshwater species richness has lagged behind efforts for the terrestrial realm, and the first attempt at mapping global freshwater ecoregions and hotspots was unveiled relatively recently (Abell et al. 2008). This is an important development because we lack confirmation on whether terrestrial and freshwater hotspots overlap (Strayer and Dudgeon 2010), and the analysis at the scale of river catchments throughout Africa suggests that such overlap is low (Darwall et al. 2011a). In addition, terrestrial vertebrates are poor surrogates for the overall freshwater diversity in a given area (Rodrigues and Brooks 2007).

A recent example of a major conflict among potential users of water is the actual boom in hydropower development, in Europe and globally. Although the utmost principle of the European Water Framework Directive (WFD) is to avoid the deterioration of the status of water bodies, we actually experience an unrestrained development in hydropower production; in particular of small-scale facilities. This rising conflict among different users of water occurs mainly because different directives are responsible for managing the different components of water (e.g., biodiversity conservation, irrigation, navigation, water quality). There is an urgent need to develop synergies among the different users, for the benefit of humans and the ecosystem (Pahl-Wostl et al. 2013b).

Knowledge of the status and condition of the biodiversity present within fresh waters provide an essential basis for making decisions that will allow sustainable management of these ecosystems. Many taxa are good indicators of environmental health. For example, the amphibiotic life cycle of dragonflies (with aquatic larvae and terrestrial adults) and their sensitivity to structural habitat quality, make them well suited for use in evaluating long-term and short-term environmental change in aquatic ecosystems and the associated riparian habitats, which are resources heavily utilized by local communities (Kalkman et al. 2008; see Text Box 4). Amphibians have been used as indicators of the general health of the ecosystem (e.g., Welsh and Ollivier 1998; Rice and Mazzotti 2004). Molluscs—as well as other macro-invertebrates—are sensitive to water quality and flow, and are potentially useful in bio-monitoring programs (Strong et al. 2008); many are also threatened with extinction (Johnson et al. 2013) although global assessments of the conservation status of, for example, freshwater snails are lacking. Global biodiversity databases such as the IUCN Red List of Threatened Species can, through the provision of information on species distributions and their sensitivity to identified threats, help to inform decisions on the potential impact of developments on freshwater ecosystems.

Rockström et al. (2009) defined a set of 'planetary boundaries' that describe a safe operating space for humanity. Bogardi et al. (2012, 2013, 2013) noted that in a few decades we may transgress those planetary boundaries for freshwater, indicating that we will have failed as an international community to establish political targets or economic incentives for change. To avoid this, we must develop policies and governance that will protect freshwater ecosystems and ensure the long-term provision of freshwater services to humans (Pahl-Wostl et al. 2013b). An important approach will be to take full account of the "nexus" between water, food and energy, as one of the most fundamental relationships and increasing challenges for society (Bogardi et al. 2012; Lawford et al. 2013a; Russi et al. 2013). While biodiversity, and particularly wetland ecosystems, are at the core of this nexus (Russi et al. 2013), freshwater ecosystems and biodiversity often fail to be considered when this nexus is discussed. Their exclusion may cause a permanent source of conflict because synergies among the various users are not exploited and consensus cannot be achieved. A possible reason for excluding biodiversity and the ecosystem as pari passu partners is the complexity and uncertainty they may add.

Next Steps to Meet Global Conservation and Management Needs

As noted above, substantial gaps in knowledge of global freshwater biodiversity still remain, and considerable research is required to provide baseline data that can be used to inform conservation initiatives and action for this imperilled biota. These data should include satellite and in situ observations, combined with procedures to combine and model these global data sets (Lawford et al. 2013b) (Fig. 17.4).

We need to ensure a better allocation of environmental flows in order to allow for sufficient hydric resources to properly support ecosystem functions while also attending human requirements (Poff and Matthews 2013), and this needs to be tied with research on how climate change will affect those allocations. Modification of flows in some regions is likely to be unavoidable, to meet essential human requirements. When this occurs, the implementation of comprehensive environmental impact assessments with recommendations as to how to mitigate the most deleterious impacts is crucial.

The need for more data is an obvious priority, but conservation biologists must also be ready to make the most of the data that are currently available, and to use these to help landscape managers make appropriate decisions. There are many

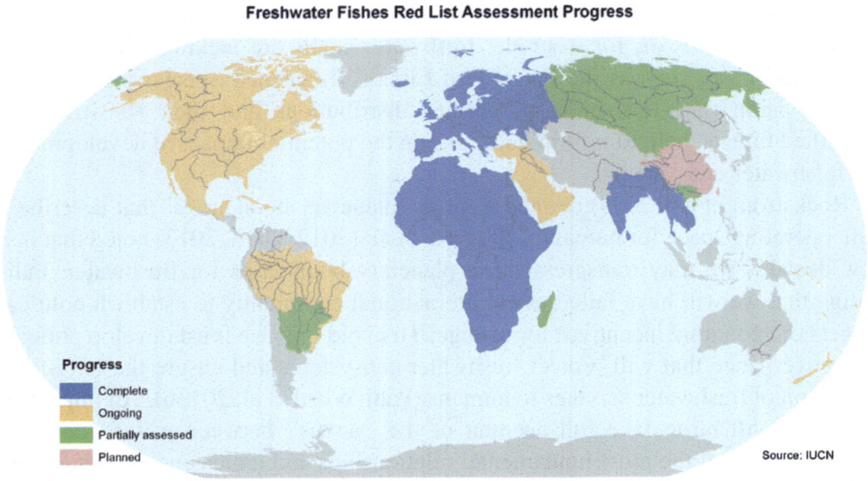

Fig. 17.4 Map showing the progress towards completion of Red List assessments for freshwater fishes in different parts of the world

excellent systems for collating biodiversity data into integrated systems that can support monitoring and measurement of change (Scholes et al. 2012; and see discussion above on the IUCN Red List). Some databases are specifically designed to collate and present ecological information, drawn from multiple data sets, to assist private and public-sector decision-makers in developing ecologically sustainable business and management practices (e.g., see Text Box 5). When developing new analytical tools for evaluating impacts on freshwater biodiversity it will be important to look carefully at the needs of the likely users. In some cases in the past, the relevant users and stakeholders have not been sufficiently engaged during the process of tool development (Morrison et al. 2010).

While awareness of the extent of threats to freshwater biodiversity has grown during the last decade, a great deal more needs to be done in order to conserve it. A major challenge we face is to raise awareness of the tremendous diversity of species living within our freshwater ecosystems, as they remain largely unseen and unvalued. The fact that most freshwater species live in a habitat that very few people explore or appreciate leaves them highly vulnerable to the impacts of the Anthropocene. Many freshwater species, some of which may be truly impressive creatures, such as *Pangasianodon gigas*, the Giant Mekong Catfish, are heading for extinction yet few people will even notice their passing. As this chapter indicates, it is often the very activities that enhance human well-being and water security which place freshwater species at risk (e.g. Vörösmarty et al. 2010). It remains a huge challenge to manage the Anthropocene global water system in a manner that will meet the water, food and energy needs of people, while allowing for sufficient semblance of natural ecosystem functions to remain in order to sustain biodiversity. For some large, iconic animals it may already be too late to reverse population declines, but it would be a travesty to permit the many freshwater species now recognized as globally threatened to follow path of the Yangtze dolphin into our history books. We already have much of the knowledge and many of the tools we need to protect freshwater biodiversity; we must now demonstrate the will to act.

Acknowledgements The authors are grateful to the Amphibian Survival Alliance and Bio-Fresh for sponsoring the symposium "Biodiversity Freshwater Ecosystems: Status, Trends, Pressures, and Conservation Priorities" at the Global Water System Project meeting on 'Water in the Anthropocene' (May, 2013), which was attended by most authors (DD, IH, JGM, KT, NT, VC) and primed the writing of this chapter. The authors also want to thank Michele Thieme (WWF-US) for comments on the manuscript. Robin Abell (WWF-US) and Jamie Pittock (Australian National University) also kindly checked the manuscript and allowed use of content from previous collaborations, including permission to use a figure first published by Abell et al. (2007). Jonathan Loh (WWF/ZSL) kindly gave access to the most current Living Planet Index figures. Alex Mauroner (Conservation International Intern, Center for Environment and Peace) adapted Tables 17.1 and 17.2 from the originals. Kurt Buhlmann provided us with a recent picture of the Kihansi Spray Toad. Ian Harrison is grateful to the Department of Ichthyology, American Museum of Natural History, New York for granting him Research Associate status, and to the staff of the Museum library for assisting in locating published materials; he is also grateful to Columbia University, New York for granting Adjunct Research Scientist Status (for Center for Environmental Research and Conservation) and External Affiliate Status (Department of Ecology, Evolution, and Environmental Biology) and allowing access the library facilities. This manuscript represents the view of individual authors and not necessarily that of the organisations they represent.

A.1 Appendix

Box 1: Balancing Development and Biodiversity Conservation

The Kihansi dam generates about 20 % of Tanzania's electricity. It is located in the Udzungwa Mountains, where the Kihansi River plunges off an escarpment. Because of its steep drop and dependable water flow, it was selected to develop a hydropower project that was started in 1994. Before the dam was completed, biological surveys of the area yielded the discovery of several species new to science, most famously the Kihansi spray toad (*Nectophrynoides asperginis*), which was endemic to a very small area of about 2 ha in the spray zone of the Kihansi Falls, the smallest distribution known for a vertebrate (Poynton et al. 1998).

As a result of these biological findings, the government agreed to let 10 % of the river flow to continue its original course—a reduction from over 16 to about 2 m^3/s (Rija et al. 2011); this reduced flow proved insufficient to maintain the mist zone that created the toad's habitat. In combination with other events, such as a one-time flushing of pesticide-rich sediments accumulated at the dam and the possible occurrence of an amphibian fungal disease (Krajick 2006), the toad's population crashed from an estimated high

Photo 17.1 Kihansi spray toad (*Nectophrynoides asperginus*) © Kurt Buhlmann

of over 20,000 individuals in 2003 to less than five individuals seen in 2004 (Channing et al. 2009). There have been no confirmed records since, and the species has been listed as Extinct in the Wild in the Red List of Threatened Species since 2009.

In 2000, some toads were collected from the field in an attempt to establish a captive breeding programme in the US (Bronx and Toledo zoos) that collaborates extensively with Tanzanian authorities. There have also been attempts to recreate the natural spray zone at the bottom of the gorge by means of an artificial sprinkler system, though it is becoming clear that this may not be sufficient, as some elements of the original ecosystem are still absent—for instance, the waterfall created continuous winds that replenished the area with wet silt (Rija et al. 2011). Since 2010, there are ongoing efforts to try to reintroduce some captive bred toads back into the spray zone of the falls, with the first ones released in 2012, but the road ahead is not easy (Khatibu et al. 2008). Millions of dollars have been spent to try to prevent this species from going extinct and change its Red List status from Extinct in the Wild back to Critically Endangered, which would be a first in recorded history. In spite of this, for many locals the dam is the source of their access to electricity that they cherish, even if it comes at the cost of a little known toad (Photo 17.1).

Box 2: Rapid Assessment (AquaRAP) Programs for Fresh Waters

Since 1996, 13 Rapid Assessment Programs have been implemented to specifically target freshwater ecosystems, focusing on surveys across watersheds or basins. The AquaRAP program has several objectives, listed by Alonso and Willinck (2011). These include increasing the priority given to conservation of freshwater systems; catalysing multinational, multidisciplinary, collaborative research on freshwater systems that includes training of students; highlighting the importance of systematic research and collections for conservation; and generating a body of reliable data about the selected watersheds.

AquaRAPs have confirmed the fact that our knowledge of biodiversity is woefully low for many parts of the world. Just in Latin America, AquaRAPs have identified 238 new basin and country records for fishes in addition to 105 species new to science. New records have also been identified for number of records for planktonic and benthic organisms, but the numbers are certainly underestimates of the total number of species, since there are often not enough taxonomists working on these groups to allow species identifications.

The conservation and management impacts of AquaRAP have been important, resulting in the creation of new protected areas, and the provision of information and advice that has been used by decision-makers (Alonso and Willinck 2011). Harrison et al. (2011) give several examples of where AquaRAP surveys have provided critical data for biodiversity assessments of African freshwater species, as well as application of information for management decisions. For example, the AquaRAP expedition to the Okavango Delta, Botswana catalyzed a process for resolving conflicts between local fishermen and sport fishermen in the delta.

Box 3: Iconic, Flagship Fishes and River Conservation

Large-bodied river fishes are particularly vulnerable to human impacts arising from overexploitation, pollution, dam construction and habitat alteration because many of them are slow growing and/or late maturing and migratory, and thus apt to encounter a variety of threats or stressors at different times and locations during their lives (reviewed by Dudgeon et al. 2006; see also Limburg and Waldman 2009). Examples include the Mekong giant catfish, the Yangtze paddlefish, African tiger-fishes, sturgeon, salmonids and a variety of other anadromous species. Many of these species have (or had) economic value which contributed to their exploitation and subsequent decline. However, this value also provides an opportunity for species protection that is predicated on the adoption of a payment for ecosystem services (PES) model. One example is provided by Everard and Kataria (2011) who describe the benefits obtained by a local community in the Himalayas of northern India from protection of a large 'flagship' fish species in the Western Ramganga River. The golden mahseer (*Tor putitora*: Cyprinidae), which may exceed 50 kg, is a favoured species for recreational angling. Along with associated cultural and wildlife tourism, angling generates income that creates incentives for protection of intact river systems by the local rural populace. They benefit economically from sustainable mahseer exploitation through catch-and-release fisheries, thereby establishing a PES market involving local people, tour operators and visiting anglers. This PES market is

Photo 17.2 Mekong giant catfish (*Pangasianodon gigas*) © Zeb Hogan

sustainable provided that people can benefit economically to a greater extent than they would through killing of fish for sale and consumption.

As Everard and Kataria (2011) explain, creation of local incentives through PES may be the most effective means for preventing destructive over-exploitation of large fishes. The Western Ramganga River model is potentially transferable to other rivers that support potential flagship fish species. It offers means of supporting regional development through involvement of riparian populations in markets for large, iconic fishes, especially where such species also have symbolic or cultural values. It must be stressed that sharing of the benefits of recreational angling markets is essential to promote self-interested resource stewardship of the type practiced along the Western Ramganga River, because without distribution of the revenues from tourism (for instance, where profits accrue to a few business operators only), local people are unlikely to have any incentive to protect freshwater ecosystems (Photo 17.2).

Box 4: Guardians of the Watershed. Dragonflies as Flagship Species for Water Quality

Dragonflies are employed successfully as indicators of ecosystem health in environmental impact assessments and monitoring programs, particularly in Australia (Bush et al. 2013) and Europe (Sahlen and Ekestubbe 2001). They can be used as environmental sentinels and as the whistleblowers for freshwater health, providing an easy tool not only for environmental impact assessments, but also for freshwater monitoring, carried out by various stakeholder groups. Using dragonflies as a flagship species—beautiful, easy to observe and positively perceived throughout—a monitoring scheme can be applied not only at the level of decision makers and conservationists, but also at the local community level.

Photo 17.3 Violet dropwing (*Trithemis annulata*) © Viola Clausnitzer

Recent projects in Angola and Tanzania, which included stakeholders from various backgrounds, have shown that the general problems of environmental health can also be explained here by using dragonflies as flagship species. Once the connection between the presence of certain species and habitat quality is understood, dragonflies can act as the guardians of the watershed—indicating the quality of the water habitat without the need of expensive or difficult tools or survey protocols (see report at www.speciesconservation.org/case-studies-projects/amani-flatwing/4044 (Photo 17.3).

Box 5: The Integrated Biodiversity Assessment Tool

The Integrated Biodiversity Assessment Tool (IBAT) for business (https://www.ibatforbusiness.org) has been developed through a partnership between UNEP-WCMC, IUCN, BirdLife International and Conservation International. IBAT is a web based decision support tool that provides planners with access to critical spatial information on conservation priorities (e.g. species, protected areas and key biodiversity areas) to inform decision-making processes with the intent of addressing any potential biodiversity risks associated with a development as early as possible. Hence, IBAT can help its users integrate biodiversity risk assessment into development plans; this reduces potentially costly impacts to critical ecosystems and supports well-informed decisions about where to invest effort in sustainable use and management of natural ecosystems. Commercial users currently support underlying data maintenance and update processes via a subscription service. This tool is currently supported by a number of private and public sector users including 25 extractive companies, and is being updated to include more specific functionality related to freshwater including direct access to data on species and sites as well as summarized indices intended to support existing water risk assessment tools in use by the private sector (e.g. WBCSD's Global and

Local Water Tools). It has been referenced by International Finance Corporation's safeguard systems and featured as a case study by the International Council on Mining and Metals (ICMM) of good biodiversity practice. A free version for non-commercial users (e.g., governments, NGOs or academics) is also available for conservation planning and research purposes (https://www.ibat-alliance.org/ibat-conservation/).

References

Abell R, Allan JD, Lehner B (2007) Unlocking the potential of protected areas for freshwaters. Biol Conserv 134:48–63

Abell R, Thieme ML, Revenga C et al (2008) Freshwater ecoregions of the world: a new map of biogeographic units for freshwater biodiversity conservation. Bioscience 58(5):403–414

Abramovitz JN (1996) Imperiled waters, impoverished future: the decline of freshwater ecosystems. Worldwatch Paper vol 128. Worldwatch Institute, Washington DC

Addams L, Boccaletti G, Kerlin M, Stuchtey M (2009) Charting our water future. Economic frameworks to inform decision-making. 2030 Water Resources Group

Alcamo JM, Vörösmarty CJ, Naiman RJ et al (2008) A grand challenge for freshwater research: understanding the global water system. Environ Res Lett 3:1–6

Allan D, Esselman P, Abell R et al (2010) Protected areas for freshwater ecosystems: essential but underrepresented. In: Mittermeier RA, Farrell TA, Harrison IJ et al (eds) Fresh water: the essence of life. CEMEX & ILCP, Arlington, pp 155–178

Allen DJ, Molur S, Daniel BA (compilers) (2010) The status and distribution of freshwater biodiversity in the Eastern Himalaya. IUCN, Cambridge, UK and Gland, Switzerland, and Zoo Outreach Organisation, Coimbatore, India

Allen DJ, Smith, KG, Darwall, WRT (compilers) (2012) The status and distribution of freshwater biodiversity in Indo-Burma. IUCN, Cambridge, UK and Gland, Switzerland

Alonso LE, Willinck P (2011) History and overview of AquaRAP. In: Alonso LE, Deichmann JL, McKenna SA et al (eds) Still counting…: biodiversity exploration for conservation—the first 20 years of the rapid assessment program. Conservation International, Arlington, VA, US, p 80–91

Anderson DM, Glibert PM, Burkholder JM (2002) Harmful algal blooms and eutrophication: nutrient sources, composition, and consequences. Estuaries 25:704–726

Balian EV, Lévêque C, Segers H, Martens K (2008) The freshwater animal diversity assessment: an overview of the results. Hydrobiologia 595:627–637

Balian E, Harrison I, Barber-James H et al (2010) A wealth of life: species diversity in freshwater systems. In: Mittermeier RA, Farrell TA, Harrison IJ et al (eds) Fresh water: the essence of life. CEMEX & ILCP, Arlington, pp 53–89

BioFresh (2013) BioFresh project: the network for global freshwater biodiversity. http://project.freshwaterbiodiversity.eu/. Accessed August 28 2013

Bogardi JJ, Dudgeon D, Lawford R et al (2012) Water security for a planet under pressure: interconnected challenges of a changing world call for sustainable solutions. Curr Opin Environ Sustain. 4:35–43

Bogardi JJ, Fekete BM, Vörösmarty CJ (2013) Planetary boundaries revisted: a view through the 'water lens'. Curr Opin Sustain 5:581–589

Broadmedow S, Nisbet TR (2004) The effects of riparian forest management on the freshwater environment: a literature review of best management practice. Hydrol Earth Syst Sci 8(3):286–305

Bush A, Theischinger G, Nipperess D et al (2013) Dragonflies: climate canaries for river management. Divers Distrib 19:86–97

Capps KA, Flecker AS (2013) Invasive fishes generate biogeochemical hotspots in a nutrient-limited system. PLoS One 8(1):e54093. doi:10.1371/journal.pone.0054093

Cardinale BJ (2011) Biodiversity improves water quality through niche partitioning. Nature 472:86–89

Cardinale BJ, Duffy JE, Gonzalez A et al (2012) Biodiversity loss and its impact on humanity. Nature 486:59–67

Cardoso P, Erwin TL, Borges PAV, New TR (2011) The seven impediments in invertebrate conservation and how to overcome them. Biol Conserv 144:2647–2655

Carrizo SF, Smith KG, Darwall WRT (2013) Progress towards a global assessment of the status of freshwater fishes (Pisces) for the IUCN Red List: application to conservation programmes in zoos and aquariums. Int Zoo Yearb 47:46–64

Channing A, Howell K, Loader S et al (2009) *Nectophrynoides asperginis*. In: IUCN 2013. IUCN Red List of Threatened Species. Version 2013.1. Accessed August 28 2013

Chao BF (1995) Anthropogenic impact on global geodynamics due to reservoir water impoundment. Geophys Res Lett 22(24):3529–3532

Chao BF, Wu JH, Li YS (2008) Impact of artificial reservoir water impoundment on global sea level. Science 320:212–214

Clausnitzer V, Kalkman VJ, Ram M et al (2009) Odonata enter the biodiversity crisis debate: the first global assessment of an insect group. Biol Conserv 142:1864–1869

Clausnitzer V, Dijkstra K-DB, Koch R et al (2012) Focus on African freshwaters: hotspots of dragonfly diversity and conservation concern. Front Ecol Environ 10:129–134

Collen B, Whitton F, Dyer EE et al (2014) Global patterns of freshwater species diversity, threat and endemism. Glob Ecol Biogeogr 23:40–51

Costanza R., d'Arge R, de-Groot R et al (1997) The value of the world's ecosystem services and natural capital. Ecol Econ 25:3–15

Culver DC, Holsinger JR (1992) How many species of troglobites are there? Bull Natl Speleol Soc 54:59–80

Cuttelod A, Seddon M, Neubert E (2011) European Red List of non-marine molluscs. Publications Office of the European Union, Luxembourg

Darwall WRT, Smith KG, Allen J et al (2009) Freshwater biodiversity: a hidden resource under threat. In: Vié J, Hilton-Taylor C, Stuart SN (eds) Wildlife in a changing world: an analysis of the 2008 IUCN Red List of Threatened Species. IUCN, Gland, pp 43–53

Darwall WRT, Holland RA, Smith et al (2011a) Implications of bias in conservation research and investment for freshwater species. Conserv Lett 4:474–482

Darwall WRT, Smith KG, Allen DJ et al (eds) (2011b) The diversity of life in african freshwaters: under water, under threat. An analysis of the status and distribution of freshwater species throughout mainland Africa. IUCN, Cambridge, United Kingdom and Gland, Switzerland

de Groot R, Brander L, van der Ploeg S, Costanza R et al (2012) Global estimates of the value of ecosystems and their services in monetary units. Ecosyst Serv 1:50–61

Diaz RJ, Rosenberg R (2008) Spreading dead zones and consequences for marine ecosystems. Science 321:926–929

Dudgeon D (2010) Prospects for sustaining freshwater biodiversity in the 21st century: linking ecosystem structure and function. Curr Opin Environ Sustain 2:422–430

Dudgeon D (2011) Asian river fishes in the Anthropocene: threats and conservation challenges in an era of rapid environmental change. J Fish Biol 79:1487–1524

Dudgeon D, Arthington AH, Gessner MO et al (2006) Freshwater biodiversity: importance, threats, status and conservation challenges. Biol Rev 81:163–182

Dunn H (2003) Can conservation assessment criteria developed for terrestrial systems be applied to riverine systems? Aquat Ecosyst Health Manage 6:81–95

Everard M, Kataria G (2011) Recreational angling markets to advance the conservation of a reach of the Western Ramganga River, India. Aquatic Conserv 21:101–108

Foster S, Chilton J, Nijsten G-J, Richts A (2013) Groundwater—a global focus on the 'local resource'. Curr Opin Environ Sustain 5:685–695
Frost DR, Grant T, Faivovich J et al (2006) The Amphibian Tree of Life. Bull Am Mus Nat Hist 297:1–370
Gaston KJ, Jackson SF, Cantú-Salazar L, Cruz-Piñón G (2008) The ecological performance of protected areas. Ann Rev Ecol Evol Syst 39:93–113
Gessner MO, Swan CM, Dang CK et al (2010) Diversity meets decomposition. Trends Ecol Evol 25:372–380
Griggs D, Stafford-Smith M, Gaffney O et al (2013) Sustainable development goals for people and planet. Nature 495:305–307
Harrison I, Upgren A, Alonso LE (2011) Chapter 8. Synthesis for all taxa. Section 8.5 Areas for further work. In: Darwall WRT, Smith KG, Allen DJ et al (eds) The diversity of life in African freshwaters: under water, under threat. An analysis of the status and distribution of freshwater species throughout mainland Africa. IUCN, Cambridge, United Kingdom and Gland, Switzerland
Heisler J, Glibert PM, Burkholder JM et al (2008) Eutrophication and harmful algal blooms: a scientific consensus. Harmful Algae 8:3–13
Herbert ME, McIntyre PB, Doran PJ et al (2010) Terrestrial reserve networks do not adequately represent aquatic ecosystems. Conserv Biol 24(4):1002–1011
Holland RA, Darwall WRT, Smith KG (2012) Conservation priorities for freshwater biodiversity: the key biodiversity area approach refined and tested for continental Africa. Biol Conserv 148:167–179
Hooper DU, Adair EC, Cardinale BJ et al (2012) A global synthesis reveals biodiversity loss as a major driver of ecosystem change. Nature 486:105–108
Horowitz P, Finlayson M (2011) Wetlands as settings for human health: incorporating ecosystem services and health impact assessment into water resource management. Bioscience 61(9):678–688
IUCN (2013) The IUCN Red List of Threatened Species 2013.2. www.iucnredlist.org. Accessed August 23 2013
Jelks HL, Walsh SJ, Burkhead NM et al (2008) Conservation status of imperilled North American freshwater and diadromous fishes. Fisheries 33:372–407
Jenkins WA, Murray BC, Kramer RA, Faulkner SP (2010) Valuing ecosystem services from wetlands restoration in the Mississippi alluvial valley. Ecol Econ 69:1051–1061
Johnson DP, Bogan AE, Brown KM et al (2013) Conservation status of freshwater gastropods of Canada and the United States. Fisheries 38:247–282
Junk WJ, An S, Finlayson HCM et al (2013) Current state of knowledge regarding the world's wetlands and their future under global climate change: a synthesis. Aquat Sci 75:151–167
Kalkman VJ, Clausnitzer V, Dijkstra K-DB et al (2008) Global diversity of dragonflies (Odonata; Insecta) in freshwater. Hydrobiologia 595:351–363
Khatibu FHA, Church DR, Moore RD et al (eds) (2008) Kihansi spray toad re-introduction guidelines. IUCN SSC Amphibian Specialist Group
Kottelat M, Freyhof J (2007) Handbook of European freshwater fishes. Kottelat, Cornol and Freyhof, Berlin
Krajick K (2006) The lost world of the Kihansi toad. Science 311:1230–1232
Lawford R, Bogardi J, Marx S et al (2013a) Basin perspectives on the water–energy–food security nexus. Curr Opin Environ Sustain 5:607–616
Lawford R, Strauch A, Toll D et al (2013b) Earth observations for global water security. Curr Opin Environ Sustain 5:633–643
Lehner B (2012) Hydrobasins: version 1.b. Global watershed boundaries and sub-basin delineation derived from hydrosheds data at 15 second resolution. World Wildlife Fund, Washington, DC
Limburg KE, Waldman JB (2009) Dramatic declines in North Atlantic diadromous fishes. Bioscience 59:955–965

Loh J, Green RH, Ricketts T et al (2005) The living planet index: using species population time series to track trends in biodiversity. Philos Trans Roy Soc B 260:289–295

McIntyre PB, Jones LE, Flecker AS, Vanni MJ (2007) Fish extinctions alter nutrient recycling in tropical freshwaters. Proc Natl Acad Sci USA 104:4461–4466

Mittermeier RA, Brooks TM, Farrell TA et al (2010) Introduction. Fresh water: the essence of life. In: Mittermeier RA, Farrell TA, Harrison IJ et al (eds) Fresh water: the essence of life. CEMEX & ILCP, Arlington, Virginia, USA, p 15–41

Molur S, Smith KG, Daniel BA, Darwall WRT (compilers) (2011) The status and distribution of freshwater biodiversity in the Western Ghats, India. IUCN: Cambridge, UK and Gland, Switzerland; Zoo Outreach Organisation, Coimbatore, India

Morrison J, Schulte P, Schenk R (2010) Corporate water accounting: an analysis of methods and tools for measuring water use and its impacts. United Nations Environment Programme, United Nations Global Compact, Pacific Institute

Ngcobo S, Jewitt GPW, Stuart-Hill SI, Warburton ML (2013) Impacts of global change on southern African water resources systems. Curr Opin Environ Sustain 5:655–666

Pahl-Wostl C, Vörösmarty C, Bhaduri A et al (2013a) Towards a sustainable water future: shaping the next decade of global water research. Curr Opin Environ Sustain 5:708–714

Pahl-Wostl C, Palmer M, Richards K (2013b) Enhancing water security for the benefits of humans and nature—the role of governance. Curr Opin Environ Sustain 5:676–684

Palmer M (2010) Beyond infrastructure. Nature 467:34–535

Palmer MA, Reidy C, Nilsson C et al (2008) Climate change and the world's river basins: anticipating response options. Front Ecol Environ 6:81–89

Pearce-Kelly P, Khela S, Ferri C, Filed D (2013) Climate-change impact considerations for freshwater-fish conservation, with special reference to the aquarium and zoo community. Int Zoo Yearb 47:81–92

Pittock J, Finlayson M, Gardner A, McKay C (2010) Changing character: the Ramsar Convention on Wetlands and climate change in the Murray-Darling Basin, Australia. EPLJ 27:401–425

Poff NL, Matthews JH (2013) Environmental flows in the Anthropocene: past progress and future prospects. Curr Opin Environ Sustain 5:667–675

Poynton JC, Howell KM, Clarke BT, Lovett JC (1998) A critically endangered new species of *Nectophrynoides* (Anura: Bufonidae) from the Kihansi Gorge, Udzungwa Mountains, Tanzania. Afr J Herpetol 47(2):59–67

Ramsar Convention Secretariat (2010) Managing groundwater: guidelines for the management of groundwater to maintain wetland ecological character. Ramsar handbooks for the wise use of wetlands, 4th edn, vol 11. Ramsar Convention Secretariat, Gland, Switzerland

Reid G McG, Contreras MacBeath T, Csatadi K (2013) Global challenges in freshwater-fish conservation related to public aquariums and the aquarium industry. Int Zoo Yearb 47:6–45

Rice K, Mazzotti F (2004) Use of amphibians as indicators of ecosystem restoration success. US Geological Service Fact Sheet 2004-3106

Rija AA, Khatibu FH, Kohi EM, Muheto R (2011) Status and reintroduction of the Kihansi spray toad *Nectophrynoides asperginis* in Kihansi gorge: challenges and opportunities. In: Proceedings of the 7th TAWIRI Scientific Conference: 2–4 December 2009, Corridor Springs Hotel, Arusha, Tanzania. (Tanzania Wildlife Research Institute) Arusha, Tanzania, p 11–20

Rockström J, Steffen W, Noone K et al (2009) A safe operating space for humanity. Nature 461:472–475

Rodrigues AS, Brooks TM (2007) Shortcuts for biodiversity conservation planning: the effectiveness of surrogates. Ann Rev Ecol Evol Syst 38:713–737

Rodrigues ASL, Andelman SJ, Bakarr ML et al (2004) Effectiveness of the global protected area network in representing species diversity. Nature 428:640–643

Russi D, ten Brink P, Farmer A et al (2013) The economics of ecosystems and biodiversity for water and wetlands. IEEP Lond, Brussels

Sahlen G, Ekestubbe K (2001) Identification of dragonflies (Odonata) as indicators of general species richness in boreal forest lakes. Biodivers Conserv 10:673–690

Scholes RJ, Walters M, Turak E et al (2012) Building a global observing system for biodiversity. Curr Opin Environ Sustain 4:1–8

Smith BD, Zhou K, Wang D et al (2008) *Lipotes vexillifer*. In: IUCN 2013. IUCN Red List of Threatened Species. Version 2013.1. www.iucnredlist.org. Accessed on 23 August 2013

Strayer DL (2006) Challenges for freshwater invertebrate conservation. J North Am Benthol Soc 25:271–287

Strayer DL (2010) Alien species in fresh waters: ecological effects, interactions with other stressors, and prospects for the future. Freshw Biol 55(Supplement 1):152–174

Strayer DL, Dudgeon D (2010) Freshwater biodiversity conservation: recent progress and future challenges. J North Am Benthol Soc 29:344–358

Strong EE, Gargominy O, Ponder W, Bouchet P (2008) Global diversity of gastropods (Gastropoda; Mollusca) in freshwater. Hydrobiologia 595:149–166

Stoch F, Galassi DMP (2010) Stygobiotic crustacean species richness: a question of numbers, a matter of scale. Hydrobiologia 653:217–234

Stuart SN, Collen B (2013) Conserving biodiversity in a target driven world. In: Collen B, Pettorelli N, Baillie JEM, Durant SM (eds) Biodiversity monitoring and conservation: bridging the gap between global commitment and local action. Wiley-Blackwell, Cambridge, U.K., p 421–438

Taylor RG, Scanlon B, Döll P et al (2012) Ground water and climate change. Nat Clim Change. doi:10.1038/NCLIMATE1744 (Published online: 25 November 2012)

Tedesco PA, Oberdorff T, Cornu J-F et al (2013) A scenario for impacts of water availability loss due to climate change on riverine fish extinction rates. J Appl Ecol. doi:10.1111/1365-2664. 12125

Tisseuil C, Cornu J-F, Beauchard O et al (2012) Global diversity patterns and cross-taxa convergence in freshwater systems. J Anim Ecol 82:365–376

Thieme ML, Turak E, McIntyre P et al (2010) Freshwater ecosystems under threat: the ultimate hotspot. In: Mittermeier RA, Farrell TA, Harrison IJ, Upgren AJ, Brooks TM (eds) Fresh water: the essence of life. Arlington, CEMEX & ILCP, pp 123–143

Thieme ML, Rudulph J, Higgins J, Takats JA (2012) Protected areas and freshwater conservation: a survey of protected area managers in the Tennessee and Cumberland River Basins, USA. J Environ Manage 109:189–199

Tomimatsu H, Sasaki T, Kurokawa H et al (2013) Sustaining ecosystem functions in a changing world: a call for an integrated approach. J Appl Ecol. doi:10.1111/1365-2664.12116

Totten MP, Killeen TJ, Farrell TF (2010) Non-dam alternatives for delivering water services at least cost and risk. Water Altern 3(2):207–230

Turvey ST, Pitman RL, Taylor BL et al (2007) First human-caused extinction of a cetacean species. Biol Lett 3:537–540

Vörösmarty CJ, Sahagian D (2000) Anthropogenic disturbance of the terrestrial water cycle. Bioscience 50:753–765

Vörösmarty C, McIntyre PB, Gessner MO et al (2010) Global threats to human water security and river biodiversity. Nature 467:555–561. http://riverthreat.net/

Vörösmarty CJ, Pahl-Wostl C, Bunn SE, Lawford R (2013) Global water, the anthropocene and the transformation of a science. Curr Opin Environ Sustain 5:539–550

Wallace JS, Acreman MC, Sullivan CA (2003) The sharing of water between society and ecosystems: from conflict to catchment-based co-management. Phil Trans R Soc Lond B 358:2011–2026

Welsh HH Jr, Ollivier LM (1998) Stream amphibians as indicators of ecosystem stress: a case study from California's redwoods. Ecol Appl 8(4):1118–1132

Zehnder AJB, Yang H, Schertenlieb R (2003) Water issues: the need for action at different levels. Aquat Sci 65:1–20

Chapter 18
Water Governance and Management Systems and the Role of Ecosystem Services: Case Study Insights—Groundwater Management in the Sandveld Region, South Africa

Kathrin Knüppe and Claudia Pahl-Wostl

Abstract Freshwater resources deliver a broad set of ecosystem services essential for human health and well-being, food and energy production, social and economic stability, and for protecting and maintaining ecosystems. The ever increasing demand of water resources often result in substantial declines in the provision of ecosystem services. The management of human and environmental water needs is therefore challenging and calls for an integrative view on ecosystem services. A shift of current water management objectives is required to ensure water security for current and future generations. This article analyzes water governance and management systems (WGMS) and highlights characteristics assumed to be crucial for adaptive and integrated management: (i) institutional settings, (ii) actor networks, and (iii) multi-level structures. To understand complex WGMS one has to link these characteristics to management performances including impacts on ecosystem services. We applied this approach to the Sandveld in South Africa focusing on actor networks and the management of ecosystem services. We indicate that a basic re-thinking of water management objectives at national and regional level according to groundwater sustainability took place. A bottom-up movement in the Sandveld developed approaches to protect and sustain groundwater resources. Nevertheless, cooperation between actors and sectors from different levels is weak which in turn provides a huge barrier for the integration of ecosystem services into groundwater policies.

K. Knüppe (✉) · C. Pahl-Wostl
Institute of Environmental Systems Research,
University of Osnabrück, Osnabrück, Germany
e-mail: kathrin.knueppe@uos.de

A. Bhaduri et al. (eds.), *The Global Water System in the Anthropocene*,
Springer Water, DOI: 10.1007/978-3-319-07548-8_18,

Introduction

Water is vital to all social and economic sectors as well as the natural resource base on which the world depends. Expanding population, growing economies, urbanization, poor water management as well as diverse uncertainties regarding global climate change are putting unprecedented pressure on the Earth's fresh water resources. To guarantee sustainable water resources management is therefore one of the central tasks of the 21st century (UNEP 2007; Bates et al. 2008).

Worldwide, the concept of ecosystem services has received attention in the sustainable management of natural resources and as a way to communicate human dependence on ecological life support systems (Odum 1996; Daily 1997; de Groot et al. 2002). Daily (1997, p 3) defines ecosystem services as "the conditions and processes through which natural ecosystems, and the species that make them up, sustain and fulfill human life." The concept embraces an ecologically based management approach and serves as an integrated model bridging societal and ecological systems (MA 2005; Loring et al. 2008).

Freshwater systems provide a diverse set of ecosystem services important to socio-economic development and aquatic as well as terrestrial ecosystems. Often these services are claimed and modified by various actors (e.g. farmers, conservationists, municipalities) which in turn produces social and ecological trade-offs (e.g. between irrigation and biodiversity) because the use of some services comes at the expense of others (Bennett et al. 2009; Raudsepp-Hearne et al. 2010). Water governance and management systems[1] have been given overwhelming emphasis to provisioning services, whereas regulating, supporting and cultural services and the requirements for maintaining them have been largely ignored. The Millennium Ecosystem Assessment (2005) developed four categories to distinguish ecosystem services: provisioning (e.g. water supply), regulating (e.g. flood attenuation), supporting (e.g. nutrient cycling) and cultural (e.g. recreation) services. This categorization serves as a functional abstraction from ecological resources to 'used services' that highlights the linkages and dependencies between these services and human well-being (Loring et al. 2008).

The ignorance of complex feedbacks between ecosystem services often led to ineffective resource governance and management of trade-offs (e.g. between irrigation and biodiversity) with long-term negative consequences for human well-being (e.g. resource access, livelihoods, food, clean air and water). Fundamental to develop and implement innovative approaches is a shift of water management objectives and practices towards an integrative perspective on ecosystem services.

[1] A distinction is made by Pahl-Wostl (2009) between resources management and governance. 'Resources management' refers to the activities of analyzing and monitoring, developing and implementing measures to keep the state of a resource within desirable bounds. The notion of 'resource governance' takes into account the different actors and networks that help formulate and implement environmental policy and/or policy instruments. Governance sets the rules under which management operates.

This article addresses certain characteristics of WGMS assumed to be crucial for integrated management: (i) institutional settings (e.g. water policy frameworks), (ii) actor networks (e.g. involvement of non-state actors), (iii) multi-level structures (e.g. coordination and cooperation between different administrative levels) (see e.g. Ostrom 2005; Pahl-Wostl et al. 2010; Huntjens et al. 2010; Krysanova et al. 2010; Knüppe and Pahl-Wostl 2011). To understand the development and behavior of WGMS as well as system changes, one has to link these characteristics to management performances. Explorative studies that have systematically investigated these linkages are relatively rare. This article addresses this research gap, in the realm of the Sandveld region in South Africa. This region is characterized by ecosystem service trade-offs between freshwater requirements for intensive irrigation and nature conservation. This includes a long-lasting conflict between the different stakeholders of this region. Supported by literature review and qualitative expert interviews a shift of management objectives and practices was analyzed and main challenges towards the integration of ecosystem services into water management were identified.

Ecosystem Services and Water Governance Challenges

Under the present day conditions discrepancies exist between human and environmental water needs. Managing water resources in a sustainable, equitable and efficient manner requires integrative perspectives on societal and ecological systems: a coupled, inseparable system of humans and nature (Folke et al. 2005; MA 2005) in which ecosystem services are conceived as a bridging part (Bennett et al. 2009) (Fig. 18.1).

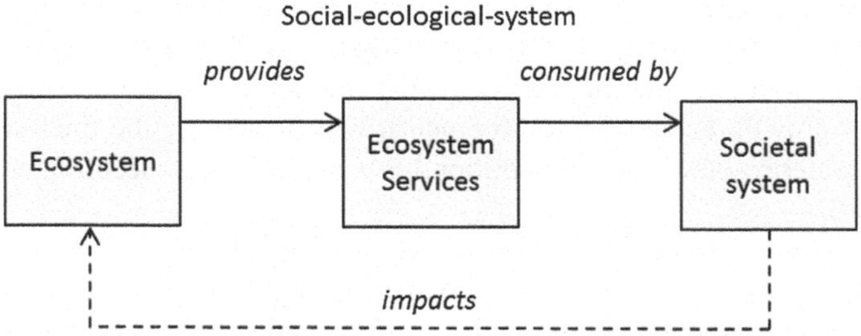

Fig. 18.1 Ecosystem services link the ecological and the societal sphere; the societal system may impact the ecosystem which in turn might influence the magnitude and quality of ecosystem services

Trade-Offs and Priorities

It is crucial to acknowledge that different ecosystem services are not independent of one another and the relationships between them may be highly non-linear (Rodríguez et al. 2006). The use of certain services often results in substantial declines in the provision of other services; in other words, they are traded-off (Holling and Meffe 1996). For example, an increase in one service such as food production or timber harvesting can negatively affect the provision of other ecosystem services, e.g. drinking water quality (Raudsepp-Hearne et al. 2010). Water related trade-offs often occur if water is removed from streams or lakes for drinking, sanitation, irrigation, and industry. These uses often conflict with other freshwater services that rely on the maintenance of stream flow or lake levels, e.g. power generation, fish production, transport, waste removal, and recreation (Rodríguez et al. 2006). It is not trivial assessing different trade-offs, but managing multiple bundles of ecosystem services simultaneously is crucial and at the same time extremely challenging (Bennett et al. 2009). An overly narrow focus on a limited set of ecosystem services (e.g. favoring water for irrigation purposes) may lead to ecological shifts with sudden losses of ecosystem services.

The analysis of trade-offs among competitive priorities is one of the core issues in natural resource research. Due to the preference of human societies for provisioning services over regulating, supporting and cultural trade-offs often arise from management choices made by humans, which can change the magnitude and mixture of ecosystem services (Foley et al. 2005; van Jaarsveld et al. 2005). More specifically, trade-offs occur when (i) management ignores the interactions between ecosystem services, (ii) knowledge and understanding of how they work is incorrect or incomplete, and (iii) when there are no specific markets for the ecosystem services in question (Rodríguez et al. 2006).

Balancing complex and conflicting demands for water among different actors and sectors is a difficult task. This is especially true for areas where water is used predominantly by a single sector (e.g. irrigated agriculture). Further concerns relate to the environmental impacts of global change (e.g. extreme climate events, population growth, economic development), which increase the pressures on the available water resource and competition for water will become stronger (Schlüter et al. 2009).

Many of these problems are associated with the failure of governance and management systems (Bakker et al. 2008; Rogers and Hall 2003). Currently, many management regimes focus on the delivery of a single service (e.g. agriculture production) while multi-sectoral and integrated water management are assumed to have a positive impact on ecosystem services. However, given the dominance of agriculture in many areas a transition to more integrated perspective on ecosystem services faces many barriers.

Integrating Ecosystem Services into Water Management

Given the historical mostly technocratic development of water management, most WGMS do not provide the structural conditions necessary to implement integrative approaches without changing certain system characteristics. For a shift in favor of integration of ecosystem services into water management the following considerations are necessary (e.g. Berkes et al. 2003; Biswas and Tortajada 2010; Irwin and Ranganathan 2008; Ostrom 2007; Pahl-Wostl 2009; Pahl-Wostl et al. 2007):

- a shift towards participatory management and collaborative decision making,
- integration of different water dependent sectors (e.g. water supply, waste water treatment, agriculture, forestry, fishery, tourism, and nature conversation),
- implementing decentralized and polycentric management approaches, which take uncertainties and unexpected events into account,
- incorporation of social-ecological-system properties and their linkages into management goals at all levels (from local to international),
- provision of free access to information and the conscious collection of data and monitoring of the state of ecosystem services.

Conceptualizing Water Governance

Water Governance Characteristics and Performance

Our research design builds upon the analytical framework developed within the Twin2Go Project.[2] The framework supports a generic but contextual diagnostic approach (Ostrom 2007) to assess the transferability of insights between similar problems and contexts. A clear distinction is made between (a) WGMS, (b) performance, and (c) ecological and societal context (Fig. 18.2) (Pahl-Wostl et al. 2012).

We extended the framework by including the role of ecosystem services and environmental hazards in water management. This enables researchers to analyze (i) the response to emerging ecosystem services trade-offs, (ii) the consideration of risks and uncertainties associated with floods and droughts, and (iii) the drivers and barriers towards adaptive and integrative water management.

The context in which a WGMS is embedded has a strong influence on the WGMS and its performance as the dependent variable. Regarding the WGMS the focus is mainly on structural characteristics whereas the performance is

[2] Coordinating twinning partnerships towards more adaptive governance in river basins supported by the European Commission under the 7th Framework Programme (2009–2011).

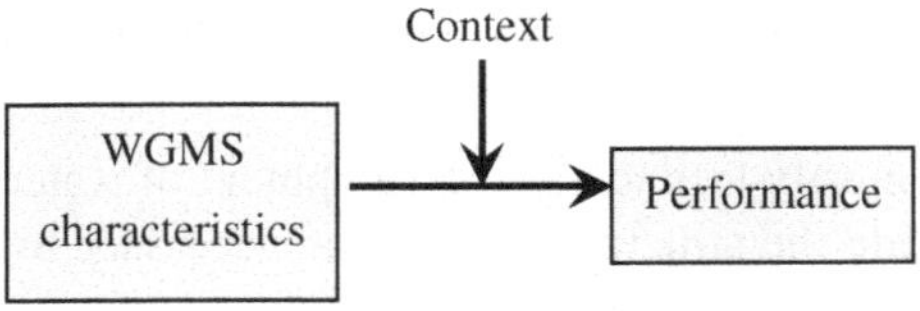

Fig. 18.2 Distinction is made between the characteristics of a WGMS, its performance and the context influencing the relationship between WGMS and performance

emphasizing the process dimensions (Table 18.1). This can include approaches and measures impacting the state of an ecosystem as well as the state of human well-being. Table 18.1 presents indicators which can be used for empirical analysis in order to understand WGMS characteristics, performance measures and context and to compare them across different case studies (e.g. river basins). Depending on different research topics the set of indicators can be changed or extended.

The outlined WGMS characteristics have the potential to strengthen the consideration and the protection of ecosystem services. Performance measures refer to the state of implementation by asking how are water resources, ecosystem services and trade-offs as well as climate change indeed being dealt with in practice? Further, performance measures allow assessing and evaluating the degree of satisfaction with the current state of the WGMS. A failure of achieving the stated goals is a sign of a non-satisfactory performance.

Data Collection and Analysis

Qualitative research was preferred because it delves into system complexities and processes by exploring where and why certain management approaches and policies are chosen, and how they influence ecosystem services. The study focused to a large extent on the perceptions, meanings and understandings of individuals in relation to their experience and beliefs in the field of water management.

As a first step, data collection was based on literature research (legal documents, publications on laws and regulations, research articles and reports) to understand general features of the ecological and societal context. In a second step, a series of expert interviews was carried out during field work in South Africa in the years 2010–2012. The number of interviewees included 22 experts chosen based upon their specific knowledge and broad experience in the field of water management and reflected various types of expertise: politics and administration, consulting, water supply, agriculture, research and nature conservation. The interviewees were asked to outline a sequence of water policy and management processes covering the past 20–25 years. This allowed us to identify shifts of water management objectives and behavior. After data collection was finalized, interview information and the findings from the literature review were transferred to a database (Microsoft Office Access)

Table 18.1 Set of indicators related to context, WGMS characteristics and performance measures

Context	WGMS characteristics	Performance measures
Ecological system	*Institutional settings*	Trade-offs between human and environmental water needs
(Variation of) water availability	Dealing with ecosystem services and environmental hazards	
	Legal frameworks	ES and trade-offs are considered/ mentioned
Variability in precipitation	Water (basin) strategies/plans	Objectives, measures, strategies (e.g. market instruments) in place
	Appropriate financing system	
Relative Water Stress Index	Market instruments for ecosystem services	Objectives, measures, strategies are implemented (positive impacts)
Degree of watershed modification	Presence of complementary informal institutions	*Societal impacts of floods and droughts*
		Economic damage
	Actor networks	Societal damage/Casualties
Expected climatic changes	Role and patterns of interactions of state and none-state actors and power relationships	*Response to climatic changes*
Societal system	Cross-sectoral cooperation	State of development of CC adaptation plans (*Ways to deal with climate variability (floods and droughts)*)
State of societal development (HDI)	Cooperation between administrative levels	Coordinated and integrated implementation of adaptation measures
	Stakeholder participation	
	Multi-level structures	
Government types	Degree of centralization	
GDP per capita	Degree of polycentricity	
Social equity (GINI index)		
Efficiency of formal institutions (CPI)		

and systematically analyzed (Knieper et al. 2010). Starting forms of the database were created containing consistent datasets and information to facilitate the description and representation of WGMS and their respective geographic, ecological, political and societal contexts. The database approach allows for a representation of dynamic management processes by developing a sequence of important processes. To analyze linkages between the characteristics of a WGMS and its performance a set of queries was developed to operationalize the indicators presented in Table 18.1. The queries provide variables, used as a function to calculate one or more relationships within and between relevant system elements (e.g. management, actors, institutions, ecosystem services). For example, queries can be calculated to identify whether actors are active at multiple administrative levels during management processes or whether the connectivity between levels within a certain management process is high or low.

Here we apply the database approach to the Sandveld region in South Africa. The Sandveld presents a typical example of the dilemma of managing ecosystem service trade-offs and the challenge to safeguard human and environmental water needs in a holistic manner.

Results: Managing Groundwater Ecosystem Services

Case Study: Sandveld, South Africa

The north western part of South Africa experiences severe periods of drought and will be most vulnerable to future climate induced water supply stress. Therefore, the Sandveld region, a sub-area of the Olifants/Doorn Water Management Area, was selected as a case study grappling with the difficulties to balance the demand placed on water resources by socio-economic development, especially for the poor, and sustaining the provision of ecosystem services.

The Sandveld consists of 4,590 km^2 of coastal plain along the west coast of South Africa, bordered by the Olifants River catchment to the north and east, the Berg River catchment to the south and the Atlantic Ocean coastline to the west (DWAF 2005). The area has some small seasonal rivers and streams (e.g. Langvlei and Verlorevlei river) important for ecological processes and functions but minor important for water supply and usages. The area features sandy and nutrient poor soils and comprises granular primary aquifers and deeper fractured rock secondary aquifers. The volume of the water stored in the Sandveld aquifer is estimated to be approximately 500 Mm^3, which is recharged by the catchment area of the Cederberg mountain ranges to the east of the Sandveld region (Conrad et al. 2005). The region experiences dry summer and wet winter conditions, similar to other regions of the Western Cape (see Fig. 18.3).

The people of the Sandveld depend on a wide range of different ecosystem services important for their well-being (Table 18.2). Agriculture is the dominant employer in the Sandveld and potato production and processing is the main economic activity, complemented by some cereal and rooibos tea production (Franke et al. 2011). Water for irrigation requires 35 million m^3/year while water for rural and urban supply solely requires 3 million m^3/year (DWA 2005). Groundwater abstraction has had a significant detrimental impact on the low flows throughout the catchment. Moreover, irrigation enhances the leaching of agri-chemicals into groundwater reserves and ecological habitats and landscapes are being threatened or fragmented due to land clearance for potato production. The core of the

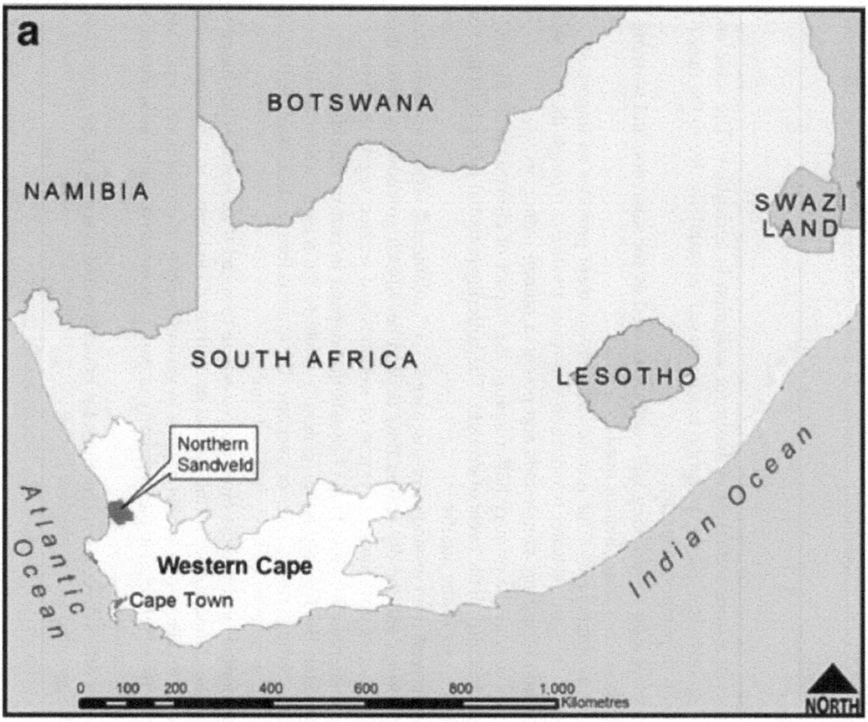

Fig. 18.3 Location of the Sandveld (Münch and Conrad 2007)

agricultural production area coincides with sensitive aquatic and terrestrial ecosystems (e.g. Cape Floristic Kingdom, Verlorenvlei RAMSAR site).

Trade-offs between ecosystem services can be ascribed to the usages/requirements of different sectors:

- Agriculture: the potato production industry depends on a huge amount of groundwater for irrigation purposes.
- Conservation: the different ecosystems and protected areas support population of coastal migrant birds, habitat for flora and fauna and regulate the overall landscape-water regime.
- Domestic water use: groundwater from the Sandveld aquifers is abstracted for supply to the few towns in the Sandveld area (Lamberts Bay, Elands Bay, Graafwater, Strandfontein and Doringbaai).
- Tourism: the tourist trade flourishes during the spring and summer months (birding at Verlorevlei, hiking in Cederberg Wilderness Area, collecting wild flowers and seafood).

Table 18.2 Ecosystem services in the Sandveld (we do not include supporting services to avoid double counting as these services are defined as '*those that are necessary for the production of all other ecosystem services*' (MA 2005)

Ecosystem services (MA 2005)	Groundwater related ecosystem services	Explanation
Provisioning '*products obtained from ecosystems*'	Irrigation	Groundwater is a store and retention basin for irrigation in agriculture. The scale and rate of groundwater use for irrigation has increased substantially due to the massive expansion in pumping capacity
	Domestic supply	Groundwater is used for drinking and cooking as well as for sanitation and washing requirements as a basic human need
Regulating '*benefits obtained from the regulation of ecosystem processes*'	Purification/waste treatment	The biological component of the groundwater environment provides an important service in the form of water purification and waste treatment through the microbial degradation of organic compounds and potential human pathogens
	Drought buffer	Groundwater acts as the primary buffer against the impact of climate variability and spatial variability in the event of drought. The buffering potential depends on the soil and rock types of the aquifer
	Base flow	Base flow derived from groundwater discharge is a fundamental service in many areas where springs and the dry-season flow depend heavily on groundwater. Base flow controls factors governing the extent of wetlands and surface vegetation types
	Flora/fauna habitat	There are numerous flora and fauna habitats that depend in part or entirely upon groundwater. Biodiversity issues generally relate to the regions where aquifers discharge through rivers, lakes or swamps. These areas form critical wildlife habitats and serve as sources of food, fuel and timber
Cultural '*non-material benefits that people obtain from ecosystems*'	Recreation/tourism	Local communities and visitors often choose where to spend their leisure time based in part on the characteristics of the natural or cultivated landscapes in a particular area
	Aesthetic beauty	Many people find beauty or aesthetic value in various aspects of groundwater dependent ecosystems, as reflected in the popularity of parks, scenic drives, and the selection of housing locations
	Education/research	Groundwater offers diverse opportunities for education and research in the context of social, economic and ecological issues

Balancing Human and Environmental Water Needs

During the last 2 decades, the management of water resources in South Africa has undergone substantial transformation (Braune 2000). With the democratization of the country in 1994 came a policy shift towards providing basic services, including water and sanitation services, to all citizens (Braune and Xu 2008).

For a long time groundwater was dealt with in isolation and was basically viewed from a technological or hydrological perspective. Since the early 2000s the management of groundwater became a topic of interest for different actors and sectors in the Sandveld. The interview results show that during the last two decades programs and measures were developed to better understand the quality and quantity of groundwater, geological formations, the storage capacity of aquifers, and the ecological integrity of the landscape-water regime. To implement sustainable irrigation practices, abstraction must be regularly monitored, wells must be registered and licensed, hydro-geological and climate data must be accessible for farmers, and a fair pricing system must be established. Although the area is receiving much attention due to its environmental uniqueness, and its significant groundwater resources, we identified many challenges which still remain for water and land managers: illegal agriculture activities, insufficient pivot irrigation systems, decreasing water quantities, increased water contamination and salt water intrusion along the coast. Over the last 5 years the industrial sector, the conservation sector and farmers have joined forces to address agricultural sustainability, conserve the remaining fragments of the biodiversity-rich land, and establish natural corridors connecting fragmented habitats.

Considering these development processes described above, we ask the following question: What are important WGMS characteristics towards integrated and sustainable provision of groundwater ecosystem services? A special research focus is put on actor networks.

Figure 18.4 summarizes most important WGMS characteristics, performance measures and context factors of the Sandveld. We found that sectoral cooperation, exchange between administrative levels and the involvement of stakeholders constitute crucial characteristics for water management in the Sandveld. These characteristics have a strong influence on the performance measure: dealing with trade-offs between irrigation and water quantity/base flow as well as between agriculture and biodiversity aspects. The ecological and societal context has a strong influence on past and current management behaviour. Since the end of the Apartheid era in 1994 South Africa's water legislation has undergone significant transformation processes including socio-economic shifts and acquired an entirely new dimension for social, economic and political reconstructions. These changes have had a significant impact on groundwater resources. Beside the overall societal context, the ecological system aspects such as the climatic conditions and the high degree of watershed and landscape modifications imply additional impacts on the performance measures

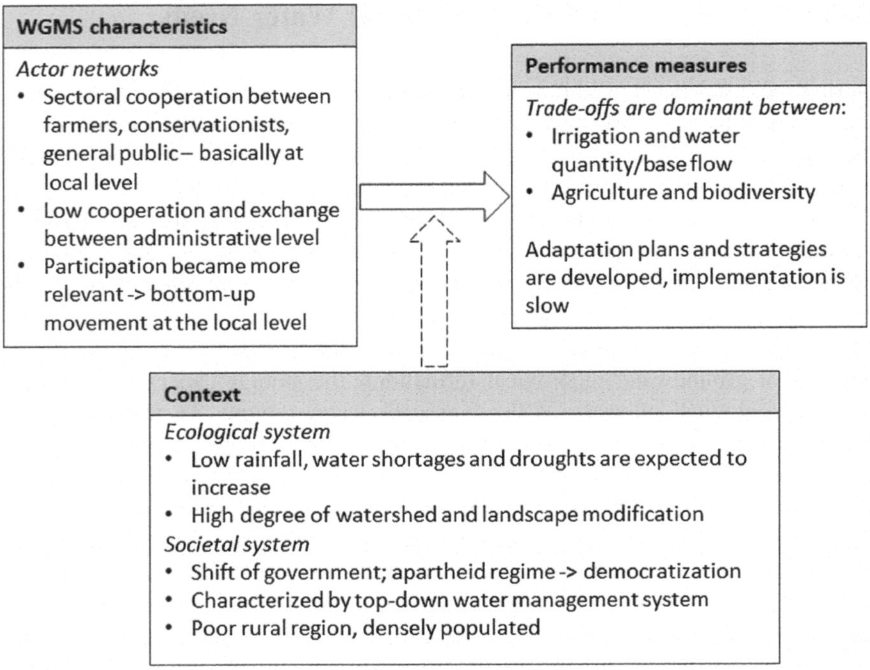

Fig. 18.4 Relationship between actor networks (WGMS characteristic) and the handling of ecosystem service trade-offs (performance measures) influenced by contextual conditions (Due to space restrictions the authors focus on a restricted number of indicators, yet the proposed approach allows other researchers to include as many indicators as relevant to a specific research question)

Discussion

Coordination and Cooperation Across Actors and Sectors

South Africa offers ground-breaking institutional water frameworks (e.g. Water Services Act of 1997, National Water Act of 1998 and the National Water Resource Strategy of 2004). Integrated Water Resources Management (IWRM) became a guiding principle to the water policy sector of South Africa including the optimal balance between social, economic and environmental water use. In this context the status of groundwater changed from one of a privately owned resource asset, coupled to property, to one of a public, or national, resource asset. Furthermore, for the first time groundwater was recognized as a part of the hydrological cycle. This new policy situation induced massive changes of administrative responsibilities and innovative regulatory instruments for groundwater assessment, planning and management, economic instruments to influence water use patterns as well as cooperative measurements at all levels.

Currently, groundwater policies are basically developed at the national level and there exists hardly any discourse or exchange of experience and knowledge among actors and sectors from different levels. Especially the cooperation and the exchange between state and non-state actors is weak or completely absent. Goals, knowledge and experience about groundwater and ecosystem services from regional/local actors are not circulated or considered. This disconnect between lower and higher levels triggers mistrust and conflicts between official authorities and people at the lower levels who have to implement policy requirements.

For a long time the groundwater management system of the Sandveld did not provide the capacity to assure effective and sustainable resource regulation and allocation. During the last 10 years, a basic rethinking of what groundwater use and protection means led the Sandveld's potato industry, conservation sector, farmers, and local municipalities to make a major attempt to integrate and mainstream ecological thinking into the agriculture sector. A bottom-up movement in the Sandveld initiated multi-stakeholder processes to develop approaches to protect and sustain groundwater resources without negative consequences for the socio-economic development of the Sandveld farmers. In other words, human and environmental water needs must be considered and managed integrative rather than in isolation and therefore, minimize trade-offs. As an outcome of this bottom-up movement the Greater Cederberg Biodiversity Corridor (GCBC) as well as Biodiversity Best Practices for Potato Production were established and implemented in the Sandveld. The implementation includes a set of regulatory provisions and measures (e.g. corridor planning, registration of water abstraction, environmental management plans). Currently, the implementation makes slowly progress and is hampered by specific WGMS characteristics. It became apparent that mostly actors and sectors located at the local or regional level were engaged in these developments and actors from the national level (including water, agriculture, and environmental authorities) were not involved. Furthermore, bottom-up programs such as the GCBC and the Best Practices for Potato Production are difficult to institutionalize at higher levels if cooperation and information exchange is absent.

Albeit the people of the Sandveld region made progress towards more sustainable groundwater management, they still lack approaches to integrate an ecosystem service perspective. The following three topics were identified as being core challenges for the integration of ecosystem services into groundwater policies both at the regional and national level.

Undervaluation of the Importance of Groundwater Resources and the Meaning for Ecosystem Services

A major complex of problems concerning the use of groundwater for domestic purposes, irrigation and nature conversation in the Sandveld exists in people's minds, which are often shaped by cultural and individual patterns. Groundwater was and still is barely recognized as being a life-essential resource (e.g. drinking and sanitation), and is only exploited during prolonged periods of drought and being important for agriculture production.

Water managers in the Sandveld as well as at higher administrative levels are often not aware of direct linkages between groundwater storage, recharge and discharge, nor of the wide variety of ecosystem services groundwater provides. Consequently, the linkages between groundwater resources, the services they provide and human well-being are not apparent and therefore not recognized by official authorities during planning, decision and implementation processes.

Shortages of Adequate Data and Expertise

Groundwater ecosystem services are characterized by a shortfall of physical, hydrogeological as well as socio-economic data related to groundwater, aquifer properties and linkages to human well-being. The latter is extremely important as it plays an important role related to poverty alleviation, health standards and social vulnerability.

It was further identified that human resources and capacities are lacking at all management levels including national and regional state offices. Important management positions remain unfilled or are taken up by people requiring further specialized training before they can deal appropriately with the challenges facing their respective positions (e.g. allocation of water licenses, monitoring of water abstraction).

Centralization of Power

Groundwater in the Sandveld is dominated by a top-down management system. The national and regional offices of the Department of Water Affairs exhibits huge lacks in terms of the structures in place for cooperation between the different political agencies, administrative levels and other stakeholders (e.g. farmers, civil society, NGOs). Decentralized management systems may lead to greater efficiency, effectiveness and equity. In general, South Africa's water sector lacks these integration structures, both in terms of the exchange between responsible administrative levels and in terms of cooperation between different sectors such as agriculture, land use planning, nature conservation, forestry and society as a whole.

Conclusion

This chapter elaborated a set of indicators (Table 18.1) relevant to understand WGMS characteristics, the performance and the overall context in relation to the management of ecosystem services. We applied relevant indicators to the Sandveld region in South Africa. In conclusion, we can summarize the following findings. First, the overall water management of South Africa went through a period of massive change. Perception of groundwater changed dramatically: for the first time groundwater was recognized as a part of the hydrological cycle and ecosystem services found access to policies and legislation (e.g. National Water Act). Albeit the very progressive water resource legislation the country is facing many difficulties by implementing innovative approaches and measures for water use and protection at the local level. As groundwater policies are basically developed at the national level there exists hardly any discourse or exchange of experience and knowledge with actors from lower levels. Participation at lower levels only occurs during informal programs and projects but is not implemented in the formal procedures at higher levels. Second, bottom-up movements in the Sandveld brought together actors with different interests in order to develop programs and strategies towards more sustainable groundwater management. Unfortunately, these developments still lack approaches to integrate an ecosystem service perspective and require some clear identification of trade-offs. Further, actors from higher levels, especially state actors, are not involved in the Sandveld programs (GCBC and Biodiversity Best Practices for Potato Production) in order to support knowledge.

A fundamental challenge for water managers all over the world is to understand the dynamics of ecosystem services and human well-being, and to develop integrated and sustainable management practices dealing with these complexities. To meet the steady and increasing demand for water they must reflect new social priorities, economic realities and environmental goals.

A current research project (*WaterNeeds*, www.waterneeds.uni-osnabrueck.de) considers these complexities by exploring the relation between WGMS characteristics, the performance and the context of individual case studies across the world. *WaterNeeds* is invented to analyze under which conditions a reframing of water management objectives in terms of an integrated perspective on ecosystem services and environmental hazards support transitions towards sustainable water governance and management.

References

Bakker K, Kooy M, Shofiania NE, Martijn EJ (2008) Governance failure: rethinking the institutional dimensions of urban water supply to poor households. World Dev 36(10):1891–1915

Bates BC, Kundzewicz ZW, Wu S, Palutikof JP (eds) (2008) Climate change and water. Technical Paper of the Intergovernmental Panel on Climate Change. IPCC

Bennett EM, Peterson GD, Gordon LJ (2009) Understanding relationships among multiple ecosystem services. Ecol Lett 12:1–11

Berkes F, Colding J, Folke C (2003) Navigating social-ecological systems: building resilience for complexity and change. Cambridge University Press, Cambridge

Biswas AK, Tortajada C (2010) Future water governance: problems and perspectives. Water Resour Dev 26(2):129–139

Braune E (2000) Towards comprehensive groundwater resource management in South Africa. IAH Congress 2000, Cape Town

Braune E, Xu Y (2008) Groundwater management issues in southern Africa—an IWRM perspective. Water SA 34(6):699–706

Conrad J, Nel J, Wentzel J (2005) The challenges and implications of assessing groundwater recharge: A case study—northern Sandveld, Western Cape, South Africa. Water SA 30(5):75–81

Daily GC (1997) Nature's services—societal dependence on natural ecosystems. Island, Washington

de Groot RS, Wilson MA et al (2002) A typology for the classification, description and valuation of ecosystem function, goods and services. Ecol Econ 41(3):393–408

Department of Water Affairs and Forestry (DWAF) South Africa (2005) Olifants/Doorn water management area: Internal Strategic Perspective. Prepared by Ninham Shand (Pty) Ltd in association with Jakoet and Associates, Umvoto Africa, FST and Tlou and Matji, on behalf of the Directorate: National Water Resource Planning. DWAF Report No P WMA 17/000/00/0305

Foley JA, DeFries R, Asner GP et al (2005) Global consequences of land use. Science 309:570–574

Folke C, Hahn T, Olsson P, Norberg J (2005) Adaptive governance of social-ecological systems. Annu Rev Environ Resour 30:441–473

Franke AC, Steyn JM, Ranger KS, Haverkort AJ (2011) Developing environmental principles, criteria, indicators and norms for potato production in South Africa through field surveys and modeling. Agric Syst 104(4):297–306

Holling CS, Meffe GK (1996) Command and control and the pathology of natural resource management. Conserv Biol 10(2):328–337

Huntjens P, Pahl-Wostl C, Grin J (2010) Climate change adaptation in European river basins. Reg Environ Change 10(4):263–284

Irwin F, Ranganathan J (2008) Action agenda for sustaining ecosystem services. In: Ranganathan J, Munasinghe M, Irwin F (eds) Policies for sustainable governance of global ecosystem services. Washington, pp 21–69

Knieper C, Kastens B, Holtz G, Pahl-Wostl C (2010) Analysing water governance in heterogeneous case studies—experiences with a database approach. Environ Sci Policy 13(7):592–603

Knüppe K, Pahl-Wostl C (2011) Requirements for adaptive governance of groundwater ecosystem services—Insights from Sandveld (South Africa), Upper Guadiana (Spain) and Spree (Germany). Reg Environ Change 13(1):53–66

Krysanova V, Dickens C, Timmerman J, Varela-Ortega C et al (2010) Cross-comparison of climate change adaptation strategies across large river basins in Europe, Africa and Asia. Water Resour Manage 24(14):4121–4160

Loring AP, Chapin FS, Gerlach SC (2008) The services-oriented architecture: ecosystem services as a framework for diagnosing change in social ecological systems. Ecosystems 11(3):478–489

MA [Millennium Ecosystem Assessment] (2005) Introduction and conceptual framework. Ecosystems and human well-being: a framework for assessment. Millennium Ecosystem Assessment, Washington

Münch Z, Conrad C (2007) Remote sensing and GIS based determination of groundwater dependent ecosystems in the Western Cape, South Africa. Hydrogeol J 15(1):19–28

Odum HT (1996) Environmental accounting: emery and environmental decision making. John Wiley, New York

Ostrom E (2005) Understanding institutional diversity. Princeton University Press, New Jersey

Ostrom E (2007) A diagnostic approach for going beyond Panaceas. Proc Natl Acad Sci USA 104:15181–15187

Pahl-Wostl C (2009) A conceptual framework for analyzing adaptive capacity and multi-level learning processes in resource governance regimes. Glob Environ Change 19(3):354–365

Pahl-Wostl C, Sendzimir J, Jeffrey P, Aerts J, Berkamp G, Cross K (2007) Managing change toward adaptive water management through social learning. Ecol Soc 12(2):30. http://www.ecologyandsociety.org/vol12/iss2/art30/

Pahl-Wostl C, Holtz G, Kastens B, Knieper C (2010) Analysing complex water governance regimes: the management and transition framework. Environ Sci Policy 13(7):571–581

Pahl-Wostl C, Lebel L, Knieper C, Nikitina E (2012) From applying panaceas to amstering complexity: towards adaptive water governance in river basins. Environ Sci Policy 23:24–34

Raudsepp-Hearne C, Peterson GD, Bennett EM (2010) Ecosystem services bundles for analyzing tradeoffs in diverse landscapes. PNAS USA 107(11):5242–5247

Rogers P, Hall AW (2003) Effective water governance. TEC Background Paper, Stockholm, Global Water Partnership

Rodríguez JP, Beard TD, Bennett EM et al. (2006) Trade-offs across space, time, and ecosystem services. Ecol Soc 11(1):28. http://www.ecologyandsociety.org/vol11/iss1/art28/

Schlüter M, Leslie H, Levin S (2009) Managing water-use trade-offs in a semi-arid river delta to sustain multiple ecosystem services: a modeling approach. Ecol Res 24(3):491–503

UNEP [United Nations Environment Programme] (2007) Global environment outlook: environment for development (GEO 4). Progress Press, Valletta

van Jaarsveld AS, Biggs R, Scholes RJ et al (2005) Measuring conditions and trends in ecosystem services at multiple scales: the Southern African Millennium Ecosystem Assessment (SAfMA) experience. Philos Trans Royal Soc B-Biol Sci 360:425–441

Chapter 19
Tackling the 'How' Question: Enabling and Enacting Practical Action for Managing the Wicked Problem of Nonpoint Source Pollution in Catchments

James J. Patterson, Jennifer Bellamy and Carl Smith

Abstract Managing nonpoint source (NPS) pollution in catchments is a 'wicked problem' and a persistent challenge in protecting the health of freshwater and marine ecosystems, and the human systems that depend on them. NPS pollution arises through complex interactions between human activities and dynamic natural systems, both spatially and across multiple levels of organisation. A key challenge is enabling and enacting purposeful and concerted collective action ('practical action') within multi-level catchment management systems. This challenge has been explored through theory-informed empirical research, involving an in-depth case study in South-East Queensland (SEQ), Australia, which is a large, complex and rapidly growing coastal region facing significant ongoing waterway health challenges. Three embedded catchment cases within the SEQ region were studied using multiple qualitative methods (semi-structured interviews, field observation, document review, feedback workshops). A conceptual framework was used to analyse 'enabling capacities' that are important for practical action, which were: prior experience and contingency; institutional arrangements; collaboration; engagement; vision and strategy; knowledge building and brokerage; resourcing; entrepreneurship and leadership; and reflection and adaptation. Practical action was found to be emergent from the combined interplay of these enabling capacities, at and across multiple levels. These findings imply that management efforts need to focus on building enabling capacities that underpin the emergence of practical action within complex, dynamic and changing situations, rather than solely on prescribing actions and targets to be achieved.

J. J. Patterson (✉) · J. Bellamy · C. Smith
School of Agriculture and Food Sciences, University of Queensland, Brisbane, Australia
e-mail: james.patterson@uqconnect.edu.au

A. Bhaduri et al. (eds.), *The Global Water System in the Anthropocene*,
Springer Water, DOI: 10.1007/978-3-319-07548-8_19,

Introduction: Nonpoint Source Pollution and the Challenge of Practical Action

Changes to global water systems emerge from the interaction of diverse factors such as human modification of hydrological flows, population pressure, land use change and climate change, which are dynamic and interlinked across multiple levels of organisation (Postel et al. 1996; Vörösmarty et al. 2010). In this context, managing catchments is vital for protecting water quality, water security, biodiversity and ecosystem services, on which humanity relies (Parkes et al. 2010; UN-Water 2011). Nonpoint source (NPS) water pollution is a particular catchment management issue that remains a major and persistent challenge to the health of freshwater and marine ecosystems, and a significant issue for sustainability and human wellbeing linked to natural water systems (Carpenter et al. 1998; UN-Water 2011).

NPS pollution refers to the cumulative impact of pollutants released to waterways (e.g. nutrients, sediments, toxicants and pathogens) from across large areas of catchments and basins, that is linked to human activities and land use change (e.g. urbanisation, deforestation and agriculture) (Carpenter et al. 1998; Gunningham and Sinclair 2005). It is highly spatially and temporally dispersed and can span rural and urban contexts, and it can include specific issues of urban stormwater runoff (Brown 2009), agricultural runoff (Moss 2008), and stream and landscape erosion (Bunn et al. 2010). This contrasts with point source pollution (e.g. from wastewater treatment plants) which involves discrete and clearly identifiable sources, a limited number of actors, and it is commonly amenable to relatively straightforward technical or regulatory intervention (Gunningham and Sinclair 2005). There have been only limited efforts to manage NPS pollution in industrialized contexts in Europe (CEC 2007), the U.S.A. (Weitman 2011), and Australia (Patterson 2014), and they have all faced substantial ongoing difficulties in practice because of the complex and cross-cutting nature of the issue. More generally, NPS pollution arguably remains an under-researched and poorly-understood issue, particularly its social and institutional dimensions.

From a catchment or basin perspective, the challenge of managing NPS pollution can be seen as a 'wicked problem' (Rittel and Webber 1973; APSC 2007; Head 2008; Smith and Porter 2010), involving complex, uncertain, multi-actor, multi-scalar, dynamic and changing situations (Patterson et al. 2013). A key challenge in this context is enabling and enacting 'practical action'; which is defined here as purposeful and concerted collective action oriented towards a local level within a multi-level resource governance system. It includes various types of management activities, such as: environmentally-focused activities (e.g. on-ground restoration and mitigation works); socially-focused activities (e.g. engagement and promoting behavior change); and institutionally-focused activities (e.g. knowledge co-generation, network-building, collaborative planning, and policy feedback) (Patterson et al. 2013). However, enabling and enacting practical action is a major challenge in catchments because it is linked to "the dynamics of social, institutional and biophysical interactions, multiple drivers of change, and patterns of behaviour of multiple actors in particular contexts that evolve over time" (Patterson et al. 2013). Nevertheless, practical action is central within broader management and governance efforts aiming to shape change in catchments (e.g. towards adaptation or transformation for sustainability), and to bridging the gap between theory and practice (Ingram 2008; Barmuta et al. 2011) in addressing persistent and increasingly pressing catchment issues. In response, this paper focuses on the need to better understand how to enable and enact practical action for managing NPS pollution in practice.

Research Approach

The research adopted a theory-informed empirical approach, which involved both theory development on 'enabling capacities' and an in-depth empirical case study. This approach was taken given the need for a systemic perspective of the complex and cross-cutting research problem, and to link conceptual understanding with practical experience in exploring how practical action can be enabled and enacted in real situations. The empirical investigation involved an embedded case study design (Yin 2009) consisting of the in-depth study of three embedded catchment cases within a single broader region (described in the following section). Multiple qualitative methods were utilised in order to draw on the rich body of practical experience in the catchment areas and the broader case study region. They include: semi-structured key informant interviews ($n = 53$); field observation; review of academic and grey literature; and feedback workshops. This approach enabled an in-depth and context-sensitive study of practical action in catchments, as well as its linkages and embeddedness within the broader multi-level regional governance system.

Based on in-depth review and synthesis of literature, a conceptual framework was developed to understand and analyse the factors that can influence the emergence of practical action, from a systemic perspective (Fig. 19.1). The framework identified a diverse range of systemic enabling capacities across multiple institutional levels that can influence practical action, namely: prior experience and contingency; institutional arrangements; collaboration; engagement; vision and strategy; knowledge building and brokerage; resourcing; entrepreneurship and leadership; and reflection and adaptation (Patterson et al. 2013). The framework reflects a complex systems perspective under which practical action is 'emergent' from enabling capacities and their interactions across multiple levels. It provides a systemic and context-sensitive mechanism for cross-case analysis to study the complex phenomenon of practical action. The framework was applied in the empirical case study to analyse enabling capacities and their cross-level interplay that are important for enabling and enacting practical action in the embedded catchment cases.

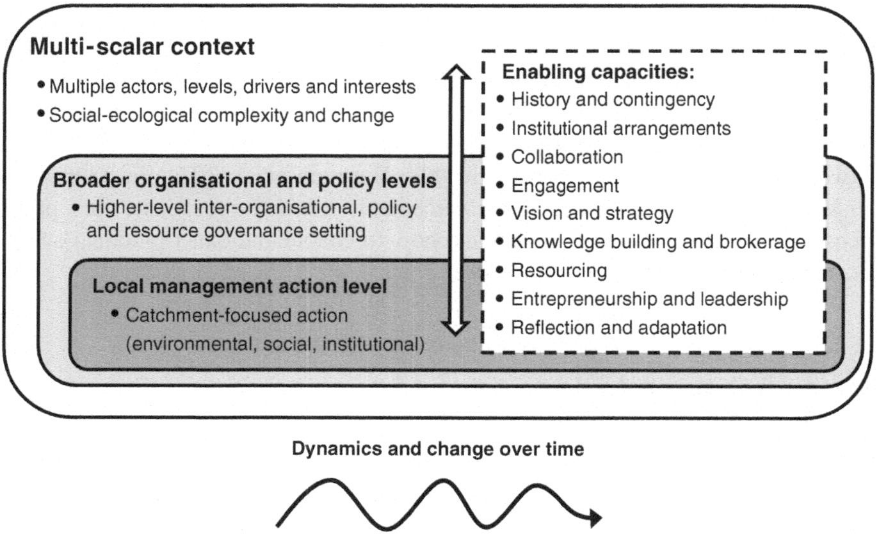

Fig. 19.1 Conceptual framework for understanding enabling capacities that can influence practical action for NPS pollution in catchments (*Source* Patterson et al. 2013)

Case Study Area: South-East Queensland, Australia

The overall case study area was South-East Queensland (SEQ), Australia (Fig. 19.2), which is a large and complex regional landscape that is also one of the most rapidly growing and urbanising coastal regions in the country (DERM 2009). The region contains a state capital city (Brisbane) and substantial surrounding urban areas, with a total population of 3 million that is expected to grow to 4.4 million over the next two decades (SEQHWP 2006; DERM 2009). It has a rich history of largely self-organised regional collaboration regarding waterway and catchment management over two decades, driven largely by concerns regarding the health of freshwater and marine ecosystems, and linked social, economic and cultural values (Abal et al. 2005; SEQHWP 2006; Bunn et al. 2010). In this context, waterway health has been recognised as a key challenge in the dynamic interplay of the regions linked social and ecological systems. Through substantial investment over several years, there have been major successes in addressing point sources of pollution, in particular sewage treatment plant effluent impacting the sensitive marine receiving environment of Moreton Bay (Fig. 19.2) (SEQHWP 2006; Bunn et al. 2010; Healthy Waterways 2013). However, ongoing monitoring of the cumulative impact of the many pressures on waterway health, across both rural and urban areas, has highlighted NPS pollution as the major ongoing and increasingly urgent challenge to waterway health in the region (SEQHWP 2006; Bunn et al. 2010; Healthy Waterways 2013).

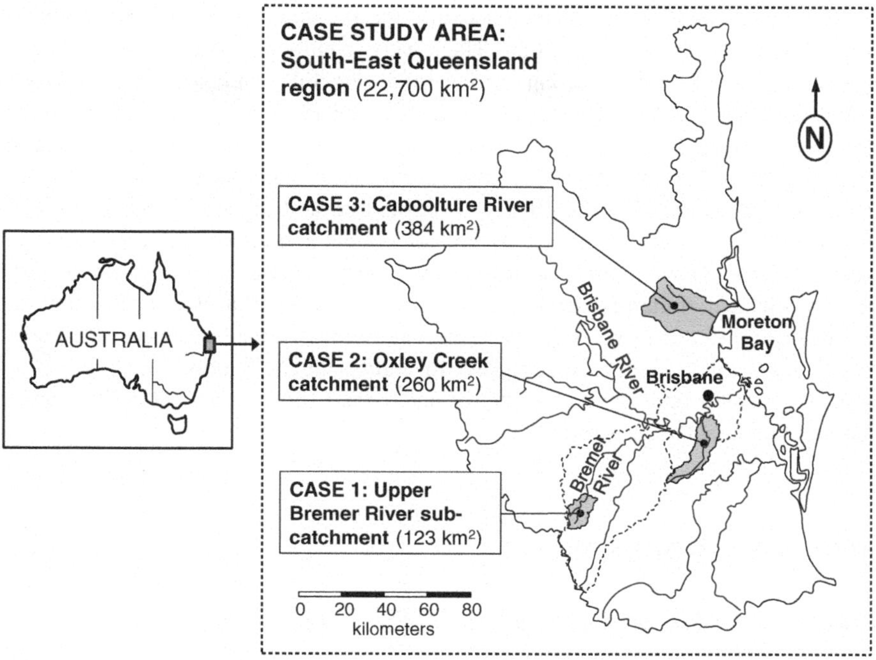

Fig. 19.2 The embedded catchment cases within the broader SEQ region case study area

Addressing NPS pollution has proven to be extremely difficult in practice, even though initiatives in SEQ are widely regarded as being among the forefront of efforts to address water quality and waterway health issues in Australia (Bunn et al. 2010; Patterson 2014). These difficulties also echo experience from other places across the world (such as Europe and the U.S.A.) regarding the challenges of addressing persistent NPS pollution issues in practice (Patterson 2014). There is a widely recognised need for practical action to address NPS pollution issues in SEQ (particularly sediments and nutrients), due to their impacts on the health of fresh, estuarine and marine waterway systems (Abal et al. 2005; Bunn et al. 2010; MJA 2011; Healthy Waterways 2013). This requires not just on-ground action (e.g. by farmers, industry, natural resource management bodies, and community groups), but also action at broader strategic/planning levels (e.g. State and Local Government, cross-sectoral partnerships, natural resource management bodies, and water utilities) because practical action at the local level is embedded within a broader evolving multi-level regional resource governance system. Therefore the analysis presented here explores efforts to foster practical action that have occurred in particular catchment cases which are believed to be among the most 'successful' in the region. It focuses on understanding how practical action to address NPS pollution can be generated, using the conceptual framework (Fig. 19.1) as the basis for analysis.

Three catchment cases embedded within the broader regional problem context were studied in-depth (Fig. 19.2). They were contrasting catchment situations spanning a mix of rural and urban settings involving different resource management initiatives, that collectively encompass a snapshot of some of the most pressing waterway-related challenges facing the region overall. These challenges include: restoring large areas of degraded streams and rural landscapes (Cases 1, 2, 3); managing and mitigating the effects of substantial and rapid ongoing population growth and urbanisation (Cases 2, 3); and adapting and transforming existing urban and industrial areas in more water-sensitive ways (Case 2, 3). All three cases involve substantial management efforts to address NPS pollution issues (sediments and nutrients) in recent years. These efforts are widely considered to be innovative and to have made significant progress in addressing NPS pollution issues, thus offering learning opportunities to better understand 'how' practical action can be enabled and enacted in practice.

Results and Discussion

Practical Action and Outcomes in the Local Cases

Different types of 'practical action' have been pursued in the three cases, including: on-ground environmentally-focused action, socially-focused action, and institutionally-focused action. Case 1 involved efforts to foster targeted on-ground catchment restoration through non-coercive (e.g. voluntary and inventive-based) mechanisms. Case 2 involved efforts to foster on-ground catchment restoration and to enhance strategic coordination among multiple actors, through a mix of non-coercive (e.g. community engagement, institutional engagement), negotiatory (e.g. negotiation among strategic interests), and coercive (e.g. regulatory and compliance) mechanisms. Case 3 involved efforts to enhance strategic coordination among multiple actors through non-coercive (e.g. institutional engagement) and negotiatory (e.g. argumentation among strategic interests) mechanisms.

The impacts of management efforts varied across the three cases, but included a mix of environmental, social and institutional outcomes. Cases 1 and 2 focused largely on addressing sources of sediment pollution, and were both anecdotally considered to have made significant progress through on-ground restoration activities, such as stream and gully stabilization, changed land use practices, and enhanced protection of riparian areas (Patterson 2014). Case 3, while focused on addressing nutrient pollution linked to existing activities and significant impending urban growth, has not as yet involved on-ground action to the same degree as the other two cases. Environmental outcomes regarding actual reductions in NPS pollution in three cases are difficult to assess due to the relatively short timeframes of the cases compared to the much longer timeframes (e.g. legacies, lags) and complex biophysical dynamics driving NPS pollution in SEQ. However, in each of the cases there was significant evidence of improved capacity to address NPS pollution. For example, across all cases a range of social and institutional outcomes were observed including: fostering of stronger landholder, community and industry engagement with catchment issues; enhanced linkages between catchment issues and urban planning within complex Local and State Government policy and regulatory arrangements; new knowledge for managing NPS pollution issues in the catchments; emergent forms of collaboration that were more locally-centred; and wider motivation and interest generated across levels of decision-making and action in the region for addressing NPS pollution. Furthermore, each case clearly provides evidence of 'learning-by-doing' among core actors over several years, as demonstrated by: re-framing of problems over time (e.g. from an initial single-outcome focus to a broader multi-outcome focus in each case); and significant adaptation of activities and strategies over time in response to new understandings (e.g. technical, social, institutional, political). These social and institutional outcomes reflect crucial forms of progress in addressing NPS pollution issues from a longer-term perspective (e.g. several years to decades), and suggest impacts across multiple levels of planning, decision-making and action.

'Enabling Capacities' for Practical Action

As part of a broader study (Patterson 2014; Patterson et al. in prep.), the conceptual framework (Fig. 19.1) was applied to analyse enabling capacities for practical action across multiple levels of organisation in each case, particularly focusing on 'local' and 'regional' institutional levels. These levels were defined contextually in each case, but included multiple spatial, jurisdictional and other institutional levels clustered either more towards a 'local' level (i.e. actors and activities coalesced at the catchment level) or more towards a 'regional' level (i.e. actors and activities coalesced at the SEQ region level). This allowed for an analytical perspective that appreciated the real 'messily nested' jurisdictional and institutional setting (Bellamy 2007), but also allowed a simplified multi-level typology for empirical analysis of enabling capacities.

The findings highlight the importance of all of the enabling capacities in the conceptual framework in each case, at both local and regional levels. However, the way in which these capacities manifested varied across the different cases, linked to: emphases on different types of practical action; different combinations of drivers for management action; and embeddedness in differing multi-level contexts. Nevertheless, there were many common patterns in enabling capacities amongst the catchment cases (Table 19.1).

History and contingency influenced management efforts in all cases, particularly prior collaborative experiences at a local level (across landholders, community groups and government). For example, a significant feature in Case 1 was a history of rural landholders collaborating with government on land management issues, as well as past experiences of catchment management associated with a

Table 19.1 Enabling capacities that influenced practical action in the SEQ case study (Patterson 2014)

Enabling capacities	Enabling capacities identified at local and regional levels	
	Local level	Regional level
History and contingency	Prior collaborative experience with catchment-related activities Local community receptivity	Rich history of self-organised collaboration in waterway management over two decades, largely focused on strategic vision and coordination
Institutional arrangements	Multi-actor platforms for interaction at catchment level	Relatively strong regulatory and policy arrangements compared to other places in Australia, however ongoing fragmentation issues Regional bridging organisations for interaction
Collaboration	Problem-focused collaboration involving multiple actors from across multiple institutional levels	Substantial collaborative experiences and culture fostered over time, but some persistent inter-organizational tensions exist
Engagement	Substantial efforts to engage relevant actors and generate buy-in and commitment	Relatively strong institutional as well as social and political engagement with waterway issues
Vision and strategy	Locally-relevant and linked to broader regional level goals and priorities	Sustained efforts to collaboratively develop regional strategies and multi-sectoral policy coordination
Knowledge building and brokerage	Incorporation of knowledge of many different actors, and efforts to foster knowledge brokerage	Major efforts at scientific knowledge-building and science-policy linkages over many years
Resourcing	Substantial resource availability from broader organizational/ policy levels (financial, people, coordination and sharing of resources)	Substantial mobilization of organizational resources through regional bridging organisations, but overall mismatch in scale of resources available and need for local ongoing resource generation mechanisms
Entrepreneur-ship and leadership	Critical importance of key individuals for facilitating collaboration and generating agency in networks	Long history and critical importance of key individuals for generating vision and fostering collaboration at regional level
Reflection and adaptation	Reflection on progress in fostering practical action over time, and ongoing adaptation of approaches (i.e. 'learning by doing')	Unique feedback mechanism of long-term regional waterway health monitoring program Shifts in problem framing and focus of action over time

local catchment group. This provided social conditions which were relatively conducive for more-contemporary initiatives (examined in this paper) than has been the case in other nearby areas within the same region. In Case 2, a strong history of community-based catchment activities provided an important contingency for more recent government-led efforts. More broadly, all the cases were embedded within the SEQ region which has a rich history of collaboration in waterways management over two decades, and provides a regional context that is relatively conducive to local-level initiatives.

Institutional arrangements and *collaboration* were fundamentally important in all cases, particularly linked to multi-actor platforms (SLIM 2004a; Kerr 2007) which fostered catchment-focused interaction among multiple actors. In Case 1, such a platform emerged largely through active local facilitation by key individuals, but in Cases 2 and 3 platforms were established by Local Government. These contrasting approaches reflect the differing emphasis of practical action in each case; Case 1 was focused primarily on on-ground management action through non-coercive mechanisms, whereas Cases 2 and 3 involved a much stronger institutional focus with non-coercive and negotiatory mechanisms. Existing collaborative groups at a regional level (focused on regional waterways management, and regional natural resource management) proved highly significant in functioning as regional 'bridging organizations' (Folke et al. 2005; Hahn et al. 2006), that fostered regional-focused interaction among a wide range of actors. *Engagement* was closely linked to *collaboration* and was especially important for generating buy-in and commitment of a range of actors at a local level in each case.

Vision and strategy was important in all cases in regards to how well-aligned goals and motivations were amongst various actors, and across local and regional levels. In each case, this alignment was enhanced over time by the re-framing of problems (Bouwen and Taillieu 2004; Brugnach and Ingram 2012) that were initially strongly framed in regional terms, and were re-framed to be more relevant to the local level (e.g. interests and concerns of actors at a local level). *Knowledge building and brokerage* was crucial in all cases, reflected through efforts to integrate multiple types of knowledge and foster knowledge brokerage at a local level (Bouwen and Taillieu 2004; Folke et al. 2005; Ison et al. 2011). For example, a significant feature of Case 1 was the active effort to foster knowledge brokerage between regional science-policy actors and local landholders in response to an earlier conflict between them, which proved pivotal to the eventual success of management activities in this case. Cases 2 and 3 also involved significant efforts to foster knowledge brokerage between multiple actors across institutional levels, particularly through the multi-actor platforms in each case. *Resourcing* was critical for supporting practical action through enabling substantial financial and organizational resources to be available from higher institutional levels in each case, although only on a relatively short-term basis.

Entrepreneurship and leadership was fundamentally important in all cases in multiple ways, including: facilitating interactions and bridging perspectives between different actors (e.g. fostering *collaboration*, and *knowledge building and brokerage*) (Folke et al. 2005; Hahn et al. 2006; Steyaert and Jiggins 2007);

generating *agency* within networks especially at the local level (Moore and Westley 2011); and enabling institutional feedback from local level experiences to the regional level (Folke et al. 2005; Meijerink and Huitema 2010). For example, in Case 1 key individuals were crucial in overcoming major collaboration difficulties and conflict in the early stages of the program, which involved facilitating interactions and bridging perspectives between different actors, re-framing problems, and promoting and 'championing' joint activities. In Cases 2 and 3, key individuals also played pivotal roles in similar ways, although more in response to the need to engage multiple strategically-important organizational actors.

Finally, *reflection and adaptation* is critical to the adaptive and evolving management activities at the local level in each case which was underpinned by mutual learning amongst multiple actors over time (Bouwen and Taillieu 2004; Folke et al. 2005; Steyaert and Jiggins 2007). This is reflected through the re-framing of problems, and enhanced mutual understanding and collaboration over time in each case. For example, in Case 1 management activities adapted significantly to the challenges of linking regional science and policy to local on-ground practice. Cases 2 and 3 adapted and evolved over time from having a strong initial focus on specified waterway health targets, to a broader appreciation of multiple forms of social and institutional change required within difficult catchment situations. Each case also involved significant emphasis on 'learning by doing' approaches at a local level, driven largely by key individuals (*entrepreneurship and leadership*).

Collectively the enabling capacities contributed to the emergence of a diversity of forms of practical action across the catchment cases. In all cases, some enabling capacities were particularly vital at the early stages (e.g. *prior experience and contingency*, *institutional arrangements*, *resourcing*), whereas others were generated over time through collective management efforts and interactions between multiple actors (*collaboration*, *engagement*, *vision and strategy*, *knowledge building and brokerage*, *entrepreneurship and leadership*, *reflection and adaptation*). This is because the enabling capacities are highly interactional in nature, arising largely through ongoing management efforts and multi-actor interactions in actual situations of practice.

While all of the enabling capacities in the conceptual framework were important, the relative importance of different capacities and interactions between them varied across the cases. Hence as well as the enabling capacities themselves, the interplay amongst these capacities is also crucial for supporting emergent and adaptive responses. This is demonstrated by the different ways in which each case responded over time to dynamic and evolving situations and many ongoing challenges in practice, in order to continue to work towards enabling and enacting practical action. Thus it would not have been possible to analyse the enabling capacities separately or deterministically, because what was most important was their interactive effect in fostering emergent and adaptive forms of practical action in particular situations.

Conclusion

A key implication of the findings presented is that practical action in catchments is 'emergent' from the systemic functionality generated collectively by the enabling capacities and their interplay at and across multiple levels. This was demonstrated in the three catchment cases where practical action emerged in different ways over time, in response to different types of complex, dynamic and evolving catchment situations, despite all being embedded within the same overall region. Hence practical action (i.e. 'purposeful and concerted collective action oriented towards the local level') is not something that can be 'delivered' or 'implemented' necessarily from outside, but rather it is generated through ongoing efforts to shape change within particular situations. This highlights the need to focus on the *process* of fostering practical action by building the capacities for adaptive and contextually appropriate responses to emerge in particular situations, rather than solely on achieving pre-defined targets. Therefore understanding 'how' to address NPS pollution is largely about improving the capacity for practical action to emerge in particular situations, which is important for achieving outcomes across a range of dimensions (e.g. environmental, social, institutional).

Seeing practical action as an emergent property aligns with the argument of Collins and Ison (2010) that integrated catchment management (ICM): "arises out of a set of practices for managing catchments in particular contexts … [where] a shift in understanding … from a deterministic goal to an emergent phenomenon requires a shift in practices away from prescription of outcomes towards theory-led process design and, ultimately, … trusting emergence". This implies a need for a focus on 'emergence' as the phenomenon to manage in 'wicked problem' catchment situations; in this instance by building enabling capacities that are important for generating practical action. While individual catchment situations will evolve and change over time, what is most important are the capacities that enable the emergence of concerted, adaptive, and ultimately effective management action (Patterson et al. 2013). The conceptual framework developed in this research provides insights into understanding the enabling capacities and their interactions that may provide the appropriate conditions for the emergence of practical action.

The findings also have significant implications for understanding the performance of management efforts to shape change in catchments. In particular, the findings imply a need for a strong focus on the *process* of fostering practical action within particular contexts (that is, the process of building enabling capacities that collectively generate practical action), rather than focusing, for example, only on achieving pre-defined targets. Management activities often focus on achieving pre-defined (usually biophysical) targets, however this may lead to timescale mismatches in our understanding of progress and change due to complexity and time lags in social-ecological situations (Cash et al. 2006; Cumming et al. 2006). Moreover, in complex systems it is not possible to 'control' outcomes directly, as the properties of an ecosystem are the emergent outcome of social, institutional and biophysical interactions and co-evolutionary dynamics (SLIM 2004b; Collins and Ison 2010; Ison et al. 2011). In this light, the findings of this research imply the need to shift the focus of management from setting and meeting fixed targets towards an endeavor that seeks to build conditions and capacities that can foster the emergence of purposeful and concerted collective action.

This paper has explored the challenge of enabling and enacting practical action for the wicked problem of NPS pollution in catchments, which is a globally-significant issue for managing natural water systems. The paper contributes to understanding and analyzing practical action in complex and dynamic multi-level catchment systems for the wicked problem of NPS pollution. The capacities framework presented has relevance for other situations facing similar challenges generating practical action to address 'wicked' water governance problems that require diverse actors to work together across multiple levels of decision-making and action, in order to enable and enact practical action at a local level.

Acknowledgments This work was supported by an Australian Postgraduate Award, a Queensland Government Smart Futures PhD Top-Up scholarship, and financial and in-kind support provided by the University of Queensland (http://www.uq.edu.au/) and Healthy Waterways Limited (http://www.healthywaterways.org). The authors gratefully acknowledge the support of these organisations.

References

Abal EG, Bunn SE, Dennison WC (eds) (2005) Healthy waterways, healthy catchments: making the connection in South East Queensland, Australia. Moreton Bay and Catchments Partnership, Brisbane

APSC (2007) Tackling wicked problems: a public policy perspective. Australian Public Service Commission, Commonwealth of Australia, Canberra

Barmuta LA, Linke S, Turak E (2011) Bridging the gap between 'planning' and 'doing' for biodiversity conservation in freshwaters. Freshw Biol 56:180–195

Bellamy J (2007) Adaptive governance: the challenge for regional natural resource management. In: Brown AJ, Bellamy J (eds) Federalism and regionalism in Australia: new approaches, new institutions?. ANU E Press, Canberra, pp 95–117

Bouwen R, Taillieu T (2004) Multi-party collaboration as social learning for interdependence: developing relational knowing for sustainable natural resource management. J Community Appl Soc Psychol 14(3):137–153

Brown LR, Cuffney TF, Coles JF, Fitzpatrick F, McMahon G, Steuer J, Bell AH, May JT (2009) Urban streams across the USA: lessons learned from studies in 9 metropolitan areas. J North Am Benthological Soc 28(4):1051–1069

Brugnach M, Ingram H (2012) Ambiguity: the challenge of knowing and deciding together. Environ Sci Policy 15:60–71

Bunn SE, Abal EG, Smith MJ, Choy SC, Fellows CS, Harch BD, Kennard MJ, Sheldon F (2010) Integration of science and monitoring of river ecosystem health to guide investments in catchment protection and rehabilitation. Freshw Biol 55:223–240

Carpenter SR, Caraco NF, Correll DL, Howarth RW, Sharpley AN, Smith VH (1998) Nonpoint pollution of surface waters with phosphorus and nitrogen. Ecol Appl 8(3):559–568

Cash DW, Adger WN, Berkes F, Garden P, Lebel L, Olsson P, Pritchard L, Young O (2006) Scale and cross-scale dynamics: governance and information in a multilevel world. Ecol Soc 11(2):8

CEC (2007) Communication from the commission to the European parliament and the council: towards sustainable water management in the European Union—first stage in the implementation of the water framework directive 2000/60/EC. Commission of the European Communities, Brussels

Collins K, Ison R (2010) Trusting emergence: some experiences of learning about integrated catchment science with the environment agency of England and Wales. Water Resour Manage 24(4):669–688

Cumming GS, Cumming DHM, Redman CL (2006) Scale mismatches in social-ecological systems: causes, consequences, and solutions. Ecol Soc 11(1):14

DERM (2009) South East Queensland natural resource management plan 2009–2031. Department of environment and resource management. Queensland Government, and SEQ Catchments, Brisbane

Folke C, Hahn T, Olsson P, Norberg J (2005) Adaptive governance of social-ecological systems. Annu Rev Environ Resour 30(1):441–473

Gunningham N, Sinclair D (2005) Policy instrument choice and diffuse source pollution. J Environ Law 17(1):51–81

Hahn T, Olsson P, Folke C, Johansson K (2006) Trust-building, knowledge generation and organizational innovations: the role of a bridging organization for adaptive comanagement of a wetland landscape around Kristianstad, Sweden. Hum Ecol 34(4):573–592

Head BW (2008) Wicked problems in public policy. Public Policy 3(2):101–118

Healthy Waterways (2013) Ecosystem health monitoring program, healthy waterways limited, Brisbane. http://www.healthywaterways.org/ehmphome.aspx. Accessed 1 Nov 2013

Ingram H (2008) Beyond universal remedies for good water governance: a political and contextual approach. Paper presented at the 6th Biennial Rosenberg water policy forum, Zaragoza, Spain, 24–27 June 2008

Ison R, Collins K, Colvin J, Jiggins J, Paolo Roggero P, Seddaiu G, Steyaert P, Toderi M, Zanolla C (2011) Sustainable catchment managing in a climate changing world: new integrative modalities for connecting policy makers, scientists and other stakeholders. Water Resourc Manage 25(15):3977–3992

Kerr J (2007) Watershed management: lessons from common property theory. Int J Commons 1(1):89–110

Meijerink S, Huitema D (2010) Policy entrepreneurs and change strategies: lessons from sixteen case studies of water transitions around the globe. Ecol Soc 15(2):21

MJA (2011) The future of our bay: the business case for managing and enhancing South East Queensland's waterways (2012–2015). Marsden Jacob Associates, prepared for: Department of Environment and Resource Management, Brisbane

Moore M, Westley F (2011) Surmountable chasms: networks and social innovation for resilient systems. Ecol Soc 16(1):5

Moss B (2008) Water pollution by agriculture. Philos Trans R Soc B: Biol Sci 363:659–666

Parkes MW, Morrison KE, Bunch MJ, Hallströmb LK, Neudoerffer RC, Venema HD, Waltner-Toews D (2010) Towards integrated governance for water, health and social-ecological systems: the watershed governance prism. Glob Environ Change 20:693–704

Patterson JJ (2014) Enabling and enacting 'practical action' for the wicked problem of managing non-point source water pollution in catchments. Ph.D. thesis, University of Queensland

Patterson JJ, Bellamy J., Smith C (in prep.) Enabling and enacting practical action in catchments: responding to the wicked problem of non-point source pollution in Coastal subtropical Australia

Patterson JJ, Smith C, Bellamy J (2013) Understanding enabling capacities for managing the 'wicked problem' of nonpoint source water pollution in catchments: a conceptual framework. J Environ Manage 128:441–452

Postel SL, Daily GC, Ehrlich PR (1996) Human appropriation of renewable fresh water. Science 271(5250):785–788

Rittel HWJ, Webber MM (1973) Dilemmas in a general theory of planning. Policy Sci 4(2):155–169

SEQHWP (2006) South East Queensland healthy waterways strategy 2007–2012: strategy overview. South East Queensland Healthy Waterways Partnership, Brisbane, Queensland
SLIM (2004a) Developing conducive and enabling institutions for concerted action, policy briefing no. 3. In: Social learning for the integrated management and sustainable use of water at catchment scale. http://slim.open.ac.uk. Accessed 28 Aug 2013
SLIM (2004b) The Role of Learning Processes in Integrated Catchment Management and the Sustainable Use of Water, Policy Briefing No. 6. In: Social Learning for the Integrated Management and Sustainable Use of Water at Catchment Scale. http://slim.open.ac.uk. Accessed 28 Aug 2013
Smith LED, Porter KS (2010) Management of catchments for the protection of water resources: drawing on the New York City watershed experience. Reg Environ Change 10(4):311–326
Steyaert P, Jiggins J (2007) Governance of complex environmental situations through social learning: a synthesis of SLIM's lessons for research, policy and practice. Environ Sci Policy 10(6):575–586
UN-Water (2011) Policy brief on water quality. United nations. Available via UN-Water. http://www.unwater.org. Accessed 28 Aug 2013
Vörösmarty CJ, McIntyre PB, Gessner MO, Dudgeon D, Prusevich A, Green P, Glidden S, Bunn SE, Sullivan CA, Reidy Liermann C, Davies PM (2010) Global threats to human water security and river biodiversity. Nature 467:555
Weitman D (2011) Regulatory and voluntary programs to control nonpoint source pollution in the United States. In: Bochove E, Vanrolleghem PA, Chambers PA, Thériault G, Novotná B, Burkart MR (eds) DIPCON 2010 Issues and solutions to diffuse pollution: selected papers from the 14th international conference of the IWA diffuse pollution specialist group, Québec, September 2010
Yin RK (2009) Case study research: design and methods. Applied social research methods series, 4th edn. SAGE Publications, Thousand Oaks

Chapter 20
Experiences with a Transdisciplinary Research Approach for Integrating Ecosystem Services into Water Management in Northwest China

Tuck Fatt Siew, Petra Döll and Hamid Yimit

Abstract Consideration of the relatively new concept of ecosystem services (ESS) in management decisions calls for a transdisciplinary research (TR) approach that aims at integration of knowledge among scientists from multiple disciplines and stakeholders from multiple sectors. In this paper, we present our experiences with the implementation of a TR approach to support the integration of ESS into land and water management under climate change in the arid Tarim River Basin, Northwest China (SuMaRiO project). Our initial TR approach focused on the execution of a stakeholder dialogue (15–20 interviews and five workshops, including participatory modeling) to integrate stakeholder knowledge with research results from SuMaRiO scientists. In the first project phase, the approach was adapted by adding a stakeholder analysis, with explicit efforts to integrate knowledge among the multidisciplinary German scientists, and between German and Chinese scientists. Two key stakeholders from the water sector, together with other representatives of governmental organizations from the sector crop production, animal husbandry, environment, and forestry, were involved in the TR process. The applied TR approach resulted in an improved understanding on issues related to land and water management as well as ESS, and a joint problem perception of stakeholders and scientists. Based on the overall perception graph and discussion with stakeholders and scientists, gaps in the present knowledge related to water and ESS were identified. Chinese stakeholders and scientists appreciated that the TR process facilitated cross-sectoral and multi-disciplinary communication and knowledge exchange. TR (including methods of knowledge elicitation and integration) needs to be continually adapted in reaction

T. F. Siew (✉) · P. Döll
Goethe University Frankfurt, Frankfurt, Germany
e-mail: tuckfatt.siew@em.uni-frankfurt.de

H. Yimit
Xinjiang Normal University, Urumqi, Xinjiang, People's Republic of China

A. Bhaduri et al. (eds.), *The Global Water System in the Anthropocene*, Springer Water, DOI: 10.1007/978-3-319-07548-8_20,

to the challenges encountered in the socio-cultural and institutional setting in the study area. Explicit efforts of network and trust building are a prerequisite for TR, in particular in China.

Introduction

The concept of ecosystem services (ESS) is increasingly recognized as a useful policy tool that can help foster sustainable development. The concept connects environmental health to human well-being and shows the benefits of conservation of ecosystems (and their services) for the development of human society. ESS are goods and services (e.g. clean water or food) that people derive from nature (Millennium Ecosystem Assessment 2005). To help achieve a sustainable use of natural capital for the development of human welfare, ESS need to be integrated into policy making as well as business decisions (TEEB 2010). In China, the integration of ESS into decision-making processes is recommended to support ecosystem management (Chen et al. 2014). The interplay between ESS and land and water management (or integrated water resources management, IWRM) has been demonstrated in a number of studies (e.g. Jewitt 2002; Nakamura 2003; Le Maitre et al. 2007; Gordon et al. 2010; Wainger et al. 2010; Siew and Döll 2012). Coupling ESS and IWRM concepts could help to overcome the conflict between "development" and "environment", i.e. between "freshwater for humans" versus "freshwater for nature". The integrative concept facilitates the identification and negotiation of trade-offs between management options. It can also be used to develop policies to align private incentives with societal objectives (Engel and Schaefer 2013). The question is how to implement ESS into practical land and water management in light of the existing "implementation gaps" (Cook and Spray 2012).

The operationalization of ESS in strategic management decisions calls for the cooperation among multiple disciplinary scientists as well as the engagement of stakeholders in an iterative process of modeling and valuing ESS (Daily et al. 2009). In this process, scientific knowledge about coupled social-ecological systems is synthesized using modeling methods, while perceptions and needs of stakeholders are taken into account. Such approach of integrating knowledge from inside and outside of academia is termed transdisciplinary research (TR) approach (Thompson Klein et al. 2001; Hirsch Hadorn et al. 2006). TR has been applied in various problem fields in Europe, North America, South Africa, and Asia to support for instance sustainable agriculture development (Vandermeulen and van Huylenbroeck 2008), regional planning (Wiek and Walter 2009), conservation planning (Steventon 2008; Reyers et al. 2010), and water management (Cain et al. 2003). The potential contribution of TR to the development of sustainable socio-economic strategies in China is recognized (Jiang 2009).

A TR approach has been conceptualized to support the integration of ESS into land and water management under climate change and uncertainty in the Tarim

River Basin, Northwest China (Siew and Döll 2012). Competing uses of limited water resources for agriculture development and nature protection between upstream and downstream users is the major problem in the arid basin. Our TR project Sustainable Management of River Oases along the Tarim River (SuMaRiO) started in 2011. We are bringing scientists from multiple disciplines and stakeholders from multiple sectors together to develop knowledge-based management strategies through a recursive process. Scientific and stakeholder knowledge is integrated using participatory methods including actor modeling (Titz and Döll 2009), actor-based modeling (Döll et al. 2013), Bayesian Network (Düspohl et al. 2012), and participatory scenario development. The aim of this paper is to describe our experiences with the implementation of our TR approach, highlighting the challenges faced and adapted research strategies.

In the next section, the current situation in the Tarim River Basin is described. We then elaborate on the execution of the TR approach in the study area. Subsequently, outcomes and challenges to the implementation of the approach are presented, and finally conclusions are drawn.

"Development" and "Environment" Trade-Offs in the Tarim River Basin

With an approximately one million km^2 and eight million inhabitants, the Tarim River Basin is the largest inland basin in China (Fig. 20.1a). The lowland part of the basin is characterized by low annual precipitation (less than 50 mm per year) and high potential evapotranspiration (more than 2,000 mm/year). The Tarim River Basin consists of four tributaries (Aksu, Hotan, Yarkant, and Kaidu-Kenqi Rivers) and the mainstem (Tarim River) which flows eastwards to the end lake Taitema (Fig. 20.1a). Glacier and snow melt feeds the tributaries. The Kenqi River is connected to the Tarim River by a constructed channel that transfers water from Bosten Lake to the lower reaches of the Tarim River. Long-term average annual river discharge flowing from the four tributaries into the mainstem Tarim is 4.7 km^3/year (Deng 2009). Three quarters of the annual discharge occurs in July and August (Thevs 2011).

Agriculture production, which depends completely on irrigation, is the key driver for socio-economic development in the water scarce river basin (Zhuang et al. 2010). Major products are cotton, grain crops and horticultural products such as pears, apricots, and walnuts. Along the tributaries, 15,000–22,500 m^3 of water is used per year for irrigating one hectare of cultivated land (Jiang et al. 2005), including the water required for leaching salt before the growing season. In the Aksu river subbasin, 5.2 km^3 of 8.4 km^3 of discharge is used for agriculture. The amount of water use in the upstream tributary basins has increased strongly over the last decades; while river discharge upstream of the Aksu oasis has significantly increased, most likely because of glacier mass losses due to anthropogenic climate change, river discharge downstream at the confluence of the Aksu into the Tarim

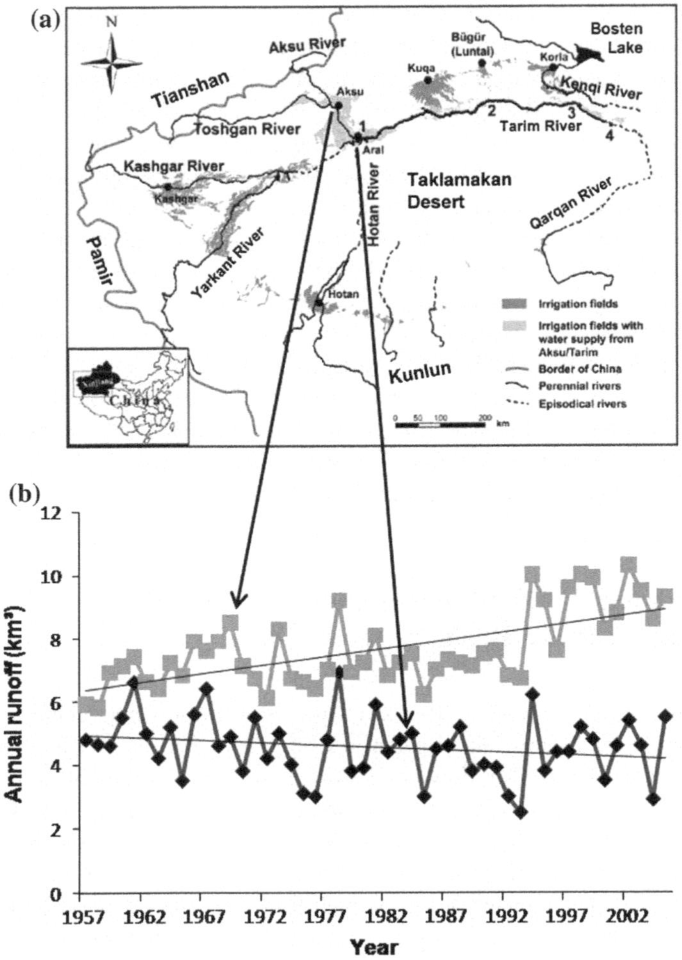

Fig. 20.1 **a** Location of the Tarim River Basin. **b** Annual runoff at the confluence of Toshgan and Aksu River (-■-) and at the Station Aral (-◆-), 1957–2005 (modified from Thevs 2011)

has significantly decreased (Tang and Deng 2010; Fig. 20.1b). In the mainstem Tarim basin, approximately one fourth of river discharge is used for agriculture and the rest by the riparian vegetation (Deng 2009).

The exploitation of land and water resources for irrigation agriculture has serious impacts on the environmental conditions in the Tarim River Basin (Shen and Lein 2005). The most prominent problem, which caught international and national attention, is the deterioration of riparian vegetation ("Green Corridor") in the lower reaches of the Tarim River and the drying up of Taitema Lake. The 436-km long "Green Corridor", which mainly consists of poplar trees (*Populus euphratica*), is the shelterbelt on both sides of the Tarim Desert Highway. It protects the road from wind and sand and therefore safeguards "the lifeline for oil and gas exploitation, traffic, and economy in Southern Xinjiang" (Zhuang et al. 2010).

To restore the "Green Corridor", the central government of China invested 10.7 billion RMB Yuan (1.29 billion USD) for a water conveyance project since 2001, under the Integrated Environment Restoration Plan (Lu et al. 2010). Water is channeled from upstream Tarim River through the diked midstream as well as from Bosten Lake via Kenqi River to the lower reaches of the Tarim River. The water diversion project has successfully recharged groundwater in the floodplain and therefore improved the quantity and quality of the riparian vegetation (Tao et al. 2008). However, it had a negative impact on Bosten Lake and the upstream water users. Water use conflicts arose between Kenqi river basin and the "Green Corridor" in 2004–2005, when the basin experienced a dry period (Tao et al. 2008).

Development, use, and management of water resources in the Tarim River Basin are guided by the principle of "four tributaries and one mainstem". Formulated based on a water quota system, an annual water allocation plan is agreed upon by the Tarim Basin Water Resources Commission (TBWRC) each year to regulate the distribution of water to different regions, different users (irrigation, industry, households, environment), and different types of farm operations (local and state farms) (Thevs 2011). TBRWC consists of a number of governmental organizations from the water, agricultural, forestry, and environmental sectors. In practice, actual water abstraction may significantly differ from the agreed upon water allocation plan. The central argument is how much water should be allocated to irrigation agriculture and to riparian vegetation,[1] which is regarded as a trade-off between "development" and "environment".

Integrating ESS concept into land and water management has been considered as a new approach to overcome the conflict between development and environment brought on by water scarcity in the Tarim River Basin (Siew and Döll 2012). River discharge generated in the upstream mountain areas should be allocated such that multiple ESS (e.g. crop production, dust retention, climate regulation, etc.) are maximized. By determining ESS, preferably in terms of monetary unit per unit of land or water used, awareness about the relative importance of ESS to policy makers can be raised (de Groot et al. 2012), decisions about allocating resources between competing uses can be better supported (Farley 2008), and the efficient use of funds for nature protection and restoration can be improved (Crossman et al. 2011).

Implementation of a Transdisciplinary Research Approach

The TR approach of SuMaRiO is illustrated in Fig. 20.2. The initial approach focused on the execution of a stakeholder dialogue comprising interviews and

---

[1] For the Tugai vegetation along the river, its regeneration and growth depend not only on the 'sufficient' amount of water received but also the right timing of water release from summer flood (Thevs 2011).

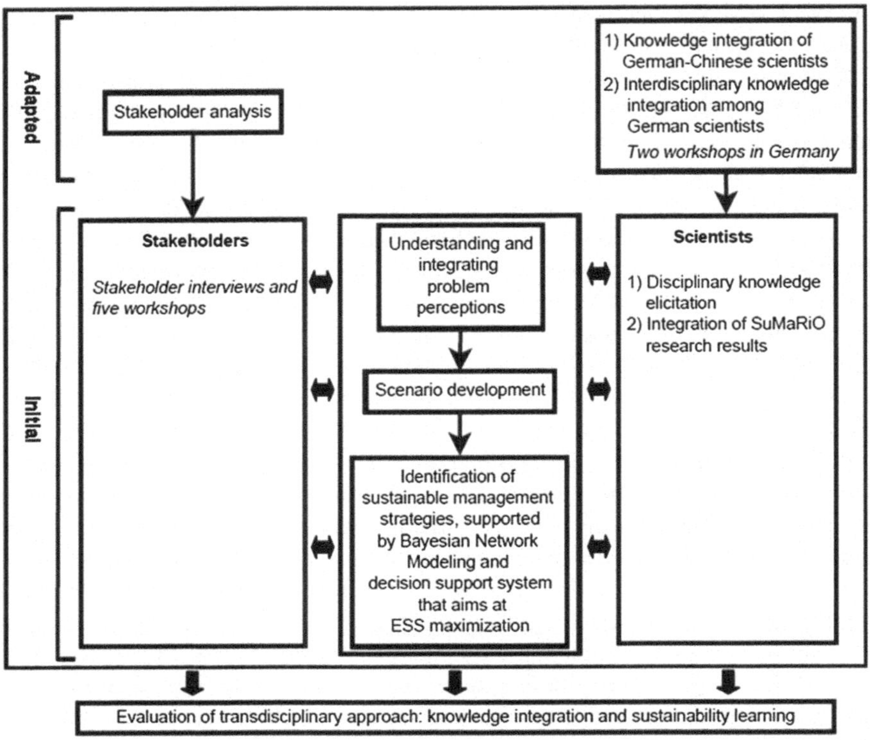

Fig. 20.2 Initial and adapted transdisciplinary approach for the support of integrating ecosystem services (ESS) into land and water management strategies in the Tarim River Basin, Northwest China. Chinese stakeholders are represented in the *box on the left* and scientists in the *box on the right*. The *box in the middle* depicts the process of integrating scientists and stakeholder knowledge on land, water, and ESS leading to the identification of sustainable management strategies. Tarim DSS (decision support system) is a software tool to be developed in the project SuMaRiO to support ESS maximization by stakeholders. The success of sustainability learning in terms of knowledge integration will be evaluated upon completion of the project

workshops. After eliciting problem perceptions of stakeholders and integration of SuMaRiO research results, scenarios under future climate and socio-economic change should then be developed. The identification of sustainable land and water management strategies is to be supported by a decision support system (Tarim DSS) that aims at maximizing identified ESS. Bayesian Networks (BNs) modeling method[2] is incorporated in the software system to account for uncertain knowledge about ESS.

The initial approach has been adapted during the first phase of the project. A stakeholder analysis was performed in addition to elicitation of problem

[2] BNs modeling method has become a core method in transdisciplinary research (Düspohl et al. 2012).

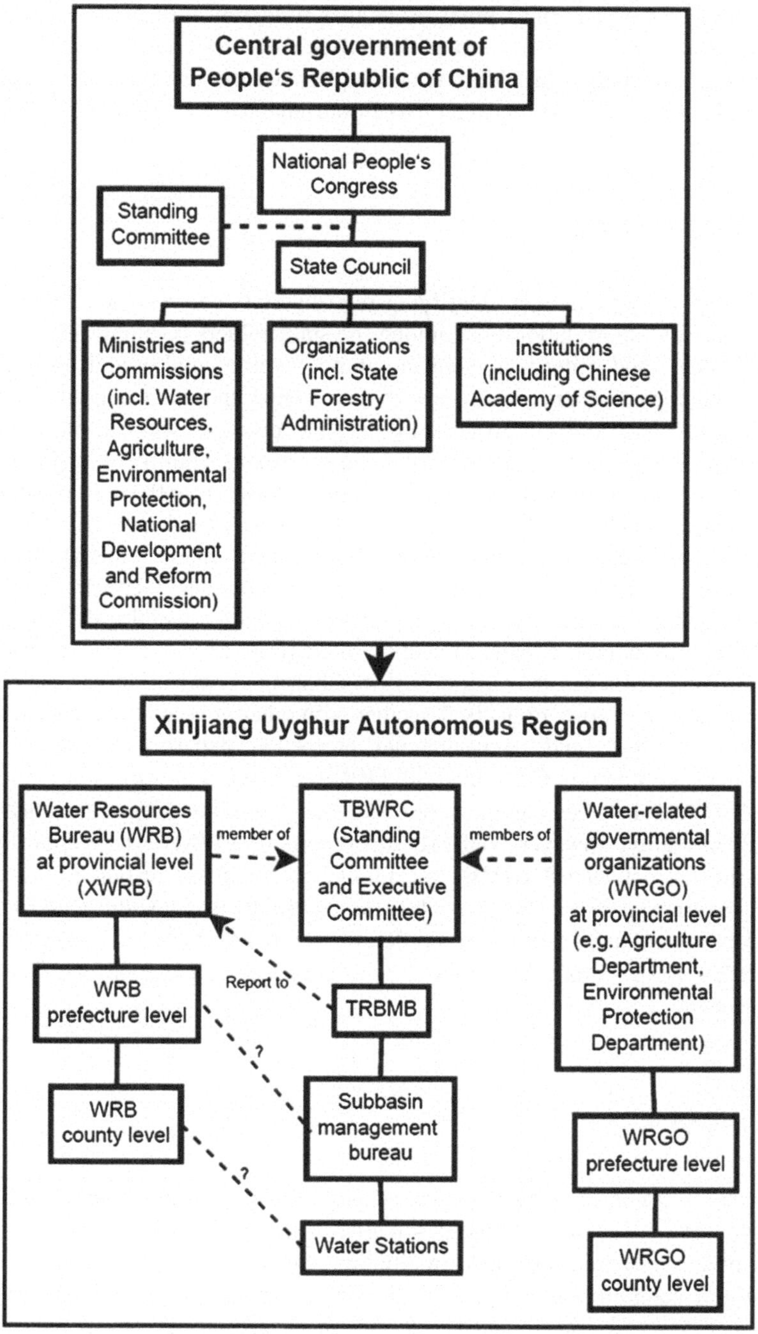

Fig. 20.3 Hierarchical structure of water management related governmental organizations in Xinjiang and China. "?" depicts unclear horizontal administrative interrelationships. (*TBWRC* Tarim Basin Water Resources Commission, *TRBMB* Tarim River Basin Management Bureau)

perceptions of German and Chinese scientists. Integration of knowledge of Chinese scientists with that of German scientists is as important as integration of stakeholder knowledge. In addition, we emphasized the integration of interdisciplinary knowledge of German scientists in our adapted approach (Fig. 20.2).

Stakeholder Analysis

Stakeholder analysis aims at identifying relevant stakeholders, their interests and agenda, and the interrelationships among the stakeholders (Grimble 1998). Stakeholders in land and water management can be governmental organizations, water user associations, farmer associations, environmental and other non-governmental organizations, and citizen groups. In our TR process, we can only involve governmental organizations from provincial, prefecture, county, or basin levels as stakeholders. Governmental organizations include those from the sector water, crop production, animal husbandry, environment, and forestry (including fruit trees). Our key stakeholders are the Xinjiang Water Resources Bureau (XWRB) at the provincial level and the Tarim River Basin Management Bureau (TRBMB). TRBMB is a basin organization who is in charge of preparing and implementing decisions made by TBWRC, in particular water allocation plans (Fig. 20.3).

TRBMB coordinates the operations of subbasin organizations as well as the execution of engineering projects. Together with other governmental organizations such as agriculture and environmental protection departments, XWRB is a member of the Standing Committee of TBWRC. Since the institutional reform in 2011, TRBMB has been empowered and has the same hierarchical status as XWRB. In reality, however, TRBMB still reports to XWRB. The responsibilities and mandates between TRBMB and administrative water resources bureaus at lower governance levels are not clear to us. Likewise, institutional functions related to water management across different sectors are also overlapping (c.f. Yan et al. 2006). Both sectoral and cross-sectoral cooperation and coordination need to be improved.

TR Process

Altogether 13 interviews were conducted in Xinjiang (9 in November 2011 and 4 in November 2012) with Chinese scientists coming from different institutions (academy and research institutes) in Xinjiang and various disciplines (hydrology, agricultural economy, ecology, and climate specialist). A causal network (perception graph) was constructed together with each interview partner during the interview which took about 2–3 h. The perception graph depicts the perspectives of interview partner with regard to the goals of land and water management, factors affecting the goal factors, possible action options that can lead to the

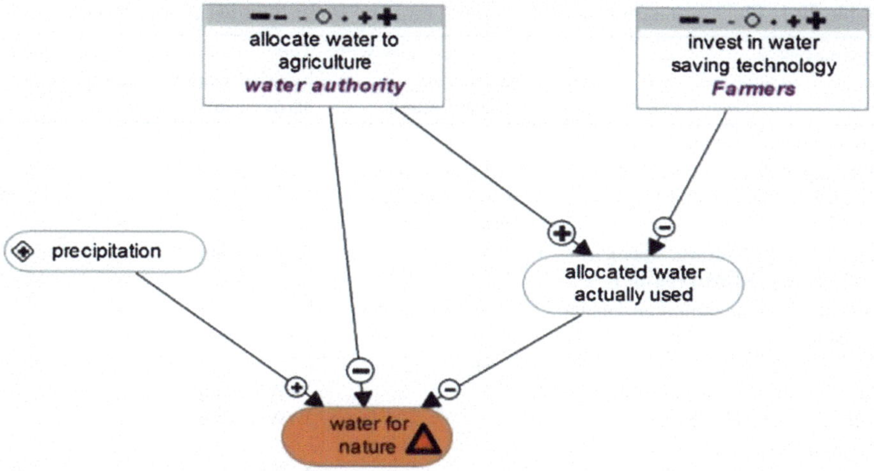

Fig. 20.4 An example of a simple perception graph depicting the problem perspective of an environmental protection organization in a fictive case of the problem field water management in arid regions. The *arrows* qualified with ⊕ or ⊖ sign in different sizes depict the intensity of the correlation between each element of the goal factor, influence factor, actions, and prospect. The perception graph expresses the perception that water allocation to agriculture by the water authority decreases strongly water for nature, and that farmers can increase water for nature somewhat by investing in water saving technology. Increasing precipitation will also lead to somewhat more water for nature (Siew and Döll 2012)

achievement of the goals, and the causal links between all these elements. After the interview, the paper-format perception graph was converted into a digital version using DANA software (http://dana.actoranalysis.com, Bots et al. 2007; Döll et al. 2013). An example of a perception graph is illustrated in Fig. 20.4.

Individual perception graphs generated from interviews in November 2011 were combined after all interviews were done, as first basis for a joint problem perception. The overall perception graph, which comprises all actions and factor mentioned in the individual graphs, was updated by integrating results gained from the four interviews conducted in November 2012. In between November 2011 and November 2012, we also constructed an overall graph representing the perspectives of German scientists within SuMaRiO project. The German overall perception graph was discussed in a SuMaRiO researcher workshop conducted in Germany in February 2012. Both German and Chinese overall perception graphs provide a basis for the development of the Tarim DSS.

The first version of German and Chinese overall perception graphs were presented at the first workshop organized in Urumqi (the capital city of Xinjiang) in March 2012. The workshop was attended by eight Chinese scientists who were interviewed and four Chinese scientists who were interviewed later in November 2012. Additionally, two representatives from TRBMB and two representatives from other governmental organizations were present at the workshop (Table 20.1). A World Café format (Welp et al. 2006) was used to encourage interactive small

Table 20.1 Chinese stakeholders and scientists involved in the transdisciplinary research

Institutions	No. of interviews with generation of perception graph	Total no. of persons participated in half-day workshops (workshop participated)
Governmental organizations		
Xinjiang Water Resources Bureau (XWRB)/provincial level	1 Informal interview (no perception graph)	1 (Workshop 2 and 3)
Tarim River Basin Management Bureau (TRBMB)/basin level	0	8 (Workshop 1, 2 and 3)
Forestry Administration/provincial level	1 Informal interview (no perception graph)	0
Agriculture Bureau/prefecture level	0	1 (Workshop 3)
Animal Husbandry Bureau/ prefecture level	0	1 (Workshop 3)
Wild Plants and Animals and Nature Reserve Management Office (under Forestry Administration)/prefecture level	0	1 (Workshop 2 and 3)
Water Conservancy Bureau Water Management Station/county level	0	1 (Workshop 1, 2 and 3)
Academy and research institutes		
Chinese Academy of Science	2	3 (Workshop 1 and 2)
Xinjiang Normal University	2	2 (Workshop 1 and 2)
Xinjiang University	4	5 (Workshop 1 and 2)
Xinjiang Academy of Forestry	1	1 (Workshop1 and 2)
Xinjiang Agricultural University	1	2 (Workshop 1 and 2)
Xinjiang Academy of Agriculture	1	1 (Workshop 2)
Xinjiang Water Resources Research Institute	1	1 (Workshop 1 and 2)
Research unit of Xinjiang Bureau of Meteorology	1	1 (Workshop 1 and 2)

group discussion among all participants. During the discussion, workshop participants were asked to formulate the most pertinent questions that need to be answered by German scientists (i.e. SuMaRiO researchers) as well as the requirements for a decision support system for land and water management. Both types of information were also collected using questionnaires, which was formulated in Chinese. Additionally, information pertaining to the current problems faced in the Tarim River Basin was also captured. Discussion in small groups was found to be an effective way of encouraging intensive interaction and exchange of ideas and information among workshop participants. In a plenary session, participants may feel reluctant to express their opinions as a sign of showing respect for high-level officials or as a way to avoid "losing face". The questionnaire was used in particular to gather information from workshop participants who did not voice their views at all in plenary or small group discussion.

Based on the results of the first workshop and the interviews with Chinese scientists before the second workshop in November 2012, the Chinese overall perception graph was modified. The updated version of the overall perception graph was presented in the second workshop and intensively discussed by workshop participants. The participants included representatives of our two key stakeholders (deputy director of TRBMB who had also participated in the first workshop, and the vice president of XWRB). Two representatives, who are in charge of nature protection at prefecture level and water management at county level, respectively, were also present. On top of that, eight Chinese scientists of the first workshop were also present. The goal of the second workshop was to come up with a final joint problem perception of stakeholders and scientists. Besides, a list of ESS relevant to the Tarim River Basin was identified by workshop participants.

In February 2013, a second German SuMaRiO researcher workshop was organized to discuss the system description of the Tarim DSS, develop storylines of two scenarios, and identify possible land and water management measures from the perspective of German scientists. The outcome of this internal workshop was presented at the third workshop in Xinjiang in March 2013. The third workshop was attended by 12 representatives of institutional stakeholders (including vice president of XWRB, deputy director and chief engineer of TRBMB, deputy director and chief engineer of Tarim Mainstem Management Bureau, and representatives from agriculture, animal husbandry, and nature protection departments). The participants provided feedback to the system description and storylines. Additionally, they identified possible land and water management measures from their perspectives, each of them individually filling out a table. These were combined with the management measures identified by German and Chinese scientists. Together with climate and socio-economic scenarios, the impacts of a combination of management measures on ESS are to be analyzed in SuMaRiO. Issues pertaining to data needs for Tarim DSS were also discussed in the third workshop.

Two more workshops are planned to be conducted in Xinjiang with participation of all selected stakeholders by 2015. As depicted in Fig. 20.2, the overall TR process with regard to sustainability learning and knowledge integration will be evaluated upon completion of the project. Throughout the process, TR activities and their outcomes are continually documented and reflected on.

Outcomes of the TR Process

Our TR process as described in the previous section is on-going. At this stage, we are able to offer an insight about the following outcomes of our TR process, which can contribute to the integration of ESS into land and water management: an improved understanding on issues at hand, a joint problem perception of stakeholder and scientists, identified knowledge gaps, and improved communication and knowledge exchange among stakeholders and scientists.

An understanding on issues related to land and water management in the Tarim River Basin from the perspectives of Chinese stakeholders and scientists was gained at the beginning of our TR process. The issues of their concerned, including socio-economic, environmental, and institutional issues, were articulated through informal exchange and formal interviews as well as by means of group discussion and questionnaires at workshops. According to Chinese stakeholders and scientists, the deficiency of the institutional arrangement is a major obstacle that prevents integrated land and water management in the Tarim River Basin. Institutional functions across different sectors are overlapping, while there is a lack of cross-sectoral communication among government institutions. On the other hand, it was perceived that downstream reaches are mostly affected by environmental and socio-economic problems such as land desertification, loss of riparian vegetation, and poverty. The understanding of the diverse issues belongs to the first step in our TR process that focuses on defining the problem taking into account different perceptions of scientists and stakeholders as well as the interests and goals of relevant stakeholders (Siew and Döll 2012).

A joint problem perception of scientists and stakeholders was obtained by integrating the knowledge of stakeholders into the overall perception graph of Chinese scientists. The graph depicts the cause-effect relationships between socio-economic and environmental factors, which are impacted by possible action options and which have impacts on achieving goals of land and water management. The graph as a causal network was used to facilitate the discussion, among others, about prevailing issues related to the allocation of water for agriculture irrigation use and water for nature use (i.e. for the restoration and protection of natural vegetation along the Tarim River, especially in the lower reaches). Causal networks visualize the structure of the present knowledge of the involved stakeholders and scientists about the complex human-environment systems and thus can be used to come up with an accepted problem definition (Welp et al. 2006).

By generating the overall perception graph and discussing it with stakeholders and scientists, gaps in the present knowledge were revealed. It was found that the issue of water allocation receives more attention as compared to, for example, water quality issues in the Tarim River Basin. The discussion also pointed out that there is a lack of investigation on the trade-off of the bundle of ESS provided by different ecosystems in the entire Tarim River Basin. Most studies in the Tarim River Basin focus on the valuation of the effect of water transfer on crop production and to a lesser extent on the growth of natural vegetation in the lower reaches of the Tarim River (e.g. Xu et al. 2008). A total of seven ESS is currently considered in the system description of SuMaRiO integrated model (Tarim DSS). A detailed description of the DSS is out of the scope of this paper.

Based on the feedback from interview partners and workshop participants, the applied TR approach contributed to improving cross-sectoral and multi-disciplinary communication and knowledge exchange. By sharing divergent perspectives on land and water management issues as well as the current development in the field of ESS, mutual understanding and learning among stakeholders and scientists have been strengthened. Nevertheless, knowledge on land and water management as well as ESS that exists at different institutions in Xinjiang (inside and outside of academia) has so far only been partially integrated in light of the challenges experienced in the implementation of our TR approach.

Challenges to Implementation of the TR Approach

The implementation of the TR approach in the Tarim River Basin has been very challenging due to several reasons. We only had access to representatives of governmental organizations. They were selected in a somewhat biased way, as most of them belong to a close network. In addition, except for the deputy director of TRBMB and the vice president of XWRB, the participating representatives of stakeholders do not have strong decision power regarding land and water management issues in the Tarim River Basin.

We planned to conduct 15–20 interviews before conducting workshops with representatives of different governmental organizations to elicit their problem perceptions. The resulted perception graph should include their goals pertaining to land and water management, factors that affect the goal factors, and possible action options that can lead to the achievement of the goals. However, formal interviews with stakeholder representatives were not possible until now. We only managed so far to exchange information with representatives from water and forest sectors informally. The main reason given was that SuMaRiO was not officially endorsed by the Chinese central government. The official recognition from a ministry in Xinjiang has, up to now, not been able to encourage the representatives to get involved in our TR process. On the other hand, stakeholder representatives did show their interest in our TR approach and the Tarim DSS. However, collaborating with foreign scientists without authorized approval from the powerful central government might probably be seen as a risk to their professional positions.

We initially intended to provide scientific support to land and water management in the Tarim River Basin, including the Aksu river basin which is subbasin of the overall Tarim basin. The Aksu contributes most of the river discharge into the Tarim mainstem. However, stakeholders in Xinjiang/China do not wish to be supported by German researchers regarding water management in the Aksu subbasin because it is a transboundary basin (Kyrgyzstan/China). Therefore, they want to restrict analysis, modeling (DSS), and the definition of management strategies to the Tarim mainstem (Fig. 20.1a). This is problematic because irrigation water in the Aksu subbasin is four times larger than in the mainstem basin and has significantly reduced river discharge into the Tarim mainstem (Fig. 20.1b).

In general, data sharing has been a sensitive issue for Chinese stakeholders and scientists, but especially problematic for transboundary basins. It seems to be impossible to provide daily discharge data for transboundary rivers to foreign researchers. A solution has been suggested by the representative of XWRB, but no official authorization can be done by XWRB alone. Some data can be obtained through alternative channels. Nevertheless, without discharge data in a temporal resolution that is required for hydrological modeling, the progress of the project has been substantially delayed.

In our research, most of the tasks could not be carried out within the planned time frame. On the one hand, our research activities (interviews and workshops) were given low priority by stakeholder representatives. On the other hand, it was not uncommon that potential workshop participants received short notices by their superiors that prevented them from participation in our workshops. As a result, workshop plans needed to be changed ad hoc, and additional time and resources were required. This poses a challenge to the design of a participatory process which should avoid the emergence of “stakeholder fatigue” (Reed 2008; Lamers et al. 2010). Flexibility is essential for doing research in China generally (van den Hoek et al. 2012).

Chinese language was used during interviews and workshops. Two German scientists involved in the TR process are native speakers of Chinese, and translation was only required for short presentation and intervention by German senior researchers. Technical terms and concepts related to land and water were well comprehended by Chinese stakeholder and scientists with whom we communicated. Misunderstandings and arguments arose as the terms “transdisciplinary research” and “ecosystem services” were introduced. The term “transdisciplinary research” was not recognized by the Chinese although it was directly adopted from a Chinese reference (Jiang 2009). In Chinese literature, “ecosystem service functions” is commonly used, deviating from the English usage. Although consensus was not achieved with regard to the use of these two core terms, the underlying concepts and applied methods were well accepted as innovative by both Chinese stakeholders and scientists.

Conclusion and Outlook

Transdisciplinary research has the potential to support the implementation of ecosystem services concept in land and water management. By getting scientists and stakeholders involved in the research process, divergent interests and perceptions on the balance of economic development, nature conservation, and human welfare can be shared. The joint generation and integration of knowledge using a transdisciplinary research approach can subsequently help derive ecosystem services-based strategies to resolve human-environment conflicts, especially in such a water-scarce and fragile environment as in the Tarim River Basin.

Some outcomes that contribute to the integration of ESS into land and water management could be obtained during the implementation of the TR approach in the Tarim River Basin. At the same time, a number of challenges were encountered. TR processes need to be adapted as new knowledge and surprises emerge. We have adapted our initial approach by adding a stakeholder analysis, and by concentrating our effort on knowledge integration between German and Chinese scientists. We have also intensified interdisciplinary knowledge integration within SuMaRiO that focuses on integration of SuMaRiO results and on the development of a decision support system. In this TR process that involves individuals from two culturally different countries, we have tailored the TR methods to suit ways of communication in the local socio-cultural and institutional setting. We have experienced that small group discussions in the form of World Café help to overcome the reluctance of Chinese workshop participants to voice their views in a large group. Making workshop participants fill out questionnaires during the workshop allows collecting specific information even from those who did not participate in the discussion. Workshops with Chinese scientists and stakeholder representatives are best conducted in Chinese language, including the materials provided.

Transdisciplinary research (including methods of knowledge elicitation and integration) needs to be continually adapted responding to the challenges encountered. For TR in China, it is a prerequisite to have a committed and eminent scientist as project partner. He/she should have influence, authority and good connections ("guanxi") to initiate network and trust building with Chinese stakeholders and scientists.

Our transdisciplinary approach is on-going. By 2015, we plan to have conducted interviews with representatives of all selected stakeholders, in addition to two more workshops in Xinjiang.

Acknowledgments This research is funded by the German Federal Ministry of Education and Research (BMBF). The authors would like to thank all Sustainable Management of River Oases along the Tarim River (SuMaRiO) project members and Chinese counterparts for their input and assistance. Special appreciation is also extended to our project partner and the team who help organize interviews and stakeholder workshops in Xinjiang. We also thank the reviewers for their valuable comments.

References

Bots PWG (2007) Analysis of multi-actor policy contexts using perception graphs. In: Lin TY (ed) Proceedings of the international conference on intelligent agent technologies (IAT'07), Los Alamitos, 2007

Cain JD, Jinapala K, Makin IW et al (2003) Participatory decision support for agricultural management. A case study from Sri Lanka. Agric Syst 76(2):457–482

Chen YY, Jessel B, Fu BJ et al (eds) (2014) Ecosystem services and management strategy in China. Springer, Berlin

Cook BR, Spray CJ (2012) Ecosystem services and integrated water resource management: different paths to the same end? J Environ Manage 109:93–100

Crossman ND, Bryan BA, King D (2011) Contribution of site assessment toward prioritising investment in natural capital. Environ Model Softw 26(1):30–37

Daily GC, Polasky S, Goldstein J et al (2009) Ecosystem services in decision making: time to deliver. Front Ecol Environ 7(1):21–28

de Groot R, Brander L, van der Ploeg S et al (2012) Global estimates of the value of ecosystems and their services in monetary units. Ecosyst Serv 1(1):50–61

Deng MJ (2009) Tarim River in China: theory and practice of water management. Science Press, Beijing (Chinese)
Döll C, Döll P, Bots P (2013) Semi-quantitative actor-based modelling as a tool to assess the drivers of change and physical variables in participatory integrated assessments. Environ Model Softw 46:21–32
Düspohl M, Frank S, Döll P (2012) A review of Bayesian networks as a participatory modeling approach in support of sustainable environmental management. J Sustain Dev 5(12):1–18. doi:10.5539/jsd.v5n12p1
Engel S, Schaefer M (2013) Ecosystem services—a useful concept for addressing water challenges? Curr Opin Environ Sustain 5:696–707
Farley J (2008) The role of prices in conserving critical natural capital. Conserv Biol 22(6):1399–1408
Gordon LJ, Finlayson CM, Falkenmark M (2010) Managing water in agriculture for food production and other ecosystem services. Agric Water Manag 97:512–519
Grimble R (1998) Stakeholder methodologies in natural resource management. Natural Resources Institute, Chatham
Hirsch Hadorn G, Bradley D, Pohl C et al (2006) Implications of transdisciplinarity for sustainability research. Ecol Econ 60:119–128
Jewitt G (2002) Can integrated water resources management sustain the provision of ecosystem goods and services? Phys Chem Earth 27:887–895
Jiang YM (2009) Transdisciplinarity as a new form of research. Zhejiang Social Sciences 1:8–15 (Chinese)
Jiang LW, Tong YF, Zhao ZJ et al (2005) Water resources, land exploration and population dynamics in arid areas—the case of the Tarim River Basin in Xinjiang of China. Popul Environ 26(6):471–503
Lamers M, Ottow B, Francois G et al (2010) Beyond dry feet? Experiences from a participatory water management planning case in the Netherlands. Ecol Soc 5(1):14
Le Maitre DC, Milton SJ, Jarmain C et al (2007) Linking ecosystem services and water resources: landscape-scale hydrology of the Little Karoo. Front Ecol Environ 5(5):261–270
Lu ZH, Zhao LX, Dai J (2010) A study of water resource management in the Tarim Basin, Xinjiang. Int J Environ Stud 67(2):245–255
Millennium Ecosystem Assessment (2005) Ecosystems and human well-being: synthesis. Island Press, Washington DC
Nakamura T (2003) Ecosystem-based River Basin management: its approach and policy-level application. Hydrol Process 17:2711–2725
Reed MS (2008) Stakeholder participation for environmental management: a literature review. Biol Conserv 141:2417–2431
Reyers B, Roux DJ, Cowling RM (2010) Conservation planning as a transdisciplinary process. Conserv Biol 24(4):957–965
Shen YL, Lein H (2005) Land and water resources management problems in Xinjiang Uygur Autonomous Region, China. Norsk Geografisk Tidsskrift—Norw J Geogr 59(3):237–245
Siew TF, Döll P (2012) Transdisciplinary research for supporting the integration of ecosystem services into land and water management in the Tarim River Basin, Xinjiang, China. J Arid Land 4(2):196–210
Steventon JD (2008) Conservation of marbled murrelets in British Columbia. In: Pourret O, Naim P, Marcot B (eds) Bayesian networks: a practical guide to applications. Wiley, Chichester, pp 127–148
Tang DS, Deng MJ (2010) On the management of water rights in the Tarim River Basin. China Water Power Press, Beijing
Tao H, Gemmer M, Song Y et al (2008) Ecohydrological responses on water diversion in the lower reaches of the Tarim River. Water Resources Research, China. doi:10.1029/2007WR006186

TEEB (2010) The economics of ecosystems and biodiversity: mainstreaming the economics of nature—a synthesis of the approach, conclusions and recommendations of TEEB. Progress Press, Mriehel (Malta)
Thevs N (2011) Water scarcity and allocation in the Tarim Basin: decision structures and adaptations on the local level. J Curr Chin Aff 40:113–137
Thompson Klein J, Grossenbacher-Mansury W, Häberli R et al (eds) (2001) Transdisciplinarity: joint problem-solving among science, technology and society. Birkhäuser Verlag, Basel
Titz A, Döll P (2009) Actor modelling and its contribution to the development of integrative strategies for management of pharmaceuticals in drinking water. Soc Sci Med 68:672–681
Van Den Hoek L, Baumgartner J, Doucet-Beer E et al (2012) Understanding the challenges and rewards of social-ecological research in China. Soc Nat Resour. doi:10.1080/08941920.2012.658985
Vandermeulen V, van Huylenbroeck G (2008) Designing transdisciplinary research to support policy formulation for sustainable agricultural development. Ecol Econ 67:352–361
Wainger LA, King DM, Mack RN et al (2010) Can the concept of ecosystem services be practically applied to improve natural resource management decisions? Ecol Econ 69:978–987
Welp M, de la Vega-Leinerta A, Stoll-Kleemannb S et al (2006) Science-based stakeholder dialogues: theories and tools. Glob Environ Change 16:170–181
Wiek A, Walter AI (2009) A transdisciplinary approach for formalized integrated planning and decision-making in complex systems. Eur J Oper Res 197:360–370
Xu HL, Ye M, Li JM (2008) The water transfer effects on agricultural development in the lower Tarim River, Xinjiang of China. Agric Water Manage 95(1):59–68
Yan F, He DM, Beth K (2006) Water resources administration institution in China. Water Policy 8:291–301
Zhuang L, Dong YS, Yin FH et al (2010) Historical evolution and the effects of ecological management in Tarim Basin, China. Chin Sci Bull 55(36):4097–4103

Part IV
Governing Water in the Anthropocene

When talking about the Anthropocene, realizing the magnitude of human impact on the Earth system, one has to come to the conclusion that a sustainable future is impossible without understanding the human drivers and dynamics of natural resource use. For water this is especially true since the "water crisis" is often described as a crisis of governance, rather than one of scarcity or technology. Water is a complex global and interconnected system, vital for human development as well as for other physical and functional systems, so that it cannot simply be treated as a sector in itself through technical and managerial panacea. These interconnections also imply the involvement and interest of a wide range of stakeholders from different parts of society and different parts of the world with divergent approaches to both water problems and their solutions. A multi-level approach to water governance is therefore inevitable. These complexities pose challenges for both water governance itself and the research dealing with it. While research on environmental governance has progressed toward adaptiveness and polycentric dynamics, embracing both the complicacies of governance regimes, there is still a pressing need for context-specific analyses that at the same time allow for sufficient comparability as well as for value-driven perspectives on water governance. Water governance describes the formal and informal interactions between actors dealing with water-related issues within certain structures. However, normatively it is about how socio-ecological systems *ought* to be managed, including ethical questions and value systems. These aspects become increasingly important when looking at the global dimension of governance, which has to address inequalities in both allocation and access to water, corresponding power relations, different valuations of water and the contestations of all of the above. Covering the wide range of complex governance issues, the section will move from empirical research at multiple scales to more normative perspectives on water governance.

The first two papers address the issue of multi-level collective action in the face of global change. Dell'Angelo et al. describe multi-level institutional arrangements in irrigation using the example of Kenya where legislation favors a decentralized system of governance. The authors describe the governance structure of community water projects and illustrate the challenges for adaptive capacity with respect to different social and environmental disturbances. When looking at local water institutions, Garrick et al. argue that the coordination of multi-level collective action in water governance has become more important in a context of global change and intensified competition for water. The paper examines adaptive federal-state relations in semi-arid regions, finding that local dilemmas of water governance are increasingly difficult to insulate from global environmental and political changes.

Moving up in scale from the subnational level, the next two papers deal with transboundary water issues. Suhardiman and Giordano address the issue of how the changing role of the nation state in the face of increasing international interdependencies shapes hydropower decision-making in the Mekong basin. The link between scientific knowledge and policy decisions in river basin organizations is the topic of the paper by Schmeier. She argues that the degree to the management and development of a river basin are based on scientifically sound decisions, depends on both the very nature of the science provided by the river basin organizations itself and the organization's institutional design.

The section concludes with three papers addressing the normative aspects of water governance. Recapitulating the development of the Millenium Development Goals on Water and Sanitation, Obani and Gupta compare the access to both water and sanitation in a human rights context. Taking into account the differences between the two in terms of international law, physical infrastructure needed, costs of the service, and willingness to pay, they call for an integrated management of water, sanitation, and hygiene.

Dellapenna takes the legal questions of water governance to a more general level and traces the evolution and characteristics of national, transnational, and international water law, their interlinkages and where they might be headed in terms of rising to the challenges of change. A critical analysis of the much-quoted "water crisis" is provided by Bruns and Frick. Using the example of Accra, Ghana, they examine how the term water crisis is being framed and how that discourse impacts on water policies. The findings suggest that research on the socio-political nature of water governance, i.e., how current and historic water governance is related to social power, and how this is shaping the crisis, is severely underrepresented. A shift to critical social science in water research is therefore essential to tackle the water crisis.

Meisch's paper deals with the normative implications of the Anthropocene concept for water governance. He argues that the objective of governance per se to create legitimacy necessitates ethical debates on how rules are being made and enforced. The integration of value discourses into water governance could, therefore, be a tool to deal with the social and political challenges of water in the Anthropocene.

Chapter 21
Multilevel Governance of Irrigation Systems and Adaptation to Climate Change in Kenya

Jampel Dell'Angelo, Paul F. McCord, Elizabeth Baldwin, Michael E. Cox, Drew Gower, Kelly Caylor and Tom P. Evans

Abstract Multilevel governance of common-pool natural resources has been shown under certain conditions to sustain resources over time even when faced with various social and environmental disturbances or shocks. In the case of irrigation systems, evidence shows that multilevel institutional arrangements that include communities in a decentralized system of governance can function better than centralized systems. Kenya has implemented a legislative framework for water governance that decentralizes many aspects of water management to local levels, resulting in a multilevel institutional regime. Community water projects are empowered to manage some aspects of water resources for irrigation and domestic use—purportedly a level at which decision-makers are better suited to adapt to local dynamics. However, climate change and population increase constantly challenge the ability of these water projects to modify rules for water allocation so that all water demands are met. In this chapter, we describe the governance structure of community water projects near Mt. Kenya and illustrate the challenges for adaptive capacity with respect to different social and environmental disturbances.

J. Dell'Angelo (✉) · T. P. Evans
The Vincent and Elinor Ostrom Workshop in Political Theory and Policy Analysis, Indiana University, Bloomington, IN, USA
e-mail: jdellang@indiana.edu

P. F. McCord · T. P. Evans
Department of Geography, Indiana University, Bloomington, IN, USA

E. Baldwin
School of Public and Environmental Affairs, Indiana University, Bloomington, IN, USA

M. E. Cox
Environmental Studies Program, Dartmouth College, Hanover, NH, USA

D. Gower · K. Caylor
Department of Civil and Environmental Engineering, Princeton University, Princeton, NJ, USA

A. Bhaduri et al. (eds.), *The Global Water System in the Anthropocene*,
Springer Water, DOI: 10.1007/978-3-319-07548-8_21,

Introduction

Rural livelihoods in many parts of the world are dramatically affected by climate variability and its corresponding impact on agricultural production. This is particularly the case in the semiarid tropics (SAT), which contain 22 % of the world's population and high concentrations of chronic poverty and inadequate food consumption (Falkenmark and Rockström 2008). Much of the vulnerability of smallholders within the SAT is driven by climate and hydrological dynamics both directly through rainfall variability and indirectly through additional human- or climate-induced processes that affect water availability. While smallholders may adjust cropping practices to adapt to changes in rainfed agricultural systems, a more vexing problem is how water governance arrangements in irrigated agricultural systems can adapt given the hydrological and social complexity in these systems.

User groups have several ways to respond to decreased water supply. For example, users may reallocate how water is distributed within the user group. Alternatively, agriculturalists may shift to different crops that are more drought tolerant or simply have lower water requirements. In areas with formal water rights that are distinct from land ownership, landowners may sell those rights and abandon agricultural land uses. In the presence of multilevel institutions, user groups may decide to enact strict quotas to limit how much water individual users can withdraw (which may or may not be equitably distributed across the user group). As is evident from these possible adaptations, some have strong institutional forcings while others are driven more by individual landowner decision making.

There are good reasons to believe multilevel institutional arrangements that include elements of local-level governance would produce more adaptive and robust governance regimes. First, there is more redundancy in nested systems (Andersson and Ostrom 2008), which provides insurance against failure. Second, there is more potential for learning because a variety of interventions can be tried in many different locations, ideally producing in the aggregate a giant field experiment from which one can learn valuable lessons about which solutions work and under which conditions. And finally, it is possible to take advantage of economies of scale for certain types of governance functions, which is difficult to do in fully decentralized systems. While many of these ideas are firmly established and widely accepted as theoretical principles of governance, surprisingly few empirical studies exist that test these ideas.

In this chapter, we describe a system of local-level water governance from research near Mt. Kenya located in an SAT region that features a social gradient transitioning from sedentary farming and relatively wealthier smallholders to pastoral tribes that have only recently begun engaging in cultivation. This area also exhibits a stark environmental gradient that transitions from higher precipitation levels near Mt. Kenya to greater water scarcity in the semiarid and arid rangelands characterized by multiple social and environmental gradients (Fig. 21.1). In this area, governance structures have been implemented that heavily emphasize local-

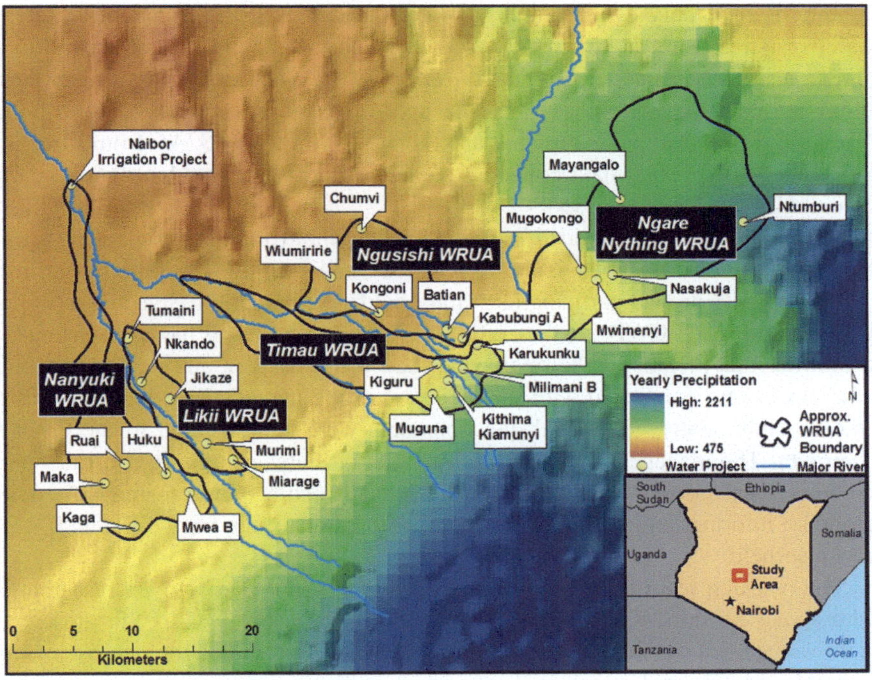

Fig. 21.1 Approximate boundaries of selected water resource users' association and communities within them in the region northwest of Mt. Kenya. Precipitation levels displayed to emphasize environmental gradient

level actors within a multilevel governance structure. These governance arrangements can be considered a prospective laboratory of institutional innovation, a governance model that potentially could be adopted in other regions of Africa. As such, we describe the governance structure in these systems and investigate the potential for adaptive capacity with respect to different social and environmental disturbances. The overall objective of this chapter is to describe the limitations and prospects for successful adaptive management in a social-ecological system that includes local-level governance within a multilevel nested institutional system. We bring attention to the legal and institutional water governance setting at the Kenyan national level and focus on community-level decision-making dynamics and the role that environmental information plays in those decisions.

Methods

This work focuses on the Upper Ewaso Ng'iro Basin in central Kenya, which extends from the northwest foot slopes of Mt. Kenya to the semiarid plains of the Laikipia plateau and then to the northern arid lowlands. We have documented the conditions of twenty-five community water projects (CWPs) from five different water resource users associations (WRUAs). Generally, WRUAs correspond to a particular river basin and coordinate the use of water resources from a single river. In this research, household-level and manager-level surveys were conducted in the Nanyuki, Likii, Timau, Ngusishi, and Ngare Nything WRUAs from June to September 2013. Total data collected consists of 850 household surveys and 71 community-level interviews.

Household surveys included sections on household members' characteristics, land and livestock assets, agricultural practices, and a central section on water use and community project participation. Households were randomly selected from the full extent of each CWP. At the community level, each CWP has a management committee that includes a chairman, vice chairman, treasurer, and secretary; group interviews were conducted with the members of the management committee as a whole. Additionally, individual questionnaires were administered to the CWP's chairman. The manager surveys addressed the attributes of the project, rules and organizational factors, trends, and management structures and practices. A focus group on information and decision making with 24 chairmen from different CWPs was also held.

The research team that administered the surveys included Kenyan researchers who spoke Swahili, Kikuyu, Kimeru, and Maasai. In every CWP, the research team was guided by members of the community who presented the purpose of the research to household members.

The analysis of the evolution of Kenya's legal water institutions is based on archival research conducted in Nairobi during June 2013 at the Kenyan National Archives and at the University of Nairobi. Additional archival research included review of strategic planning documents from the Ministry of Water and Irrigation and the Water Resources Management Authority offices in Nairobi. Original documents were also supplemented with secondary source research. In this study, we use qualitative tools such as qualitative content analysis, as well as summary statistics to demonstrate the governing and adaptation strategies within the Upper Ewaso Ng'iro Basin.

Local-Level Governance and Irrigation Systems

A substantial body of work has examined community-based irrigation systems (Coward 1979; Ostrom 1992; Shivakoti and Ostrom 2002). Much of this work overlaps with a broader literature on community-based natural resource management generally. A central focus in this literature has been investigation of the conditions that enable a group of irrigators to coordinate their activities to effectively run the system when each has his or her own information, incentives, and interests.

Prominent among such conditions has been the structure of the irrigation governance system itself. Ever since Karl Wittfogel (1957) argued that the complexity of irrigation systems necessitates a hierarchical top-down form of management, scholars have debated whether centralized or more decentralized systems are more effective. Much of this debate focuses on whether decentralized systems can effectively avoid the tragedy of the commons popularized by Hardin (1968). Many scholars and practitioners have used Wittfogel and Hardin to argue that this is unlikely.

Empirical studies have challenged Wittfogel's and Hardin's theories, resulting in a shift toward decentralization and more participatory approaches to irrigation management (Shivakoti and Ostrom 2002; Svendsen and Nott 2000). This movement is a response to noted failures in highly centralized systems and the documentation of many (but by no means all) successful, community-based, decentralized systems (Coward 1979; Lam 1998; Mabry 1996; Ostrom 1992; Tang 1992; Wade 1988).

Even in decentralized, community-based systems, however multiple levels of institutional organization play an important role (Coward 1977, 1979; Cox 2014; Geertz 1959; Siy 1982). Coward (1977) discusses how many large, indigenous irrigation systems are composed of subcomponents, and how this social decomposition subdivides the physical system as well. Coward uses his own observations and several others' (Geertz 1967; Taillard 1972; Thavaraj 1973) to support this argument. A particular social subunit corresponds to a particular geographic extent of the irrigation system. Social subunits interact to build up multiple levels of organization, which then correspond to successively larger geographic extents.

Other factors have been found to affect the ability of community-based systems to function well. These include the presence of: (1) proportionality between costs and benefits experienced by actors; (2) accountable monitoring; (3) graduated sanctions; and (4) conflict resolution mechanisms (Ostrom 1992). Attributes of the irrigators themselves also are important. Coward (1977) argues that accountable leadership is a standard theme for long-lasting, indigenous irrigation systems. Wade (1988) comes to a similar conclusion, while Lam (1998) describes the importance of leaders in traditional Taiwanese irrigation systems, and Siy (1982) discusses the consistent presence of a set of officials that runs each long-lasting *zanjera* community irrigation system in the Philippines, a set that includes a president, secretary, and treasurer. Finally, scholars have found that, with all else equal, small- to medium-sized groups are generally better at sustainably managing an irrigation system, having fewer transaction costs and more effective reputation-building mechanisms (Araral 2009; Ostrom et al. 1994; Wade 1988).

Water Governance Legislation in Kenya, Before and After the 2002 Water Act

Kenya presents a particularly challenging context for water management. Population growth is placing increasing pressure on water resources that are already very limited. Much of the country is too arid for rainfed agriculture, but irrigated agriculture is possible in some catchments. It is in this context that local-level

Table 21.1 Governance reform highlights

Key governance functions	Main actor pre-2002	Main actor post-2002
Permit issuance	National water authorities	Watershed WRMA offices
Allocation within water user groups	Local leaders within water schemes	Local leaders within water schemes
Allocation between water user groups within a watershed	N/A	Watershed WRMA offices, in consultation with WRUAs
Monitoring permit compliance	Local water bailiffs	Shared by WRMA, WRUAs, water schemes
Imposing limits on water use during drought	National water authorities	WRUAs
Conflict resolution and prevention	N/A	WRUAs
Water conservation activities	N/A	WRUA- WRMA partnerships (varies by WRUA)

water management has arisen in order to cope with the competing demands for water resources (urban/municipal, domestic, and agriculture/livestock). Until the 2002 Water Act, Kenya had not substantially changed its water allocation policies since the 1950s, when Kenya was still under colonial rule (Nilsson and Nyanchaga 2009). The number of acres under irrigation increased rapidly post-independence, however, and by the 1980s, Kenya's water officials were only able to sporadically implement and enforce its water policies (e.g., Isiolo Water District 1982; Machakos Water District 1983; Meru Water District 1981), effectively creating a vacuum of formal water policies. Kenya reformed its water laws, and over the following decade the country has begun to implement a new and polycentric framework for water governance. Local-level governance of water resources now functions within the context of national-level formal legal structures related to water rights described in greater detail below (see Table 21.1).

Water Governance Before 2002

Prior to the 2002 Water Act, water allocation in Kenya was governed by a permit system dating to the colonial era (Nilsson and Nyanchaga 2009). The program allowed riparian landowners to abstract water, provided any such uses would not harm downstream users. The basic permit program had been modified only slightly in the 1970s, when the role of local officials was enlarged and many local Water Bailiff Offices were established (Government of Kenya 1972; Nilsson and Nyanchaga 2009).

The permit program had the features of a centrally administered, top-down system. Permit applications were passed through both local and national officials, creating a byzantine approach; permit approvals could take several years from start to finish. Illegal abstraction was common. Moreover, throughout the early 1980s, many water bailiffs lacked the staff and transportation needed to issue permits, detect illegal use, and monitor permit holders' compliance with permit terms. During times of drought, government officials sometimes called on upstream users to limit their water use, but without adequate ability to enforce compliance, these efforts were not enough to protect downstream users' water rights, and downstream users often were left without sufficient water (Isiolo Water District 1982; Machakos Water District 1983; Meru Water District 1981).

At the local level, groups of water users often worked together to install, maintain, and manage water diversion systems that were shared among community members. Between schemes within a watershed, however, the limited government presence gave rise to an essentially ruleless state. Individual water schemes—usually communities of smallholders practicing subsistence agriculture or small-scale cash cropping—lacked the resources to work with other water schemes in a given watershed to develop common rules for water governance, and the national government lacked the resources to issue and enforce permits that would protect water users' rights against illegal abstractors. As a result, upstream users—often wealthier, large-scale farmers—were de facto beneficiaries of a system with little polycentric and decentralized governance.

Post-2002 Reforms

With the 2002 Water Act, Kenya departed dramatically from the colonial-era permit scheme. One of the Act's major tasks was to clarify the allocation of regulatory authority between local, regional, and national actors. Under the reforms, the central government's Ministry of Water and Irrigation (MWI) retained responsibility for formulating policy and issuing regulations designed to maintain water quality and availability throughout Kenya's river basins. Regulatory activities such as permit issuance, however, were devolved to the watershed level, through six newly designated catchment areas, each with a regional office of Kenya's Water Resources Management Authority (WRMA), a state corporation under MWI responsible for watershed-wide permits and planning (Government of Kenya 2002, 2007).

The key innovation of the 2002 Water Act and subsequent regulations was perhaps recognizing the rights of water resource users' associations (WRUAs) to create forums for community water projects and other users in a given water body to communicate, implement policies, monitor water usage, and prevent or resolve conflicts. WRUAs are voluntary civic organizations whose creation is encouraged and funded by WRMA. Legally, WRUAs are required to include all water users in a particular watershed, and must establish a participatory constitution that includes both downstream and upstream users, as well as any other relevant stakeholders in the water body (Government of Kenya 2007).

One of the key tasks of a WRUA is to bring members together to agree on a water-rationing schedule that would be imposed in times of drought. When a body of water is experiencing low levels, the catchment area WRMA office announces the need to impose rationing according to the schedule pre-determined by the WRUA. WRUAs can also take an active role in watershed management, assisting WRMA with monitoring water levels and training members about water conservation. These activities are typically formalized, and sometimes funded, via formal memoranda of understanding (MOUs) with WRMA (WSTF and WRMA 2009). Thus, while WRUAs are voluntary civic organizations, they can be given a formal role in water governance, with each WRUA's specific responsibilities and activities determined on an ad hoc basis according to local needs and capabilities.

To date, over 400 WRUAs have been formally recognized by WRMA, and over 200 of these groups have formalized subcatchment management plans with WRMA. WRUAs have developed following a decentralization principle. Representation, a corollary of decentralization, is a constitutional aspect of WRUAs. Increased representation produces a higher level of integration of the different stakeholders who share the same water source. This is considered by the stakeholders we interviewed to be a fundamental basis for both preventing and resolving conflicts. The literature also shows initial evidence that suggests the WRUAs have been successful in reducing conflicts in some of Kenya's watersheds (Aarts and Rutten 2012). Moreover, bringing the practical aspects of the management of the resource closer to the users has financial benefits as much as it potentially makes monitoring efforts easier, although quantitative monitoring of water resources (e.g. streamflow) may be beyond the means of local-level actors. Some scholars have also suggested that WRUAs may not be an ideal institutional arrangement to meet some end-users' needs, particularly the rural poor and pastoralists (Mumma 2005; Robinson et al. 2010).

Local-Level Water Governance Near Mt. Kenya

The area around Mt. Kenya constitutes one of the major natural "water towers" in the country. Annual rainfall exceeds 1,000 mm in higher-elevation areas (Notter et al. 2007) where land-use restrictions have been implemented to protect forested areas and limit soil erosion. Farther from the mountainous area rainfall levels decline to the point where agriculture is only sustainable through irrigation from streams carrying water away from the mountain. A land-use gradient follows this precipitation gradient, with agriculturalists located in mid-slopes and areas farther from the mountain primarily characterized by pastoral land uses. However, where water availability limitations are high, an increasing number of pastoralists are experimenting in a pastoral-to-agricultural transition.

Rainfall, and thereby agriculture, is highly seasonal, spanning two distinct rainy periods. The first period, known as the Long Rains, lasts from March through May while the second period, known as the Short Rains, lasts from October through December. During intervening months, rainfall is still present, particularly during

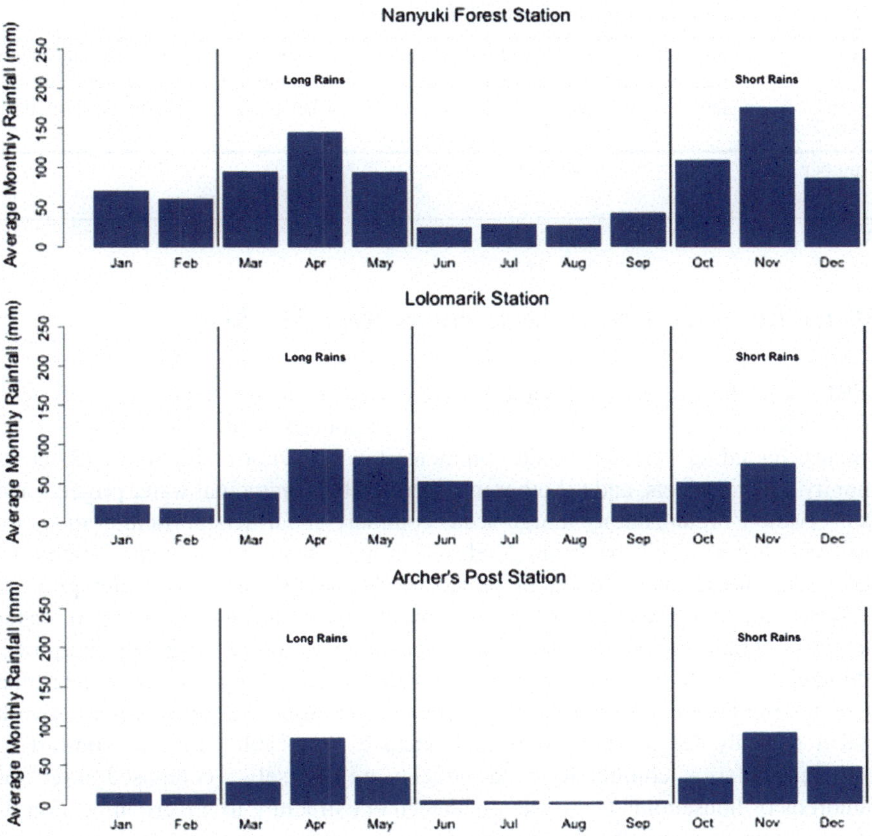

Fig. 21.2 Average monthly rainfall for three gauge stations in the Ewaso Nyiro basin

the summer, but is generally more sporadic and less intense (Fig. 21.2). In addition to this seasonal variability, the region experiences large interannual changes in total rainfall. This presents a particular challenge for those agriculturalists who cannot rely on streams and reservoirs for irrigated agriculture during years with poor rainfall.

Some data indicate little change in total annual rainfall but a statistically significant increase in storm intensity and a decrease in storm frequency (Franz et al. 2010). Although this result suggests that the future rainfall amounts may not change quickly, it also implies that water available for agriculture will decline as greater storm intensities generally produce a higher ratio of runoff to soil infiltration. This trend is not present in all relevant rainfall station records, however, pointing to the possibility of heterogeneous local-level changes that further complicate water management planning in the region.

Table 21.2 Conditions for establishing and maintaining water project membership

	Pay initial membership fee	Pay monthly maintenance fee	Attend water project meetings	Contribute labor to maintain the water project	Have an affiliation with a specific tribe or ethnicity
Number of households	725	744	732	725	2

Note $N = 749$

Water Resource Users' Associations Near Mt. Kenya

WRUAs in the Mt. Kenya region are composed of distinct types of actors who have WRMA-issued permits for a specified amount of water from a single river/stream, including (1) large-scale commercial farmers/horticulturalists, (2) community water projects, and (3) other members such as municipal water projects and conservation organizations. Large-scale commercial farmers/horticulturalists are primarily flower farms exporting goods to European markets, although some farms have large areas under wheat or potato production. Community water projects (CWPs) are composed of groups of households connected to a piped water network within the community for domestic use and/or irrigation purposes. Membership with a water project is usually obtained by paying a one-time membership fee to contribute to the expense of building the piped water network and a monthly fee to support its maintenance (see Table 21.2). In contrast to commercial farms, community water projects are themselves composed of tens or hundreds of households and have a management structure to govern allocation of water resources within the community. Commercial farms and CWPs are the most common types of WRUA members, but some WRUAs have additional types of members, as in the case of an urban area that relies on a river for municipal water supply.

WRUA management committees follow a "one head, one vote" structure with all members having an equal voting weight. In the case of a CWP, the process is bottom up and starts with the election of the individual CWP's Management Committee by the households in the project general meeting. This is different from the CWPs, where there is a system of democratic representation for the election of the committee members; in the case of the commercial farms' members, there is no committee and the chairman is usually the owner or CEO of the farm. Although the representation process is consistent among the different WRUAs investigated, the composition of a WRUA management committee varies as a reflection of the different shares of commercial members and community members in that WRUA. In Ngusishi, for example, the majority of WRUA members are commercial farms, whereas they are the minority in the Likii WRUA.

A process of partial bottom-up representation aims at including members in the rules-crafting process (Fig. 21.3). The WRUA Management Committee is the legislative body of the WRUA. In this arena, the members discuss and negotiate

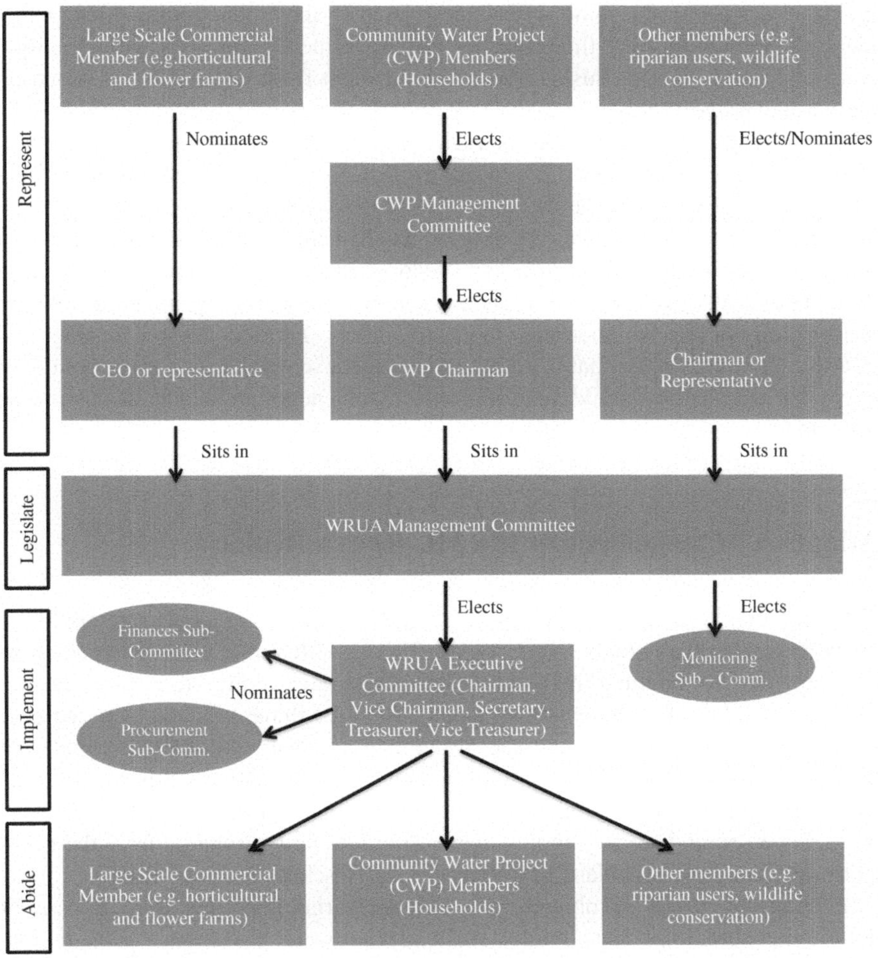

Fig. 21.3 Likii water resource users association governance structure and process

the overarching institutional arrangements (rules and strategies) to be implemented. The Management Committee is also responsible for electing the WRUA's Executive Committee, which is the main governing body of the association and composed of a chair, a vice-chair, a secretary, a vice-secretary, and a treasurer. WRUA management and operation requires a yearly budget that is funded by the members with annual fees. The fees are homogenous among the same types of members in the WRUA, and the commercial large-scale farms consistently pay a higher fee compared to the community water projects.

Community water projects pay a monthly amount relative to the amount of water that reaches the community. Each water project has a set allocation that is documented by permits maintained by the WRMA. Intake pipes of a specific dimension placed upstream carry water to individual water projects through a gravity-fed system. In some cases, these intakes have functioning meters that measure water flow over time, but not all intake meters are always functioning. A critical component of this system is the payment from water projects based on their water usage.

In contrast, households within a water project pay a regular fee for the maintenance of the pipe network of the water project. However, households do not pay a usage fee that varies by the amount of water they use. There are no meters at the household level, so water projects have no mechanism to charge households based on their water usage. This arrangement reduces incentives for water conservation at the household level although there are non-governmental organizations that are testing household-level metering in nearby areas. Farmers have a history and tradition of water access that is not tied to payment schedules, and transitioning to a pay-by-use system would require considerable negotiation and discussion to implement.

Adaptive Management in the Mt. Kenya Region?

Total population in the Upper Ewaso Ng'iro River basin increased from 50,000 in 1960 to 5,00,000 in 2000 (Ngigi et al. 2007), with rapid growth expected to continue due to ongoing regional development and inmigration from nearby locations. At the subcatchment level, population growth has resulted in an increase in river water abstraction (e.g., see Table 21.3 for Nanyuki WRUA population figures). From 1997 to 2004, the total number of river abstraction points in three subcatchments within the study area more than doubled (Liniger et al. 2005). This increase corresponds to an ongoing decrease in low-flow discharge from the Likii, Burguret, Nanyuki, and Naro Moru rivers (see Aeschbacher et al. 2005).

Population growth combined with climate variability constitutes a case of double exposure, imposing complex challenges for water governance in the region. In essence, periods of peak river water supply will fail to align with periods of peak demand, further increasing the prevalence of water scarcity (Barnett et al. 2005).

Challenges and uncertainties such as those presented here suggest that adaptive resource management (Cosens and Williams 2012) may be necessary to effectively allocate water resources in the future and support critical ecosystem services. Folke et al. (2005) describe this form of management as that which seeks the outcome of system resilience, or, put differently, a system that is able to absorb natural or human perturbations and continue to maintain essential functioning. Huitema et al. (2009) recognized four institutional prescriptions for adaptive water governance. First, resource management should be polycentric, meaning that multiple centers of control exist. Second, the public should be able to participate in management of the resource by supplying information to resource managers, engaging in co-decision making with managers, and even administrating components of the management program. Third, in complex socio-ecological systems

Table 21.3 Population growth in the Nanyuki WRUA

Sub-location	2002 population	2007 population	2012 population
Marura	85	Missing information	1,000
Kahurura	4,860	13,298	22,365
Equator	Missing information	12,936	19,285
Gathiuru	881	1,146	1,442

Source Village Chiefs

where information is limited and uncertainty and predictability high, resource managers should develop an approach that recognizes the limits of knowledge, seeks production and organization of new information, and values experimental learning. Finally, water management should take place at the river basin/watershed level. These four institutional prescriptions can either be in tension with each other or overlap. It is therefore important to understand them both separately and in conjunction as constitutive characteristics of an adaptive management system.

Polycentricity, Multilevel Governance, and Adaptive Capacity

Kenya's water governance post-2002 decentralization has resulted in a polycentric system that creates the *potential* for increased adaptive capacity. Multiple centers of control certainly exist, with MWI formulating policy and issuing regulations, WRMA carrying out regulatory activities at the regional level, WRUAs reconciling conflicts within subcatchments and implementing water management policies at the subcatchment level, and the community water projects acting as the local governing body for individual water users. This hierarchical system provides some autonomy to local-level actors for location-specific policy development and regulation of water use. Equally important to the idea of multiple centers of control in a polycentric system is that of interdependence between the levels (Cole 2011). Through interdependence and recognition by multiple levels of authority, legitimacy of rule is gained (see Table 21.4). WRUAs achieve this legitimacy by registering and signing MOUs with WRMA (WSTF and WRMA 2009). WRMA consults with WRUAs on water management issues, and WRUAs are eligible for funds that may help achieve water security for their users.

Public participation appears to be a strength of water governance in WRUAs in the Mt. Kenya region. Within the localized water projects, members are encouraged to attend regular project meetings. The frequency of these meetings varies by water project but most meet 2–4 times per year. Issues such as streamflow levels, complaints from project members, addition of new members, and election of new committee members are discussed (see Table 21.5). A high level of public participation is facilitated at the WRUA level as well as through the WRUA management committee chairs who have been elected to manage the individual water projects.

Table 21.4 CWP management committee members perception of WRMA

	Does WRMA play an important role in the management of water within your community/scheme or nearby communities/schemes?		Does WRMA respect the authority of your community/scheme to determine and implement its own rules for water and land use?	
	Yes	No	Yes	No
Likii WRUA $N = 11$	6	5	8	3
Nanyuki WRUA $N = 5$	5	0	5	0
Ngusishi WRUA $N = 5$	3	2	5	0
Ngare Nything WRUA $N = 5$	1	4	5	0
Timau WRUA $N = 5$	4	1	4	1

Table 21.5 Household participation in water project activities by WRUA

	Did a household member vote during the most recent water project election?		In the last year, how many times have you attended community meetings on water issues?			
	Yes	No	Never	Once	2–5 times	6 or more times
Likii WRUA	142	2	0	17	124	3
Nanyuki WRUA	152	0	7	9	99	36
Ngusishi WRUA	114	5	5	11	92	8
Ngare Nything WRUA	143	2	1	1	109	24
Timau WRUA	178	5	3	2	160	8

Rationing takes place between and within water projects during dry periods, typically from January to late March and from the end of July to late October. When river water levels become low, WRMA may mandate rationing within the WRUAs. The WRUAs then determine which water projects are to receive water on each day of the week following a rationing schedule approved by WRMA. The rationing schedules across all WRUAs appear to have been consistently applied over time despite variable growth and demand for water resources at the water project level. For example, while one water project has stayed at ~40 members over time, another water project has grown from 250–850 members—yet the water allocated to both communities and rotation schedules have not changed. As a result, households in the latter community typically receive water only four times a month during the dry season.

In other words, at the community/village water projects level, there is a second layer of water rotation schedules (intra-CWP household rotation). This schedule is usually established with the advice of external technicians and engineers who calibrate the rotation plan for the entire network. This becomes a structural program that can stay in place consistently for years with no changes. The intra-CWP household rotation schedules are determined based on the starting conditions of the water project pipes network (such as number of lines, size of the pipes, number of households) and are not adaptive to changing conditions. Unless there are some structural pipe/network modifications, it is rare that the intra-CWP household rotation schedules are changed.

Observations on the Role of Data and Information

Particularly important, but under investigated in the literature on common-pool resource governance, is the role that information (and data that lead to actionable information) plays in adaptive management in fast-changing social-ecological system contexts. Ostrom's (1990) design principles for common-pool resource systems, for example, refer only marginally to information. The role of information is, to a certain degree, represented in the design principles that refer to 'Collective-choice arrangements' and 'Monitoring.' In the first case, direct access and low-cost information is what gives local users a comparative advantage and makes local knowledge effective. In the second case, information is fundamental for community members to develop and implement provision rules (Cox et al. 2010). Nevertheless, for a deeper understanding of the interface between social action dynamics and cognitive processes associated with collective information, it is necessary to enter other fields of theory. Collective cognition theory (Bar-Tal 2000; Levin and Higgins 2001; Levine and Smith 2013; Mesmer-Magnus and DeChurch 2009) provides insights and can potentially expand the conventional understanding of CPR and collective-action theory. A central theme in CPR research that builds on the work of Olson (1965) and other authors (Araral 2009; Ostrom et al. 1994; Wade 1988) is about the characteristics of groups of users. In this field, outcomes are usually understood as a consequence of structures while subtler but internal processes, such as the cognitive dimension of information, are not explored. In collective cognition theory, the motivational and epistemic dimensions of collective information processes are investigated and provide counterintuitive insights (Levine and Smith 2013; McGrath et al. 2000).

In this research, we observed that information has a potentially critical role in the governance of water rationing schedules among WRUA members and intra-CWP household water-use rotation programs. Our preliminary observation in the Mt. Kenya region is that while the governance structure in place mirrors the principles that have been shown to contribute to sustained common-pool resource management (Ostrom 1990), we see little overt evidence of constitutive aspects of adaptive management such as learning from experience and integrating and producing new information for decision making. For example, at both the WRUA and CWP levels, we have found that when water rationing is enacted, the same rotation schedule has been implemented over time. There are three possible explanations that are not mutually excludable: (1) the imbalance between water demand and supply has not become acute enough to require modification to rotation schedules,

Table 21.6 Frequency of information discussed at management committee meetings

	What data and information are regularly presented at committee meetings?			
	Nanyuki $N = 3$	Ngusishi $N = 5$	Ngare Nything $N = 5$	Timau $N = 5$
Intake readings for project	xx	xxxx	xxxxx	x
Intake readings for entire WRUA or other projects	x	x	x	0
Measured water level in river	xx	xxx	xxxxx	xx
Informal perceptions of water levels in river	xxx	xxx	xxxxx	xxxxx
Amount of recent rainfall	x	xxx	xx	x
Potential for future rainfall	xx		x	x
Whether water project is over allocation of self-assessment	0	x	0	x
Complaints by members about low flow	xxx	xxxxx	xxx	xxxxx
Complaints by members about other members	xxx	xxxxx	xx	xxxx
Overloading of lines or branches	xxx	xxx	xxx	x

(2) managers lack the information necessary to evaluate the imbalance between demand and supply at the community level, and (3) individual managers' motivations might be disconnected from the collective goal of equal water sharing.

In the case of water rationing among WRUA members, the information available is not gathered and organized systematically. Often the decisions on rationings are based only on observational measurements of the river levels (see Table 21.6). Moreover, once the schedule is organized, it is likely to stay in place for the entire dry season with little adaptation to changing inputs. In the case of the intra-CWP household rotations, the combination of both the absence of up-to-date information and the inflexible adoption of standardized practices is far from the ideal for adaptive management. Also, intra-CWP household rotations have had very few modifications since the creation of the piped networks and water availability is not measured among the different households. For example, there is evidence that sharp demographic transformations in communities and villages are not reflected in modifications to the household rotation schedules. For this reason, a key element for improving the adaptive capacity of water management would be to increase WRUAs' and CWPs' efforts to gather, organize, and integrate data and information in their management activities.

Technical and financial constraints, however, prevent improvements in adaptive capacity. For example, scientists working on water flow measurements in the Likii River basin have declared that all the measurement equipment placed there in the last 15 years of research has been systematically stolen or vandalized. Among the CWPs, technical problems arise when attempting to measure water at the household level. In all CWP networks, the water is unfiltered and measurement gadgets at the household taps are technically impracticable because they inevitably get clogged with sediment. Moreover, the dynamic of gathering information faces not only technical, financial, and practical obstacles; it also strongly reflects and, at the same time, is influenced by social and political factors. We have noticed that at the CWP level, there can be a certain degree of resistance to systems of measurements that might reveal water availability inequalities among the members of the same CWP. This resistance can also be connected to the fear that household members have of measurement tools being a system of control of their activities and that might not be formally authorized or might be related to the fear that measurement devices will lead to new, additional pay-as-you-use water charges. These preliminary observations reveal that the role of information has a unique position in explaining community-level governance in a context of complex and adaptive socio-ecological systems. However, in order to deepen the analysis, it is necessary to go beyond the conventional understanding of collective-action theory and look at the internal processes of collective cognition.

Conclusion

We found that multilevel governance is prevalent in the governance of water resources near Mt. Kenya but by one metric (adapting rotation schedules to heterogeneous changes in water demand) there is little evidence of adaptation in the water governance process. One important aspect of the low level of adaptation is the limited capacity to gather and organize updated data and information and to transfer this to the decision-making level in a flexible way. We have highlighted how decisions are often implemented without incorporating new data and information. While considerable progress has been made in implementing governance structures to support adaptive resource management, a lack of actionable data and information limits the ability to do so. In future research, it will be important to go beyond the surface of governance structures and understand the internal dynamics of decision-making including perceptions of downstream and upstream water use and the departure of these perceptions from actual use.

Acknowledgments We gratefully acknowledge support from the U.S. National Science Foundation (grant SBE1115009).

References

Aarts J, Rutten M (2012) Will community based water management solve Africa's water problems? The performance of water user associations in the Upper Ewaso Ng'iro River basin, Kenya. In: Presented at the Embattled Spaces-Contested Orders Conference, VAD Association for African Studies in Germany, Cologne, Germany, May 30 June 2. https://cmsdata.iucn.org/downloads/aartsvadewasonn.pdf

Aeschbacher J, Liniger H, Weingartner R (2005) River water shortage in a highland-lowland system. Mt Res Dev 25(2):155–162

Andersson K, Ostrom E (2008) Analyzing decentralized resource regimes from a polycentric perspective. Policy Sci 41(1):71–93
Araral E (2009) What explains collective action in the commons? Theory and evidence from the Philippines. World Dev 37(3):687–697
Barnett TP, Adam JC, Lettenmaier DP (2005) Potential impacts of a warming climate on water availability in snow-dominated regions. Nature 438(7066):303–309
Bar-Tal D (2000) Shared beliefs in a society: social psychological analysis. Sage, Thousand Oaks
Cole DH (2011) From global to polycentric climate governance. Clim Law 2:395–413
Cosens BA, Williams MK (2012) Resilience and water governance: adaptive governance in the Columbia River basin. Ecol Soc 17(4):3. http://dx.doi.org/10.5751/ES-04986-170403
Coward EW Jr (1977) Irrigation management alternatives: themes from indigenous irrigation systems. Agric Adm 4(3):223–237
Coward EW Jr (1979) Principles of social organization in an indigenous irrigation system. Hum Organ 38(1):28–36
Cox M, Gwen A, Sergio V (2010) A review of design principles for community-based natural resource management. Ecol and Soc 15 (4):38. http://www.ecologyandsociety.org/vol15/iss4/art38
Cox M (2014) Applying a social-ecological system framework to the study of the Taos valley irrigation system. Hum Ecol 42(2):311–324. doi:10.1007/s10745-014-9651-y
Falkenmark M, Rockström J (2008) Building resilience to drought in desertification-prone savannas in Sub-Saharan Africa: the water perspective. Nat Resour Forum 32(2):93–102
Folke C, Hahn T, Olsson P, Norberg J (2005) Adaptive governance of social-ecological systems. Annu Rev Environ Resour 30:441–473
Franz KJ, Butcher P, Ajami NK (2010) Addressing snow model uncertainty for hydrologic prediction. Adv Water Resour 33:820–832
Geertz C (1959) Form and variation in Balinese village structure. Am Anthropol 61:991–1012
Geertz C (1967) Tihingan: a Balinese village. In: Koentjaraningrat RN (ed) Villages in Indonesia. Cornell University Press, Ithaca, pp 210–243
Government of Kenya (1972) The water (General) rules
Government of Kenya (2002) The water act
Government of Kenya (2007) The water resources management rules
Hardin G (1968) The tragedy of the commons. Science 162(3859):1243–1248
Huitema D, Mostert E, Egas W, Moellenkamp S, Pahl-Wostl C, Yalcin R (2009) Adaptive water governance: assessing the institutional prescriptions of adaptive (co-) management from a governance perspective and defining a research agenda. Ecol Soc 14(1):26. http://www.ecologyandsociety.org/vol14/iss1/art26/
Isiolo Water District (1982) Annual report to the water appeals board. Obtained from the Kenya national archives, June 2013
Lam WF (1998) Governing irrigation systems in Nepal: institutions, infrastructure, and collective action. ICS Press, Oakland
Levine JM, Higgins ET (2001) Shared reality and social influence in groups and organizations. In: Butera F, Mugny G (eds) Social influence in social reality: promoting individual and social change. Hogrefe and Huber Publishers, Bern, Switzerland, pp 33–52
Levine JM, Smith E (2013) Group cognition: collective information search and distribution. In: Carlston D (ed) Oxford handbook of social cognition. Oxford University Press, New York
Liniger H, Gikonyo J, Kiteme B, Wiesmann U (2005) Assessing and managing scarce tropical mountain water resources: the case of Mount Kenya and the semiarid Upper Ewaso Ng'iro basin. Mt Res Dev 25(2):163–173
Mabry JB (ed) (1996) Canals and communities: small-scale irrigation systems. University of Arizona Press, Tucson
Machakos Water District (1983) Annual report to the water appeals board. Obtained from the Kenya national archives, June 2013
McGrath JE, Arrow H, Berdahl JL (2000) The study of groups: past, present, and future. Pers Soc Psychol Rev 4(1):95–105

Meru Water District (1981) Annual report to the water appeals board. Obtained from the Kenya national archives, June 2013

Mesmer-Magnus JR, DeChurch LA (2009) Information sharing and team performance: a meta-analysis. J Appl Psychol 94:535–546

Mumma A (2005) Kenya's new water law: an analysis of the implications for the rural poor. In: Presented at the African water laws: plural legislative frameworks for rural water management in Africa, Johannesburg, South Africa, 26–28 Jan 2005

Ngigi SN, Savenije HHG, Gichuki FN (2007) Land use changes and hydrological impacts related to up-scaling of rainwater harvesting and management in Upper Ewaso Ng'iro basin, Kenya. Land Use Policy 24:129–140

Nilsson D, Nyanchaga EN (2009) East African water regimes: the case of Kenya. In: Dellapenna JW, Gupta J (eds) The evolution of the law and politics of water. Springer, Dordrecht, the Netherlands, pp 105–120

Notter B, McMillan L, Viviroli D, Weingartner R, Liniger HO (2007) Impacts of environmental change on water resources in the Mt. Kenya region. J Hydrol 343(3–4):266–278

OLSON M (1965) The Logic of collective action. Harvard University Press, Cambridge

Ostrom E (1990) Governing the commons: the evolution of institutions for collective action. Cambridge University Press, New York

Ostrom E (1992) Crafting institutions for self-governing irrigation systems. ICS Press, Oakland

Ostrom E, Gardner R, Walker J (1994) Rules, games, and common-pool resources. University of Michigan Press, Ann Arbor

Robinson LW, Sinclair JA, Spaling H (2010) Traditional pastoralist decision-making processes: lessons for reforms to water resources management in Kenya. J Environ Planning Manage 53(7):847–862

Shivakoti GP, Ostrom E (eds) (2002) Improving irrigation governance and management in Nepal. ICS Press, Oakland

Siy RY (1982) Community resource management: lessons from the zanjera. University of Philippines Press, Quezon City

Svendsen M, Nott G (2000) Irrigation management transfer in Turkey: process and outcomes. In: Groenfeldt D, Svendsen M (eds) Case studies in participatory irrigation management. World Bank institute, Washington, pp 27–88

Taillard C (1972) L'irrigation dans le nord du Laos: L'exemple du bassin de la Nam Song à Vang Vieng. In: Michel M (ed) Etudes de Géographie Tropicale Offertes à Pierre Gourou. Mouton, Paris, pp 241–256

Tang SY (1992) Institutions and collective action: self governance in irrigation. ICS Press, San Francisco

Thavaraj SH (1973) The necessity of terminal facilities for water management at the terminal level. Paper presented at the national seminar on water management at farm level, Alor Setar, Keoah, Malaysia

Wade R (1988) Village republics: economic conditions for collective action in South India. Cambridge University Press, Cambridge, UK

Water Services Trust Fund (WSTF) and Water Resources Management Authority (WRMA) (2009). Water Resource Users Association Development Cycle (WDC). Nairobi, Kenya. http://wstf.go.ke/index.php?option=com_docman&task=cat_view&gid=5&Itemid=496

Wittfogel K (1957) Oriental despotism: a comparative study of total power. Yale University Press, New Haven

Chapter 22
Transboundary Water Management in Federal Political Systems: A Story of Three Semi-arid Rivers

Dustin Garrick, Lucia De Stefano, Jamie Pittock and Daniel Connell

Challenges and Opportunities for Water Management in Semi-arid Federal Rivers

Federalism has increasing influence on river basin management across diverse geographic and political economic contexts, ranging from Australia and the US to India and Iraq (Garrick et al. 2013). Federal countries divide authority across territorial and national governments, which presents a classic governance test to manage conflicts and spread risk in shared waters. Federal rivers lie at the intersection of two traditions of research on collective action in the water commons—one focused on user self-organisation and the other on the geopolitics of international rivers. The coordination of local and multi-level collective action in water management has become more important in a context of global change and intensified competition for water. Local dilemmas are increasingly difficult to insulate from global change, and subnational water conflicts have overtaken international disputes as the largest share of water conflicts. Multi-level water governance dilemmas are particularly challenging in semi-arid regions vulnerable to climate variability and change; the distribution of benefits and risks for shared waters in such settings is already the source of tensions between upstream and downstream sub-national jurisdictions.

This chapter assesses the evolution and design of transboundary (interstate) water allocation institutions to manage 'difficult hydrology' (comparably low

D. Garrick (✉)
McMaster University, Hamilton, Canada
e-mail: dgarrick@mcmaster.ca

D. Garrick
University of Oxford, Oxford, UK

L. De Stefano
University of Madrid, Madrid, Spain

J. Pittock · D. Connell
Australian National University, Canberra, Australia

A. Bhaduri et al. (eds.), *The Global Water System in the Anthropocene*,
Springer Water, DOI: 10.1007/978-3-319-07548-8_22,

annual runoff, high variability) in semi-arid federal rivers prone to drought extremes. The comparative analysis identifies common institutional design features across the three basins: the Colorado River (USA/Mexico), Ebro (Spain) and Murray-Darling (Australia). In the next section we briefly review the challenge of difficult hydrology. In the third section, we introduce the basins and emphasise three trends in interstate water allocation reform in response to history, current and projected climate hazards and drought extremes.

Elazer (1987) identifies three defining elements of federal systems: a written constitution, non-centralisation and a territorial division of power (p. 157). Even unitary regimes have elements of decentralisation and devolution, which makes the distinction between federalism and unitary systems a continuum. Bednar (2011) defines federalism in terms of the geopolitical division, distribution of authority between independent state and national governments, and capacity to make binding laws for citizens falling within these overlapping jurisdictions. Beyond these defining elements, federal countries vary along a range of attributes, such as the degree of decentralisation, fiscal policy and age of the first federal constitution (and hence evolution of state-federal relations), along with other geographic and economic characteristics. This chapter considers the federal arrangements to reduce and share risks associated with drought extremes in semi-arid rivers facing 'difficult hydrology'.

Difficult Hydrology: A Shared Challenge for Semi-Arid Rivers

Semi-arid federal rivers are expected to face 'difficult hydrology' (Grey and Sadoff 2007): low mean annual runoff and high inter-annual and seasonal variability. Rivers in three regions—Southeast Australia, Spain and the Western US—confront such conditions.[1] The Colorado and Murray-Darling basins are large multi-jurisdictional rivers that experience comparably low mean annual runoff and high interannual variability. Sustained droughts affect both regions, and both are projected to experience future reductions in mean annual runoff. To cope, they have constructed high storage capacity in both absolute terms and per capita. Therefore, the Colorado and Murray-Darling have attributes associated with the difficult hydrology challenge. The Ebro is not a clear case of a semi-arid river with difficult hydrology but has a high degree of irrigation dependence (in terms of the percentage of basin area being irrigated) which magnifies the effects of projected reductions in runoff.

[1] The Rio Grande of the US/Mexico and most rivers of Southern Spain all share these attributes. However, economic development further distinguishes these basins. While South Africa's federal rivers confront similar hydroclimatic risks, for example, the country has achieved comparably lower levels of economic development.

Spatial and temporal variability in runoff also underpins upstream-downstream asymmetries. For example, the upstream state of Colorado (in the Colorado River basin) contributes higher mean annual runoff than the downstream state of Arizona due to its snowmelt-driven hydrograph; California, on the other hand, contributes a negligible proportion of the basin's runoff yet holds an entitlement to almost a third of the average annual runoff. Drought events entail uneven levels of exposure to risk across territories in these federations, requiring institutional arrangements to manage disputes between upstream states and downstream states. The comparative assessment traces the institutional evolution of interstate water allocation arrangements to explore factors contributing to or hindering robustness to difficult hydrology and jurisdictional complexity.

Interstate Water Allocation in Semi-arid Federal Rivers

How have large, semi-arid federal rivers spread hydroclimatic hazards and managed allocation disputes between multiple jurisdictions? What elements of institutional design contribute to the robustness of interstate water allocation institutions to climate hazards, and particularly droughts? This section briefly introduces the three river basins and then considers trends in institutional design in response to historic, current and projected hydroclimatic hazards. The analysis of trends in institutional design shows the presence of at least three common features in addition to context-specific elements tailored to circumstances in each basin: proportional allocation rules, institutional flexibility to update historic agreements, and river basin organisations to balance self-governance with basin-wide coordination.

Colorado River

The Colorado River straddles seven states in the US and two in Mexico (629,100 km^2) and has supported extensive irrigation development (5.5 million acres of irrigated agriculture), hydropower production, and rapid urban growth for up to 40 million people in the major population centres of the Western US (US Bureau of Reclamation 2012). Long-term average annual flows are approximately 18.5 billion m^3 with an extreme annual low and high in 1977 and 1984, respectively. Upstream reservoirs store up to 4 years of the basin's annual mean runoff to buffer against climate variability and sustained drought conditions, which are a prominent feature of the instrumented and paleoclimate hydrologic records (Woodhouse et al. 2006). The once-vast delta ecosystem has declined due to the combination of upstream reservoirs and diversions. Projected climate change impacts include decreases in runoff, earlier snowmelt runoff and more severe droughts.

Ebro River

The Ebro River (85,362 Km^2 or 17 % of Spain) crosses nine autonomous regions and has a small portion of its territory in France and Andorra. It is managed by the Spanish government through the Ebro River Basin Authority. Average natural runoff is 14.62 billion m^3/year, with a decrease of 11 % during the past two decades. Climate change projections point to a significant decrease in average runoff (Quiroga et al. 2011) and suggest a hotter climate, with increases in prolonged droughts (Bovolo et al. 2010). Ebro's water resources support the irrigation of about 800,000 ha, livestock breeding, energy production and water supplies for a sparsely populated territory (32.3 inhabitants per km^2). Most users withdraw water from 187 reservoirs having a total capacity of 7.49 billion m^3. The Ebro delta hosts a high-value ecosystem that is heavily affected by the decrease in water and sediment flows due to upstream water development and is threatened by projected climate change impacts on coastal dynamics (Sánchez-Arcilla et al. 2008). Persistent pollutants from historical mining and industrial activity, and organic pollution from agricultural and urban areas are also sources of concern.

Murray-Darling River

The Murray-Darling Basin (MDB) lies within the jurisdictions of four states, a territory and the federal government and totals over a million square kilometres. It is the country's most productive agricultural region generating 65 % of irrigated agriculture and a gross value of $15 billion in 2005–2006. The average annual outflows to the sea at the Murray Mouth have declined from 12.23 billion m^3/year (prior to water resource development) to 4.73 billion m^3/year (after upstream development) (CSIRO 2008). Decadal-scale droughts are a recurrent feature, to which Australia has responded by constructing reservoirs that can store the equivalent of an average year's runoff. Overallocation to irrigation and the impact of infrastructure has resulted in extensive environmental degradation, including flood plain forest death, salinization, thermal pollution, acidification, invasion of exotic species, barriers to species' migration and loss of biodiversity (Pittock et al. 2010). Confronted by climate change predictions of more extreme floods and droughts and in anticipation of intensified competition, policy makers are struggling to develop effective institutions to manage uncertainties and overcome interstate tensions (Connell and Grafton 2011).

Institutional Design and Change in Interstate Water Allocation

Section 'Difficult Hydrology' showed that these three basins share similar characteristics in terms of difficult hydrology, particularly low mean annual runoff, high interannual variability and significant storage capacity to buffer against seasonal and interannual variability. These hazards strain already contested interstate water allocation agreements when combined with overallocation or intensified competition for water (principally for irrigation, hydropower and, more recently emerging, urban and environmental needs). Efforts to improve water use efficiency, especially in irrigation, are also a feature common to the three basins. In this context, three trends in institutional design have emerged: a transition to proportional allocation rules; emergence of multi-layered river basin arrangements for planning, conflict resolution and joint monitoring; and new flexibility to adjust historic allocation patterns. Proportional interstate allocation rules (based on a share of the available water rather than fixed volumes) have perceived advantages for managing climatic variability, which mirrors findings for international waters. The presence of interstate river basin organisations aims at ensuring effective local, state and federal involvement and therefore balancing self-governance and basin-wide coordination. The basins require flexibility to adjust historic allocation agreements without risk of defection by individual states or costly court action.

Property Rights Systems for Interstate Water Allocation: Proportional Rules

Historic property rights systems featured fixed allocations and priority systems underpinning asymmetries in water supply and use across state jurisdictions. Many of these rules were devised under assumptions of stationarity, even as infrastructure was planned to buffer seasonal and some inter-annual variability. These allocation rules have not proven robust to climate extremes and prolonged droughts, which revealed an uneven distribution of risk across (and within) jurisdictions, particularly when combined with mounting competition among a growing number and type of water users and stakeholders. Fixed allocations to deliver water for downstream states have provided security but limit flexibility to adjust allocation rules to share risks during extended droughts, imposing residual risk on upstream states. However, the Murray-Darling and Colorado confirm the trend toward proportional allocation rules as climate hazards and competition intensify, particularly among upstream states and during extended droughts.

Colorado River The Colorado River Basin is governed by a complex mix of more than 100 laws, court decisions, operational guidelines and technical rules known as the Law of the River (Garrick et al. 2008). The 1922 Colorado River Compact and the 1928 Boulder Project Act established a fixed allocation for downstream states. The legal framework required 'upper division' states (Wyoming, Colorado, Utah and New Mexico) to deliver 92.5 billion m^3 to the 'lower division' states (Arizona, California and Nevada) over a rolling 10-year period. It formally allocated an equivalent volume to the upper division states, as well as 1.85 billion m^3 annually for Mexico under a subsequent 1944 international treaty. The 10-year accounting period acknowledged interannual climate variability. However, the fixed allocation left the upper basin states with residual flows and disproportionate exposure to hydroclimate hazards. The lower division states received fixed allocations, which has contributed to comparatively higher levels of interstate conflicts, particularly between California and Arizona. The legal framework further divided the upper division entitlement on a proportional basis among the upper division states of Wyoming, Colorado, Utah and New Mexico; this proportional allocation rule has limited conflict among these states.

Ebro River In 1926 the Spanish government created the Ebro River Basin Authority (RBA) to manage the basin with the participation of irrigators. During Franco's dictatorship (1939–1975) the powerful central government determined water allocation to users (individual or collective water rights) and executed it through the construction of large water infrastructure. With the 1978 democratic Constitution, Spain became a quasi-federal country, with 17 regions having broad powers and their own parliament. The Constitution established that interregional rivers like the Ebro would be managed by the central government through its RBAs. The 1985 Water Act for the first time admitted representatives of regions into some of the RBA boards and committees, with participation quotas proportional to the regions' territory and population shares in the basin. According to the 1985 water act water uses should be regulated through River Basin Management Plans (RBMPs), which allocate water volumes to basin subsystems sharing regulation and distribution networks ('exploitation systems') and to specific user groups (irrigators, industries, etc.) within each subsystem. Individual or collective water rights are nested in these subsystems, where annual allocation quotas to rights holders are defined in user-based RBA bodies based on annual precipitation and available water volumes.

Murray Darling Evolving state-federal water management institutions have created a mixed legacy. The initial agreement in 1915 only included the three southern basin states and the federal government, codified a minimum annual volume (fixed allocation) of water for South Australia as the most downstream state, and was focused on enabling use of water for irrigation and transport (Connell 2007).

The extreme variability of flows was recognised in the initial River Murray Waters Agreement that was incorporated in identical legislation adopted by the three southern MDB states and the Commonwealth in 1914/15. Although complex, the basic formula developed at that time has been included in all subsequent agreements (Connell 2007). In non-drought periods New South Wales and Victoria are required to provide a designated volume to South Australia. The two upriver states, New South Wales and Victoria, then share equally whatever is left in the storages (proportional allocation). In addition, they are entitled to all water in their tributaries flowing into the Murray.

The water allocation system between states and to individual irrigators is based on the principle of proportions of the water actually available each season. During serious drought, 'special accounting' rules are implemented, and the three southern states are each entitled to an equal share of two upriver storages. A countervailing pressure comes through a great variety of administrative arrangements that stem from the understandable desire of irrigators (and the subsequent political pressure) to lock in supplies when water is scarce. To the degree that this pressure is successful, the forced reduction is concentrated in that portion of the flow left for the environment (Connell 2007). Until recently environmental water was poorly defined and largely met from water left over after consumptive use diversions. The 2012 Basin Plan under the federal Water Act 2008 now legally defines and increases proportional water allocations for the environment, effectively increasing flows to the most downstream state.

River Basin Governance Arrangements: Multi-layered Planning, Monitoring and Conflict Resolution

Recent interstate water allocation responses to hydroclimate hazards, overallocation or intensive river regulation have involved a mixture of hard (binding) and soft (deliberative) decision-making forums to govern multi-jurisdictional tradeoffs—exemplified by the Murray-Darling Basin Plan, the Colorado River Basin Study, and the Ebro River Basin planning (most recently under the European Union Water Framework Directive, EU WFD). The federal government plays the role of catalyst of cooperation among states, using federal funds or the commitment to build new water infrastructure to foster that cooperation. These initiatives include new interstate cooperation to monitor water supply and use, and to manage conflicts through multi-stakeholder forums at nested user association, state and interjurisdictional levels. This marks a departure from the high cost, zero-sum conflict resolution processes in early periods of the Colorado River, or the recent protests of the proposed Ebro River inter-basin transfer from Northeast to Southern Spain.

Colorado River Since the early 2000s, an unprecedented dry period has prompted a range of basin planning and shortage sharing reforms. The 2007 shortage sharing agreement was the result of an environmental impact assessment undertaken through a multi-level river basin planning process coordinated by the states and federal government. The assessment evaluated management alternatives negotiated by the seven US basin states to establish operational reservoir management criteria for triggering shortage conditions and sharing the associated risks. The resulting agreement included supply augmentation (re-regulating reservoirs), demand management (conservation programmes) and interstate water storage agreements, including the 'intentionally created surplus' programme for states to store unused water allocations to buffer against future droughts. These mechanisms relied on new institutional linkages in information gathering and basin planning. The Lower Colorado River states agreed to divide shortage risks between Arizona and Nevada before California experienced reductions. Mexico's allocation was unaffected until the November 2012 passage of Minute 319 (an amendment) of the international water treaty between the US and Mexico. The agreement reduces deliveries to Mexico only when the US experiences operational shortages under the 2007 agreement described above.

Ebro River Although autonomous regions since 1985 are represented in the River Basin Authority boards, allocation decisions are still largely controlled by a rather close community of users and developers (Hernández-Mora et al. 2013). In 1992, Aragon was the first region to make explicit its claims over water through the Aragon Water Pact (AWP), a list of more than 20 new hydraulic works that would allow for doubling Aragon's irrigated surface. In 1998 the RBMPs of all the Spanish basins—including the Ebro—were approved. In 2001 the central government approved the National Hydrological Plan (NHP), which deals with interbasin issues. Both the Ebro River Basin Management Plan and the National Hydrological Plan incorporate the AWP water works. The NHP also proposed the transfer of 1 billion m^3/yr from the Ebro to other basins. This project triggered fierce opposition, mainly in Aragon and Catalonia. Even though it was cancelled in 2004, it marked a tipping point in the evolution of power balance between regions and the central government.

Murray-Darling Recognition of environmental degradation and limits to water resources led to a new Murray-Darling Basin Agreement in 1992. A consensus-based commission was established by the governments to administer jointly agreed programs. Water allocations were capped, and a market was established to enable seasonal or permanent trade in entitlements between water users and across state borders (Connell 2007). The limitations of the lowest-common-denominator commission governance structure resulted in the federal government using indirect constitutional powers to centralise governance with the Water Act 2008 and the subsidiary Basin Plan and Authority.

Historical Continuity: Moving Beyond Stationarity in Interstate Water Allocations

Interstate water allocation tensions have been strongly influenced by path dependency of past property rights systems (fixed allocations) and associated infrastructure. However, hydroclimatic hazards and drought events have combined with other driving forces of global and regional change to trigger interstate allocation reforms to achieve the proportional sharing mechanisms and expanding multi-layered decision venues, as outlined above. For example, international agreements and emerging recognition of downstream and basin-wide environmental needs has spurred basin-wide negotiations in the Murray-Darling (Ramsar Convention on Wetlands), Ebro (EU WFD) and Colorado (Minute 319).

Colorado River The 2009 Secure Water Act established a national basin study programme to assess supply-demand imbalances under projected scenarios of climate change. In 2010, the Colorado River Basin became a pilot area for this study programme with a cost-share between the federal and state governments to assess supply and demand, system reliability metrics and options for balancing supply and demand through 2060 (US Bureau of Reclamation 2012). The information-sharing mechanisms and system-modelling efforts have demonstrated the reliance on system-wide basin planning to reform interstate water allocation in response to historic and projected hydroclimate hazards. The shortage sharing agreement has led to bilateral negotiations with Mexico to address intractable international shortage sharing, as well as restoration of the delta ecosystem through interstate and international allocation agreements. Minute 319 authorised incipient efforts to reconnect the river with its delta through a range of restoration projects and reallocation agreements.

Ebro River The 2000 European Union Water Framework Directive set new challenges that the Spanish government began facing only in 2004, after the Ebro transfer repeal. In terms of water allocation, the WFD entails opening a new 6-year planning cycle and adds a new layer of complexity to allocation, as water uses should be compatible with the achievement of good status of all waters. In May 2012 the new Ebro RBMP was issued for public consultation, after strenuous negotiations over the in-stream flows in the Ebro delta (in Catalonia), whose maintenance is considered by many to be at odds with the current and planned upstream regulation. The new RBMP includes Aragon's water claims and 'water reserves' for other regions in the Ebro, to be executed through new hydraulic works. The new RBMP acknowledges projected climate change impact on runoff but fails to formulate adaptation strategies or to reconsider the planned water projects. Users seek water supply security mainly through increased water use efficiency and through lobbying for new dams. In 2007 the RBA approved a Special Drought Plan that established drought measures in each local 'exploitation system', while leaving basin-wide issues to ad hoc negotiations.

Murray-Darling Backed by a funding package of AUD \$14.7 billion[2] over 17 years, which includes AUD \$3.1 billion to purchase water entitlements for the environment, the Basin Plan adopted in late 2012 is meant to take account of all current and emerging issues (Connell and Grafton 2011). The plan shifts responsibility for high-level policy to the national government, leaving the MDB state governments responsible for implementation. These institutional changes are being widely resisted by the states. In Australia financial power rests overwhelmingly with the federal government, which has been repeatedly frustrated by state governments determined to use their greater water knowledge and administrative capacity to promote their own goals (Connell 2007). In preparation of the Basin Plan, conflicts arose between state governments, with the upstream states resisting loss of water for their farmers. By contrast, South Australia threatened to take the Plan to the High Court if there were insufficient environmental water allocations. Policies for adaptation to climatic variability and change are focused on water markets, environmental flows, infrastructure for greater water efficiency

[2] AUD \$1 = USD \$0.96 as of October 28th 2013.

Table 22.1 Interstate water allocation and difficult hydrology: a three-basin comparison

	Interstate apportionment		Drought provisions		River basin decision venues	
	Lower states	Upper states	Lower states	Upper states	Historic	Current
Colorado River, USA and Mexico	Fixed	Proportional	Fixed priority[a]	Proportional	Interstate compact; courts	Basin planning; environmental assessments
Ebro River, Spain	Fixed[b] Allocation by water exploitation systems		Priority by use Ad hoc negotiations when needed		Centralised decision, or unilateral claims	River basin management plans; courts
Murray-Darling River, Australia	Fixed	Proportional	Proportional (special accounting)		Inter-governmental agreements	Federal authority; Basin plan

Notes

[a] Arizona bears the brunt of shortage risk due to its junior priority under the 1968 Colorado River Basin Project Act; Nevada and Mexico share relatively small proportions of lower division shortages

[b] Established by River Management Plans. Aragon and Catalonia have asserted unilateral claims for irrigation use and downstream deliveries for the Ebro Delta, respectively

and iterative planning. A broader range of complementary adaptation measures would spread risks (Pittock and Finlayson 2011).

Table 22.1 summarize key features of institutional arrangements and trends in the three case studies.

Concluding Remarks

We examined semi-arid federal rivers with common climate and governance risks in interstate water allocation. The polycentric governance arrangements in semi-arid federal rivers are not static and instead have adapted by renegotiating the balance of devolved decision making and federal coordination. Interstate water allocation reforms have evolved to establish proportional sharing, drought provisions to balance security for downstream states and flexibility for upstream states, and a portfolio of hard (binding) and soft (deliberative) river basin decision-making forums to balance self-governance and basin-wide coordination to manage a broadening set of interacting risks. Drought and environmental demands have been catalysts for interstate water allocation reforms, unlocking historic path dependencies and upstream/downstream asymmetries in risk sharing. The case studies illustrate that federal-state relations are not static, but evolve as risks and institutions interact and change; therefore the right balance between levels will shift, and is increasingly impacted by policy changes at international and supra-national levels.

Acknowledgments This chapter is adapted from the case study section (Sect. 5) of Garrick et al. 2013.

References

Bednar J (2011) The political science of federalism. Annu Rev Law Soc Sci 7:269–288. doi:10.1146/annurev-lawsocsci-102510-105522

Bovolo CI et al (2010) Climate change, water resources and pollution in the Ebro Basin—towards an integrated approach. In: Barcelo' D, Petrovic M (eds) The Ebro River Basin: the handbook of environmental chemistry. Springer, Berlin, pp 295–329

Connell D (2007) Water politics in the Murray-Darling Basin. The Federation Press, Leichardt

Connell D, Grafton RQ (2011) Water reform in the Murray–Darling Basin. Water Resour Res 47:W00G03. doi:10.1029/2010wr009820

CSIRO (2008) Water availability in the Murray–Darling Basin report. Sustainable yields project. CSIRO Publishing, Canberra

Elazar D (1987) Exploring federalism. University of Alabama Press, Tuscaloosa

Garrick D, Jacobs K, Garfin G (2008) Models, assumptions, and stakeholders: planning for water supply variability in the Colorado river basin. J Am Water Resour Assoc 44, 381–398. doi:10.1111/j.1752-1688.2007.00154.x

Garrick D, De Stefano L, Fung F, Pittock J, Schlager E, New M, Connell D (2013) Managing hydroclimatic risks in federal rivers: a diagnostic assessment. Phil Trans R Soc A 371:20120415. http://dx.doi.org/10.1098/rsta.2012.0415

Grey D, Sadoff CW (2007) Sink or swim? Water security for growth and development. Water Policy 9:545–571. doi:10.2166/wp.2007.021

Hernández-Mora N, del Moral L, La Roca F, La Calle A, Schmidt G (2013) Interbasin water transfers in Spain. Interregional conflicts and governance responses. In: Schneier-Madanes G (ed) Globalized water. Springer, Dordrecht, Germany

Pittock J, Finlayson CM (2011) Freshwater ecosystem conservation in the Basin: principles versus policy. In: Grafton Q, Connell D (eds) Basin futures: water reform in the Murray-Darling Basin. ANU E-press, Canberra, pp 39–58

Pittock J, Finlayson CM, Gardner A, McKay C (2010) Changing character: the Ramsar convention on wetlands and climate change in the Murray-Darling Basin, Australia. Environ Plann Law J 27:401–425

Quiroga S, Garrote L, Iglesias A, Fernández-Haddad Z, Schlickenrieder J, de Lama C, Sánchez-Arcilla A (2011) The economic value of drought information for water management under climate change: a case study in the Ebro basin. Nat Hazards Earth Syst Sci 11:643–657. doi:10.5194/nhess-11-643-2011

Sánchez-Arcilla A, Jiménez JA, Valdemoro HI, Gracia V (2008) Implications of climatic change on Spanish Mediterranean low-lying coasts: the Ebro delta case. J Coastal Res 24:306–316. doi:10.2112/07A-0005.1

US Bureau of Reclamation (2012) Colorado River Basin water supply and demand study

Chapter 23
Legal Plurality in Mekong Hydropower: Its Emergence and Policy Implications

Diana Suhardiman and Mark Giordano

Abstract The changing role of the state and the increased participation of non-state actors has blurred the meaning of international affairs and highlighted overlapping power structures at international, national, and local levels. This paper illustrates how these power structures shape the hydropower decision making landscape in one of the world's most dynamic transboundary basins, the Mekong. Using the Lao PDR as a case study, we highlight how international donors' influence in the overall shaping of national policy and legal frameworks, the state's positioning of hydropower development as the main source of revenue, and the emerging importance of private sector actors manifested in overlapping rules and legal plurality in hydropower decision making. While legal plurality reflects the inherently contested terrain of hydropower, it also highlights the importance of power geometries and the scale dynamics in hydropower governance. The growing role of non-state actors may be interpreted as a reduction in state decision making power, but it may also be seen as a means for the state to take advantage of competing interests, in this case receiving both donor funding and private capital. If international donors expect national government agencies to promote meaningful application of internationally defined socio-environmental safeguards, they need to create space for critical discussion and move beyond the current standardized approach in promoting sustainable hydropower development.

D. Suhardiman (✉)
International Water Management Institute, Southeast Asia Regional Office,
P.O. Box 4199, Vientiane, Lao PDR
e-mail: d.suhardiman@cgiar.org

M. Giordano
Edmund a. Walsh School of Foreign Service, Georgetown University,
37th & O Streets NW, Washington, DC 20057, USA
e-mail: mg1382@georgetown.edu

A. Bhaduri et al. (eds.), *The Global Water System in the Anthropocene*,
Springer Water, DOI: 10.1007/978-3-319-07548-8_23,

Introduction

The changing role of the state in international relations has been widely discussed in political science (Migdal 1988, 2001; Schulte-Nordholt 2003; Scott 1987) and geography literature (Harvey 1989; Cox 1997; Escobar 2001). Authors have described how the state has been hollowed out (Jessop 2004) and how particularizing and universalizing tendencies beyond the state now interact in a process of glocalization (Robertson 1995; Swyngedouw 1997).

This change in understanding is also evidenced in thinking on transboundary waters (Furlong 2006; Sneddon 2003). Such work emphasizes the importance of 'power geometries' (Norman and Bakker 2009), as shaped by state and non-state actors, and 'scale dynamics' (Feitelson and Fischhendler 2009) at multiple overlapping levels of water governance. It highlights significant changes in states' roles, power, and influence, bringing to light the emergence of other important actors in transboundary water governance.

Bakker (2010) has placed political geography in the wider context of political economy to discuss how nation-states have coped with these changes as related to natural resource management. Using neo-liberalism and market environmentalism perspectives as her entry points, she showed that the state can actually sustain its power through reregulation, and that increased participation of non-state actors (both private and local) does not necessarily result in reduced state power. This reregulation, which concerns the redrawing of the formal boundaries of states' mandates, entails a shift in power structure benefiting some groups and disadvantaging others.

While this research has shown how states sustain their power through reregulation and other means, there has been little structural analysis on the factors that drive states' decisions to redefine or change their roles and how these decisions are linked to wider power geometries. Bakker's (2010) analysis provides valuable insights on how states can maintain their power amidst global, national, and local pressures. However, we still have little understanding of the decisive factors behind the decision to reregulate and how this decision might be linked to the way states view their position within wider power structures.

We expand this discussion in a developing country context by examining the role of international donors and private sector actors in driving state decisions to redefine their role in natural resource management, shedding light on the link with wider power geometries. To do this, we analyze the power interplay between the state, international donors, and the private sector, using the example of hydropower development in the Mekong region. We illustrate how this power interplay manifests in overlapping and contradictory legal orders in hydropower decision making, creating so-called legal plurality, and the policy implications for hydropower development. Taking the Lao PDR as our case study, we illustrate how legal plurality emerges, driven by international donors' interference in national decision making, the national government's interest in promoting economic growth through rapid hydropower development, and the emerging importance of private sector actors as the financial drivers of hydropower development. We conclude that to ensure meaningful application of socio-environmental safeguards desired by donors for sustainable hydropower development, it is crucial for them to understand the power geometries and the scale dynamics in hydropower governance.

Legal Plurality: The Intermingling of Global and National Norms

The notion of pluralism in natural resource management had been widely discussed and analyzed by legal scholars[1] (Spiertz 2000; Benda-Beckmann et al. 1996; Spiertz and Wiber 1996). The concept of legal pluralism highlights the role of law as a discourse, a process and element of social change at local and national levels (Moore 1978). Responding to the process of globalization and how this affects the overall shaping of the role of law and nation state on the world stage, the concept of global legal plurality emerged in 2000s (Agnew 2005; Berman 2009; Randeria 2007). Today, it includes a variety of institutions, norms, and processes located and produced at various scales, reconfiguring the overall notion of law, state, and territoriality (Randeria 2007). As stated by Santos (2006, p. 45): *'alongside local and national legal orders, supranational legal orders are emerging, which interfere in multiple ways with the former. Sub-national legal plurality acts in combination with supranational legal plurality'.*

Scholars of legal plurality have focused on a variety of issues important in shaping the overall process of state repositioning including multi-directional interaction of local, national and international norms (Santos and Garavito 2005); dialectical legal interactions (Burke-White 2004); non-state international law making (Ahdieh 2004); conflicts of law (Resnik 2008); and the disaggregation of the state (Singer 2003). Building on this body of work, we illustrate the ambiguous role of the state as both central and marginal (Randeria 2007). Deriving from this notion of ambiguity, we present the role of the state and its policy and legal instruments as an inherently 'contested terrain' (Randeria 2007; Santos 2002). As stated by Berman (2009, p. 236): "We can conceive of a legal system as both autonomous and permeable; outside norms affect the system but do not dominate it fully".

In Laos, legal plurality is rooted in international donors' policy interventions in land, water, and environmental management, through the incorporation of internationally defined socio-environmental safeguards as part of national policies and legal frameworks. Equipped by their power to fund, donors ensure the adoption of international guidelines and protocols for sustainable hydropower development

---

[1] Legal pluralism as a concept emerges as a response to the legal centralist conception of law. As stated by Griffiths (1986, p. 4): *"Legal pluralism is the fact. Legal centralism is a myth, an ideal, a claim, an illusion".*

(e.g. Environmental Impact Assessment or EIA, resettlement action plan) by national governments. In practice, however, our case study illustrates how the incorporation of internationally defined socio-environmental safeguards can contradict the state's positioning of hydropower development as the main source of revenue and manifest in overlapping policies and legal frameworks. We illustrate how the intermingling of global and national norms in hydropower development has resulted in policy inconsistency and institutional discrepancy related to land, water, and the environment.

Research Methodology

The line of analysis and arguments presented in this paper are derived from in-depth case study research (Burawoy 1991; Yin 1994), conducted from May 2010 to June 2011, focused on understanding the overall power interplay in Mekong hydropower development (including Cambodia, Vietnam, Thailand, and China) in general, and in the Lao PDR in particular. It looks at the way hydropower development is governed by multiple actors (various segments of the state, international donors, private sector actors), and how this manifests in the current state of legal plurality. To understand the emergence of legal plurality and its policy implications, we look at (1) how international donors incorporate their agenda into national policies and legal frameworks; (2) how this incorporation relates to existing national policies and legal frameworks and the overall institutional set up in hydropower development; and (3) how private sector actors position themselves as ad hoc decision makers in hydropower development through 'forum shopping'[2] (Benda-Beckmann 1981).

To understand international donors' role and influence in hydropower decision making, we discuss donor-driven policies and legal frameworks, and how they are formulated as a means to incorporate socio-environmental safeguards into national hydropower decision making.

To understand the institutional set up of hydropower development, we mapped the national-level (sectoral) decision-making landscapes (Aligica 2006; Azevedo 1997). We interviewed key actors from the Ministry of Planning and Investment (MPI) which has direct links to private developers, Ministry of Energy and Mines (MEM) as the agency in charge for hydropower development, Ministry of Natural Resources and Environment (MoNRE) as the government agency assigned to review the Environmental Impact Assessment (EIA) of each hydropower project, National Land Management Authority (NLMA) as the agency responsible for national land use planning in relation to land concession agreement, Ministry of Agriculture and Forestry (MAF), as well as private developers and civil society groups. The resulting

[2] Through 'forum shopping' private sector actors select and choose legal frameworks and policies that are in line with their interests.

institutional mapping includes organizational analysis of relevant government agencies and other key actors in hydropower sector development.

Finally, we reviewed existing procedures for hydropower development project design, construction, and operation, focusing on the role and involvement of private sector actors. These procedures include the formulation of Environmental Impact Assessments (EIA), Resettlement Action Plans (RAP), Power Purchase Agreements (PPA), and land concession agreements. In addition, we interviewed staff from a variety of hydropower companies to understand their experience and gain insights into hydropower project formulation and implementation.

Mekong Hydropower

Mekong hydropower is developing rapidly (MRC report 2009), responding to growing regional demand for electricity, export-led economic growth, expanding domestic consumer markets, and facilitated by the emerging importance of private sector financing[3] (Middleton et al. 2009). Laos is at the forefront of this development. Nationally, hydropower development is perceived as the state's primary means to promote economic growth and achieve development targets through industrialization and domestic market development and, importantly, as a means for government revenue generation. Regionally, international financial institutions such as the Asian Development Bank (ADB) present Laos' hydropower potential as an integral part of the ADB's regional power trade plan and emphasize the country's potential role as the battery for Southeast Asia (ADB 2009). Currently, there are 99 dams planned in addition to 17 already under operation (MRC report 2009). Rapid hydropower development in Laos cannot be viewed in isolation from the role of neighboring countries (e.g. China, Thailand and Vietnam) with regard to their capacities as hydropower dam developers, investors and/or power purchasers. In 2013, for example, the energy sector was the largest recipient of foreign investment, over US$ 1,565 million, and China and Vietnam were the main sources of funding (Vientiane Times, 21 October 2013). Moreover, some 90 % of power generated in Laos is now and is planned for export to its neighboring countries especially Thailand and Vietnam (same source).

Hydropower development in general has been met with resistance from NGOs, environmentalists, international agencies and others who are concerned with its potential social and environmental impacts. Partially addressing this concern, major international donors including the World Bank, Australian Aid (AusAid), Government of Finland, Swedish International Development Agency (SIDA) and Danish International Development Agency (DANIDA) have focused their efforts

[3] Unlike before, hydropower projects are built and operated by private developers in collaboration with key government agencies, with little or no involvement from the international financial institutions (IFIs) such as the World Bank and the Asian Development Bank.

on promoting sustainable hydropower development in the region, primarily through the incorporation of socio-environmental safeguards into national policies and legal frameworks. In Laos, this incorporation is most apparent in the formulation of a number of national policies and legal frameworks (see below).

The incorporation of socio-environmental safeguards into national policies and the establishment of the Ministry of Natural Resources and Environment (MoNRE) formally represent the state's willingness to adopt internationally defined standards and procedures. Yet, the incorporation of socio-environmental safeguards into national policies and legal frameworks is neither sufficient to synergize environmental and other sectoral ministries' roles in policy formulation and implementation, nor amend existing policies and legal frameworks which are not in line with those newly adopted. Moreover, the policy adoption and the establishment of new government bodies to promote sustainable hydropower development do not automatically result in well-functioning government bodies to implement, monitor, evaluate and enforce the resulting laws and policies.[4] Instead, the result is the creation of multiple, sometimes conflicting legal orders and overlapping authoritative and operational boundaries between agencies in charge of policy formulation and implementation. In other words, the result is legal plurality.

Legal Plurality: Its Emergence and Policy Implications

This section discusses how the intermingling of donors' preferences, government development priorities, and private actors' interests result in legal plurality in hydropower decision making and its policy implications. First, we discuss the formulation of donor-driven national policies and legal framework. Second, we highlight the existing policy inconsistency and institutional discrepancy in land, water, and environmental management. Third, we discuss the role of private sector actors as ad hoc decision maker in hydropower development.

Donor-Driven Policies and Legal Frameworks

International donors' interest in promoting sustainable hydropower development in Laos is most apparent in the formulation of the: (1) National Policy on Sustainable Hydropower (2009); (2) guidelines on EIA review (2011); (3) Government Decree on Resettlement, Compensation and Grievance Procedure for Project Affected People (2005); and (4) draft National Water Resources Policy (2010).

---

[4] See also Carruthers and Halliday (2006) on the question of global convergence versus national divergence in legal frameworks and practices.

The formulation of the National Policy on Sustainable Hydropower (NPSH) (2009) originates from the World Bank's objective to translate lessons from the Nam Theun 2 project into national policies to promote sustainable hydropower development. Projected as an overarching legal framework to promote social and environmental protection in hydropower development, the policy highlights the need to include environmental impact assessment as an integral part of hydropower decision-making processes, and to recognize the rights of local populations affected by hydropower projects. Moreover, it urges the need for information disclosure for any reports related to feasibility study, mitigation planning and monitoring of hydropower projects.

Complementing the NPSH, the guidelines on EIA review (2011) and the Government Decree on Resettlement, Compensation and Grievance Procedure for Project Affected People (2005) were formulated under the leadership of MoNRE with support from SIDA and DANIDA. The guideline introduces in detail each step that should be taken by private developers to ensure their compliance with environmental protection. It outlines the procedure for social and environmental management monitoring.

The Government Decree on Resettlement, Compensation and Grievance Procedure for Project Affected People (2005) defines principles, rules and measures to mitigate adverse social impacts and to compensate damages that result from involuntary acquisition or repossession of land and fixed or movable assets, including change in land use, restriction of access to community or natural resources affecting community livelihood and income sources (see article 1). Moreover, it obliges project developers to prepare different types of assessment: initial social assessment, land and assets acquisition assessment, and social impact assessment, as well as land acquisition and compensation report, and a resettlement action plan.

In line with the above defined policies and legal frameworks the National Water Resources Policy (2010) was drafted to promote the application of Integrated Water Resources Management to coordinate relevant government agencies and private sector actors. Formulated under the leadership of MoNRE with support from the Asian Development Bank, the draft policy assigns MoNRE itself, Lao National Mekong Committee (LNMC), and River Basin Committee (RBC) as the responsible government agencies in charge for water resources coordination.

Legal Plurality in Hydropower Decision Making: Policy Inconsistency and Institutional Discrepancy

Despite considerable achievement with regard to the incorporation of social and environmental safeguards into national policies and legal frameworks, existing policies on land-water-environmental management give a contradicting standpoint on how hydropower development should be directed through either sectoral or cross-sectoral approaches. The principle of Integrated Water Resources Management incorporated in the draft National Water Resources Strategy (2009) contradicts the sectoral approach under the Water Resources Law (1996), which assigns the responsibility to use available water resources to an individual sector ministry.[5] Similarly, both the Electricity Law (2010) and the Ministry of Energy and Mine's (MEM) decision-making chart place the full authority to direct hydropower decision making to MEM, and appoint it as the sole government agency in charge of each stage of hydropower development processes (i.e. design, review, feasibility, construction, monitoring, evaluation). This is in contrast with the National Policy on Sustainable Hydropower which urges the integration of hydropower development into the overall water resources development plan, as well as the guideline on EIA review and the Government Decree on Resettlement, Compensation and Grievance Procedure for Project Affected People which highlight the need to involve relevant ministries in hydropower decision-making processes.

Inconsistent policy in land, water and the environment is also reflected in institutional discrepancy between government agencies assigned with policy implementation task. Theoretically, National Land Management Authority (NLMA) and Ministry of Agriculture and Forestry (MAF) are responsible for land use planning and are supposed to be involved in all types of land concession negotiation, including for hydropower development. However, it is unclear how land concession for hydropower should be negotiated between private investors and MEM or MAF, or NLMA, as the Land Law (1999) combines both an integrated and sectoral approach towards land management. The law incorporates the task of land management planning, of which land concession forms an integral part. At the same time it categorizes land use types in line with sectoral ministries' areas of development (such as use of industrial land under the MEM or former Ministry of Industry and Handicraft, forest and agricultural land under MAF). Legally, the law assigns NLMA the role and responsibility for overall land management planning (through land zoning) but gives sector ministries the responsibility to regulate land use in accordance to their sectoral development activities (through land categorization).

Private Sector Actor as Ad Hoc Decision Maker

Private sector actors are the financial driver for hydropower sector development. At present, both Electricite du Laos (EdL), a public, state owned company and the Ministry of Energy and Mines (MEM) rely primarily on Independent Power Producers (IPPs) or private investors to finance hydropower development in the

[5] Currently, the Government of Laos is in the process of renewing its Water Law. Yet, how the new law will address the issue of sectoral versus integrated water resource management remains unclear.

country. According to the National Power Development Plan (2010), around 85 % of hydropower generation in Laos is owned by IPPs as of 2010, and this share is expected to increase. In line with this plan, about 90 % of power generation capacity (mainly for export to neighboring countries) will be developed by IPPs over the next 10 years. In theory, EdL can shape and reshape private investors' modes of operation, and in its capacity as a state owned company, EdL formally contributes a fraction of all hydropower investment. In practice, however, this share is relatively small (often less than 10 % of the total investment). While one may think that EdL could hardly influence private investors' codes of conduct in hydropower development (given their relatively small financial contribution to the overall investment), as a state owned company, EdL could still help the GoL in conducting its hydropower development policies especially with regard to sharing these policies to private investors. In practice, however, EdL's interaction with private developers is derived mainly from their interest to achieve the defined targets in hydropower development.

Equipped with the power to fund hydropower project formulation and implementation, private investors become ad hoc decision makers in hydropower development. Technically, private investors can only proceed with proposed projects after receiving an approval from the relevant government agencies. In practice, they are fully entitled to decide on how to proceed with the projects, without having to wait for MEM's approval of their feasibility study, or regardless of EIA review outcome from Ministry of Natural Resources and Environment (MoNRE). As said by one of the power company directors: "The government does not have to approve/disapprove the feasibility study. It is completely the company's decision whether or not to proceed with the project. In short, as long as the company is willing to invest in it and take all the potential risks, they can proceed regardless of the result of the feasibility study" (interview with a power company director, November 2010).

Moreover, as existing policy inconsistency and institutional discrepancy in hydropower decision making do not provide a clear strategy or guideline on how the government should monitor and evaluate private sector actors' conducts in each stage of hydropower development, this allows private sector actors to choose government agencies and legal rules which fit their interests, regardless of whether these interests are in line with the overall notion of sustainable hydropower development. Private developers would also sign land concession agreements with MEM, referring to MEM's full authority to direct hydropower development as stated in the Electricity Law (2010), partially ignoring other relevant ministries' role in national land use planning. As stated by NLMA official: "In practice, land concession agreement for hydropower is mainly negotiated between private developers and MEM with no involvement from NLMA" (interview with NLMA official, March 2011). Similarly, private developers approached local authorities (district governments) for their direct approval for the defined RAP, without consulting with MoNRE staff at national level, circumventing MoNRE's responsibility of arranging consultations with local population as regards the resettlement plan.

Conclusions

Our Mekong case study illustrates the role of international donors and private sector actors in shaping the Government of Laos' decision to redefine its role in natural resource management. Responding to international donors' pressure to promote sustainable hydropower development and relying on private sector actors as the financial driver for hydropower sector development, the state decided to adopt internationally defined guidelines and protocols as part of its national policies and legal frameworks, while continuing to promote rapid hydropower development and shaping the role of private sector actors as ad hoc decision makers for project planning and approval. The simultaneous existence of three different rule systems defines a state of legal plurality.

Our analysis on the emergence of legal plurality and its policy implications for hydropower decision-making in the Lao PDR provides two main findings.

First, it shows how the state, or at least some segments of it, can sustain its power in national level decision-making even amidst policy interventions and pressures from international donor agencies. Together with the emerging importance of private sector actors in hydropower development, the current state of legal plurality has blurred the boundary between the state as de-jure regulatory body and private sector actors as de facto decision makers. This is most apparent from the way the state informally shifted its responsibility in hydropower project formulation and implementation to private sector actors. While this shift in 'responsibility' may imply the state's declining power in hydropower development, we believe that it also helps the state to maintain its power and allows it to 'appear' and 'disappear' in accordance with its interests, using uncertainty and unpredictability of rule enforcement as escape hatches.

To ensure private hydropower finance while at the same time maintaining its reputation and good relations with donors, the Lao government has, for instance, positioned the Ministry of Energy and Mines (MEM) as the sole decision maker in formal hydropower development and given them the task and responsibility of promoting rapid hydropower development. Unlike the Ministry of Natural Resources and Environment (MoNRE) which has no particular targets in terms of Environmental Impact Assessment (EIA) review and approval, MEM's task and responsibility is focused on achieving investment targets from hydropower development as road mapped in the government's national socio-economic development plan 2011–2015. In this light, the existing legal plurality should not be misinterpreted as a sign of legal authorities' indeterminate, obscure or malleable function. Rather, it reveals the state's governing rationale as regards its natural resource management, and how this rationale is partially shaped by the hydropolitical situation within the basin.

Second, it highlights the policy implications of the existing legal plurality and how this reduces the actual significance of international donors' efforts to incorporate socio-environmental safeguards into national policies and legal frameworks. There is a tendency to think that the existing policy inconsistency and institutional discrepancy in hydropower decision making occurs primarily due to the government's lack of capacity to integrate its sectoral policies and planning. As a response to this interpretation, current discussions by donors on hydropower governance primarily emphasize the need to streamline hydropower policies and integrate the government agencies in charge of its implementation. While lack of capacity both in terms of technical expertise for the case of EIA review and lack of personnel in the case of RAP monitoring and enforcement is certainly an issue, it is merely a symptom of a much greater problem rooted in the political importance of hydropower development.

What needs to be addressed is not just building capacity and encouraging cross-sectoral coordination within and between government agencies but also creating space for critical discussion[6] to identify how best to pursue growth and sustainable hydropower development. International donor agencies such as the World Bank, ADB, SIDA, and AusAid have taken turns in providing financial support for MoNRE staff capacity building with regard to capacity building and in initiating cross-sectoral coordination. Yet, this capacity building alone cannot make up for MoNRE's relative absence at the provincial and district level. It is very likely that the state will use revenue from hydropower to improve its capability to build more dams, rather than to financially support MoNRE in terms of staff capacity for conducting EIA reviews or in changing hydropower practice. Unless the issue of prioritization in state budgeting is included as part of discussions on 'capacity building' programs, donor-funded training programs will continue to have little significance in increasing the current standard of EIA review, monitoring, and evaluation or the practice of hydropower development.

References

Asian Development Bank (2009) Building a sustainable energy future. The Greater Mekong Subregion

Agnew J (2005) Sovereignty regimes: territoriality and state authority in contemporary world politics. Ann Assoc Am Geogr 95(2):437–461

Ahdieh RB (2004) Between dialogue and decree: international review of national courts. N.Y. Univ Law Rev 79(6):2029–2163

Aligica PD (2006) Institutional and stakeholder mapping: frameworks for policy analysis and institutional change. Public Organ Rev 6:79–90

Azevedo J (1997) Mapping reality: an evolutionary realist methodology for the natural and social science. SUNY Series in the Philosophy of the Social Science, New York

Bakker K (2010) Neoliberalizing nature? Market environmentalism in water supply in England and Wales. Ann Assoc Am Geogr 95(3):542–565

Berman PS (2009) The new legal pluralism. Annu Rev Law Soc Sci 5:225–242

[6] Here space becomes a social product, which represents the embodiment of social relations and how they shape and transform the overall process of power interplay.

Burawoy M (1991) The extended case method. In: Buroway M (ed) Ethnography unbound: Power and resistance in the modern metropolis, pp 271–287. University of California Pres, Los Angeles
Burke-White WW (2004) International legal pluralism. Michigan J Int Law 25(4):963–979
Carruthers B, Halliday T (2006) Negotiating globalization: globalization and the construction of Insolvency regimes in East Asia. Law Soc Inquiry 31:3
Cox K (1997) Spaces of globalization: reasserting the power of the local. Guilford, New York
Escobar A (2001) Culture sits in places: reflections on globalism and subaltern strategies of localization. Polit Geogr 20:139–174
Feitelson E, Fischhendler I (2009) Spaces of water governance: the case of Israel and its neighbors. Ann Assoc Am Geogr 99(4):728–745
Furlong K (2006) Hidden theories, troubled waters: international relations, the territorial trap and the Southern African Development Community's transboundary waters. Polit Geogr 25(4):438–458
Griffiths J (1986) What is legal pluralism? J Legal Pluralism 24:1–55
Harvey D (1989) The condition of postmodernity. Blackwell, Oxford
Jessop B (2004) Hollowing out the 'nation state' and multilevel governance. In: Kennet P (ed) A handbook of comparative social policy. Edward Elgar, Cheltenham, pp 11–25
Mekong River Commission (2009) Hydropower sector review for the joint basin planning process. Basin Development Plan Program phase II. Vientiane
Middleton C, Garcia J, Foran T (2009) Old and new hydropower players in the Mekong region: agendas and strategies. In: Molle F, Foran T, Kakonen M (eds) Constested waterscapes in the Mekong region: hydropower, livelihoods and governance. Earthscan, London, pp 23–54
Migdal J (1988) Strong societies and weak states: state-society relations and state capabilities in the third world. Princeton University Press, Princeton
Migdal J (2001) State in society: studying how states and societies transform and constitute one another. Cambridge University Press, Cambridge
Moore S (1978) Law and social change: the semi-autonomous social field as an appropriate subject of study. In: Moore S (ed) Law as a process: an anthropological approach. Routledge, London, pp 54–81
Norman ES, Bakker K (2009) Transgressing scales: water governance across the Canada-US borderland. Ann Assoc Am Geogr 99(1):99–117
Randeria S (2007) The state of globalization: legal plurality, overlapping sovereignties and ambiguous alliances between civil society and the cunning state in India. Theory Cult Soc 24(1):1–33
Resnik J (2008) Foreign as domestic affairs: rethinking horizontal federalism and foreign affairs preemption in light of translocal internationalism. Emory Law J 57(1):31–92
Robertson R (1995) Glocalization: time-space and homogeneity-heterogeneity. In: Featherstone M, Lash SM, Robertson R (eds) Global modernities: theory, culture, society. Sage Publications, London and New Delhi, pp 25–44
Santos BD (2006) The heterogeneous state and legal pluralism in Mozambique. Law Soc Rev 40(1):39–76
Santos BS (2002) The heterogeneous state and legal pluralism in Mozambique. Law Soc Rev 40(1):39–76
Santos BS, Rodriguez-Garavito CA (2005) Law and globalization from below: towards a cosmopolitan legality. Cambridge University Press, Cambridge
Schulte-Nordholt H (2003) Renegotiating boundaries: access, agency and identity in post-Suharto Indonesia. Bijdragen tot de Taal Land en Volkenkunde 159(4):550–589
Scott J (1987) Weapons of the weak: everyday forms of peasant resistance. Yale University Press, US
Singer PW (2003) Corporate warriors: the rise of the privatized military industry. Cornell University Press, Ithaca
Sneddon C (2003) Reconfiguring scale and power: the Khong-Chi-Mun project in northeast Thailand. Environ Plann 35:2229–2250

Spiertz J, Wiber MG (1996) The role of law in natural resource management. VUGA, The Hague
Spiertz J (2000) Water rights and legal pluralism: some basics of a legal anthropological approach. In: Bruns B, Meinzen-Dick R (eds) Negotiating water rights. Intermediate Technology Publications, London
Swyngedouw E (1997) Neither global nor local: "Glocalization" and the politics of scale. In: Cox K (ed) Spaces of globalization: reasserting the power of the local. Guilford, New York, pp 137–166
Vientiane Times (2013) Ministry of Planning and Investment records investment value in excess of US$ 3.04 billion. 13 October 2013
von Benda-Beckmann K (1981) Forum shopping and shopping forums: dispute processing in a Minangkabau village in West Sumatra. J Legal Pluralism 19:117–159
von Benda-Beckmann F, von Benda-Beckmann K, Spiertz J (1996) Water rights and water policy. In: Spiertz J, Wiber MG (eds) The role of law in natural resource management. VUGA, The Hague, pp 77–99
Yin RK (1994) Case study research: designs and methods. Applied social research methods series no. 5. Thousand Oaks and Sage Publications, London and New Delhi

Chapter 24
International River Basin Organizations Lost in Translation? Transboundary River Basin Governance Between Science and Policy

Susanne Schmeier

Abstract Successfully governing water resources requires sound scientific understanding of the watercourse and the challenges it is facing. This is particularly important in transboundary watercourses due to the additional layer of complexity added when water resources transcend the politico-administrative boundaries of nation states. River Basin Organizations (RBOs) established for addressing such transboundary challenges have, however, not always been successful in linking scientific knowledge to policy decisions concerning the sustainable development of the river basins. This has implications for the overall effectiveness of river basin governance and the long-term sustainable development of the watercourse. This chapter analyzes the science-policy-link in RBOs. It finds that the strength of the science-policy-link varies considerably across RBOs, depending not only on the scientific knowledge provided by the RBO but also on the design of the RBO as well as the mechanisms it provides to its members for addressing the basin's challenges.

Introduction

International River Basin Organizations (RBOs) have been established around the world to address water-related problems that emerge due to the transboundary nature of many of the world's watercourses. Not all of them have lived up to the expectation that they would ensure the long-term sustainable development of their watercourse. Instead, many RBOs struggle with developing and implementing effective measures for the sustainable management of shared water resources,

This paper represents the opinion of the author and is not necessarily representative of the position of GIZ.

S. Schmeier (✉)
Deutsche Gesellschaft für internationale Zusammenarbeit (GIZ), Eschborn, Germany
e-mail: schmeier@transnationalstudies.eu

A. Bhaduri et al. (eds.), *The Global Water System in the Anthropocene*,
Springer Water, DOI: 10.1007/978-3-319-07548-8_24,

leaving a considerable number of watercourses in a situation in which their resources are overexploited.

These shortcomings are due to a number of factors determining the effectiveness of river basin governance, including, for instance, the issue about which riparian states disagree, the geographical distribution of power among riparians, or the number of riparian states involved in an RBO (Bernauer 1997; Marty 2001; Dombrowsky 2007; Schmeier 2013). In addition to these rather well-researched determinants of water resources governance, science—and its linkages to as well as its influence on policy—necessarily is an additional key factor, especially in systems as complex as shared watercourses. However, while the link between science and policy has been addressed extensively by scholars studying (international) environmental politics (Hegger et al. 2011; Miles et al. 2002; Cash et al. 2003; Roux et al. 2006; McNie 2007; Van den Hove 2007; Vogel et al. 2007; Koetz et al. 2011; Gupta et al. 2012), our understanding of the linkages between science and policy in the specific issue-area of water resources management remains limited. Most research focuses on local or national water resources management (Poff et al. 2003; Molle 2004; Cleaver and Franks 2008; McDonnell 2008) and very few studies address internationally shared watercourses (Blatter 2001; Quevauviller et al. 2005; Friend 2009; Weller and Popovici 2011).

This chapter therefore focuses on the link between science and policy in the management of shared watercourses—with a particular emphasis on RBOs as the institutionalized means for jointly addressing water resources use and protection. It asks to what extent science influences policy in the context of RBOs and is particularly interested in the factors potentially determining whether and to what extent science can influence policy. After briefly introducing an analytical framework for analyzing the science-policy-link (SPL), the chapter discusses experiences in four RBOs—the International Commission for the Protection of the Danube River (ICPDR), the International Commission for the Protection of the Rhine (ICPR), the Lake Victoria Fisheries Organization (LVFO) and the Mekong River Commission (MRC). It finds that the strength of the SPL varies considerably across RBOs, depending on (1) the nature of science provided to policy makers and (2) the design of an RBO and the mechanisms it provides for water resources governance.

Linking Science to Policy in Transboundary Water Resources Management

While acknowledging the different notions of science in existing research on the linkages between science and policy (e.g. McNie 2007; van den Hove 2007; Vogel et al. 2007; Koetz et al. 2011), science is defined as the product of a comprehensive analytical process, guided by accepted approaches and methods of a specific discipline, which addresses a specific issue at stake with the aim to generate knowledge and provide answers to a clearly defined question, when addressing water issues, most often aiming at understanding the characteristics of a

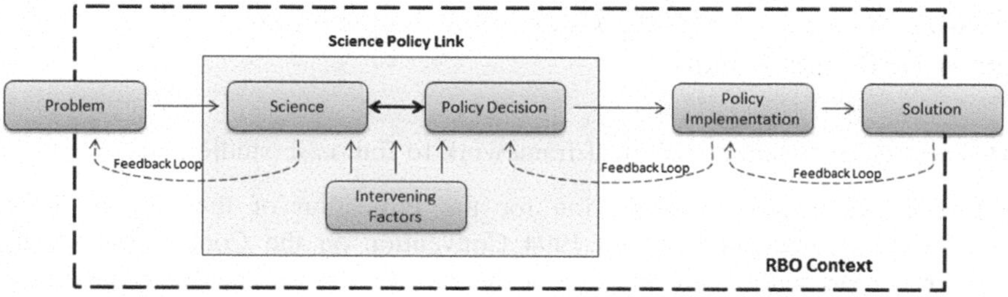

Fig. 24.1 The SPL in RBOs

watercourse, including the challenges it faces due to environmental change, and developing approaches for achieving a certain desired state of the respective watercourse.

Policy describes a policy decision taken by policy-makers concerned with a specific problem and the implementation of this policy decision through specific measures and activities. In water resources management, this refers to actions taken by water policy makers in a river basin that most often aim at addressing water-specific challenges in order to improve the state of the watercourse and/or the benefits riparians gain from it.

The interaction between science and policy with the aim to address a certain problem on the basis of a sound understanding of the problem and its potential solutions through political decision-making and the implementation of such decisions by political action is captured by the SPL (see Fig. 24.1). In many shared watercourses where institutionalized cooperation exists, the SPL is set within the framework of RBOs. The link between science and policy occurs throughout the water resources management cycle: science provides a basis for understanding the conditions of a watercourse in all its complexity, including data on the river's hydrology, the basin's aquatic and land-based ecosystems, the basin's climate, the use of its different water by different user groups as well as the dependence of riparian populations on these resources; science provides input for understanding and assessing environmental and socioeconomic challenges a basin is facing; science is required for developing policy options and for deciding which ones to implement; and science is required for monitoring the effectiveness of measures implemented to address a certain water-specific problem and thus the effectiveness of the water resources management process.

Given the embeddedness of the SPL in the broader RBO context, the SPL is influenced by the design of the RBO itself. Among the many institutional design elements of RBOs (Bernauer 1997; Marty 2001; Dombrowsky 2007; Schmeier 2013), the way decisions are taken within an RBO, the mechanisms through which data and information is shared and the state of the river is monitored are expected to matter most. The next section focuses on the SPL with its intervening factors in more detail, also providing cursory empirical evidence for how the different factors determine whether and to what extent science influences policy in different RBOs.

Strong Science = Strong Decisions? Experiences from Different Basins

This section applies the analytical framework to four case studies:

(1) The International Commission for the Protection of the Danube River (ICPDR), mandated by the 1994 Convention on the Cooperation for the Protection and Sustainable Use of the Danube River (Danube River Protection Convention) and its commitment to "improve the current environmental and water quality conditions of the Danube River" (Art. 2 Danube Convention) to work on water quality but also other ecological as well as flood management challenges;
(2) The International Commission for the Protection of the Rhine (ICPR), on the basis of the 1999 Convention on the Protection of the Rhine (Rhine Convention) committed to "the sustainable development of the Rhine ecosystem" (Art. 2 Rhine Convention) and mainly addressing water quality challenges arising from the intensive economic use of the basin;
(3) The Lake Victoria Fisheries Organization (LVFO), mandated by the 1994 Convention on the Establishment of the Lake Victoria Fisheries Organization (LVFO Convention) to work on fisheries management challenges with the aim of "restoring and maintaining the health of the ecosystem and ensuring the sustainable development for the benefit of present and future generations" (LVFO 2005a, p. 17); and
(4) The Mekong River Commission (MRC), addressing numerous issues in the Mekong River Basin—ranging from fisheries management or the assessment of large infrastructure schemes to river basin planning or navigation—on the basis of the 1995 Agreement on the Cooperation for the Sustainable Use of the Mekong River Basin (Mekong Agreement) and members' commitment to "cooperate and promote […] the sustainable development, utilization, conservation and management to the Mekong River Basin water and related resources" (Preamble, Mekong Agreement).

The extent to which these RBOs link the science they produce to policy decisions they are supposed to guide varies considerably: The SPL is very strong in the ICPDR and the ICPR where science and policy interact intensively in order to develop solutions for challenges the basins are facing.

The development of a Rhine climate change adaptation strategy provides an interesting example. In 2007, the Conference of Rhine Ministers charged the ICPR to provide members with a scenario study of the Rhine under climate change conditions. Accordingly, an Expert Group (EG KLIMA) was set up that brings together climate change experts from ICPR members, including both policy makers and academics. They came up with three key products—(1) a literature review evaluating existing knowledge on climate change in the Rhine River Basin (ICPR 2009a), (2) an analysis of potential hydrological scenarios under climate change conditions (ICPR 2011), and (3) an analysis of the effects of hydrological regime changes on the Rhine ecosystem (ICPR 2013a). Based on those analyses, an adaptation strategy that guides members in their adaptation actions is being developed. These examples demonstrate a strong SPL in which science and policy complement each other, with challenges analyzed by science and acknowledged by policy makers and solutions then being developed by science and implemented by policy makers.

A similarly strong link can be found in the ICPR's work on reintroducing the salmon to the Rhine. On the basis of scientific analyses on the state of the salmon in the Rhine, a strategy called Rhine Salmon 2020 (ICPR 2004) as well as a more recent master plan for migratory fish (ICPR 2009b) have been developed and a number of measures implemented—including the legal bindingness of ensuring fish passability in the Rhine that is tied to ICPR members' commitment to the EU Water Framework Directive (EUWFD). These measures considerably improved the passability of the Rhine for salmon (as well as other species). Scientific monitoring of the state of the salmon could then show that in recent years, nearly 7,000 salmon returned to the Rhine (ICPR 2013b). Here, the close interaction between science and policy has ensured a veritable improvement of the ecological state of the river that would otherwise not have been possible.

The SPL is slightly weaker yet still important in the case of the LVFO, where strong science on fisheries management is generated and used by decision-makers, yet often depends on externally financed, commissioned and conducted research instead of building riparian science which remains comparatively weak. For instance, a comprehensive monitoring for the introduced Nile perch has been implemented under the framework of the LVFO that collects comprehensive data and develops management recommendations (Bucceri and Fink 2003). These recommendations are largely implemented by LVFO members, however constrained by gaps in financial and technical capacity in national fisheries departments as well as enforcement bodies.

On the other hand, the SPL is particularly weak in the MRC. Although a lot of highly valuable scientific knowledge is generated by the MRC, it is not translated into policy decisions. A prominent example is the case of the Xayaburi hydropower project which underwent MRC's process of Notification, Prior Consultation and Agreement (PNPCA) in 2011: Based on comprehensive analyses by a large number of well renowned researchers, both the MRC's Strategic Environmental Assessment of Hydropower on the Mekong Mainstream (MRC 2010a) and the Xayaburi-specific PNPCA Review Report (MRC 2011a) found that environmental impacts (including transboundary ones) were to be expected "even after the project related mitigation measures have been introduced" (MRC 2011a, p. 96) and therefore recommended that "decisions on mainstream dams should be deferred for a period of 10 years" (MRC 2010a, p. 24). This assessment was shared by downstream countries with Vietnam similarly demanding that "the Xayaburi Hydropower Project as well as all other planned hydropower projects on the Mekong mainstream [should] be deferred for at least 10 years" (MRC 2011b, p. 1). Nonetheless, Laos went ahead with the construction of the Xayaburi project based on a unilateral policy decision, threatening the sustainable development of the entire river basin. Hence, in spite of a large amount of scientifically sound knowledge being available through the MRC, the influence on policy makers (especially in Laos and Thailand, the latter one financing the project) was close to irrelevant. For the MRC, lacking knowledge is not the problem. Instead, there is a gap between the available science and what policy makers make of it.

The following paragraphs focus on the different aspects of the SPL in the different RBOs, including the RBO's institutional design characteristics identified as decisive for linking science to policy in the analytical framework. This helps explaining why the strength of the SPL varies so dramatically across the different river basins and RBOs studied and developing responses to the respective shortcomings in order to strengthen the SPL.

The Nature of Scientific Knowledge: Meeting Policy-Makers Needs

First, the type of scientific input itself can be expected to matter for whether and to what extent science is translated into policy. The type of scientific input and, in particular, its targetedness on the issues at stake in the basin varies significantly across the RBOs studied[1]: 47 % of the ICPDR's publications focus directly on water quality, the most important challenge the basin is facing, followed by another 20 % of publications addressing governance and institutional issues, most often related to water quality objectives and the implementation of related policy measures (Fig. 24.2). A similarly strong focus on the issues at stake is found in the LVFO (Fig. 24.3), where 40 % of all documents focus specifically on fisheries and another 30 % on institutional matters related to implementing specific fisheries policies (such as the "National Guidelines for Beach Management Units", LVFO 2005b). Moreover, LVFO's products are generally presented in a policy-oriented way with limited scientific jargon and in the form of recommendations or implementation plans, directly targeting policy-makers in their decision-making roles (such as, for instance, the "Implementation and Financing Plan for the Strategy to reduce the Impact of HIV/Aids on Fishing Communities", LVFO 2005c). It is hence ready to be taken up by policy-makers without requiring any translation.

In the case of the MRC, on the other hand, no clear issue focus can be observed (Fig. 24.4). The issues addressed most often are basin planning and fisheries (with 28 and 27 % of all documents respectively)—not necessarily the most important challenges the basin is facing. Hydropower is only addressed in less than 4 % of

[1] The analysis of the scientific reports of all RBOs was conducted by compiling a list of all technical publications of the four RBOs and coding them according to the specific issue-area they cover as well as the type of advice they provide (e.g. consultant studies, reports, action plans, recommendations, etc.). A full list of all documents, including coding and analysis, is on file with the author.

Fig. 24.2 The distribution of functional issues—ICPDR

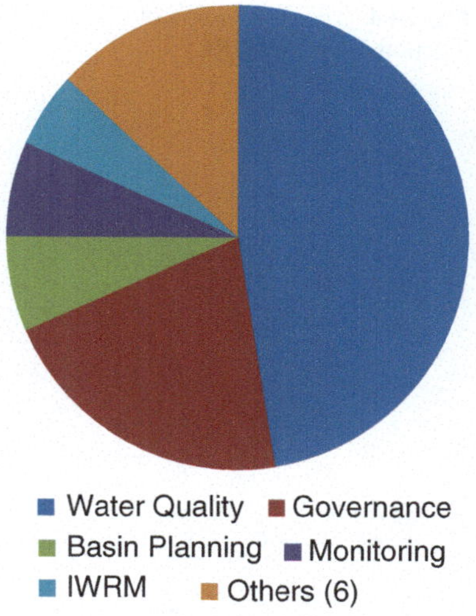

Fig. 24.3 The distribution of functional issues—LVFO

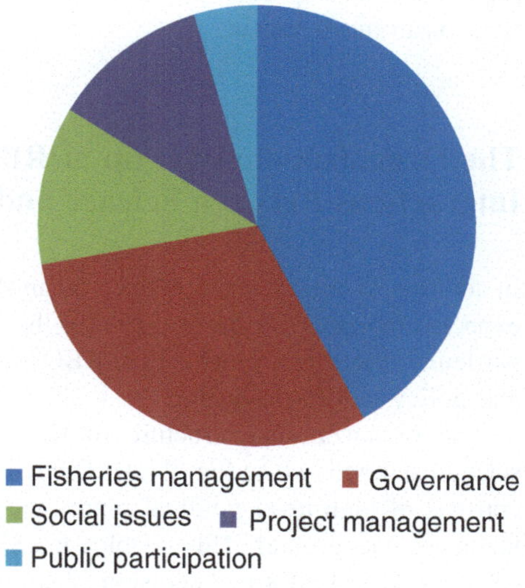

MRC's scientific outputs—particularly counterintuitive given the important of hydropower developments for the overall health of the basin. Nearly all hydropower-specific documents were published in recent years (especially since the MRC's Strategic Environmental Assessment of the Mekong Mainstream Dams) when hydropower developments became particularly ambitious in the Mekong River Basin—indicating some degree of adaptiveness of the MRC to newly emerging challenges in the basin. Moreover, most documents are highly scientific in nature and do not provide easy and straightforward policy recommendations.

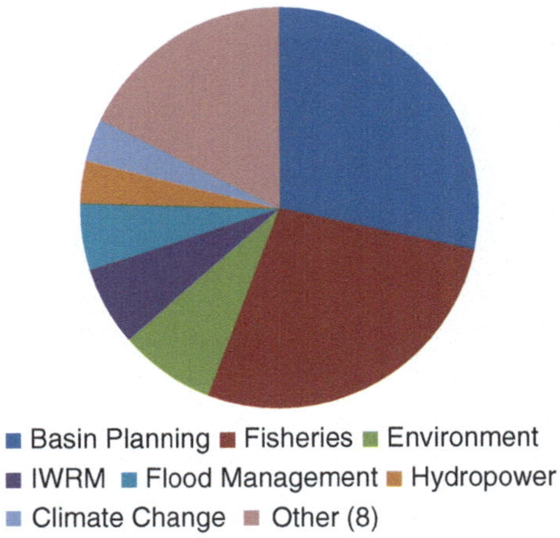

Fig. 24.4 The distribution of functional issues—MRC

The targetedness of scientific outputs of an RBO—both in terms of addressing the right topics and presenting them in a policy-oriented manner—can thus be regarded as an important prerequisite for science guiding policy in the process of river basin management.

The Organizational Set-up of RBOs: Facilitating Interactions Between Science and Policy

In addition to the science produced by an RBO, the set-up of the RBO itself can be expected to matter for the strength (or the weakness) of the SPL. This concerns, in particular, the extent to which the RBO's organizational structure allows scientists and policy-makers to interact.

The organizational structure of the ICPDR facilitates interactions between science and policy (see Fig. 24.4). The ICPDR's governing body, the Commission, consists of high-level political representatives from ICPDR members yet with a technical background. This enables the Commission's members to make technically sound and informed decisions that are at the same time politically mandated. Input to these decisions is provided by the ICPDR's Expert Groups (EGs). They consist of representatives of ICPDR members who are technical experts in specific water resources management topics. If deemed necessary, they can be complemented by external experts from research institutions. The EGs are managed by Technical Experts (TEs) at the ICPDR Secretariat. Such direct integration of scientific knowledge into the political planning and decision-making process on the basis of a strongly science-oriented organizational structure ensures a strong SPL in the RBO. The ICPR relies on a similar organizational set-up—however with an even smaller Secretariat and an even larger reliance on both scientists and policy-makers in its member states (Fig. 24.5).

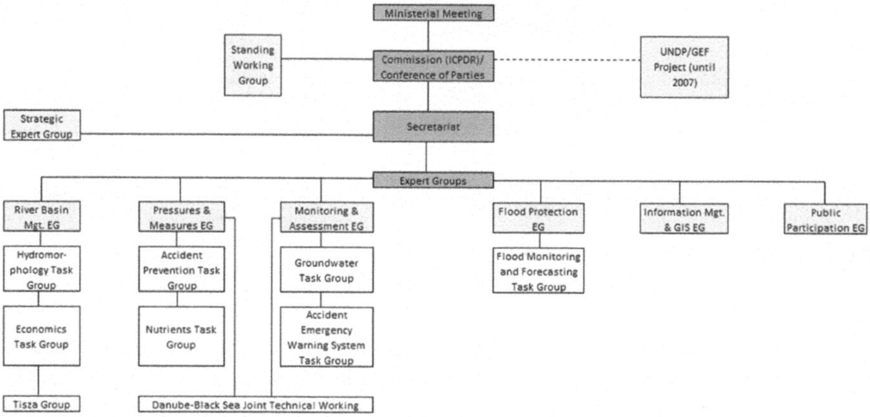

Fig. 24.5 The organizational structure of the ICPDR

Fig. 24.6 The organizational structure of the LVFO

A similarly close relationship between science and decision-making is ensured by the LVFO's organizational structure (see Fig. 24.6): The Scientific Committee is specifically designed for providing input into the fisheries management process. Consisting of the Heads of Department of member states' key research institutes in fisheries, it provides key scientific guidance for the identification of research requirements, the development of standard operating procedures, the collection, analysis and dissemination as well as the review of research results and their inclusion into policy recommendations. In addition, national research institutes are involved through the LVFO's National Committees. This ensures that policy decisions on how to manage the lake's fish resources are strongly guided by a strong interaction between science and policy. Moreover, it provides national research institutes with funding being acquired by the LVFO that enables research otherwise not possible.

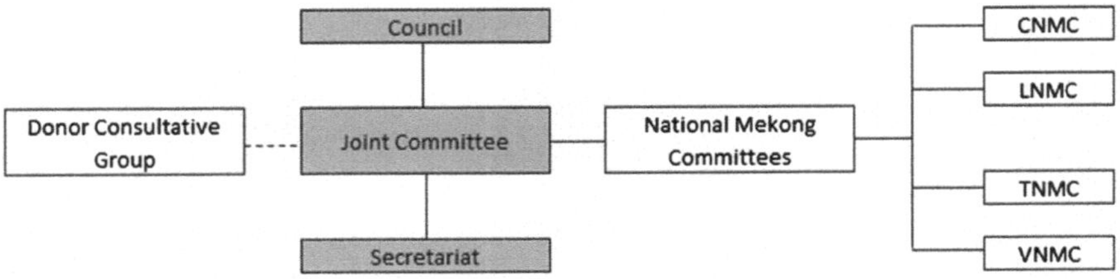

Fig. 24.7 The organizational structure of the MRC

The organizational structure of the MRC is different (see Fig. 24.7). Here, no direct link between the high-level governing bodies and the science generated for decision-making exists. Instead, all scientific input is generated through the Secretariat which itself has rather weak links to those making decisions about the river basin's future. This link is further strained by the role of MRC's National Mekong Committees (NMCs), constituting bottlenecks of information flow and thus reinforcing a situation in which science developed at the regional level rarely reaches those in member states who actually make decisions on how to use the river's resources.

The Influence of Water Resources Management Mechanisms

One of the water resources management mechanisms that can strengthen or weaken the SPL is the *decision-making* within an RBO. In the case of the ICPDR, decision-making is particularly open to scientific voices. ICPDR Commission meetings are open to observers and clear guidelines (ICPDR 2005) define the role of observers. Observers such as the Danube Environmental Forum, an association of more than 150 environmental NGOs in the basin, the International Association for Danube Research or the Regional Center for Central and Eastern Europe are not only allowed to participate in meetings, but they can also voice their opinion during the discussion and comment on statements made by the Heads of Delegations from ICPDR member states. They thus have an intermediate influence on the discussion even without being granted formal voting rights. In the MRC, on the other hand, decision-making is a rather closed business and, moreover, often takes place even outside of pre-defined decision-making forums under the auspices of the MRC. As a consequence, decisions are often taken in an intransparent manner and unilateral or bilateral considerations prevail over the well-being of the entire basin. This is illustrated, for instance, by the case of the Xayaburi hydropower project where a decision under the framework of the MRC was first delayed and later completely evaded and Laos instead engaged in bilateral negotiations with its neighboring countries whose protests against the project silenced completely since

2013 (Vientiane Times 2013a, b). Recent developments on the Don Sahong hydropower project in Southern Laos exhibit similar patterns, with negotiations between Laos and Cambodia apparently taking place outside of the MRC and thus under little influence of MRC's science (RFA 2013).

Taking informed decisions requires the availability of sound *data and information*. The way how data and information is managed within RBOs is therefore another important determinant for the influence science can have on policy. Regular exchange of data, for instance via an RBO-based database, can considerably improve the link between scientific data and information and policy decisions taken on the basis of a better understanding of the river.

In the LVFO, for instance, members are obliged to provide data on a wide range of fisheries-related aspects, including fisheries-relevant laws and regulations, data on fish landings, fish catch rates and stock assessments as well as the socioeconomic development of fisheries communities. This data is assessed through LVFO-related research projects and shared with the other RBO members, ensuring that all riparians possess the same information when taking decisions that concern the entire basin. A recent and very powerful example is the cooperation between LVFO members on cracking down illegal fishing—based on shared information on illegal fishing activities in each of the countries under the SmartFish Project by LVFO together with its partners (FAO 2013) and on joint policy actions. This demonstrates how strong scientific information can provide the basis for policy decisions and implementation (such as joint enforcement activities).

An important element for gathering data and information on the state of the river as well as on the effectiveness of river basin management mechanisms to guide policy makers is *monitoring*. The ICPDR has developed a comprehensive monitoring system, the so-called Trans-National Monitoring Network (TNMN) which monitors water quality along a number of monitoring stations on the mainstream and on tributaries (see Fig. 24.8). It focuses on 30 pre-defined parameters that are measured in the same way at all stations. While the information is gathered by each member state, it is submitted to the Secretariat, which summarizes findings in the TNMN Yearbook (e.g. ICPDR 2009) and informs the governance level of the ICPDR on achievements in restoring water quality throughout the basin. This guides policy makers in their decisions on what policy measures to implement to overcome persisting challenges.

The MRC has also established a wide-ranging monitoring network for the Lower Mekong Basin that gathers information on the basin's hydrology, ecology, climate or biodiversity (see Fig. 24.9). Monitoring results inform various MRC publications such as the State of the Basin Report (MRC 2010b) and are made available to members. Data gathering and analysis is highly sophisticated, relying on MRC's Procedures for Data and Information Exchange and Sharing (PDIES) as well as a comprehensive MRC Information System (MRC-IS) and highly developed analytical and modeling skills within the MRC Secretariat. The dissemination of this knowledge to the member states and their respective line agencies, thus going beyond the bottlenecks of the NMCs, is, however, limited. Consequently, knowledge that could inform better decisions does not reach those taking the

Fig. 24.8 Monitoring stations for the TNMN for the Danube River Basin (ICPDR 2009, p. 8)

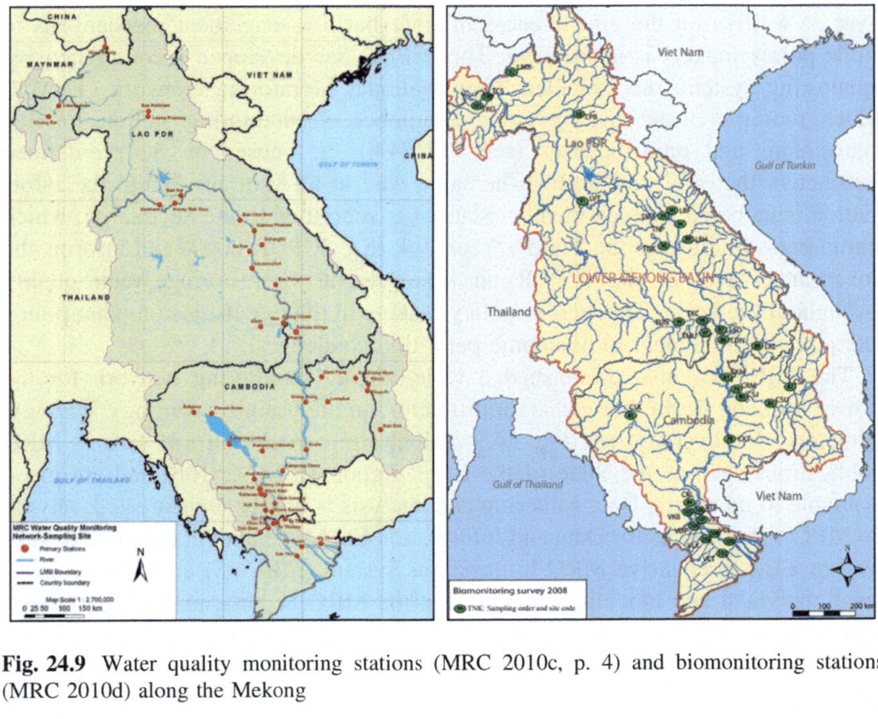

Fig. 24.9 Water quality monitoring stations (MRC 2010c, p. 4) and biomonitoring stations (MRC 2010d) along the Mekong

decisions. This underlines, again, that for the MRC it is not so much the availability of scientific data and information but the willingness of members to integrate this knowledge into their decision and policy making that impedes an effective SPL.

Conclusion

This analysis shows that the SPL varies considerably across RBOs—while the ICPDR and the ICPR have been particularly successful in translating river basin management science into policy, the MRC continues to struggle with ensuring that decisions concerning the management and the development of the Mekong River are based on scientifically sound decisions that ensure the long-term sustainable development of the basin. This can, to a large extent, be explained by the nature of the science provided by the RBO itself as well as by the institutional design of the RBO and the mechanisms it has at hand for translating science into policy.

While a number of research gaps on the SPL remain and a myriad of other factors can potentially determine the SPL as well, the analysis has allowed for drawing some conclusions that provide guidance to policy makers on how to strengthen the say of science in policy-making for international water resources management. They include, most importantly, efforts in (1) clarifying the RBO's role as a provider of sound, impartial and scientific knowledge, (2) ensuring the relevance of the RBO's science with regard to the most pressing challenges the basin and hence its policy-makers are facing, (3) adapting the RBO's organizational set-up to the required interactions between scientists and policy-makers, (4) building accessible data and information management and sharing systems, and (5) establishing approaches and systems for monitoring and evaluating change in the basin.

References

Agreement on the Cooperation for the Sustainable Development of the Mekong River Basin, 5 April 1995, Chiang Mai, Thailand

Bernauer T (1997) Managing international rivers. In: Young O (ed) Global governance. Drawing insights from the environmental experience. MIT Press, Cambridge, pp 155–195

Blatter J (2001) Integrative Symbole und regenerative Normen bei der Institutionenbildung. Erkenntnisse vom Gewässerschutz am Bodensee. Zeitschrift für Internationale Beziehungen 8(1):5–40

Bucceri A, Fink M (2003) Case study Lake Victoria. In: ETH seminar: science and politics of international freshwater management 2003/04, Zürich, Switzerland

Cash D, Adger N, Berkes F, Garden P, Lebel L, Olsson P, Pritchard L, Young O (2003) Scale and cross-scale dynamics: governance and information in a multilevel world. Ecol Soc 11(2), online version

Cleaver F, Franks T (2008) Distilling or diluting? Negotiating the water research-policy interface. Water Altern 1(1):157–176
Convention on the cooperation for the protection and sustainable use of the Danube River (Danube River Protection Convention), 29 June 1994, Sofia, Bulgaria
Convention on the establishment of the Lake Victoria fisheries organization (LVFO Convention), 30 June 1994, Kisumu, Kenya
Convention on the Protection of the Rhine (Rhine Convention), 12 April 1999 in Bern, Switzerland
Dombrowsky I (2007) Conflict, cooperation and institutions in international water management: an economic analysis. Elgar, Cheltenham
FAO (2013) The smartfish programme—implementation of a regional fisheries strategy for the eastern-southern Africa and indian ocean region. http://www.fao.org/fishery/smartfish/en
Friend R (2009) Fishing for influence: fisheries science and evidence in water resources development in the Mekong Basin. Water Altern 2(2):167–182
Gupta A, Andresen S, Siebenhüner B, Biermann F (2012) Science networks. In: Biermann F, Pattberg P (eds) Global environmental governance revisited. MIT Press, Cambridge, pp 69–93
Hegger D, Lamers M, Van Zeijl-Rozema A, Dieperink C (2011) Knowledge co-production in climate change adaptation projects: what are the levers of action? In: Paper presented at the 2001 earth system governance conference, Fort Collins, CO, 17–20 May 2011
ICPDR (2005) Guidelines for participants with consultative status and for observers to the ICPDR. Vienna, Austria, 26 April 2005
ICPDR (2009) Water quality in the Danube River basin—2009. TNMN Yearbook 2009. Vienna, Austria
ICPR (2004) Rhein. Lachs. 2020, Koblenz, Germany
ICPR (2009a) Analysis of the state of knowledge on climate changes so far and on the impact of climate change on the water regime in the Rhine watershed. Literature evaluation. Report No. 174, Koblenz, Germany
ICPR (2009b) Masterplan Wanderfische. ICPR Report No. 179, Koblenz, Germany
ICPR (2011) Study of scenarios for the discharge pattern of the Rhine. Report No. 188, Koblenz, Germany
ICPR (2013a) Aktueller Kenntnisstand über mögliche Auswirkungen von Änderungen des Abflussgeschehens und der Wassertemperatur auf das Ökosystem Rhein und mögliche Handlungsperspektiven. Report No. 204, Koblenz, Germany
IPCR (2013b) Lachs kann an rund 480 Hindernissen im Rheineinzugsgebiet wieder flussaufwärts wandern. Koblenz, Germany, http://www.iksr.org/index.php?id=191&ignoreMobile=1&tx_ttnews%5Btt_news%5D=813&cHash=907d488c97b22804fc6921a3faa47bd7
Koetz T, Farrell K, Bridgewater P (2011) Building better science-policy interfaces for international environmental governance: assessing potential within the intergovernmental platform for biodiversity and ecosystem services. Int Environ Agreem. doi:10.1007/s10784-011-9152-z (online version)
LVFO (2005a) The institutional structure and programs of the organization. Entebbe, Uganda 2005
LVFO (2005b) National guidelines for beach management units. Entebbe, Uganda
LVFO (2005c) Implementation and financing plan for the strategy to reduce the impact of HIV/Aids on fishing communities—2005–2015. Entebbe, Uganda
Marty F (2001) Managing international rivers. Problems, politics and institutions. Peter Lang, Bern
McDonnell R (2008) Challenges for integrated water resources management: how do we provide the knowledge to support truly integrated thinking? Int J Water Resour Dev 24(1):131–143
McNie E (2007) Reconciling the supply of scientific information with user demands: an analysis of the problem and review of the literature. Environ Sci Policy 10:17–38
Miles E, Underdal A, Andresen S, Wettestad J, Skjaerseth J, Carlin E (eds) (2002) Environmental regime effectiveness. Confronting theory with evidence. MIT Press, Cambridge
Molle F (2004) Nirvana concepts, narratives and policy models: insights from the water sector. Water Altern 1(1):131–156

MRC (2010a) MRC strategic environmental assessment of hydropower on the Mekong mainstream. Final Report, Vientiane, Lao PDR, MRCS
MRC (2010b) State of the Basin Report 2010. Vientiane, Lao PDR
MRC (2010c) The Mekong River report card on water quality. Vientiane, Lao PDR
MRC (2010d) Report on the 2008 biomonitoring survey of the lower Mekong River and selected tributaries, MRC technical paper No. 27. Vientiane, Lao PDR
MRC (2011a) Procedures for notification, prior consultation and agreement (PNPCA). Proposed xayaburi dam project—Mekong River. Prior consultation project review report. 24 March 2011, Vientiane, Lao PDR
MRC (2011b) Procedures for notification, prior consultation and agreement form of reply to prior consultation—replying state: The socialist republic of Vietnam, submitted to the special MRC joint committee meeting. 19 April 2011, Vientiane, Lao PDR
Poff L, Allan D, Palmer M, Hart D, Richter B, Arthington A, Rogers K, Meyer J, Stanford J (2003) River flows and water wars: emerging science for environmental decision making. Frontiers Ecol 1(6):298–306
Quevauviller P, Balabanis P, Fragakis C, Wydert M, Oliver M, Kaschl A, Arnold G, Kroll A, Galbiati L, Zaldivar J, Bidoglio G (2005) Science-policy integration needs in support of the implementation of the EU water framework directive. Environ Sci Policy 8:203–2011
Radio Free Asia (RFA) (2013) Laos confirms 'preparatory' work on controversial Dam Project, 31 July 2013
Roux D, Rogers K, Biggs H, Ashton P, Sergeant A (2006) Bridging the science-management divide: moving from unidirectional knowledge transfer to knowledge interfacing and sharing. Ecol Soc 11(1), online version
Schmeier S (2013) Governing international watercourses. River Basin organizations and the sustainable governance of internationally shared rivers and lakes. Routledge, London
Van den Hove S (2007) A rational for science-policy interfaces. Futures 39(7):1–19
Vientiane Times (2013a) Work powers ahead on Xayaburi Dam, 23 January 2013
Vientiane Times (2013b) MRC countries not opposed to Xayaburi Dam, 25 January 2013
Vogel C, Moser S, Kasperson R, Dabelko G (2007) Linking vulnerability, adaptation, and resilience science to practice: pathways, players and partnerships. Glob Environ Change 17:349–364
Weller P, Popovici M (2011) Danube River basin management—rationale and results. How to link science, as the basis for policy. River Syst 20(1–2):103–109

Chapter 25
The Human Right to Water and Sanitation: Reflections on Making the System Effective

Pedi Obani and Joyeeta Gupta

Abstract The Millenium Development Goal (MDG) on water has been more successful than the MDG on sanitation. Does this have implications for the human right to sanitation? This chapter argues that there are key differences between access to water and sanitation in terms of the legal content of both, the physical infrastructure needed, the costs of the service, cost recovery, and the differences in the preparedness of people to pay for this service which may in some cases make this right an 'imposed right'. These differences may lead one to argue that in different circumstances it may be more appropriate to talk of a combined right or separate rights, respectively. Given that there are unhygienic alternatives to sanitation services, there may be a need to include another key element into the right to sanitation, namely: to provide people a better knowledge of the need for sanitation services and to explain why this is seen as both a right and a responsibility. This is essential to making the system effective and economically viable, as only if people understand why this right has been created will it be possible to make people pay for it.

This chapter covers ongoing PHD research financed by the NUFFIC Netherlands Fellowship Professional Project No. CF 162/2012.

P. Obani (✉) · J. Gupta
Department of Integrated Water Systems and Governance, UNESCO-IHE
Institute for Water Education, Delft, Netherlands
e-mail: pedi.obani@gmail.com

P. Obani
Faculty of Law, University of Benin, Benin, Nigeria

P. Obani · J. Gupta
Governance and Inclusive Development, Amsterdam Institute for Social Science Research, University of Amsterdam, Amsterdam, Netherlands

A. Bhaduri et al. (eds.), *The Global Water System in the Anthropocene*,
Springer Water, DOI: 10.1007/978-3-319-07548-8_25,

Introduction

Drinking water and sanitation are public goods, i.e. the use of these services by one should not diminish the use by others and no one should be excluded from their use (i.e. non-rival and non-exclusive). By their very definition, the market is generally inadequately able to provide for public goods and this calls for some sort of state intervention. Such state intervention can be articulated in terms of state responsibility or in terms of human rights. At the international and national level, there has been a choice to articulate this governance initiative in terms of human rights; this automatically means that the state but also other actors may be held responsible for implementing these rights.

The human rights approach to the provision of water and sanitation services is, hence, appropriate because it elevates access to water and sanitation services to the status of a legal entitlement, and makes it the primary responsibility of national governments to respect, protect, and fulfil the right in favour of their populations (Albuquerque 2012); promotes good governance,[1] including accountability and sustainable solutions (London and Schneider 2012); helps create contextually relevant standards for minimum access to water and sanitation services (Albuquerque 2012); gives a voice to the voiceless (Sinden 2005, 2009; Gearty 2006; London and Schneider 2012); and stimulates the setting up of institutions to provide services and adjudicate where these rights are violated (London and Schneider 2012). Despite the shortcomings in the human rights approach—it does not guarantee implementation (Donoho 2006; Nkhata and Mwenifumbo 2010), may not take environmental issues into account (Stallworthy 2008), and may not account for locally perceived priorities (Joshi et al. 2011)—, the recognition of access to water and sanitation as a human right is a necessary step for an adequate living standard for everyone, irrespective of someone‘s social or economic status (Albuquerque 2010).

Following a gradual process of adopting international law instruments in this field, in 2010, the UN General Assembly (UNGA) and the UN Human Rights Council (HRC) passed two resolutions (UN General Assembly 2010; UN Human Rights Council 2010) acknowledging the human right to water and sanitation, and reiterating that water and sanitation are essential to the realisation of all human rights. Before then, water and sanitation were also included in the Millennium Development Goals (MDG). These targets aim at halving the number of people without access to water and sanitation by 2015 (WHO/UNICEF Joint Monitoring Programme for Water Supply and Sanitation 2006). The increased recognition of water and sanitation as a MDG and a human right has resulted in close and regular monitoring of the level of access to improved drinking water sources and improved

---

[1] The principles of good governance include independence, openness and transparency, accountability, effectiveness, clarity of purpose, legitimacy, legality, efficiency, integrity and the rule of law.

sanitation facilities, by various UN bodies, international organizations, NGOs, and government agencies. The official records highlight inequities in the levels of access to and the cost of improved drinking water and sanitation between different regions, and within population groups within countries (United Nations 2013; World Health Organization and UNICEF 2013). They show that although between 1990 and 2011, 2.1 billion people gained access to improved potable water services, only 1.9 billion people gained access to improved sanitation facilities (World Health Organization and UNICEF 2013). Around the world, lack of access to water and sanitation still affects approximately 768 million and 2.5 billion people, respectively and 1 billion people still engage in open defecation using buckets, plastic bags, water bodies, open land, and other public spaces due to lack of adequate toilets (World Health Organization and UNICEF 2013). Therefore, in the long run, the MDG target on water has proved to be more successful than the target on sanitation.

Given the differences in success rates between meeting access to water and sanitation goals, this chapter therefore explores three main questions: (a) Is the human right to sanitation (HRS) different from the human right to water (HRW)? (b) What are the contextual issues that should be taken into account to enhance the implementation of HRS? (c) What are the implications of such contextual issues for the effective implementation of the HRS?

The first section Differences Between HRS and HRW analyses the differences between the HRS and HRW and some contextual issues, while the following section Implications for Enhancing the Effective Implementation of the Human Right to Sanitation addresses the implications of these differences for enhancing the effective implementation of HRS.

Differences Between HRS and HRW

This section examines the differences in the legal content and the physical infrastructure required for HRS and HRW. It also addresses some of the important issues which affect implementation, such as the cost of services, cost recovery, and the reasons why people are unprepared to pay for HRS as opposed to HRW.

Differences in Legal Content

HRS and HRW have evolved through three parallel but independent routes, of implicit recognition (when the right to water and sanitation services was seen as implicit in the broader right to life and health), explicit recognition (when treaties and other instruments explicitly recognized this right), and an independent but combined recognition (when the right emerged independently of other rights in documents). This evolution has taken place in various international human rights declarations, international treaties, national laws, and court cases. In August 2007, following decision 2/104 of the United Nations High Commissioner for Human Rights, the report from the High Commissioner for Human Rights on the scope and content of the relevant human rights obligations related to equitable access to safe drinking water and sanitation under international law expressly stated that

> It is now the time to consider access to safe drinking water and sanitation as a human right, defined as the right to equal and non-discriminatory access to a sufficient amount of safe drinking water for personal and domestic uses… to sustain life and health.

This mandate covers water supply for sanitation, only in a limited sense, but it does not expressly mention other sanitation services such as waste collection, disposal and treatment. Subsequent declarations, such as the 2010 UNGA and UNHRC resolutions, have neither clarified nor further expanded the meaning of HRS. Furthermore, although drinking water quality standards exist at the international level, and this is used as a baseline in many national water policies, there are no similar global quality standards for safe sanitation beyond ensuring that toilets are technically and hygienically safe for personal and domestic use.[2] Beyond toilets, the meaning of sanitation also encompasses access to "wastewater facilities and services that ensure privacy and dignity, ensuring a clean and healthy living environment for all" (COHRE 2008, p. 17). With regards to the MDG, target 7.C explicitly refers to access to safe drinking water, the safety criterion is only implied in the case of sanitation in terms of "hygienically separating human excreta from human contact" (World Health Organization and United Nations Children's Fund 2013, p. 12), but this does not go far enough to include wider aspects of personal hygiene, or even the safe collection, treatment, disposal, or re-use of human excreta in order to ensure a healthy living environment (Albuquerque 2010); while the official records only monitor access to "improved" water sources and sanitation facilities" (World Health Organization and United Nations Children's Fund 2013). The adoption of these rights has given further impetus to the implementation of the MDGs as many countries have supported this right and are gradually trying to implement them. However, it is important to note that while many developing countries already had a position on the human right to water, most of them did not have a similar position on the human right to sanitation and so the base line for effectiveness differs between the two. For example, the United Kingdom officially recognised sanitation as a human right only in 2012, but this excludes "the collection and transport of human waste" (Foreign and Commonwealth Office 2012). In general the HRW implies access to a certain quantity of water (i.e. minimum amount per person per day at a reasonable distance from the location of an individual), quality (i.e. safety in terms of health issues), accessibility (in terms of access for all ages and groups), and affordability

[2] The existing guidelines for sanitation mainly recommend the number of people who may use the same toilet in public places such as schools and offices (Workplace (Health, Safety and Welfare) Regulations 1992: The Sphere Handbook), and technical requirements for the safety of sanitation facilities.

(in terms of the costs of water). Many of these criteria have been further articulated (Albuquerque 2010, 2012). Such articulation includes, for instance, that people are expected to contribute to the cost of meeting their water and sanitation needs, but the cost should not be more than 5 % of the household's income (UN-Water Decade Programme on Advocacy and Communication, Water Supply and Sanitation Collaborative Council 2010). However, while the HRS implies access in terms of safety and proximity to improved sanitation at affordable costs, the actual articulation of these standards is significantly less well developed.

Cultural and Infrastructural Differences

In addition to the differences in the legal content, there are also cultural and infrastructural differences between the HRS and the HRW. First, while access to water supply and its benefits for well-being are freely discussed and openly demanded as a necessity for humans, the equally important and closely linked subject of sanitation is often seen as a taboo subject for public discussion (Black and Fawcett 2008), reducing the demand for such facilities and the political need for action. This may also affect the location of facilities.[3] There is no alternative to safe drinking water for personal and domestic use, while there is an alternative to improved sanitation services—namely open defecation. As a result, there is a stronger demand for drinking water facilities whereas the demand for sanitation is more latent, especially in low income settings (Department for International Development 2007). This suggests that sanitation may be perceived as an imposed right, particularly in cultures where the subject is not openly broached for discussion.[4] Hence, policies tend to emphasize reducing the collection time and travel distance to access improved sources of water supply. This has resulted in the establishment of drinking water supply systems that are not always linked to sanitation systems.

Second, drinking water and sanitation systems can have independent infrastructures as where ground water is funnelled for drinking water services, and sanitation services just link the sanitation system to a disposal regime; or an integrated system which recycles the water from sanitation systems into drinking water. In the short term, integrated systems may be more expensive than non-integrated systems (see Sect. Cost Recovery). But in order to ensure that the service meets the needs of the users, there is need for effective stakeholder participation in the design of these systems (Patkar and Gosling 2011).

---

[3] For instance, in some parts of Madagascar, locating a toilet inside a house or on one's land is a local taboo; nationwide over 10,000 people die annually from sanitation-related diseases (IRIN 2012).

[4] This could also account for the emphasis on sanitation marketing, particularly among the poor who often have other competing basic needs, such as shelter and feeding, which may rank higher in their scheme of priorities.

Third, the quantity of water needed for drinking is relatively limited compared to what is required for sanitation and hygiene, especially in the developed world when a standard western style toilet is conceived. While the world's population that lacks access to clean water is located mainly in developing countries, and use only about 5 l of water daily (United Nations Development Programme 2006), the recommended minimum daily requirement for drinking water is 7.5 l per person, excluding the quantity needed for other health and well-being purposes such as food preparation (Howard and Bartram 2003; World Health Organization 2003). In comparison, the daily average quantity of water used in flushing toilets in developed countries amounts to about 50 l per person (United Nations Development Programme 2006). However, the latter does not need to meet potable water standards! This raises issues of cost, of the use of scarce water resources, and of the kind of sanitation system that is contextually appropriate in different parts of the world.

Cost of Service

The global benefit-cost ratios calculated in terms of return per dollar invested is significantly higher for sanitation, followed by sanitation and water supply combined, and then water supply alone (Hutton and World Health Organization 2012, Hutton 2013). Utilising the standard water supply and sanitation coverage definitions, the WHO and UNICEF Joint Monitoring Programme for Water Supply and Sanitation estimated the global benefit-cost ratios calculated in terms of return per dollar invested as 5.5 for sanitation, 4.3 for sanitation and water supply combined, and 2.0 for water supply alone (Hutton and World Health Organization 2012, Hutton 2013). The cost of providing drinking water is also significantly lower than the costs of providing sanitation services, although there are differences in the various estimates depending on the definition of access to water supply and sanitation, and the type of technology selected (Toubkiss 2006) and the extent to which environmental issues are also taken into account. The estimated cost of providing universal access is USD 35 billion per year for sanitation and USD 17.5 billion for drinking-water, over a period of 5-years from 2010–2015 (Hutton 2013). This can be related to the significantly lower level of access to sanitation that has been observed among the poorest households, as measured by wealth quintiles, and among poor regions such as East Asia, Latin America and the Caribbean, West Asia, South-East Asia, Caucasus and Central Asia, South Asia, North Africa, Oceania, and sub-Saharan Africa (Hutton and World Health Organization 2012; WHO and UNICEF Joint Monitoring Programme for Water Supply and Sanitation 2012). However, in some cases the high cost and lack of access to sanitation facilities may be attributed to high technical specifications which make the facilities unaffordable. For instance, a recent study conducted in Cambodia showed that although there was a strong demand for toilets, there was also a preference for a sophisticated design which cost as much as USD 150 compared to simpler but hygienic options that cost less than USD 10 (Salter 2008). Similarly, the 'toilet wars' being waged in South Africa are also partly driven by discontent over the technical design of the facilities that are provided by the government in low income areas (Zille 2013; Robins 2013).

Cost Recovery

The right level of costing or tariff is crucial to ensure the sustainability of water and sanitation services. Ten years after the Dublin Statement on Water (1992) declared that water should be recognised "as an economic good", privatization and full cost recovery for water and sanitation services were advocated at the World Summit on Sustainable Development in 2002 (Barlow 2009). However, the provision of water and sanitation services is capital and infrastructure intensive and requires long-term investments (Baumann and Boland 1998). Poor households in developing countries often discount the future heavily, and as a result they cannot provide the long-term financial commitments and huge capital investments required (Poulos and Whittington 2000). This makes the service available only to those who can afford to pay and has therefore generated wide-spread criticism and precipitated a water crisis in many poor parts of Latin America and Africa (Barlow 2009). This is also especially true where these households are located in remote regions requiring even higher service costs.

If one of the rationales for full cost recovery is to promote efficient use and prevent wastage of water resources, this is not supported by the evidence from developed countries where a larger amount of water is often used for showers, sanitation, and washing than is available for the average household in developing countries, and at a relatively lower cost too. For instance, the estimated quantity of water required, based on a developed country plumbing system, is: (a) 11.4–18.9 l per flush for toilets (b) 11.4–18.9 l per minute for running tap water required for brushing the teeth, washing dishes, etc. (c) 18.9–26.5 l per minute for showering, and (d) 132.6–189.3 l per bath (The World Bank Group 2003).[5] This exceeds by far the quantity of water used by an average slum dweller for a whole day; people who live more than 1 km away from a water source often have less than 5 l of unsafe water available for use per day (World Health Organization and United Nations Children's Fund 2005). In addition, while an average poor household in developing countries spends between 9 and 20 % of its income on water as illustrated in Table 25.1 (The World Bank Group 2003), by contrast in developed countries, the expenditure of households on water and sanitation is much less in relation to income.[6]

---

[5] The figures are calculated based on a metric conversion rate of 1 gallon equal to 3.7854 l.

[6] For instance in France, water and sanitation services only represent 1.25 % of average household income and there are provisions made for social tariffs to benefit the poor.

Table 25.1 Percentage of poor households income spent on water in developing countries

Region	City/Country	Percentage of poor households income spent on water
East Africa	Ukunda, Kenya	9
Sub-Saharan Africa	Addis Ababa, Ethiopia	9
Asia and Pacific	Bangladesh	11[a]
Sub-Saharan Africa	Onitsha, Nigeria	18
North and Central America and the Carribean	Port-au-Prince, Haiti	20
North and Central America and the Carribean	Port of Spain, Trinidad and Tobago	20

[a] This is the amount spent on fuel for boiling drinking water
Source World Bank (1993); The World Bank Group (2003)

Differences in People's Preparedness to Pay for Water and Sanitation Services

It appears that people may be more prepared to pay for water than for sanitation services for a number of reasons. First, people may be constrained to pay for safe drinking water because it a necessity and non-substitutable and alternative sources of water (e.g. bottled water) may be more expensive, but they may choose not to pay for sanitation and instead opt for open defecation (OD).

Second, the poor quality and unreliable nature of services from utilities has been recorded as affecting the preparedness of customers to pay for both water and sanitation services (Whittington et al. 1990; World Bank 1993). Despite the high rate of water coverage in urban areas especially, the quality of service still remains a concern. In the case of sanitation, in addition to availability, toilets have to be hygienically maintained in order to ensure their use and to avoid diseases. Health care savings account for between 5 and 13 % of total global economic benefits of water and sanitation, estimated at USD 260 billion annually (Trémolet and Rama 2012). However, the maintenance of public toilets is an unpleasant task, and in some societies this may be avoided by people leaving public toilets dirty and unusable. This may also push people to more readily engage in OD (BBC 2008; Pappas 2011). In India, personal observation has revealed that many public toilets are simply not used because of the poor maintenance of these toilets, the task of cleaning toilets is left to specific castes, most people do not leave the toilet as clean as they found it and clearly people are not willing to pay enough for their maintenance. However, gender may play a determining role in who is willing to pay. As women face a higher risk of physical assault, loss of dignity, and exposure to diseases as a result of engaging in OD than do men, they may be more willing to pay for sanitation services but this will only be relevant where they are able to influence economic decisions in the household to cater for their needs. Age is also an important factor affecting people's ability to engage in OD; children may find it easier to engage in OD, while adults, especially the elderly may be discouraged by physical, and health reasons, as well as social factors.[7]

A third important reason is the level of individual ownership and responsibility for access to water and sanitation. These may be affected by various factors, including: psychological factors (Rosenquist and Emilia 2005), political history and governance structure (Njoh and Akiwumi 2011), and land tenure[8] (Scott et al. 2013). On the one hand, people may be more willing to take ownership for the provision of safe drinking water because it is seen as a necessity. On the other hand, providing subsidies for toilets may not automatically solve the problem if people do not prioritize toilets and divert the subsidy for other uses.[9] Hence, the internalization of the knowledge and changes in mindsets and behavior at all levels are very important for successful sanitation programmes (Hickling and Bevan 2009); people need to realize that: "they are eating one another's shit" (Chambers 2009, p. 11) in environments where open defecation is practiced.

Implications

HRS and HRW are different in terms of legal content, physical infrastructure requirements, and the cost of implementation, which is significantly higher for sanitation. The human rights approach imposes a legal duty on States to provide adequate services, and requires that no one should be deprived of access as a result of their inability to pay (UN-Water Decade Programme on Advocacy and Communication, Water Supply and Sanitation Collaborative Council 2010, UN-HR, OHCHR, UN-HABITAT et al. 2010). The cost of sanitation services especially needs to be a small percentage of the income of the poor in order to be affordable, because of the options of OD or stay at home—for women, girls and the elderly. This would require innovation to ensure both technical safety, hygiene, and acceptability at an affordable cost for the potential users. WHO uses a 3 % standard of income for water services (Howard and Bartram 2003)—but whether this also applies to sanitation is unclear.

---

[7] In 2007, Help the Aged, UK, published the results of a survey which showed that 52 % of respondents did not go out as often as they wanted due to the fear of not finding a toilet to use (Help the Aged, *Nowhere to Go: Public Toilet Provision in the Uk* (London: Help the Aged 2007).

[8] In India for instance, home owners pay a fixed price for water and sanitation services which is often quite low while non-home owners in urban areas have to pay per glass of waters or per visit to the toilet. Furthermore, slum dwellers may be unwilling to invest in the provision of sanitation infrastructure due to the imminent fear of forced eviction (personal observation).

[9] For instance, in the 1980s in India, the government project of providing subsidised toilets failed as a result of lack of use; people preferred instead to return to open defecation that they were accustomed to (Stephanie Pappas, "With 7 Billion People, World Has a Poop Problem" http://www.livescience.com/16713-7-billion-people-world-poop-problem.html (accessed 7 August 2013)).

While some governments have responded by developing social tariffs to protect access to water and sanitation services for the poor, social tariffs alone cannot mitigate other forms of vulnerability and marginalization which affect women, children, the physically disabled, minorities such as indigenous people, and migratory populations, for instance. Hence, the human rights approach cannot guarantee the sustainability of water and services without adequate funding and, as such, service providers need to find innovative ways of maintaining a balance between cost recovery and protecting access for the poor, marginalized and vulnerable. The human rights approach must also address other factors which affect people's preparedness to pay, in order to secure a sustainable financial base for service provision.

Implications for Enhancing the Effective Implementation of the Human Right to Sanitation

Given the differences between the human rights to water and sanitation highlighted in the previous section, this section analyses whether the rights should continue to be recognised as combined or independent rights. The section also examines whether the provision of sanitation services should be driven by supply or demand.

Arguments for and against a Combined Right

There are three clear options for formulating the human rights to water and sanitation, namely: as a combined right that is implemented in an integrated manner, as a combined right with differentiated implementation, or as two separate rights, implemented separately. The similarities between water supply and sanitation, the use of water for both services, and the relative cost-effectiveness of an integrated system (see above) support a *prima facie* case for calling for integrated water supply and sanitation. The further need for integration between the water supply and sanitation sector and other water consuming sectors has also led to the adoption of integrated water resources management (IWRM) approaches which seek integrated and affordable solutions for efficiently managing the competing demands for the limited water resources available (Butterworth and Soussan 2001).

If water and sanitation are provided in an integrated manner and are properly managed, this may lead to the recovery of essential nutrients such as phosphorus from wastewater and other benefits for human health, the environment, and the economy (Montangero and Belevi 2008; Cordell et al. 2009, Kvarnström et al. 2011). Although in the short term such integration may prove to be more expensive than simply providing drinking water alone, adequate sanitation is necessary to ensure the long term sustainability of the water supply as well as contributing to human health. Therefore, in countries where the human right to water and sanitation is yet to be recognised, it may still be advisable to combine both to ensure the sustainability of water supply and for the issue of sanitation to benefit from the wide acceptance of drinking water supply in the development discourse, at least until such time when the cultural inhibitions against sanitation have been overcome. Furthermore, in situations where water and sanitation are already recognised as human rights, it is also necessary to recognise their distinct normative requirements for better effectiveness. For instance, in the case of sanitation there are additional normative requirements such as cultural and social acceptability for different population groupings which may not be equally important in determining drinking water quality.

The Role of Supply and Demand

There are now many interventions focused on creating a ''demand'' for drinking water and sanitation services due to the belief that people in areas with the least access may be reluctant to pay for access to formal water and sanitation services. Nevertheless, such arguments neglect the issue of "low-level" performance equilibrium which commonly affects public utility services in developing countries where the users have low confidence in the ability of the public utility to deliver a high quality of service and as such they commonly accept low performance from the public utility in return for a small tariff or non-payment for services (Nickson and Franceys 2003). The lack of demand for improvements in the quality of service and the poor funding eventually create a vicious cycle of poor performance by the utilities and lack of willingness to pay for unreliable services on the part of users. It therefore follows that sanitation programmes which are solely dependent on government or donor funding often fail at the end of the funding period. To be sustainable, sanitation services and new technology must be reliable and responsive to human needs and expectations in order to avoid the vicious cycle of low-level performance equilibrium.

Implications

The recognition of sanitation as a human right in combination with water, for instance as contained in the UNGA and HRC Resolutions of 2010 and in the Uruguay Constitution, offers a range of benefits for the integrated management of water, sanitation, and hygiene because all three services are intricately connected. For instance, the disconnection of water services as a result of the inability to pay also impacts on water dependent sanitation and hygiene (Albuquerque 2009). Furthermore, the unsafe disposal of untreated waste may also lead to the contamination of groundwater (Albuquerque 2010). However, in practice, the combination of water and sanitation has often led to greater emphasis on potable water and the neglect of sanitation; for instance, only about one third of the total aid targeted at the water and sanitation sector is actually used for sanitation even though more effort is required in this area (World Health Organization 2010). The options remain to recognise sanitation as a separate right but this may be more difficult due to enormous costs and cultural constraints. However, in all cases, it is necessary to differentiate the normative content of HRS from the HRW and develop distinct indicators for monitoring and evaluating progress with implementation. It is also important to ensure that sanitation facilities are suitable, acceptable, and capable of reliable service delivery in order to ensure sustainability.

Conclusion

There are fundamental differences between HRW and HRS which require further unbundling of the normative content of the latter, in line with the needs and expectations of the poor, vulnerable, and marginalised, rather than anticipating demand. This is essential to ensuring universal acceptability and effective implementation of HRS. Given that the poor, vulnerable, and marginalized have a less hygienic alternative to sanitation services, it may be necessary to ensure that this right (a) also includes a public awareness component to make people (those who lack the service but also more broadly others in society in order to create the demand for such services) understand why the right is important. (b) If this right is to be effectively implemented, it is also necessary to ensure that following the public awareness of the significance of this right, the specific needs of the local populations are taken into account, possibly through public consultation. (c) An integrated approach to water and sanitation appears to be more economic than separate infrastructures for both, but is more expensive than simply providing drinking water. This may however be a short-term concern. It will in the long-term be essential to close the sanitation-supply cycle to ensure the sustainability of the water resource. It may also be possible to reduce the costs of joint infrastructure if a separation is made between the kind of water quality needed for drinking and cooking and for sanitation services. (d) Clearly the provision of water supply and sanitation needs to be financially viable. Although the human rights approach does not prescribe any particular economic model for the provision of facilities and services, financially able users can reasonably be expected to pay for the cost of their water and sanitation services. However, the cost must not affect the household's capacity to acquire other essential goods and services, including education, food, health services, and housing (UN-Water Decade Programme on Advocacy and Communication, Water Supply and Sanitation Collaborative Council 2010). This calls for a further articulation of the percentage of income that can be spent on sanitation services. Some sort of system of cross-subsidies will be necessary to ensure universal access. (e) Finally, it is critical that the system also takes environmental issues into account and includes technological innovation (for instance, composting dry toilets) that may help to close cycles in an innovative and contextual way, thus reducing costs, while perhaps recycling the scarce phosphorous in human excreta. In the ultimate analysis, the system needs to be economically viable (either directly or indirectly) and environmentally sustainable in order for the human right to sanitation to be effectively implemented. Legal scholars should not just stop at the normative adoption of the principle, but they should understand their subject in order to provide clearer guidance about how this right can be actually implemented.

References

Albuquerque CD (2009) Promotion and protection of all human rights, civil, political, economic, social and cultural rights, including the right to development: report of the independent expert on the issue of human rights obligations related to access to safe drinking water and sanitation United Nations Human Right Council

Albuquerque CD (2010) Report of the independent expert on the issue of human rights obligations related to access to safe drinking water and sanitation. United Nations General Assembly

Albuquerque CD (2012) On the right track back: good practices in realising the rights to water and sanitation. Texttype, Lisbon

Barlow M (2009) Blue covenant: the global water crisis and the coming battle for the right to water. New Press, New York

Baumann DD, Boland JJ (1998) The case for managing urban water. In: Baumann DD, Boland JJ, Hanemann WM (eds) Urban water demand management and planning. Mc-Graw-Hill, New York, pp 1–30

BBC (2008) School toilets 'Bullying Fears'. http://news.bbc.co.uk/2/hi/uk_news/education/7213977.stm. Accessed 7 Aug 2013

Black M, Fawcett B (2008) The last taboo: opening the door on the global sanitation crisis. Earthscan, London

Butterworth J, Soussan J (2001) Water supply and sanitation & integrated water resources management: why seek better integration? Prepared for WHIRL project workshop on 'water supply & sanitation and watershed development: positive and negative interactions', Andhra Pradesh, India, 5–14 May 2001

Chambers R (2009) Going to scale with community-led total sanitation: reflections on experience, issues and ways forward. IDS Practice Paper 1. IDS, Brighton

COHRE, UN-HABITAT, WATERAID, SDC (2008) Sanitation: a human rights imperative. COHRE, Geneva

Cordell D, Drangert JO, White S (2009) The story of phosphorus: global security and food for thought. Global Environ Chang 19:292–305

Department for International Development (2007) Sanitation policy background paper: water is life, sanitation is dignity. DFID, London

Donoho DL (2006) Human rights enforcement in the 21st century. Bepress Legal Series. http://law.bepress.com/expresso/eps/1282. Accessed 2 Aug 2013

Foreign & Commonwealth Office (2012) UK recognises right to sanitation. www.gov.uk/government/news/uk-recognises-right-to-sanitation. Accessed 1 Aug 2013

Gearty CA (2006) Can human rights survive? the Hamilyn Lectures 2005. Cambridge University Press, Cambridge

Help the Aged (2007) Nowhere to go: public toilet provision in the UK. Help the Aged, London

Hickling S, Bevan J (2009) Scaling up CLTS in Sub-Saharan Africa. In: Bongartz P (ed) Tales of shit: comunity-let total sanitation in Africa. IIED, London

Howard G, Bartram J (2003) Domestic water quantity, service level, and health. World Health Organization, Geneva

Hutton G (2013) Global costs and benefits of reaching universal coverage of sanitation and drinking-water supply. J Water Health 11(1):1–12

Hutton G, World Health Organization (2012) Global costs and benefits of drinking-water supply and sanitation interventions to reach the MDG target and universal coverage. World Health Organization, Geneva. http://whqlibdoc.who.int/hq/2012/WHO_HSE_WSH_12.01_eng.pdf. Accessed 1 Aug 2013

IRIN (2012) Madagascar: addressing toilet taboos to improve sanitation. Available through IRIN News. http://www.irinnews.org/report/95136/madagascar-addressing-toilet-taboos-to-improve-sanitation. Accessed 7 Aug 2013

Joshi D, Fawcett B, Mannan F (2011) Health, hygiene and appropriate sanitation: experiences and perceptions of the urban poor. Environ Urban 23(1):91–111

Kvarnström E, McConville J, Bracken P et al (2011) The sanitation ladder—a need for a revamp? J Water San Hyg Dev 1:3–12

London L, Schneider H (2012) Globalisation and health inequalities: can a human rights paradigm create space for civil society action? Soc Sci Med 74:6–13

Montangero A, Belevi H (2008) An approach to optimise nutrient management in environmental sanitation systems despite limited data. J Environ Manag 88(4):1538–1551

Nickson A, Franceys R (2003) Tapping the market: the challenge of institutional reform in the urban water sector. Palgrave Macmillan, New York

Njoh AJ, Akiwumi FA (2011) The impact of colonization on access to improved water and sanitation facilities in African cities. Cities 28(5):452–460

Nkhata MJ, Mwenifumbo AW (2010) The more things seem to change, the more they stay the same: the strained matrix for the protection and enforcement of human rights in Africa. E Afr J Peac Hum Rights 16(2):139–215

Pappas S (2011) With 7 billion people, world has a poop problem. http://www.livescience.com/16713-7-billion-people-world-poop-problem.html. Accessed 7 Aug 2013

Patkar A, Gosling L (2011) Equity and inclusion in sanitation and hygiene in Africa: a regional synthesis paper. A catalytic working paper for the 3rd AfricaSan conference supporting the technical session 'reaching the unserved: equity and inclusion in Africa', Kigali, July 2011

Poulos C, Whittington D (2000) Individuals'rates of time preference in developing countries: results of a multi-country study. Environ Sci Technol 43(8):1445–1455

Robins S (2013) Toilet wars: politicisation of human waste. Cape Times, 27 Sept 2013, p 9

Rosenquist D, Emilia L (2005) A psychosocial analysis of the human-sanitation nexus. J Environ Psychol 25(3):335–346

Salter D (2008) Identifying constraints to increasing sanitation coverage: sanitation demand and supply in Cambodia. Water and Sanitation Program, Phnom Penh

Scott P, Cotton A, Khan MS (2013) Tenure security and household investment decisions for urban sanitation: the case of Dakar, Senegal. Habitat Int 40:58–64

Sinden A (2005) In defense of absolutes: combating the politics of power in environmental law. Iowa Law Rev 90:1405–1512

Sinden A (2009) Power and responsibility: why human rights should address corporate environmental wrongs. In: Mcbarnet DJ, Voiculescu A, Campbell T (eds) The new corporate accountability: corporate social responsibility and the law. Cambridge University Press, Cambridge, pp 501–528

Stallworthy M (2008) Understanding environmental law. Sweet & Maxwell, London

The Dublin statement on water and sustainable development (1992) International conference on water and sustainable development, Rio de Janeiro, June 1992

The World Bank Group (2003) Development education program's DEPweb. http://www.worldbank.org/depweb/english/modules/environm/water/textq.html. Accessed 30 Jul 2013

Toubkiss J (2006) Costing MDG target 10 on water supply and sanitation: comparative analysis obstacles and recommendations. World Water Council, Quebéc

Trémolet S, Rama M (2012) Tracking national financial flows into sanitation, hygiene and drinking-water. WHO, Geneva

UN-Water Decade Programme on Advocacy and Communication, Water Supply and Sanitation Collaborative Council (2010) The human right to water and sanitation: media brief. UNW-DPAC, Zaragoza

UN-HR, OHCHR, UN-HABITAT et al (2010) (The) Right to water, fact sheet no 35. OHCHR, Geneva

UN General Assembly Res. 64/292, U.N. Doc. A/RES/64/292 (2010)

UN Human Rights Council Res. 15/9, U.N. Doc. A/HRC/RES/15/9 (2010)

United Nations (2013) The millennium development goals report 2013. United Nations Publications, New York

United Nations Development Programme (2006) Human development report 2006, beyond scarcity: power, poverty and the global water crisis. Palgrave Macmillan, New York

Whittington D, Okorafor A, Okore A et al (1990) Strategy for cost recovery in the rural water sector: a case study of Nsukka District, Anambra State, Nigeria. Water Resour Res 26(9):1899–1913

WHO & UNICEF Joint Water Supply Sanitation Monitoring Programme (2006) Meeting the MDG drinking water and sanitation target : the urban and rural challenge of the decade. World Health Organization, Geneva

WHO & UNICEF Joint Water Supply Sanitation Monitoring Programme (2012) Progress on drinking water and sanitation: 2012 update. UNICEF, New York

Workplace (Health, Safety and Welfare) Regulations (1992) The sphere handbook

World Bank (1993) Water resources management. World Bank, Washington D.C

World Health Organization (2003) The right to water. World Health Organization, Geneva

World Health Organization (2010) UN-Water Global Annual Assessment of Sanitation and Drinking Water 2010. World Health Organization, Geneva

World Health Organization & UNICEF (2013) Progress on sanitation and drinking water-2013 update. World Health Organization, Geneva

World Health Organization, United Nations Children's Fund (2005) Water for life: making it happen. Joint Monitoring Programme for Water Supply and Sanitation, Geneva

Zille H (2013) The politics behind the "toilet wars". Politics, Web. http://www.politicsweb.co.za/politicsweb/view/politicsweb/en/page71619?oid=383760&sn=Detail&pid=71619. Accessed 2 Aug 2013

Chapter 26
Patterns of Water Law

Joseph W. Dellapenna

Abstract Water law through the centuries has conformed to a limited set of patterns, in part because of the characteristics of the resource and in part because of the migration of water laws from society to society. In order to describe these patterns, this chapter summarily traces the evolution and characteristics of national, transnational (regional), and international water law, how they are related, and where they might be headed.

Introduction: What Counts as Law?

Law applies to water usage at all levels, from informal regulations by small or large communities through local, state/provincial, national, regional, and international or global law. Despite the seemingly infinite variations of these different bodies of law, water law (the law applicable to the management or use of water) actually fits a limited number of patterns. While some of this derives from the nature of the resource itself, other features reflect the spread of laws by various means from place to place and time to time. This chapter seeks to describe the characteristic features of these patterns and the processes by which they were disseminated across the globe.

Readers who come from countries with highly developed formal legal systems are likely to have a firm idea of what 'law' means and how it works, typically involving a legislatively created, highly determinate rule enforced by courts and police. This notion of law, called 'legal positivism', focuses attention on 'positive' law, law that is formally enacted and formally enforced. A leading legal positivist, Austin, defined law as 'the command of a sovereign enforced by a sanction'

J. W. Dellapenna (✉)
212 St. George's Road, Ardmore, PA 19003, USA
e-mail: dellapen@law.villanova.edu

A. Bhaduri et al. (eds.), *The Global Water System in the Anthropocene*,
Springer Water, DOI: 10.1007/978-3-319-07548-8_26,

(Austin 1998, p. 133). The Austinian paradigm is not an adequate notion of what law is and how law operates, a point perhaps best expressed by Goodhart: 'It is because a rule is regarded as obligatory that a measure of coercion may be attached to it; it is not obligatory because there is coercion' (Goodhart 1953, p. 17).

'Law' refers to an organic mechanism whereby certain claims of right are socially established as collectively enforced norms and other claims of right are denied such status. These norms might be formally established and enforced through legislatures, courts, and executives, or informally established and enforced through custom and informal, but often highly effective, collective action. When normative judgments are accepted as law, few will violate the norms, and those who do will pay a higher price than someone who violates a mere social or moral convention: The price might well be exposure to official coercion, while in informal settings enforcement might range from social pressure to exclusion from the group to corporal punishment or even death. What then is the function of formal law, 'law on the books'? History teaches us that informal law functions successfully when persons in a particular community know the others in the community and what they are doing, each depends on the others for wide ranging social support, and each realizes that overreaching too far or too often will cost them the social support needed for survival or thriving. As societies become larger, social interaction becomes less personal and the complex mutual reciprocities that ensure compliance with purely customary rules break down. Formal law allows adequate certainty and predictability of right and obligation when informal or customary law is no longer adequate (Dellapenna 2000). This was as true of Hammurabi's Babylon as it is in modern Europe (Glenn 2010).

Certainty and predictability are important values, particularly for one seeking to make firm plans for the future, but they are not the only consequences that societies resort to formal law. One consequence that seems to follow regularly from the development of systems of formal law is that it ensures that the state itself abides by the law. Yet societies change. The problem confronting lawyers and judges is to mediate the resulting tension between the need for stability and certainty with the need for flexibility and change to accommodate new social realities (Cardozo 1921). Too little flexibility and change and formal law loses touch with social realities. Too much flexibility and change makes planning and legal control impossible.

Lurking behind any discussion of law is the question of how effective such regimes actually are. An effective legal regime cannot be created simply by decree, or by importing a foreign model that works well in the country where it originated. The law in every country is 'path-dependent', a result of what has gone before as well as what is sought for the future. At the extreme, formal law may play little or no real role in structuring social relations or resolving disputes (Dellapenna 1997). In each society, one must learn who the lawyers and judges are, to whom they are connected, and what their role in the state and the economy is. A judiciary or other dispute resolution process functions effectively only when it is embedded in the structures of social, political, and economic power. Yet embedding might serve only to entrench existing power structures to the disadvantage of innovators or the poorly connected.

With a concept of law that includes both formal and informal norms and institutions of varying degrees of effectiveness in mind, a society (of people, of communities, or of states) is never without law, but law can take a myriad of forms and express highly varied content. We must not overstress formal legal structures applicable to water except when they actually reflect how water is managed and how disputes over water are resolved. In many ways this notion of law mirrors the on-going discussion on governance, where there is a shift from centralized, top-down, hierarchical approaches to more diffuse systems of rulemaking in society (Gupta 2011). This is also in line with discussions on global administrative law where scholars have found that international law emerges not just from legislative and judicial processes, but also from executive and administrative actions (Krisch 2006). These introductory remarks allow us to understand how pervasive and varied water law is even while searching out patterns of consistency across societies. If we find such consistencies, the consistencies, and not the variations, demand explanation.

The Beginnings of Formal Water Law

Today, water laws are found around the world as local customs and regulations, national legislation, regional agreements, and global treaties, together creating a complex legal governance framework for water. The framework is a result of historical processes. Given the broad concept of law indicated above, there cannot be a society without water law of some sort, and formal water laws are found in the earliest human civilizations. So central was the need to regulate water in these early civilizations that Wittfogel concluded that this need drove the emergence of basin-wide or other hydraulically focused empires in early civilizations (Wittfogel 1981).

In Mesopotamia, archaeologists have uncovered numerous records of contracts and legal cases. Codes of laws inscribed on steles set forth early water law, including the *Code of Hammurabi* (1738 BCE) (King 1910, pp. 53–56). These early laws indicate communal management, although the actual provisions of the Mesopotamian codes were limited to liability for flooding a neighbour's fields (Kornfeld 2009, pp. 29–33). The ancient Hindu *Arthashastra* (*ca.* 300 BCE) (Rangarajan 2000) are similarly limited, providing that water belonged to the king but authorizing private uses on payment of a tax so long as the private actor maintained the infrastructure, with severe penalties for injuring another water user (Cullet and Gupta 2009, p. 160). The slightly later *Laws of Manu* (*ca.* 200 BCE) in India are similar (Cullet and Gupta 2009, p. 159). The *Law of Moses* (*ca.* 1000 BCE), as developed and extended by rabbinical scholarship, remained focused on a few simple rules regarding rights to use water and the duty to protect its purity (Laster et al. 2009).

These water laws developed in a highly contextual manner reflecting the history, geography, and political systems of the countries concerned. Early water laws exhibit certain recurring patterns. Some of these are purely cultural, reflecting the predominant forms of social structure of a time and place. Most commonly in ancient times, laws were presented as divinely revealed. Other features reflect the nature of the resource and patterns of use. Thus the right to use water is variously granted to owners of riparian land (land contiguous to a water source) or because of temporal priority in using the water (first in time, first in right) (Scott and Coustalin 1995). The riparian approach generally required a sharing of the water, while the priority approach often did not. Some cultures would mix the two principles, while others gave preferences to particular types of use (e.g., irrigation vs. municipal uses). And from the beginning, the laws addressed questions of pollution as well as the allocation to particular uses such as prohibitions of allowing cattle to defecate in flowing water. These ancient water laws tended to be most developed in arid or semi-arid regions. As in arid and semi-arid regions today, the resulting water laws emphasized allocation rather than pollution (Teclaff 1985).

How Water Law Systems Spread Across the Planet

The nature of water resources and the nature of the uses of the resource to some extent provide a measure of unity to patterns of water law, along with a continuing debate about which legal approach is best (Dellapenna 2008b; Trelease 1974). The purely social, or jurisprudential, features of water laws create a possibility of passing water law even regarding features that do not simply reflect the nature of the resource or its uses from one society to another through one or more of several processes. These have included: (1) the spread of civilizations or cultures (Kornfield 2009); (2) the spread of religion when laws are considered divinely revealed (Naff 2009; Laster et al. 2009); (3) conquest and colonization, including the spread and decline of Communism (Cullet and Gupta 2009; Farias 2009; Kidd 2009; Kotov 2009; McCay and Marsden 2009; Nilsson and Nyanchaga 2009; van der Zaag 2009); (4) the widespread codification of law in the nineteenth century (Watson 1993); (5) the rise of engineering and epistemic communities (Gupta 2009); (6) the spread of environmentalism (Zellmer 2009); and (7) the "second wave of globalization" (a wave of global integration set off after 1950 with the freeing of trade and accelerated by the end of the Cold War), with new water laws often promoted by aid agencies, development banks, and UN agencies (Dellapenna 2008b; Gupta 2003). These various influences can co-exist, while pre-existing institutions and laws often persist, resulting in a process that Francis Cleaver has termed "bricolage" (Cleaver 2012). By bricolage, Cleaver means an uneven blending of old practices and norms with new practices and norms. Such institutional and legal bricolage involves a constant renegotiation of norms and the reinvention of tradition.

The result today is almost 200 different national water law systems, each with country specific characteristics. These systems are composed of overlapping and contradictory elements derived from the above processes. Many nations have residual indigenous laws that conflict with water laws imposed by colonial regimes or imported from 'more advanced' systems, all subject to attempts at water law reform deriving from international legal standards or the prevalent thinking of epistemic communities (Cullet and Gupta 2009; Farias 2009; Kidd 2009; Nilsson and Nyanchaga 2009; van der Zaag 2009). This leaves multiple systems of water law competing for application (Cullet and Gupta 2009; Gupta and Leenderste 2005; Nilsson and Nyanchaga 2009).

One can find some communities applying indigenous law to manage their water resources even without formal legal recognition, while other communities in the same state apply formal law left by a colonial regime, and yet other communities in the state apply markets or otherwise embrace whatever legal thinking appears most modern. The resulting pluralism could be positive, recognizing interests that cannot be aggregated in universalist approaches (Krisch 2006, p. 248), or negative, fragmenting interests and policies and breaking down legal structures. Recent efforts to integrate different regulations into a comprehensive water code sometimes succeed for better (as in Israel, Laster and Livney 2009) or worse (as in Russia, Kotov 2009). In other cases, they founder on the resistance of those who are committed to earlier regimes. Examples of successful resistance include Brazil (Farias 2009), East Africa (Nilsson and Nyanchaga 2009), and India (Cullet and Gupta 2009).

Contemporary Patters of Water Law at the Local or National Level

The nearly 200 national water legal systems define the right to use water according to only a few possibilities (Gupta and Dellapenna 2009). Thus the right to use water might be defined be in terms of the relationship of the use to the water source: (1) based on the location of the use (a riparian connection); (2) the timing of the use (a temporal or seasonal priority system); or (3) the nature of the use (preferences for the most socially important uses). Rights to use water are often characterized as a kind of property, which allows a different typology: (1) common property (the resource is used freely by those with lawful access, without collective decision making); (2) private property (defined water rights are allocated to particular users with considerable control over 'their' water); or (3) community or public property (water is managed jointly by those entitled to share the resource) (Dellapenna 2010; Ostrom 1990).

Each type of property right must recognize to some extent the public nature of water resources, and therefore even in the most thoroughly privatized water property regime there will be regulations to: (1) enforce the property or water right regime, (2) protect the resource from pollution or degradation, and (3) promote or preclude markets.

The Evolution of Water Law at the Level of Transnational Regions

In a sense, transnational regional water law systems are as old as the earliest recorded formal water laws in the form of early hydraulic empires (Wittfogel 1981). Examples include early China, Egypt, India, and Mesopotamia. These legal systems generally imposed rules on certain limited questions of water management while deferring to local customs or laws for day-to-day decisions. Such hybrid regimes operated for centuries unless the imperial system became strong enough to displace indigenous law completely (*see*, e.g., Kotov 2009).

The demise of most empires in the twentieth century did not mean the end of transnational regional systems. Instead, in the twentieth century, states often voluntarily created transnational water law systems. The European Union's European Water Framework Directive of 2000 is now the leading example of such a transnational water law, although this is embedded in a system of transnational law covering a broad range of issues (Aubin and Varone 2004; Canelas de Castro 2009). Another type of transnational system is the growing number of river basin organizations and water commissions (Conca 2006; Merrey 2009). Although river basin bodies and water commissions rarely have strong transnational law-making functions, they are increasingly part of the growing system of international administrative law (Farrajota 2009).

One way to conceive of transnational regional water law is that sovereignty is sacrificed for the greater good of all the parties concerned. A better way to conceive of such transnational water law is that states are realizing their sovereignty by expressing it through cooperative transnational institutions. Either way, such institutions seem likely to become more common and more effective.

The Evolution of Globally Applicable International Water Law

Although international water agreements go back at least 800 years, true international water law developed only in the last two centuries. International law in general provides an institutional framework, with rules for treaty making and interpretation and means for dispute resolution. International law empowers international actors by legitimating their claims, but it also limits the claims they are allowed to make (Dellapenna 2008a). International water law is found in numerous treaties (e.g., UN 1997) and in customary international law.

Customary international law develops through states making claims and counterclaims until they agree on what the law requires (Danilenko 1993, pp. 75–82). Identifying customary law is informal and challenging. Customary international water law evolved largely through water treaties, beginning in the late eighteenth century. The treaties focus first on freeing navigation, then (because of the industrial revolution in the nineteenth century) on water allocation, and finally on cooperative or joint management regimes in the twentieth and twenty-first centuries (Dellapenna 1994).

Contemporary customary international water law resembles the common principles underlying national water laws, including recognizing rights in riparian or aquifer states, considering temporal priority to some extent, and emphasizing the nature of and need for particular uses. These principles often take on different colorations when applied to an incompletely organized community of states. Customary international water law primarily includes three principles: (1) limited territorial sovereignty over national waters (requiring states to consider the needs of other riparian states) (Dellapenna 2001); (2) the no-harm principle (derived from the Roman law maxim, *sic utero tuo ut alineium non laedes*—'Use not your property so as to injure the property of another') (Dellapenna 2008a); and (3) the obligation to settle disputes peacefully. Some states claim historic rights, i.e., the right to use the quantity of water they have been using for a significant period of time (Brunnée and Toope 2002). These principles emerged through a dialectic process where the claim of absolute territorial sovereignty competed with claims of absolute integrity of state territory. Examples abound (Dellapenna 1996, 2001). Perhaps the best known is the dispute (at the turn of the nineteenth century) between Mexico and the United States which took about a decade to negotiate to a sharing agreement (McCaffrey 1996). Today, limited sovereignty prevails, expressed as the principle of equitable utilisation (ILA 1966, art. IV; ILA 2004, art. 12; UN 1997, art. 5), i.e., the need to share international waters equitably (fairly) (Dellapenna 1996).

The codification of the customary international water law effectively began with the International Law Association's approval of the *Helsinki Rules on the uses of international rivers* (ILA 1966). The UN General Assembly then asked the International Law Commission to codify international water law based in large part on the *Helsinki Rules*. The result was the UN Watercourses Convention, approved by a vote of 103–3 on 21 May 1997 (UN 1997). Thus far 30 states have ratified, leaving the convention 5 short of entering into force. Still the convention is an authoritative reflection of existing customary water law (Gabçikovo-Nagymaros Case 1997, p. 140) and influencing regional law in Southern Africa, South Asia, and Europe (Farrajota 2009; van der Zaag 2009).

The UN Watercourses Convention adopts the principles of limited sovereignty (equitable utilisation), no significant harm, and peaceful resolution of disputes, with great emphasis on the procedures to be followed. Its approval is important for showing that the principle of limited sovereignty is not inconsistent with the principle of 'permanent sovereignty' of states over their natural resources approved by the General Assembly some 35 years earlier (UN 1962). The convention is a limited framework for structuring negotiations. Although it includes environmental values and some modern ideas about water governance, arguably it was out-of-date when it was adopted for it scarcely refers to legal developments in the environmental, human rights, and investment arenas since the *Helsinki Rules*.

Environmental concerns are not entirely absent from the UN Watercourses Convention, but they appear only in the most general terms and only in terms as transboundary issues. These limits highlight the most basic problems with the UN Watercourses Convention—its attempt to address transboundary water issues in isolation from other, intimately connect water issues (such as groundwater) (UN 2008). Perhaps its most glaring omission is the lack of any mention the right of the public to participate in decision making regarding transboundary water resources, although perhaps more important is the failure to consider the extent to which modern international law speaks to environmental and resource issues within states and not just in transboundary contexts. Both of these points become critical with the emergence (after the UN Watercourses Convention was completed) of an increasingly well established human right to water (UN 2010).

International water agreements provide sources of law for participating states as well as for inferring a developing customary international law. A major regional and increasingly globally relevant source of water law is the 1992 UN Economic Commission for Europe Convention on Transboundary Watercourses (UNECE 1992). This treaty obliges parties to prevent, control, and reduce transboundary impacts and to use the waters in an ecologically sound and rational way, to conserve water resources, and to protect the environment. While it embraces the principle of equitable utilization, its emphasis is on environmental protection—the 'no harm' side of the equation. There are hundreds of other bilateral and multilateral international water agreements (Oregon State University 2002) which together give rise to a body of international customary law that sets basic standards even for water resources not covered by an international agreement (Dellapenna 2001). International adjudication of water disputes is another rich area of legal development (Gabçikovo-Nagymaros Case 1997; Castillo-Laborde 2009).

The most recent effort to codify all this body of law is the *Berlin Rules on Water Resources,* approved unanimously by the International Law Association in 2004 to replace the *Helsinki Rules* (ILA 2004). The *Berlin Rules* integrate insights from environmental, humanitarian, human rights, and resource law. Where appropriate, these comprehensive rules cover all national and international fresh waters and related resources (the aquatic environment) and thereby penetrate national jurisdiction. The rules include the principles of public participation, the obligation to use best efforts to achieve conjunctive and integrated management of waters, and the duties to achieve sustainability and to minimize environmental harm. They identify the rights and duties of states and persons, the need for environmental impact assessments, and rules relating to extreme situations including accidents, floods, and droughts. The *Berlin Rules* are grounded in existing law interpreted in light of evolving changes in global water law.

Groundwater traditionally has been neglected by national and international water law (Cassuto and Sampaio 2013). The *Berlin Rules* (ILA 2004, Chap. VIII) provided the first attempt at a comprehensive codification of the customary international law of groundwater. The UN Law Commission has subsequently adopted draft articles on transboundary aquifers that were noted but not approved by the UN General Assembly (UN 2008).

A second wave of globalization has washed over the planet in the last 60 years or so. This wave has supported neoliberal dominance, challenging concepts of sovereignty underlying traditional international law and further marginalizing states (van Creveld 1999). Economists and others have strongly advocated markets based on a private property in water resources as the best way to manage water (Griffin 2006), generating considerable controversy about the utility of markets (Dellapenna 2008b; Griffin 2006). In any event, markets need strong regulation, leading to bilateral and multilateral agreements on trade and investment (e.g., WTO 1994). The neo-liberal approach and enhanced private sector participation in water management has inspired a reaction in the form of a human rights approach that pierces the veil of sovereignty to protect access rights for the most vulnerable in society (UN 2010; Gupta et al. 2010).

Conclusion

Despite talk of 'water wars', water resources tend not to be a key reason for conflict (Kalpakkian 2004). Instead, at the national, regional, and international levels water law has served to mediate conflict and resolve disputes. Yet after 5000 years, water law remains tied to old models that, at least at a general level, can be traced back to the earliest extant historical records. As noted above, the laws for allocating water to particular uses, broadly speaking, fall into three distinct patterns (or mixtures thereof) that are found very early in the historical record: (1) based on the location of the use (a riparian connection); (2) the timing of the use (a temporal or seasonal priority system); or (3) the nature of the use (preferences for the most socially important uses). The resulting right to use water are often characterized as property rights, which in turn can be characterized in one or three ways (although some systems also mix these): (1) common property (under which all who have lawful access are allowed to use the resource freely, without collective decision making); (2) private property (under which persons holding defined water rights are allowed considerable control over 'their' water so long as they remain within the specified right); or (3) community or public property (under which water is managed jointly by those entitled to share the resource) (Dellapenna 2010; Ostrom 1990). Each type of water law regime recognizes to some extent the public nature of water resources, and therefore even in the most thoroughly privatized water property regime there will be regulations to: (1) enforce the property or water right regime, (2) protect the resource from pollution or degradation, and (3) promote or preclude markets.

Today, many challenges exist worldwide to water management and to water law. One result has been the emergence and strengthening of both transnational (regional) law and of general international law addressed to water resources. In many respects these bodies of law are still in their formative stages and no one suggests that either body of law can, or should, fully displace national or local water law regimes. These supranational regimes (regional as well as international) both in many ways reflect the same concerns as embodies in national water law patterns, but generally with much less well developed institutions for applying the law and resolving disputes.

Communities at all levels face global water problems such as access, sanitation, pollution, ecosystem destruction, and changing flow regimes as a result of dams, other human activities, and the increasingly disrupted climate. Governance systems themselves are in a state of flux (Gupta 2011). In the future, the 'global public good' characteristics of water, its ecosystem services, and its links to energy, food, and climate are likely to gain prominence (Kaul et al. 1999), further challenging traditional notions of sovereignty. Some might see law—local, national, transnational (regional), and international—as an impediment to coping adequately with the water needs of the coming century. Conceiving of rights to use water as property rights in itself introduces a kind of rigidity that can make it more difficult to introduce change into the legal structure of water use. As issues of water governance become very technical, technocratic solutions may lead to growing formal and informal administrative law and governance in the water field, some of which might be adopted through international development cooperation processes but without a formal international legal consensus. Champions of markets as a water management tool in particular often see existing water laws as an impediment to the successful operation of markets for raw water, i.e., water not yet abstracted from its natural sources (Brandes and Nowlan 2009).

History shows, however, that water law is able, if slowly, to rise to the challenge of change. While there is an on-going shift in the locus of governance (van Creveld 1999), there have been only limited shifts in the rules to guarantee legality, legitimacy, accountability, transparency, and the rule of law. Against this background, water law is slowly moving forward through regional agreements, administrative frameworks, and joint water management bodies at all levels of governance from community up to global levels. Legal systems, however slow their development, have the authority of history behind them and may ultimately provide the vehicle for problem solving and conflict resolution in the twenty-first century. Meanwhile, as global governance grapples numerous difficult issues, water law will figure prominently in the results as water management systems and social justice processes struggle to cope with tomorrow's needs.

References

Aubin D, Varone F (2004) The evolution of European water policy. In: Kissling-Näf I, Kuks S (eds) The evolution of national water regimes in Europe: transitions in water rights and water policies. Kluwer, Dordrecht, pp 49–86

Austin J (1998) The province of jurisprudence determined. In: Hart HLA (ed) Hacket Publishers, Indianapolis

Brandes OM, Nowlan L (2009) Wading into uncertain waters: using markets to transfer water rights in Canada—possibilities and pitfalls, 19. J Environ Law Pract 19:267–287

Brunnée J, Toope SJ (2002) The changing Nile basin regime: does law matter? Harvard Int Law J 43:105–159

Canelas de Castro P (2009) European community water policy. In: Dellapenna JW Gupta J (eds) The evolution of the law and politics of water, Springer, New York, pp 227–244
Cardozo NB (1921) The nature of the judicial process. Yale Univ Press, New Haven
Cassuto DN, Sampaio RSR (2013) Hard, soft & uncertain: the Guarani Aquifer and the challenges of transboundary groundwater. Colorado J Int Environ Law Policy 24:1–41
Cleaver F (2012) Development through bricolage: rethinking institutions for natural resources management. Routledge, New York
Conca K (2006) Governing water: contentious transnational politics and global institution building. MIT Press, Cambridge, MA
Cullet P, Gupta J (2009) India: the evolution of water law and policy. In: Dellapenna JW, Gupta J (eds) The evolution of the law and politics of water. Springer, New York, pp 157–173
del Castillo-Laborde L (2009) Case law on international watercourses. In: Dellapenna JW, Gupta J (eds.) The evolution of the law and politics of water. Springer, New York, pp. 319–335
Danilenko GM (1993) Law-making in the international community. M. Nijhoff, Dordrecht
Dellapenna JW (1994) Treaties as instruments for managing internationally shared water resources: restricted sovereignty vs. community of property. Case West Res J Int Law 26:27–56
Dellapenna JW (1996) Rivers as legal structures: the examples of the Jordan and the Nile. Nat Resour J Law 36:217–250
Dellapenna JW (1997) The lesson of the triple-twisted pine: plum blossoms on mountain peaks and the future of the rule of law in Hong Kong. Vanderbilt J Transnat Law 30:637–673
Dellapenna JW (2000) Law in a shrinking world: the interaction of science and technology with international law. Kentucky Law J 88:809–883
Dellapenna JW (2001) The customary international law of transboundary fresh waters. Int J Global Environ Issues 1:264–305
Dellapenna JW (2008a) International water law in a climate of disruption. Mich State J Int Law 17:43–94
Dellapenna JW (2008b) Climate disruption, the Washington consensus, and water law reform. Temple Law Rev 81:383–432
Dellapenna JW (2010) Global climate disruption and water law reform. Widener Law Rev 15:409–445
Farias PJL (2009) Brazil: the evolution of the law and politics of water. In: Dellapenna JW, Gupta J (eds) The evolution of the law and politics of water. Springer, New York, pp 69–86
Farrajota MM (2009) International cooperation on water resources. In: Dellapenna JW, Gupta J (eds) The evolution of the law and politics of water. Springer, New York, pp 337–352
Glenn HP (2010) Legal traditions of the world: sustainable diversity in law, 4th edn. Oxford University Press, Oxford
Goodhart AL (1953) English law and the moral law. Stevens, London
Griffin RC (2006) Water resource economics: the analysis of scarcity, policies, and projects. MIT Press, Cambridge, MA
Gupta J (2003) The role of non-state actors in international environmental affairs. Heidelberg J Int Law 63:459–486
Gupta J (2011) Developing countries: trapped in the web of sustainable development governance: performance, legal effects and legitimacy. In: Dilling O, Herberg M, Winter G (eds) Transnational administrative rule-making: performance, legal effects and legitimacy. Hart Publishing, Oxford, pp 305–330
Gupta J, Ahlers R, Ahmed I (2010) The human right to water: moving toward consensus in a fragmented world. Rev Eur Community Int Environ Law 19:294–305
Gupta J, Dellapenna JW (2009) The challenge for the twenty-first century: a critical approach. In: Dellapenna JW, Gupta J (eds) The evolution of the law and politics of water. Springer, New York, pp 391–410
Gupta J, Leenderste K (2005) A multi-level, dynamic approach to water law and policy. In: Amaya OD, Milla A, Gracia MDP (eds) Incorporation of the IWRM principles in legal frameworks in Latin America: experiences and lessons learned. Universidad Externado de Colombia, Bogota, pp 17–37

Kalpakkian J (2004) Review of identity, conflict and cooperation in international river systems. Ashgate, Aldershot

Kaul I, Grunberg I, Stern M (1999) Global public goods: international cooperation in the 21st century. Oxford University Press, Oxford

Kidd M (2009) South Africa: the development of water law. In: Dellapenna JW, Gupta J (eds) The evolution of the law and politics of water. Springer, New York, pp 245–261

King LW (1910) The code of Hammurabi (trans.). British Museum, London. http://www.yale.edu/lawweb/avalon/medieval/hamframe.htm

Kornfeld I (2009) Mesopotamia: a history of water and law. In: Dellapenna JW, Gupta J (eds) The evolution of the law and politics of water. Springer, New York, pp 21–36

Kotov S (2009) Russia: historical dimensions of water management. In: Dellapenna JW, Gupta J (eds) The evolution of the law and politics of water. Springer, New York, pp 139–156

Krisch N (2006) The pPluralism of global administrative law. Eur J Int Law 17:247–278

Laster R, Aronovsky RD, Livney D (2009) Water in the Jewish legal tradition. In: Dellapenna JW, Gupta J (ed) The evolution of the law and politics of water. Springer, New York, pp 53–66

Laster R, Livney D (2009) Israel: the evolution of water law and policy. In: Dellapenna JW, Gupta J (eds) The evolution of the law and politics of water. Springer, New York, pp 121–137

McCaffrey SC (1996) The Harmon doctrine one hundred years later: buried, not praised. Nat Resour J 36:965–1007

McKay J, Marsden S (2009) Australia: the problem of sustainability in water. In: Dellapenna JW, Gupta J (eds) The evolution of the law and politics of water. Springer, New York, pp 175–189

Merrey DJ (2009) African models for transnational river basin organizations in Africa: an unexplored dimension. Water Altern 2:183–204

Naff T (2009) Islamic law and the politics of water. In: Dellapenna JW, Gupta J (eds) The evolution of the law and politics of water. Springer, New York, pp 37–51

Nilsson D, Nyanchaga EN (2009) East African water regimes: the case of Kenya. In: Dellapenna JW, Gupta J (eds) The evolution of the law and politics of water. Springer, New York, pp 105–120

Oregon State University (2002) Atlas of international freshwater agreements. UNEP, Nairobi

Ostrom E (1990) Governing the commons: the evolution of institutions for collective action. Cambridge University Press, Cambridge

Rangajaran LN (2000) The Arthashastra (ca. 300 BCE) (trans.). Penguin, New Delhi

Scott A, Coustalin B (1995) The evolution of water rights. Nat Resour J 35:821–979

Teclaff L (1985) Water law in historical perspective. W.S Hein, Buffalo, NY

Trelease FJ (1974) The model water code, the wise administrator, and the goddam bureaucrat. Nat Resour J 14:207–229

van Creveld M (1999) The rise and decline of the state. Cambridge University Press, Cambridge

van der Zaag P (2009) Southern Africa: evolving regional water laws and policies. In: Dellapenna JW, Gupta J (eds) The evolution of the law and politics of water. Springer, New York, pp 245–261

Watson A (1993) Legal transplants: an approach to comparative law, 2nd edn. University of Georgia Press, Athens, GA

Wittfogel KA (1981) Oriental despotism: a comparative study of total power, reprint edn. Vintage Books, New York

Zellmer S (2009) United States: emergence of environmental considerations. In: Dellapenna JW, Gupta J (eds.) The evolution of the law and politics of water. Springer, New York, pp. 205–223

Case

Gabçikovo-Nagymaros Case (Hungry/Slovakia) (1997) Judgment of 25 Sept 1997, 1997 ICJ no. 92

Other Legal Materials

EU Water Framework Directive (2000) EC Directive 2000/60/EC

International Law Association ('ILA') (1966) The Helsinki rules on the uses of the waters of international rivers. In: Report of the fifty-second conference of the international law association, International Law Association, London

International Law Association ('ILA') (2004) The Berlin rules on water resources. In: Report of the seventy-first conference of the international law association, International Law Association, London

UN (1962) Permanent sovereignty over natural resources, GA res. 1803 (XVII), 17 U.N. GAOR Supp. no.17 at 15, UN Doc. A/5217

UN (1997) UN convention on the law of non-navigational uses of international watercourses', UNGA Res. A/51/49. http://untreaty.un.org/ilc/texts/instruments/english/conventions/8_3_1997.pdf

UN (2008) The law of transboundary aquifers, UNGA Res/63/124. http://daccess-dds-ny.un.org/doc/UNDOC/GEN/N08/478/23/PDF/N0847823.pdf?OpenElement

UN (2010) The human right to water and sanitation, UNGA/Res/64/292. http://www.un.org/ga/search/view_doc.asp?symbol=A/RES/64/292

UNECE (UN Economic Commission for Europe) (1992), Convention on the protection and use of transboundary watercourses and international lakes. http://www.unece.org/fileadmin/DAM/env/water/pdf/watercon.pdf

WTO (World Trade Organization) (1994) Agreement establishing the world trade organization, 1867 UNTS 154. http://www.wto.org/english/docs_e/legal_e/04-wto_e.htm

Chapter 27
The Notion of the Global Water Crisis and Urban Water Realities

Antje Bruns and Fanny Frick

Abstract The global water crisis is often alluded to in scientific papers and geo-political discourse. However, the lack of a proper definition of what the term 'water crisis' means has been routinely overlooked, as much as are the reasons why it is assumed to be truly global in nature. Such generalisations and simplifications in both science and policy alike may lead to governance responses that are not fully applicable. In the following chapter we examine the relation between initially framing a problem (water crisis), and introducing policies and management principles that reflect such a situation. We will start by exploring the emergence of global water crisis in the 1990s. We then contrast these findings by examining how the urban water crisis in Accra, Ghana has worsened over time, although there is enough water to go round. We conclude with a plea that crucial socio-political perspectives within hydrology be reinforced, since these are the very factors—occurring within different spatio-temporal scales—that are often overlooked in research into water-related global change.

The Emergence of the Water Crisis and Its Implications for Water Governance

The notion that water is in crisis emerged with the publication of two key books in the early 1990s: in 1992, the Worldwatch Institute published Sandra Postel's *Last Oasis Facing Water Security*. One year later, Peter Gleick edited a review of the state of the world's freshwater system entitled *Water in Crisis*. Ever since these global crisis claims were made, awareness of a threat to water security and the so-called 'crisis' itself has increased steadily among both researchers and

A. Bruns (✉) · F. Frick
Humboldt-Universität zu Berlin, Unter Den Linden 6, 10099 Berlin, Germany
e-mail: antje.bruns@geo.hu-berlin.de

A. Bhaduri et al. (eds.), *The Global Water System in the Anthropocene*, Springer Water, DOI: 10.1007/978-3-319-07548-8_27,

politicians. The necessity of rethinking what is actually meant by the term 'water crisis' is often overlooked. Instead, many divergent phenomena have come to be described as part of its definition (Linton 2004; Trottier 2008; Srinivasan et al. 2012). Threats to water security are predominantly described as consequences of growing demand and environmental degradation (e.g. Vörösmarty et al. 2010). Later assessments have exhibited differing views as to the specific drivers of the crisis (Srinivasan et al. 2012).

The problem with any unreflecting use of such language is that there is a lack of neutrality in the descriptions upon which discourse is based. The opinions become modified and even biased owing to local custom, when, for example, defining a scientific research question, when developing policy responses, or in journalistic reporting. Since these practices are not isolated, but involve interactions between various actors, previous assumptions are constantly being reproduced, particularly in the established settings of interpretive communities (cf. Fairclough 1992; Bourdieu 2001; Adger and Benjaminsen 2001; Belina and Dzudzek 2009). This means that wherever a problem has been determined to exist, the solution is also to a large extent a predetermined one: as Linton (2004) points out, discourse on the global water crisis as initiated by Postel (1992) and Gleick (1993) was based entirely on the research of a small group of Soviet scholars. While the overall crisis talk soon began to be reproduced, the scientific basis that initially gave rise to the definition of a global water crisis was hardly ever questioned (Linton 2004). What is more, critical alerts to uncertainties that had been made by Gleick in his publication became overlooked over time (ibid.).

Broader discourse on the environmental crisis, a precedent of water crisis, gave rise to a series of restrictive environmental policies into the 1990s. Despite the measures taken to control environmental degradation, however, deterioration continued, and the so-called "crisis" grew worse. By the turn of the millennium, assessments of the low effectiveness of environmental policies led to a paradigm shift in politics, from regulatory law to cooperative governance. The Water Framework Directive in Europe is a prime example of this shift (Bruns 2010). In practice, however, the same old preconceptions as to what the issues are, still tend to dominate, because alternative ways of assessing a problem are seldom part of any planned shifts in governance (Manuel-Navarrete et al. 2009; Bruns 2010; Stoett 2012). Yet what has been rather under-examined in governance research thus far are cross-scale interactions (Agrawal 2001) and questions of power (Swyngedouw 2009; Kenis and Raab 2008).

As European experiences while implementing integrated management approaches have shown here, participative and cooperative governance processes cannot simply be enforced, but tend rather to emerge from the interaction of actors at different scales that depend on stakeholders' norms and values (Bruns and Gee 2009; Bruns 2010). Since the participative approach invited NGOs and private stakeholders to become involved in decision-making processes, this paradigm also raises concerns as to the legitimacy of these decisions (Bruns 2010). The shift in

Table 27.1 Numbers and origin of articles analysed

	Web of science	Scopus	Total articles analysed
Articles on the urban water crisis	27	54	70
Articles on the global water crisis	23	100	83

Some articles appeared in both search engines and are not counted twice; and some articles were thematically unsuitable

regulating water resources away from top-down management to more integrated management methods became, nevertheless, a universal paradigm that has influenced water policy in countries around the globe (Shiva 2003; Bakker 2009). There is hence reason enough to believe that global discourse on the water crisis is far more related to Western water management paradigms (which is inspired itself by modern hydrology) than to specific water problems in the so-called Global South.

Methods

According to the school of critical discourse analysis, discourse can be considered a social practice, reflecting the (re)production of social relations of power and domination. Hence discourse analysis can be applied as a method for analysis of societal conditions (Fairclough 1992; Belina and Dzudzek 2009). This type of discourse analysis looks at how texts are used in the production of hegemonic definitions and ideologies. A critical discourse analysis was performed within our urban water crisis case study of Accra, Ghana, based on a literature review (context analysis) and a textual analysis of key policy documents addressing the crisis in Accra. In order to interpret the influence of scientific framings of the water crisis in the (re-)production of urban water realities in Accra, the production of discourse on the global and the urban water crisis was assessed by comparing narrative elements across 153 peer-reviewed scientific articles (analysis of text production processes and interpretations). The corpus of analysis was defined in searches for *global water crisis* and *urban water crisis* in the titles of original research articles. To identify these articles the search engines Scopus and Web of Science were used. A coding scheme was developed to analyse the occurrence of narrative elements covering the following aspects (i) methods (empirically based on primary versus secondary data, or on conceptual methods), (ii) scale of analysis (household, city, regional, global) and (iii) proposed solutions. Narrative elements were identified mainly from the abstracts of the articles, since it is there that the argument is most concise. When analysing them, the four-eye principle was adopted in order to ensure methodological consistency in qualitative assessment (Table 27.1).

Urban Water Realities: The Case of Accra (Ghana)

Water governance policy in Accra was shaped considerably by early post-developmental government strategies adopted in the 1960s. After independence, the Ghanaian government's focus on water infrastructure was placed on rural areas and on poor neighbourhoods within urban areas. The aim was to reverse the urban bias in infrastructure provision that had dominated colonial planning. Provision of infrastructure was seen as a key factor in socio-economic development, and planning this without foreign intervention became symbolic of independence and modernity (Bohman 2010). In the early 1960s, the first government body to manage urban water supply and sanitation, the Ghana Water and Sanitation Corporation (GWSC), was set up, its objective being to guarantee "the provision, distribution, and conservation of the supply of water in Ghana for public, domestic and industrial purposes; and the establishment, operation and control of sewerage systems for such purposes" (Ghana Water and Sewerage Corporation Act 1965).

Among the largest infrastructure projects early in independence was the Akosombo Dam on the Volta River—the main source of fresh water in the region. Designed to supply Ghanaian industry and households with power (Volta River Development Agency Act 1961), the dam has produced one of the largest artificial lakes in the world (Bohman 2010, p. 36) and has brought multiple environmental and socio-economic impacts that have had unexpected effects on the wider region (Gyau-Boakye 2001; Karley 2009). Unintended consequences, such as changes in rainfall patterns, caused an accelerated migration to the city in the early years of independence (Karley 2009). As the urban population rapidly expanded, the extension of water supplies and distribution in urban areas was put high on the agenda in the government's seven year development plan 1963–1970 (Bohman 2010). Pricing policies did not receive the necessary consideration, and GWSC operated without achieving cost recovery.

Subsidies to support sector development were first granted to the government by the international community (via the World Bank) in 1969 and 1974, paving the way for international influence in the design of Ghana's water governance throughout the following decades (Bohman 2010). Early international community recommendations were to decentralise GWSC's responsibilities and to restructure the tariff system (Bohman 2010). Water infrastructure and service levels deteriorated during economic recession in the 1970s and 80s, and by 1990, a third of the facilities had broken down, with the rest operating below capacity (Fuest and Haffner 2007; Bohman 2010). In response to the economic crisis, the international community pressured Ghana to broaden its markets in the late 1980s. By means of structural adjustment programs (SAP) implemented to meet requirements set by the IMF/World Bank, the water sector was restructured to facilitate service provision by the private sector and local government. While urban water management remained under the authority of national government, rural water and sanitation was delegated to local authorities (Fuest and Haffner 2007; Bohman 2010).

This devolution coincided with an international wave of direct investment in Accra, likewise facilitated by SAP. Combined with obscure land ownership, the sudden pressure for economic development fuelled an informal market in land plots (Yeboah 2000; Owusu 2011). The city grew largely unchecked (Accra's population almost doubled from 863,000 in 1980 to 1,674,000 in 2000 (UN Habitat 2011)), and many plots were sold without adequate water and sanitation infrastructure (Yeboah 2000). The devolution of water authority was hindered by a lack of financial resources in local government (Bohman 2010). As a consequence, an increasingly segregated urban pattern emerged—described as the "dual structure of the city"—from the "uneven spatial distribution of potable water supply in different residential sectors within the city [with a sharp contrast between] planned high and medium-class residential areas on the one hand and low-class residential areas which constitute the rest of the residential areas" (Songsore and McGranahan 1993, p. 19). Meeting the water challenge was no longer a public health concern, but had more importantly become an economic interest among competing international investors. Plans for private sector participation were contested by a national coalition of NGOs arguing that a contract would favour "the interest of private water operators rather than Ghanaian water consumers" (Bohman 2010, p. 107).

Despite efforts made by the government of Ghana and supporting donors, the institutions established were unable to handle Accra's uncontrolled growth: In 2002, the proportion of urban Ghanaians with access to potable water was equal to, if not lower than it had been before the sector reforms. Unstable water supply had become a constraint on economic activity and on international investment in the country, and was therefore thought a major limiting factor in achievement of Ghana's vision of becoming a middle income country by 2020 (Fuest and Haffner 2007).

In 2005, a private operator (Aqua Vitens Rand, AVRL) was selected for a five-year management contract with GWCL to supply urban water, in contrast to the initial plans for private sector engagement in the water sector which had anticipated a lease contract. The shift in the design of private sector participation followed international investors' declining interest in the water sector. It reflects, moreover, a shift when considering "the possible potentials, risks, and gains from private sector involvement among investors, governments, and donor agencies" that followed the global "1990s wave of privatisation" (Bohman 2010, p. 118, 10). Objectives of the agreement between Ghanaian government and AVRL were the expansion of urban water supply, protecting the access to and affordability of water supply for low income consumers, cost recovery, adequate investment flows, and integration of the private sector (Darteh et al. 2010). However, the contractor's activities remained largely without effect: at the end of the first decade of the 21st century, "only 51% of the population has direct access to utility water supply services", while the rest of the population relies on informal connection to the GWCL system, tanker services, or alternative systems in communities on Accra's outskirts (Adank et al. 2011, p. vii). Another paradox remains: those relying on informal water supplies have to pay higher prices than those employing the formal system (ibid.). Today, Accra's water is managed through the Ghana Urban Water Company Limited, a sub-agency of GWCL that was created in absence of a successor to AVRL. While the Volta river discharge is more than enough to serve the city, "even when considering a potential drop in river flow caused by climate change and increased use of water upstream in the basin" (Adank et al. 2011, p. vii), adequate infrastructure and services to secure access to water and sanitation in Accra are lacking (ibid.).

In Bohman's analysis (2010) of sector reforms until 2005, she finds that under international influence, the water sector in Ghana has been reformed by a "shift from […] 'filling the gap' [through subsidies] to 'managing scarcity' [through staff development and integrated management approaches]" (ibid., p. 116), triggered in turn by a shift away from public health concerns and towards the broader concern of socio-economic development. This is mirrored in the World Bank's description of water as "the silent crisis in Ghana" (World Bank 2007, p. 13). The Bank continues to consider its interventions an achievement towards economically viable urban water management: "the Bank has actively engaged government and other stakeholders over the last 10 years, and this dialogue has resulted in the introduction of the private sector in the management of the urban water utility" (World Bank 2007, p. 47). In Ghana's National Water Policy (2007), the World Bank description of water crisis as a threat to economic development is upheld (Ghana National Water Policy 2007, p. 9). The transferal of guiding principles of donors' agendas into National Water Policy is evident in its promotion of selected integrated water resource management (IWRM) principles, including an emphasis on the private sector's role in providing water services that is repeated throughout the document: "The Government of Ghana is determined to halt the falling trends in water supply coverage and quality and resume a programme of expansion and improvement. This requires consistently high levels of investment and increasingly private (local and foreign) sources" (Ghana National Water Policy 2007, p. 30).

It becomes clear from the Accra study that water management and governance have been repeatedly restructured since the middle of the 20th century. Only 12 years after independence, the international community became involved in water policy design. Throughout this period, restructuring has been strongly encouraged and supported by international organisations (such as the World Bank) and companies. One could say the city has become a playground for international organisations testing modern paradigms of development assistance in the donors' darling of Sub-Saharan Africa. Today, since restructuring has shown little impact, companies are withdrawing their interest in investing in the system.

The lack of practical instruments for pro-poverty and sustainable water resource management (as stated, e.g., by Groenfeldt 2010) may originate in the plagiarising of donor strategies without assessment specific social and political contexts (because reforms were implemented without adequate participation: Fuest and Haffner 2007). Most importantly, the particular roles of informal structures in specific local conditions (such as the land tenure system) have long been disregarded.

Looking at the history of Accra's water governance since independence, it appears that no alternatives to Western paradigms for safeguarding water security currently enter policy design.

The Notion of a Water Crisis in Academia

Since global ecopolitics is generally based on scientific research (Stoett 2012; Turner and Robbins 2008) it is worthwhile to look at academic literature published on the topic, as is done in this chapter. Within the urban water situation in Accra—influenced by Western paradigms as described—two questions are of major interest: firstly, by whom and how is the water crisis in internationally policy-shaping discourses framed? And, secondly: Is there a difference in the framings of the global and the urban water crisis, respectively?

A brief look at research areas contributing to study of the water crisis reveals that the hard core of natural sciences (Geology, Environmental Sciences, Water Resource Studies and Engineering) leads the field in expertise. By contrast scientific disciplines which engage with people, cultures and the institutions they build to regulate the resources are clearly in the minority. This is surprising and even more so since principal themes accompanying the water crisis are of societal and political nature—as acknowledged by the majority of the analysed articles (see Table 27.2) (Fig. 27.1).

Research on both urban and global water crisis is informed to a large extent by secondary empirical data. Primary research is mainly carried out at the city or household level, while many articles explore the global water crisis on a very generic, more conceptual level. We found that studies on the global water crisis tend to generalise, up-scale and transfer their findings, in an attempt to provide knowledge beyond the specific analysis. A separation between disciplines is thus detectible, in methods and data that accompany different spatial scales of analysis. This divide may also explain why governance related aspects are predominantly discussed on the city level, and large scale technical and universal managerial solutions tend to dominate studies of the global water crisis.

Review of worldwide research into the urban aspect of global water crisis reveals that the scale employed in many articles as to where the crisis occurs is poorly defined. While the scale of the approach is quite clear in research referring to urban crises, the exact nature of global crises remains unclear. The poor attention to scale in global change research has already been addressed by O'Brien (2011), who stressed that scales function as an ordering system between different scientific disciplines, and thus may hinder full understanding of global environmental change. If the global water crisis is in fact the aggregate of local, individual water crises, then regional to global datasets are suitable neither for studying the crises, nor indeed for deriving governance solutions. In terms of water security,

Table 27.2 Summary of key results on framings of the 'urban water crisis' and the 'global water crisis' in peer-reviewed academic literature

Item	Urban water crisis	Global water crisis
Employed methods	Most frequent method: interpretation of secondary source empirical data (28/67), followed by primary source empirical data (19/67) and conceptual approaches (17/67); least often: lab-based approaches (7/67)	Most frequent: secondary source empirical data (39/83) and conceptual assessments (29/83). Least often applied: modelling and primary source data (11/67 each)
Household level	Six of the seven articles looking at household level were based on primary source empirical data	
Scale of analysis	Most frequent scale of analysis: city level (44/69)	Supra-national or non-specific scales dominate (33 and 34 out of 84, respectively)
Focus of analysis	Many articles (more than half—38/69) referred to more than one aspect of a water crisis. The most abundant framing was related to socio-economic problems (40/69), followed by policy-related challenges (36/69) and engineering and infrastructure (28/69)	The global water crisis was framed (analysed) most often in the context of policy issues (34/84), followed by socio-economic issues (31/84) and infrastructure/engineering challenges (26/84)
Type of solution	Primary source empirical data-based analyses predominantly led to governance recommendations (8/18)	Technical recommendations outweighed other types of recommendations slightly
	Analyses involving modelling approaches most often led to recommendations with regard to management	

such research does not address questions of environmental justice, or the dichotomies between rural and urban areas that are key to cross-scale governance (cf. Stoett 2012).

Science-Policy Interface

Key global environmental governance policies and frameworks refer to well-cited academic literature, such as the body of work analysed in this study. The frameworks and guidelines influencing national and local water policies are identical. As water governance in Accra shows, many countries and cities are obliged to integrate international frameworks in local decision-making, particularly when they depend on financial assistance from international donors. At the same time, neither Accra as a city nor Ghana as a whole have had any freedom to experiment with strategies of water governance without international intervention. Besides

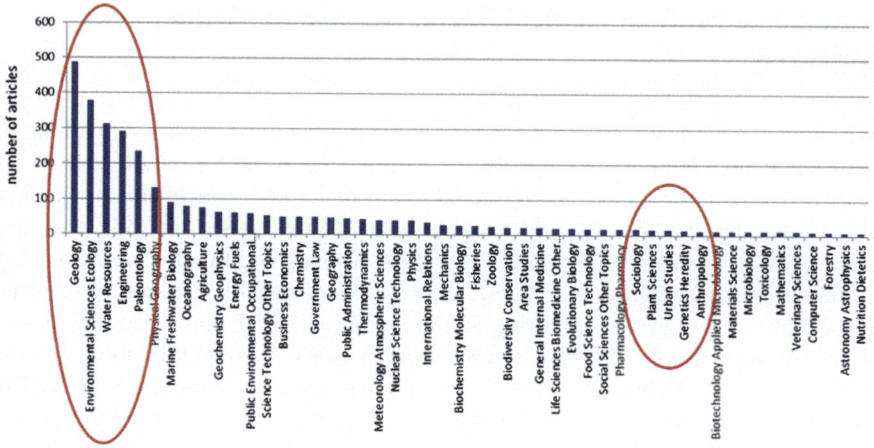

Fig. 27.1 Research areas of articles exploring the water crisis. Data from a web of science search

this, local research has long been strongly influenced by Western scholars (and continues to be so, as the brain drain mainly to Europe and the United States is ongoing (Takyi and Addai 2003)). While these internationally influenced frameworks, technical and management interventions have largely failed, local coping strategies that have been developed were not meaningfully included in national water governance strategies. So, in the case of Accra, the reproduction of discourse appears mostly unidirectionally top-to-bottom, from global to national to local level.

The above hints at a crucial gap in the interface between science and policy: Accra demonstrates several challenges that are peculiar to urban water governance. The sheer amount of research done on urban water crisis indicates that scholars are very much aware of the particular features of urban water, whilst only a limited number appear to assess the urban water crisis using primary data. Findings are often generalised. As a result, policy-makers either have to collect their own data, or rely on blueprinting solutions from elsewhere without adequate localised knowledge. Review of policy documents about Accra reveals no meaningful alternative to the dominant framing of the water crisis. Critical studies of water governance in Accra do exist (for instance, Yeboah 2006; Fuest and Haffner 2007; Adank et al. 2011), yet water governance policy shows that such research has had little to no influence on policy decisions on the national and city level.

Instead, our study revealed simplistic interpretations and reproduction of the crisis narrative in water policies inspired by managerialism. Both scholars and policy-makers alike have failed to take adequately into account water's multiple facets as "socio-physical constructions that are actively and historically produced, both in terms of social content and physical-environmental qualities" (Swyngedouw 2009, p. 56).

Towards a Shift to Critical Social Science in Water Research

We showed that mostly Western scholars, large international donors, and alliances between them claim to be capable of defining the water crisis as well as of knowing how best to solve it (by the adoption of IWRM principles). In shaping and promoting the universal discourse on the 'water crisis' in this way, the international water research and developing assistance community has become a 'hydro-hegemon' (Zeitun and Warner 2006), unintentionally implementing strategies that often trigger competition and exploitation rather than cooperation—as shown in the example of Ghana's water governance system. While policies were reframed and governance structures revised, clear definitions and goals of IWRM were however often lacking, or the Western paradigms involved stood in conflict with existing values, norms and practices (see also Agyenim and Gupta 2012). As a consequence, reforms were only partially implemented, and sometimes even triggered adverse change, instead of bringing about true integration and participative development of the water sector.

Although the water crisis is in fact a governance crisis (Pahl-Wostl et al. 2010) which was also stated in many articles analysed in this study, studies of the socio-political nature of water are rare and even more importantly those available tend to gloss over the power of hegemonic narratives (e.g. Sivipalan et al. 2012).

The overall lack of critical socio-scientific studies in water research, their restriction to smaller scales of analysis in contrast to global studies on the water crisis, and a dearth of truly interdisciplinary cross-scale studies contribute to a fragmented understanding of coupled socio-hydrological cycles. We therefore suggest reframing traditional hydrology and strengthening critical perspectives political ecology can offer. Such a new framing of society-water-interactions in the Anthropocene would disentangle how current and historic water governance is related to social power and how this is shaping the 'water crisis'.

References

Adank M, Darteh B, Moriarty P, Osei-Tutu H, Assan D, van Rooijen D (2011) Towards integrated urban water management in the Greater Accra Metropolitan Area. Current status and strategic directions for the future. SWITCH/Resource Centre Network Ghana, Accra

Adger W, Benjaminsen T (2001) Advancing a political ecology of global environmental discourses. Dev Change 32:681–715

Agrawal A (2001) Common property institutions and sustainable governance of resources. World Dev 29(10):1649–1672

Agyenim JB, Gupta J (2012) IWRM and developing countries: implementation challenges in Ghana, Physics and Chemistry of the Earth, vol 47–48, Parts A/B/C, pp 46–57. ISSN 1474-7065, http://dx.doi.org/10.1016/j.pce.2011.06.007

Bakker K (2009) Water. In: Castree N, Demeritt D, Liverman D, Rhoads B (eds) A companion to environmental geography. Wiley-Blackwell, New York, pp 515–532

Belina B, Dzudzek I (2009) Diskursanalyse als Gesellschaftsanalyse-Ideologiekritik und Kritische Diskursanalyse. In: Glasze G, Mattissek A (eds) Handbuch Diskurs und Raum. Theorien und Methoden für die Humangeographie sowie die sozial-und kulturwissenschaftliche Raumforschung. Transcript Verlag, Bielefeld, pp 129–151

Bohman A (2010) Framing the water and sanitation challenge: a history of urban water supply and sanitation in Ghana 1909–2005. Dissertation, Umeå University

Bourdieu P (2001) Das Politische Feld. UVK VertragsgesellschaftmbH, Konstanz

Bruns A (2010) Governance im Küstenraum. Dissertation, Christian-Albrechts-Universität zu Kiel

Bruns A, Gee K (2009) From state-centered decision-making to participatory governance. GAIA 2:150–157

Darteh B, Adank M, Kwabena SM (2010) Integrated urban water management in Accra: institutional arrangements and map. SWITCH, Accra

Fairclough N (1992) Discourse and text: linguistic and intertextual analysis within discourse analysis. Discourse Soc 3(2):193–217

Fuest V, Haffner SA (2007) PPP—policies, practices and problems in Ghana's urban water supply. Water Policy 9(2):169–192

Ghana National Water Policy (2007) Ministry of Water Resources, Works and Housing (eds), Accra

Gleick P (1993) Water in crisis: a guide to the world's fresh water resources. Oxford University Press, New York, Oxford

Groenfeldt D (2010) Viewpoint-the next nexus: environmental ethics, water policies, and climate change. Water Altern 3(3):575–586

Gyau-Boakye P (2001) Environmental impacts of the Akosombo dam and effects of climate change on the lake levels. Environ Dev Sustain 3:17–29

Karley N (2009) Flooding and physical planning in urban areas in West Africa: situational analysis of Accra, Ghana. Theor Empirical Res Urban Manag 13(4):25–41

Kenis P, Raab J (2008) Politiknetzwerke als Governanceform: Versuch einer Bestandsaufnahme und Neuausrichtung der Diskussion. In: Schuppert GF, Zürn M (eds) Governance in einer sich wandelnden Welt. VS Verlag für Sozialwissenschaften, Wiesbaden, pp 132–148

Linton J (2004) Global hydrology and the construction of a water crisis. Great Lakes Geogr 11(2):1–13

Manuel-Navarrete D, Pelling M, Redclift M (2009) Coping, governance, and development: the climate change adaptation triad. Environment, politics and development working paper series18. Department of Geography, King's College London

O'Brien K (2011) Responding to environmental change: a new age for human geography? Prog Hum Geogr 35(4):542–549

Owusu G (2011) Urban growth, globalization and access to housing in Ghana's largest metropolitan area, Accra. Paper presented at the 4th European conference on African studies, Uppsala, 15–18 June 2011

Pahl-Wostl C, Holtz G, Kastens B, Knieper C (2010) Analyzing complex water governance regimes: the management and transition framework. Environ Sci Policy 13(7):571–581

Postel S (1992) The last oasis: facing water scarcity, 3rd edn. Earthscan, London

Shiva V (2003) Der Kampf um das blaue Gold: Ursachen und Folgen der Wasserverknappung. Rotpunktverlag, Zürich

Sivapalan M, Savenije HHG, Blöschl G (2012) Socio-hydrology: a new science of people and water. Hydrol Process 26:1270–1276. doi: 10.1002/hyp.8426

Songsore J, McGranahan G (1993) Environment, wealth and health—towards an analysis of intraurban differentials within the Greater Accra Metropolitan-Area, Ghana. Environ Urban 5(2):10–34

Srinivasan V, Lambin EF, Gorelick SM, Thompson BH, Rozelle S (2012) The nature and causes of the global water crisis: syndromes from a meta-analysis of coupled human-water studies. Water Resour Res 48(10):1–16

Stoett PJ (2012) Global ecopolitics: crisis, governance, and justice. University of Toronto Press, North York, Ontario
Swyngedouw E (2009) The political economy and political ecology of the hydro-social cycle. J Contemp Water Res Educ 142(1):56–60
Takyi BK, Addai I (2003) Demographic processes, economic growth, and socio-economic developments in Ghana, 1960–2000. In: Tettey WJ, Puplampu KP, Berman BJ (eds) Critical perspectives in politics and socio-economic development in Ghana. Brill, Leiden, pp 223–273
Turner BL, Robbins P (2008) Land-change science and political ecology: similarities, differences, and implications for sustainability science. Ann Rev Environ Resour 33(1):295–316
Trottier J (2008) Water crises: political construction or physical reality? Contemp Politics 14(2):197–214
Vörösmarty CJ, McIntyre PB, Gessner MO, Dudgeon D, Prusevich A, Green P, Glidden S, Bunn SE, Sullivan CA, Reidy Liermann C, Davies PM (2010) Global threats to human water security and river biodiversity. Nature 467(7315):555–561
World Bank (2007) Ghana—country assistance strategy. The Worldbank, Washington D.C.
Yeboah I (2000) Structural adjustment and emerging urban form in Accra, Ghana. Afr Today 47(2):61–89
Yeboah I (2006) Subaltern strategies and development practice: urban water privatization in Ghana. Geogr J 172(1):50–65
Zeitoun M, Warner J (2006) Hydro-hegemony—a framework for analysis of trans-boundary water conflicts. Water Policy 8(5):435

Chapter 28
The Need for a Value-Reflexive Governance of Water in the Anthropocene

Simon Meisch

Abstract The paper reflects on the conditions for a value-reflexive governance of water as a tool to contribute to Sustainable Development within the Anthropocene and to deal with the social and political challenges along the way. Its contribution consists in integrating value discourses in sustainable water governance. These value discourses are necessary and unavoidable. While integrating more stakeholders in problem solving and knowledge production leads to more value disputes, it will at the same time strengthen the legitimacy of water governance. A value-reflexive governance aims to make visible the values underlying scientific and political concepts, and to treat value conflicts in an ethically informed and structured way. The paper discusses conceptual considerations and critically assesses the Anthropocene concept and deals with challenges to sustainable water governance. It then argues that the approach of value-reflexive governance might be useful for water governance and shows what questions need to be considered conceptually: What are values? What is the contribution of ethics to a value-reflexive governance of water? What is the relationship between good and value-reflexive governance? As a result a pragmatically concept of values and a more value-reflexive stance to values in governance is suggested.

The Anthropocene Concept

The term 'Anthropocene' is meant to describe the geological time period that has been dominantly shaped by humans since the beginning of the Industrial Revolution in the late 18th century (Crutzen 2002; Zalasiewicz et al. 2011). There

S. Meisch (✉)
Junior Research Group "Ethics of Science in the Research for Sustainable Development", International Centre for Ethics in the Sciences and Humanities, University of Tuebingen, Wilhelmstr. 19, 72074 Tuebingen, Germany
e-mail: simon.meisch@uni-tuebingen.de

A. Bhaduri et al. (eds.), *The Global Water System in the Anthropocene*, Springer Water, DOI: 10.1007/978-3-319-07548-8_28,

is an ongoing discussion of the scientifically detected manifestations that led to the classification of the Anthropocene as a new geological epoch (ICS 2013). Yet, regardless of contested issues, global changes such as rapid expansion of mankind, climate change, release of greenhouse gases and toxic substances, transformation of landscape by humans, loss of biodiversity etc. are evident (Rockström et al. 2009; Zalasiewicz et al. 2010).

Conceptually, 'Anthropocene' encompasses more than empirical knowledge about characteristics of geological eras. In addition to its scientific content, the concept also encompasses a normative component calling for social and political action in order to counteract global environmental changes and their detrimental effects on humans and nature. The concept describes and evaluates states of the world and fields of action and makes normative statements. It "[stresses] the enormity of humanity's responsibility as stewards of the Earth" (Crutzen and Schwägerl 2013). A business-as-usual strategy is no longer possible, because continuing with the status-quo would be "detrimental or even catastrophic for large parts of the world" (Rockström 2009, p. 472).

'Anthropocene' can therefore be described as a 'thick moral concept' or 'epistemic-moral hybrid'. Thick moral concepts comprise descriptive as well as normative and evaluative elements, i.e. facts and value judgements, which are closely intertwined in common language usage ('thick') but which nevertheless can be separated analytically (Dietrich 2004, p. 21; cf. also Ricken 2003, pp. 62–65). This hybrid raises meta-ethical as well as epistemological objections. Meta-ethical objections refer to the 'is-ought problem' (or 'naturalistic fallacy') according to which it is impossible to deduce normative statements about what *ought* to be from descriptive statements about what *is*. The mere fact of global environmental changes does not itself create obligations for action unless there is a normative concept such as justice claiming that detrimental effects on human life and nature are to be prevented. The epistemological objection is related to the meta-ethical one. In conventional understanding, science is about objective facts. As the Anthropocene concept is intertwined with normative statements, its scientific soundness might be questioned (Autin and Holbrook 2011). The argument made here is that there are scientific concepts (such as Biodiversity or Anthropocene) that evolve and can be understood against the background of a certain normative theory. Yet, they are neither naturalistic fallacy nor political ideology. In epistemological terms, they can be described as 'epistemic-moral hybrids' (Potthast 2000, 2010). Regarding scientific concepts in such a way initiates an interdisciplinary discourse on concepts themselves, their meanings and implications. More precisely, it enables reflection on underlying norms and values of the Anthropocene.

Though the Anthropocene concept increasingly resonates within political and public debates, it is still very much science-driven. Therefore, it is no surprise that scientists are ascribed a prominent task in problem solving and guiding society towards environmentally sustainable management during the era of the Anthropocene (Crutzen 2002). According to Crutzen (2002), this involves scientific and engineering measures on all scales up to internationally accepted, large-scale geo-engineering projects (for a critical view, cf. Ott and Baatz 2012). What could be understood as self-authorisation of science (cf. Lövbrand et al. 2009) is not a subject to be further discussed here. While acknowledging that science has an important role to play in solving the major future challenges, other social actors and their values also contribute to the solution. The paper turns next to issues that must be considered in order to address value pluralism in dealing with water problems within the Anthropocene.

Challenges to Sustainable Water Governance in the Anthropocene

It has been mentioned above that the Anthropocene concept has to be understood against a normative background. Since the Brundtland report in 1987, the concept of Sustainable Development (SD) has become the political strategy to balance the needs of humans and nature. The concept's normative core is the idea of inter- and intra-generational justice in the face of decreasing natural resources, the ecosystems' limited capacity to absorb human emissions, and the ongoing environmental destruction. Briefly stated, SD dictates that humans are obligated to ensure that everyone has the opportunity to live a self-determined and good life. In this endeavour, we proceed in such a way that the natural basis necessary to live such a life is at least retained (and in the best case extended) for all contemporary and future humans (Meisch 2013; Voget-Kleschin 2013). On a normative level, action in the Anthropocene refers to SD.

In many ways, water plays a crucial role in the transformation process to SD. Even without the global environmental changes associated with the Anthropocene, constituting and enforcing a human right to water constitutes a major task in itself. Access to and allocation of water is unequal and often unfair due to regional scarcities, changing consumption patterns and political entitlement (Kowarsch 2011). In the Anthropocene, human action has affected water and water systems in many ways—ranging from ocean circulation, to river hydrology, to coastal zones, and to local and regional projects (dams, channels) (Crutzen 2002; Rockström 2009). Besides social and political factors, anthropogenic change puts additional stress on water resources and supply and thereby further complicates enforcing a human right to water (Parry et al. 2007; Bals et al. 2008; Rockström et al. 2009). However, as water is essentially connected to many other SD issues (e.g. food, health and sanitation, social development etc.), it can also be seen as an ideal starting point to address many sustainability issues and to cope with global change in the Anthropocene (Steduto and Kuylenstierna 2009). Research platforms such as the Integrated Water Resources Management (IWRM) or the Water, Energy and Food Security Nexus explicitly point in this direction.

In recent years, many approaches in the water sciences and management have emerged. Yet, while scientific debates develop steadily and substantially, progress in solving real world problems is lagging behind expectations: Non-SD endures, water problems escalate and decisions remain urgent (Ingram 2011; Ostrom 2008). Politics must be brought back into water governance by directly considering value conflicts. This claim implies changing present practices e.g. by paying more attention to policy implementation and power relations. An essential component of this politicised water governance is water ethics. Water ethics deal with human actions and social institutions that affect water. In particular, water ethics aims to reflect normative and evaluative claims with regard to existing water practices and institutions and tries to assess whether they contribute to solutions that conform to the normative right or the evaluative good (cf. Düwell et al. 2011).

Coping with value conflicts that with regard to water are likely to happen (Ingram 2011; Schmidt 2010) faces ethical challenges that will be considered here. As many different values are ascribed to water, more politicised water governance is most likely to be in need for a value-reflexive governance in order to deal with competing values (Ingram 2006; Groenfeldt and Schmidt 2013; Groenfeldt 2013). Paying more attention to the value dimension of politics opens up new opportunities. For instance, there is an agreement within the political and scientific community that contextual solutions to water problems are necessary (Ostrom 2007; Pahl-Wostl and Toonen 2009a). One consequence can be seen in a greater participation of communities in developing water research issues and in implementing possible findings (Ziegler and Ott 2011; Ingram 2011). Explicit debate on values allows for a low-threshold entrance into ethical debates on sustainable water governance. Therefore, facing the ethical dimension, in which the value-dimension plays a central part, is not only vital for successful water governance but also an opportunity for involving people into ethical debates on water governance and water science. The discussion turns to the premises and conditions for value dialogue. It argues for politicising water governance and reflects about conceptual preconditions of a value-reflexive governance of water: it deals with the concept of value and its relationship to ethics and compares good and value-reflexive governance.

Value-Reflexive Governance of Water

Water Governance: Bringing Politics Back In

Though numerous scientific, technological and policy approaches to cope with water issues have been pursued, actual results seem to be disappointing. Reasons for failure are diverse and cannot be discussed in detail here (for an excellent overview, cf. Ingram 2011). The paper focuses instead on water governance and water ethics as essential parts of a solution. After being neglected by water scholars and practitioners, the political dimension must be seen as an indispensable component for sustainable innovations in water science and governance (Ingram 2011; Pahl-Wostl and Toonen 2009a; Hoppe and Wesselink 2011). Ignoring the political aspect of water and leaving it unmanaged also raises serious socio-ethical questions. For instance, unmanaged water might leave the world's poorest worst off because it is likely more vulnerable to natural disasters or subject to informal elite rule and corruption ('iron law of oligarchy') (cf. Meisch et al. 2012, p. 414).

It is reasonable to distinguish management and governance as many of the previous shortcomings and failures can be traced back to the confusion of the two. While management strives for effectiveness and efficiency, governance intends to create legitimacy (Pahl-Wostl and Toonen 2009b). Making water policies more efficient and effective does not make them more legitimate and socially accepted when distributional effects or cultural factors are ignored (Ingram 2011; Ostrom 2008). Sustainable water governance needs to fulfil several tasks: unfolding and settling value differences; finding legitimate policy solutions; dealing with uncertainty and surprise due to climate change; and finally finding ways and means for policy implementation. With regard to reforms of water policies, it seems to be undisputed that panaceas or universal solutions that are supposed to fit all situations independently of time and space are destined to fail. Contextualised solutions are needed (Ingram 2011; Ostrom 2007, 2008). Central factors in contextualized solutions are: attracting public attention in an area with values at stake; generating engagement and support; engaging social movements; making water governance an issue of politics and not only of expert circles; overcoming (formal and informal) bureaucratic path dependency; and finding means for policy implementation contexts (Hoppe and Wesselink 2011; Huitema and Mejerink 2007; Ingram 2011). Transformation to SD is an eminently knowledge-driven process. However, it has been argued that it needs to be a different science and knowledge from what we know now, namely one that is more credible, trusted and legitimate (Ingram 2011; Funtowicz and Ravetz 1993). Inter- and intra-disciplinarity as well as participation should feature more prominently (Ziegler and Ott 2011). The claim for contextual problem solving encompasses other scientific methods or models, more robust and reliable technologies (Ingram 2011; Hoppe and Wesselink 2011) and value-reflexive governance of water.

The Concept of Value

Conceptualising 'value' is notoriously difficult (Schnädelbach 1983). However, addressing the value dimension of governance requires a basic conceptual understanding. Values have to be distinguished from preferences and attitudes, as they are not on the same analytical level (Meisch and Potthast 2011, pp. 8–14) even though all three terms refer to evaluations in one way or another. In conceptualising values, it is reasonable to take a pragmatic approach and to avoid the philosophical difficulties of value-philosophical traditions with strong metaphysical claims (Joas 2001, 2008). In short, values originate in contexts of action and are connected to specific experiences in which people evaluate actions, institutions, contexts etc. with regard to their moral desirability. One can also state that practical contexts (such as science) are always value-laden. In the course of time, these evaluations become detached from specific situations. Values as nouns (such as transparency, participation etc.) then become the reference points for evaluations. As values originate in experiences, their meaning might to some degree vary in different contexts. Attitudes and preferences are related to values in different ways. Attitudes are the tendency to evaluate, not the evaluation itself. As mental sets, attitudes determine the way people evaluate a stimulus object. Preferences are part of a comparative value concept that compares objects according to some value. Therefore, they need one or more values as criteria to build preference lists. Both attitudes and preferences can be evaluated on the basis of values. While attitudes and preferences influence and structure actions, the terms themselves are normatively undetermined. Reference to values allows for ethical considerations such as argumentation, reflectivity and justification (Meisch and Potthast 2011; Joas 2008).

Thus, values must be regarded as reference points for evaluations, and, as such they work as ideals or criteria for evaluating actions, persons, institutions, things, attitudes, preferences, norms, etc. as good or bad. Values are emotionally and rationally binding, and give long-term orientation and motivate for action. They also encompass an active and passive as well as rational and emotional element (Beck et al. 2012; Meisch et al. 2012). This approach to values has several advantages. First, it takes into account that actors already have concrete and strong beliefs about their values. It is applicable by persons with different moral backgrounds in different contexts. Second, it acknowledges the situation of several and heterogeneous accepted values within a value community (i.e. freedom, wealth, etc.). Descriptions of a pluralistic society with a plurality of values are empirically undeniable and broadly accepted. Third, it offers a value theory neither claiming the eternal existence of fixed values independent of time and space nor paving the way for value relativist or value subjectivist positions. Instead, historically contingent values can be employed as valid and binding for a certain given time or at least for certain societies. Even if in pluralistic backgrounds, the acceptance of certain justification models is difficult, one must not underestimate the consensual acceptance of basic values, which gain a quasi-objective status. But, as values are generated in dynamical interactions between individuals and society, the question arises how it is possible to identify a substantial value system of a social group. This would be necessary if we wanted to solve the problem of competing values in a certain context (Beck et al. 2012).

Values and Ethics

Ethics that is aware of a plurality of values within and between societies can develop mechanisms that allow citizens to bring their values into ethical debates and thereby to participate in the finding of socially robust innovations (Funtowicz and Ravetz 1993). It is necessary to pay attention on two value-ethical claims:

First, values can be regarded as low-threshold entry to ethical debates. Including the value dimension in coping with situations of high uncertainties allows to better address citizens and stake-holders, who want to understand, support and participate consciously and deliberately in water governance. It can be expected that value dialogue contributes to politicised water governance as water issues get more easily on the political agenda, stakeholders can be better mobilised and more intensively included in policy making (Ingram 2011; Huitema and Mejerink 2007). Second, ethical debates cannot be restricted to mere value talk. Value debates need an ethical framing that enables participation as well as a philosophically sound reflection on (conflicting) values. The approach of value-reflexive governance has three aims. It wants first to contribute to a democratisation of sciences, second to make implicit value commitments explicit and third to allow for rational discourses on values. The last point is important because there are strands in moral philosophy that regard values as purely subjective and non-argumentative. The value-ethical core of a value-reflexive governance therefore does not only take stock of the values involved but also develops philosophical tools to deal with value conflicts and participatory mechanisms to deal with value conflicts in social contexts.

Ethics has different tasks in solving value conflicts. First, it contributes to the understanding of what is actually meant by a given value. Depending on a value's genesis, there are very likely different and conflicting interpretations of the same value. These have to be made transparent and open for dialogue. Ethics might as well examine different value interpretations with regard to their moral rightness or goodness. Second, ethics helps to handle conflict between different values. As mentioned earlier, water is an issue where many different values are at stake and value conflict seems likely. An ethics of values offers philosophical frames (e.g. the concept of inter- and intra-generational justice) to deal with conflicting values and it also addresses the need to set up norms and regulations in water governance (Meisch et al. 2012; Immergut 2011).

Toward a Value-Reflexive Governance: More than Good Governance

When suggesting the concept of value-reflexive governance, one needs to clarify its relationship to the concept of good governance as on the surface both concepts might appear identical. The idea of good governance first evolved within the value context of the World Bank, which developed principles that could determine the allocation of loans to developing countries and that had a strong anti-corruption bias. The genesis of good governance conceptions within the context of an international economic institution does not delegitimize the concept as such. However, as explained before, concepts have a factual and a value side that are intertwined. Therefore, one has to be aware of implicit normative and evaluative statements, make them transparent and open for rational public discourse. While the concept of good governance is mainly concerned with norms that are meant to guide governance processes, which values are at play often remains implicit. This observation leads to two main further questions to be addressed in the processes of water governance: which values form the basis of normative statements about how good water governance should be? Whose values are meant to guide governance (Meisch et al. 2012; Immergut 2011)?

An interesting example is the concept of Integrated Water Resources Management (IWRM) that has become a major point of reference for discourses in water management. It is understood to be “a process which promotes the coordinated development and management of water, land and related resources in order to maximise economic and social welfare in an equitable manner without compromising the sustainability of vital ecosystems and the environment” and which also comprises notions of good governance (GWP 2010). Although it was meant to be a science-based and impartial conception of water management, it nevertheless makes explicit and implicit moral judgements on the world water situation and social relationships with regard to water (Scmidt 2010, pp. 7–8; Biswas 2008) by explicitly referring to ethical concepts such as welfare or sustainability and by implicitly promoting a Utilitarian framework (Kowarsch 2011, pp. 47–48). It has been mentioned above that scientific practise is value-based and that scientific conceptions in the form of epistemic-moral hybrids make factual as well as normative and evaluative judgements. Both have to be made transparent for discourse and need ethical justification.

In this situation, a value-reflexive governance aims for two objectives. First, it makes explicit underlying values of good governance concepts and the processes by which they became guiding imperatives. With this, one can avoid imposing specific value systems on social contexts. Secondly, it offers solutions which ensure that open, transparent and more inclusive governance not only allows more social actors to express their values but also that those values can be translated into policy programmes. In contrast to the broader concept of good governance, the concept of value-reflexive governance stresses the point of sensitivity in regard to participants’ values in governance processes. Ensuring that all stakeholders’ values will be voiced and heard within governance processes does not say anything about how to deal with values, let alone value conflicts that are most likely going to arise (Meisch et al. 2012).

What does this mean for water governance and the water sciences? The scientific system considers itself a self-regulating social system. However, politics and business play an important role in shaping it. Both spheres interact. A value-reflexive governance of water opens up dialogue on underlying values. At the same time, water sciences take part in solving social problems and are therefore actors in governance processes. While traditionally the sciences have been assigned with the role of contributing a standing knowledge to deal with concrete problems, this (self) perception changed. Fixed, cure-all solutions failed in concrete social contexts because governance processes overlooked value dimensions of people affected by political and technical solutions. In value-reflexive governance, water sciences and practitioners do not contribute to societal requests by providing fixed knowledge but by developing context-specific solutions to problems with a specific time-space dimension. This requires an understanding of values involved and suggestions how to deal with value conflicts. In that sense, value-based governance needs, among others, conceptual clarification as well as extensive deliberation on the ethical norms and decisions to be made in water governance.

Conclusion and Outlook

In the era of Anthropocene, humanity is faced with massive challenges. Water is one of the key elements for sustainable development claiming an autonomous and good life for contemporary and future generations. While human activity has put stress on water, addressing the water issue can help to bring about SD on many scales and policies. So far, the record of water management is mixed, which can, among others, be traced back to the neglect of the political dimension of water. Finding solutions to water problems is not only about technology to be applied by experts, it is also a highly normative undertaking with values at stake. The paper suggested conceptual considerations for a value-reflexive governance as a response to the present shortcomings in water governance. The proposed approach intends to make values visible and accessible for rational discourse and at the same time suggests value discourses as a low-threshold entry for stakeholders to ethical debates of water governance.

Next research steps need to assess existing ethical tools such as, among others, the "Ethical Matrix" (Kaiser et al. 2007) and apply it to the context of water governance. In this endeavour, there are many tasks ahead which cannot all be discussed here. What is clearly needed is developing appropriate tools for different scales that also consider how they can be placed within respective processes of policy formation and implementation. Their main task would be bringing together different forms of knowledge with an ethically sound process of reflection and decision making.

Acknowledgements The paper builds on previous work and results of the EU project "Value Isobars—The Landscape and Isobars of European Values in Relation to Science and New Technology" (FP7 31004480; http://www.value-isobars.eu/). I acknowledge the provision of important discussion points from project partners and especially thank Thomas Potthast, Matthias Kaiser and Helen Ingram. The full responsibility for the content, style, errors and inaccuracies is mine.

References

Autin W, Holbrook J (2011) Is the Anthropocene an issue of stratigraphy or pop culture? GSA Today 22:60–61. doi:10.1130/G153GW.1

Bals C, Hameling S, Windfuhr M (2008) Climate change, food security and the right to adequate food. Study Published by Bread for the World and Germanwatch. Stuttgart

Beck R, Meisch S, Potthast T (2012) The value(s) of sustainability within a pragmatically justified theory of values: considerations in the context of climate change. In: Potthast T, Meisch S (eds) Climate change and sustainable development. Ethical perspectives on land use and food production. Wageningen Academic Publishers, Wageningen, pp 49–54

Biswas A (2008) Integrated water resources management: is it working? Water Resour Dev 24:5–22

Crutzen P (2002) The geology of mankind. Nature 415:23

Crutzen P, Schwägerl C (2013) Living in the anthropocene: towards a new global ethos. environment 360. http://e360.yale.edu/feature/living_in_the_anthropocene_toward_a_new_global_ethos/2363/. Accessed 17 Aug 2013

Dietrich J (2004) Ethisch-Philosophische Grundlagenkompetenzen: ein Modell für Studierende und Lehrende. In: Maring M (ed) Ethisch-Philosophisches Grundlagenstudium. Ein Studienbuch. LIT, Münster, pp 15–33

Düwell M, Hübenthal C, Werner MH (2011) Einleitung. In: Düwell M, Hübenthal C, Werner MH (eds) Handbuch ethik. Metzler, Stuttgart, pp 1–23

Funtowicz S, Ravetz J (1993) Science for the post-normal age. Futures 26:568–582

Groenfeldt D (2013) Water ethics. A values approach to solving the water crisis. Routledge, London

Groenfeldt D, Schmidt J (2013) Ethics and water governance. Ecol Soc 18. doi:10.5751/ES-04629-180114

GWP [= Global Water Partnership] (2010) What is IWRM? http://www.gwp.org/en/The-Challenge/What-is-IWRM/. Accessed 29 Dec 2013

Hoppe R, Wesselink A (2011) If post-normal science is the solution, what is the problem? Sci Technol Human Values 36:389–412

Huitema D, Mejerink S (2007) Understanding and managing water transitions: a policy science perspective. Paper presented to the First International Conference on Adaptive and Integrated Water Management, Basel, Switzerland, 12–15 Nov 2007. http://www.newater.uni-osnabrueck.de/caiwa/data/papers%20session/G1/springer01-Meijerink.pdf. Accessed 8 Sep 2012

ICS [= International Commission on Stratigraphy] (2013) Subcommission on Quaternary Stratigraphy, Working Group on the 'Anthropocene'. http://quaternary.stratigraphy.org/workinggroups/anthropocene/. Accessed 29 Dec 2013

Immergut E (2011) Democratic theory and policy analysis: four models of "Policy, Politics and Choice". der moderne staat 1: 69–86

Ingram H (2006) Water as a multi-dimensional value: implications for participation and transparency. Int Environ Agreements 6:429–433

Ingram H (2011) Beyond universal remedies for good water governance: a political and contextual approach. In: Garrido A, Ingram H (eds) Water for food in a changing world. Routledge, London, pp 241–261

Joas H (2001) The genesis of values. University of Chicago Press, Chicago

Joas H (2008) Value generalization. Limitations and possibilities of a communication about values. J Bus Econ Ethics 9:88–96

Kaiser M, Millar K, Thorstensen E et al (2007) Developing the ethical matrix as a decision support framework: GM fish as a case study. J Agric Environ Ethics 20:65–80

Kowarsch M (2011) Ethical targets and questions of water management. In: Kowarsch M (ed) Water management options in a globalised world. Proceedings of an international scientific workshop (20–23 June 2011, Bad Schönbrunn), p 38–49. http://www.hfph.de/igp/proceedings2011. Accessed 29 Dec 2013

Lövbrand E, Stripple J, Wiman B (2009) Earth system governmentality. Reflections on science in the anthropocene. Glob Environ Change 19:7–13

Meisch S (2013) Green food consumption: whose responsibility? In: Röcklinsberg H, Sandin P (eds) The ethics of consumption. The citizen, the market and the law. Wageningen Academic Publishers, Wageningen, pp 160–165

Meisch S, Potthast T (2011) Towards a pragmatically justified theory of values for governance. Conceptual analysis of values, norms, preferences and attitudes. EU-Project Value Isobars.

Work package 1 Deliverable 2 (Mar 2011). http://www.value-isobars.no/filestore/WP1_2_Conceptualanalysisofvaluesnormspreferencesandattitudes.pdf. Accessed 29 Dec 2013

Meisch S, Beck R, Potthast T (2012) Towards a value-reflexive governance of water. In: Potthast T, Meisch S (eds) Climate change and sustainable development. Ethical perspectives on land use and food production. Wageningen Academic Publishers, Wageningen, pp 413–418

Ostrom E (2007) A diagnostic approach for going beyond panaceas. Proc Natl Acad Sci 104:15181–15187

Ostrom E (2008) Sustainable development of common-pool resources. http://www.indiana.edu/~workshop/colloquia/materials/papers/ostrom_paper1.pdf. Accessed 9 Sep 2012

Ostrom E (2010) A multi-scale approach to coping with climate change and other collective action problems. Solutions 1:27–36

Ott K, Baatz C (2012) Domains of climate ethics: an overview. In: Potthast T, Meisch S (eds) Climate change and sustainable development. Ethical perspectives on land use and food production. Wageningen Academic Publishers, Wageningen, pp 23–28

Pahl-Wostl C, Toonen T (2009a) Sustainable water governance in times of global change. A major challenge for the scientific and policy communities. IHDP Update Issue 3:26–30

Pahl-Wostl C, Toonen T (2009b) Global water governance: Quo Vadis? Global Water News 8:8–10

Parry M, Canziani O, Palutikof J et al (2007) Climate change 2007: Impacts, adaptation and vulnerability. Contribution of working group II to the fourth assessment report of the IPCC. Cambridge University Press, Cambridge, pp 23–78

Potthast T (2000) Bioethics and epistemic-moral hybrids: Perspectives from the history of science. Biomedical Ethics 5:20–23

Potthast T (2010) Epistemisch-moralische Hybride und das Problem interdisziplinärer Urteilsbildung. In: Jungert M, Romfeld E, Sukopp T et al (eds) Interdisziplinarität. Theorie, Praxis, Probleme. WBG, Darmstadt, pp 173–191

Ricken F (2003) Allgemeine ethik. Kohlhammer, Stuttgart

Röckström J, Steffen W, Noone K et al (2009) A safe operating space for humanity. Nature 46:472–475

Schnädelbach H (1983) Philosophie in Deutschland 1831–1933. Suhrkamp, Frankfurt/Main

Schmidt J (2010) Water ethics and water management. In: Brown P, Schmidt J (eds) Water ethics. Foundational readings for students and professionals. Island Press, Washington, pp 3–15

Steduto P, Kuylenstierna J (2009) Climate change, energy and food security, economic development—in the end, it all trickles down to water. Climate change policy and practice. http://climate-l.iisd.org/guest-articles/climate-change-energy-and-food-security-economic-development-in-the-end-it-all-trickles-down-to-water/. Accessed 9 Sep 2012

Voget-Kleschin L (2013) Employing the capability approach in conceptualizing sustainable development. J Hum Dev Capabilities 14:483–502

Zalasiewicz J, Williams M, Steffen W et al (2010) The new world of the anthropocene. Environ Sci Technol 44:2228–2231

Zalasiewicz J, Williams M, Haywood A et al (2011) The anthropocene: a new epoch of geological time? Phil Trans R Soc 369:835–841

Ziegler R, Ott K (2011) The quality of sustainability science: a philosophical perspective. Sustain Sci Practice Policy 7:31–44

The manufacturer's authorised representative in the EU is Springer Nature Customer Service Centre GmbH, Europaplatz 3, 69115 Heidelberg, Germany. If you have any concerns regarding our products, please contact ProductSafety@springernature.com

Printed and bound by CPI Group (UK) Ltd, Croydon, CR0 4YY
30/09/2026
02225602-0015